# NONAQUEOUS
## ELECTROCHEMISTRY

# N O N A Q U E O U S
## ELECTROCHEMISTRY

EDITED BY
## DORON AURBACH
*Bar-Ilan University*
*Ramat-Gan, Israel*

## CRC Press
Taylor & Francis Group
Boca Raton London New York

CRC Press is an imprint of the
Taylor & Francis Group, an **informa** business

CRC Press
Taylor & Francis Group
6000 Broken Sound Parkway NW, Suite 300
Boca Raton, FL 33487-2742

First issued in paperback 2019

© 1999 by Taylor & Francis Group, LLC
CRC Press is an imprint of Taylor & Francis Group, an Informa business

No claim to original U.S. Government works

ISBN-13: 978-0-8247-7334-2 (hbk)
ISBN-13: 978-0-367-39957-3 (pbk)

**Visit the Taylor & Francis Web site at
http://www.taylorandfrancis.com**

**and the CRC Press Web site at
http://www.crcpress.com**

# Preface

Nonaqueous systems are becoming important in modern electrochemistry for the following reasons:

1. High energy density batteries are a crucial need as modern electronics progresses and moves more and more toward miniaturization. Many novel battery systems require the use of nonaqueous electrolyte systems because important new developments in this field are based on active metal electrodes (eg, Li and Li-C intercalation compounds).
2. The use of nonaqueous systems increases the range of applicability of electrochemical techniques to many organic and inorganic compounds that are either unstable or insoluble in water. The electrochemical window of many nonaqueous systems is much wider than that of water.

In the literature of recent years, we indeed see more and more publications on the study and development of novel nonaqueous high energy density battery systems, nonaqueous electro-organic synthesis, and mechanistic studies. This parallels worldwide efforts to commercialize nonaqueous lithium and lithium-carbon batteries. Hence, it is important to gather from time to time the knowledge accumulated in these areas, to update the literature and provide the increasing number of people working in this field with a comprehensive compendium on the practical and theoretical aspects of nonaqueous electrochemistry.

This book focuses on three types of nonaqueous systems—liquid electrolyte solutions, ionically conducting polymers, and molten salts—with emphasis on the more commonly used liquid systems. It provides a review of a variety of important and useful nonaqueous systems in terms of properties, purification procedures, and handling.

The intrinsic reactions of many important solvents, salts, and common contaminants on both active and nonactive electrodes are described and discussed. This provides the reader with a comprehensive view of the true electrochemical window of these systems. The spectroelectrochemical tools that can be applied to the study of these systems (with the emphasis on in situ techniques) are reviewed.

Important applications of nonaqueous systems in the fields of electrosynthesis, mechanistic research by electrochemical techniques, and batteries are described and demonstrated.

This book is directed at not only electrochemists but also scientists in other discipines who need to work with nonaqueous systems, including organic and inorganic chemists (in the aspects of synthesis and mechanistic research), energy physicists, and material scientists. In addition to fundamental aspects that remain invariant, great effort has been made in this book to cover the most recent accumulated knowledge.

*Doron Aurbach*

# Contents

# Contributors

**Doron Aurbach**  Department of Chemistry, Bar-Ilan University, Ramat-Gan, Israel

**George E. Blomgren**  Energizer, Westlake, Ohio

**Larry A. Dominey, Ph.D.**  Vice President, Research and Development, OMG Americas, Inc., Cleveland, Ohio

**Ioan Galasiu**  Roumanian Academy, Institute of Physical Chemistry, Bucharest, Roumania

**Rodica Galasiu**  Roumanian Academy, Institute of Physical Chemistry, Bucharest, Roumania

**Yosef Gofer**  Department of Chemistry, Bar-Ilan University, Ramat-Gan, Israel

**Shinichi Komaba\***  Department of Applied Chemistry, School of Science and Engineering, Waseda University, Shinjuku-ku, Tokyo, Japan

---

\* *Current affiliation:* Department of Applied Chemistry and Molecular Science, Faculty of Engineering, Iwate University, Morioka, Japan

**Xingjiang Liu**   Department of Applied Chemistry, School of Science and Engineering, Waseda University, Shinjuku-ku, Tokyo, Japan

**Tetsuya Osaka**   Department of Applied Chemistry, School of Science and Engineering, Waseda University, Shinjuku-ku, Tokyo, Japan

**Daniel A. Scherson**   Department of Chemistry, Case Western Reserve University, Cleveland, Ohio

**Jomar Thonstad**   Norwegian University of Science and Technology, Trondheim, Norway

**Idit Weissman**   Department of Chemistry, Bar-Ilan University, Ramat-Gan, Israel

**Arie Zaban**   Department of Chemistry, Bar-Ilan University, Ramat-Gan, Israel

# NONAQUEOUS ELECTROCHEMISTRY

# 1
# Nonaqueous Electrochemistry: An Overview

**Doron Aurbach and Idit Weissman**
*Bar-Ilan University, Ramat-Gan, Israel*

## I. INTRODUCTORY REMARKS

Defining and describing nonaqueous electrochemistry as a specific, separated branch in modern electrochemistry means that we take aqueous electrochemical systems as the reference area. Since the essence of electrochemistry is the passage of electric current in chemical systems, at least in part by ion transport and migration, electrolyte solutions based on water are ideal media for electrochemical reactions. The highly polar water (dielectric constant of about 78 at 25°C) dissolves a large variety of salts and thus forms highly ionically conducting solutions of low resistivity. However, water solutions have several limitations as electrochemical media:

1.  There are too many chemical substances of potentially important electrochemical reactivity which are insoluble in water (mostly organic compounds).
2.  Water as a protic solvent is highly reactive with a variety of electrode materials and compounds of interest, such as active metal (alkaline, alkaline earth), organic salts which undergo fast hydrolysis, basic or acidic compounds, complexes, etc.
3.  The applicability of aqueous solutions is limited to temperatures in which these media are in the liquid state. Hence, their temperature window is about 100°C, which may be too narrow for many uses, especially in areas related to energy storage and conversion (batteries, fuel cells).

4. The electrochemical window of water is too narrow. On a thermodynamic basis the potential difference between the cathodic ($H^+$ + $e^- \rightleftharpoons \frac{1}{2}H_2$) and the anodic ($\frac{1}{2}O_2 + 2H^+ + 2e^- \rightleftharpoons H_2O$) limiting reactions is only 1.299 volts. This potential window may be extended by a kinetic control, such as high overvoltage for the cathodic reaction, using certain electrode materials (e.g., mercury, lead). However, too many important compounds and substances are reduced or oxidized at potentials far beyond the potentials in which water is electrolyzed.

Consequently, there is an increasing interest in alternative (or additional) electrolyte systems in which the above limitations can be removed. Providing the answers to the above limitations of aqueous solutions and the study, matching, and classification of alternative/additional electrolyte systems is the essence of nonaqueous chemistry. An overview of this branch of modern electrochemistry should include the presentation and classification of the important nonaqueous electrolyte systems, indicating their basic features, advantages, and limitations, as well as a short review of their typical applications.

The important nonaqueous electrolyte systems may be divided into five groups:

1. Liquid solutions based on organic and inorganic solvents
2. Liquid systems based on molten salts (high temperature and ambient temperature)
3. Ionically and electronically conducting polymers
4. Solutions based on liquefied gases (inorganic, compressed liquids at ambient temperatures)
5. Conducting solids (doped oxides and glass)

These five types of electrolyte systems are described briefly in this chapter. The first three groups of systems, which are the most common and important, are treated in detail in this book.


## II. LIQUID ELECTROLYTE SOLUTIONS

## A. General

Nonaqueous liquid electrolyte solutions may be divided into subgroups according to several criteria based on the differences among the various polar aprotic solvents. The first division can be between protic or polar aprotic nonaqueous solvents and nonpolar solvents. In polar aprotic and protic nonaqueous systems, conductivity is achieved by the dissolution of the electrolytes and the appropriate charge separation of the dissolved species, allowing for their free migration under the electrical field. In nonpolar systems the conductance mechanism may be more complicated and is based on cluster formation and hopping of the charge carriers from cluster to cluster.

In this chapter, as well as in the others dealing with nonaqueous liquid systems, major attention will be devoted to the polar aprotic systems because these are the most important and useful in modern electrochemistry. Another classification relates to the division between organic and inorganic polar aprotic solvents. In addition, liquid electrolyte systems may be classified according to their usage. For instance, as discussed later, for lithium battery applications, the relevant solvent groups are mainly alkyl carbonates, ethers, and esters with Li salts as electrolytes, while for basic studies of electrochemical reactions, tetra-alkyl ammonium salt solutions in solvents such as acetonitrile or *N,N*-dimethyl-formamide may be more adequate. Understanding the fundamentals of nonaqueous electrochemical liquid systems requires discussion of the following issues:

1. Review of a large variety of solvents and salts which are relevant to modern electrochemistry and the presentation of their properties.
2. Methods and criteria for evaluating the polarity of these systems.
3. Properties of solutions, solubility of salts, their degree of dissociation, and the resulting conductivity of nonaqueous solutions. In this respect, the effect of the solvent mixture and various possible conducting mechanisms should be discussed.
4. Electrochemical windows of nonaqueous liquid electrolyte solutions and what determines them.
5. Experimental aspects: purifying, drying, environmental aspects, safety, cell configurations, and the use of appropriate reference electrodes.
6. Stability of electrodes in liquid electrolyte solutions, corrosion problems. Especially interesting is the electrochemistry of active metals.
7. Typical applications
   a. criteria for the choice of nonaqueous electrochemical systems for electrochemical applications
   b. nonaqueous batteries
   c. study of highly sensitive reagents
   d. study of supercritical systems
   e. electroorganic synthesis.

The above points are dealt with briefly in this chapter and in greater detail in subsequent chapters.

## B. Polar Aprotic Solvents

Scheme 1 provides names and structure formulas of the most important organic and inorganic polar aprotic solvents used in modern electrochemistry. Table 1 summarizes some of the physical properties of several solvents appearing in

ETHERS

Diethyl ether (DEE)                    CH₃CH₂OCH₂CH₃

Tetrahydrofuran (THF)                  

2-Methyl-THF (2Me-THF)                 

Diethoxyethane                         CH₃CH₂OCH₂CH₂OCH₂CH₃

1-3 Dioxolane                          

ALKYL CARBONATES

Ethylene carbonate (EC)                

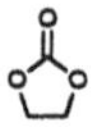

Propylene carbonate (PC)               

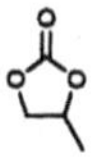

Dimethyl carbonate (DMC)               CH3OCOCH3

Diethyl carbonate (DEC)                CH3CH2OCOCH2CH3

Ethyl methyl carbonate (EMC)           CH3CH2OCOCH3

ESTERS

Methyl formate (MF)                    HCOCH3

γ-Butyrolactone (BL)                   

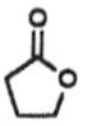

Methylacetate                          CH3COCH3

**(a)**

**Scheme 1**   (a and b) The most important organic polar aprotic solvents in electrochemistry. (c) The most important inorganic nonaqueous solvents in electrochemistry.

OTHERS

| | |
|---|---|
| Acetonitrile | $CH_3CN$ |
| Nitromethane | $CH_3NO_2$ |
| N-N dimethyl formamide | |

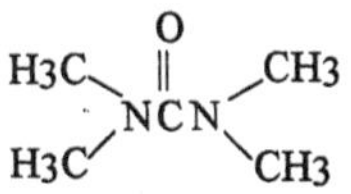

| | |
|---|---|
| Dimethyl sulfoxide | |

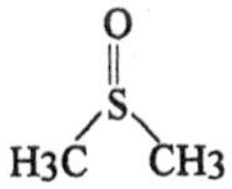

| | |
|---|---|
| Sulfolane | |

| | |
|---|---|
| Methylene chloride | $CH_2Cl_2$ |

**(b)**

| | |
|---|---|
| Thionyl chloride | |

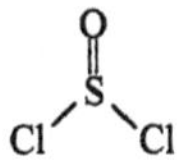

| | |
|---|---|
| Sulfuryl chloride | |

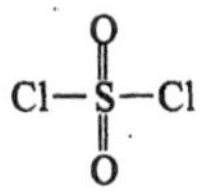

| | |
|---|---|
| Sulfur Dioxide | $O=S=O$ |
| Ammonia | $NH_3$ |

**(c)**

Scheme 1, based on Refs. 1–5. The criteria for the choice of solvents depend on the following factors:

1. The reactivity of the solvent toward the electroactive species and the electrode material. In this respect, the acidity ($pK_a$) or basicity of the solvent may be important.
2. The polarity, and hence the ability to dissolve salts and provide appropriate conductivity.
3. The electrochemical window of the solvent (which means the voltage domain with no solvent reaction). As discussed below, this parameter, however, may relate strongly to the salt chosen [6]. It should be emphasized that the electrochemical window of these systems relates strongly to their purity. For instance, trace oxygen and water are highly reactive

**Table 1** Physical Properties of Some Solvents

| Solvent | Dielectric constant ($\varepsilon$) | Density (g/cm$^{-3}$) | BP (°C) | MP (°C) | Viscosity (cP) | Dipole moment[a] (Debye) |
|---|---|---|---|---|---|---|
| Organic Solvents | | | | | | |
| Diethyl ether (DEE) | 4.265 | 0.70768 | 34.60 | −116.2 | 0.224 | 1.18 |
| Tetrahydrofuran (THF) | 7.39 | 0.880 | 65.0 | −108.5 | 0.46 | 1.71 (B) |
| 2-Methyl-THF | 6.24 | 0.848 | 80 | — | 0.457 | — |
| 1,3-Dioxolane | 7.13 | 1.0600 (20°C) | 78 | −95 | 0.589 | 1.47 |
| Ethylene carbonate (EC) | 89.6 (40°) | 1.3218 (39°) | 248 | 39–40 | 1.85 (40°) | 4.80 (B) |
| Propylene carbonate (PC) | 64.4 | 1.19 | 241 | 18.55 | 2.53 | 5.21 (B) |
| Dimethyl carbonate (DMC) | 3.12 | — | 90 | 3 | 0.585 | — |
| Diethyl carbonate (DEC) | 2.82 | 0.9693 | 127 | −43 | 0.748 | 0.90 |
| Methyl formate (MF) | 8.5 (20°C) | 0.9741 | 31.50 | −99 | 0.3298 | 1.77 (B) |
| γ-Butyrolactone (BL) | 39.1 | 1.13 | 202 | −43 | 1.75 | 4.12 (B) |
| Methylacetate | 6.68 | 0.9279 | 56.9 | −98.05 | 0.364 | 1.61 |
| Acetonitrile | 35.95 | 0.7768 | 81.6 | −45.7 | 0.3409 | 3.94 |
| Nitromethane | 35.94 | 1.1312 | 101.2 | −28.6 | 0.694 | 3.50 |
| N,N-dimethyl formamide | 36.71 | 0.9443 | 158 | −61 | 0.796 | 3.86 |
| Dimethyl sulfoxide | 46.45 | 1.0955 | 189.0 | 18.55 | 1.991 | 3.96 |
| Sulfolane | 43.26 (30°C) | 1.2619 (30°C) | 287.3 | 28.86 | 10.284 (30°C) | 4.7 |
| Methylene chloride | 8.93 | — | 39.7 | −95.1 | 0.413 | 1.14 |
| Inorganic Solvents | | | | | | |
| Thionyl chloride | 9.25 (20°C) | 1.629 (20°C) | 75.8 | −104.5 | 0.603 (20°C) | 1.38 (B) |
| Sulfuryl chloride | 9.15 (22°C) | 1.6570 | 69.4 | −54.1 | 0.674 | 1.795 |
| Sulfur Dioxide | 1.41 (20°C) | 1.3695 | −10.02 | −75.46 | 0.291 | 1.60 |
| Ammonia | 23.9 | 0.747* 10$^{-3}$ | −33.33 | −77.74 | 0.1660 | 1.47 |

[a] Measured in the gas phase unless marked (B), which means measured in benzene.

in these systems, and their reduction may interfere with the pure electrochemical behavior of the system, thus limiting the electrochemical window of the solvent-salt system [7,8]. Hence, the possibility of obtaining *highly* pure material is also an important criterion for the choice of solvent.

4. Safety aspects such as toxicity and flammability.
5. Physical aspects such as volatility.

The first criterion for the choice of salts that can be used with respect to these solvents relates to the possibility of obtaining appropriate conductivity. Basically, in order to obtain good solubility and charge separation in solution, the charge density of the ions has to be low, and thus bulky anions and/or cations have to be chosen. Hence, the cations may be $Li^+$, $Mg^{2+}$, or $Ca^{2+}$ for battery applications (based on these active metals as anodes), and $Na^+$, $K^+$, or tetraalkylammonium $(R_4N^+)$ for general application. Since many applications dictate small cations to be used (as for Li batteries), the commonly used anions are characterized by bulky structures, and thus include $I^-$, $Br^-$, $ClO_4^-$, $AsF_6^-$, $PF_6^-$, $BF_4^-$, $CF_3SO_3^-$, $(CF_3SO_2)_2N^-$, $(CF_3SO_2)_3C^-$, $BR_4^-$ (R = alkyl or aryl), $SCN^-$, $RCO_2^-$.

Basically, the final choice of the cation has to relate strictly to the application. The presence of cations such as $Li^+$ or $Na^+$ in solutions may lead to precipitation of insoluble surface films or noble metal electrodes and thus interfere with the basic electrochemical behavior of many redox couples on nonactive metal electrodes in polar aprotic solvents [9]. The use of tetraalkyl ammonium salts eliminates this problem because the thermodynamics of insoluble salt precipitation on electrodes differs in the presence of these bulky cations from that developed in the presence of cations of alkaline or alkaline earth metals [6–9].

Another example is Li battery application. Here, the key factor relates to the surface chemistry of lithium developed in solutions, as discussed in Chapter 6. Thus, the choice of the salt anion may depend primarily on its reactivity on the active metal surface [10–11].

## III. EVALUATION OF SOLVENT POLARITY

A key requirement for solvents in electrochemical systems is their ability to form conductive electrolyte solutions. The possibility of dissolving salts and separating ions in solution depends on the polarity of the solvent. A primary measure for the polarity of solvents can be properties such as the dielectric constant (Table 1) or dipole moment, which influences electrostatic interactions of solvents with solutes. However, these parameters are not sufficient for an appropriate evaluation of solvents for electrochemistry. The crucial problem with their use is that the solvating power of a solvent is a fairly complex quantity which depends on various factors, such as solvent basicity, acidity, polarity, structure, etc., whose relative contributions may vary considerably with the substrate under consideration. In other words, different reactions may respond differently to the same changes in solvent environment.

The basic requirement for the development of a more generally applicable solvent concept is the need to try to separate the various factors responsible for the solvating power of a solvent. It is important to find criteria for the solvents' character that can be correlated not only to salt solubility and apparent conductivity but also to the impact of the solvents on the thermodynamics and kinetics of the electrochemical reactions. There are several approaches to defining a typical solvent property that can represent its polarity and be correlated to the thermodynamics and kinetics of reactions conducted in its solutions (i.e., a linear free-energy relationship). A comprehensive review of such approaches by Reichardt [12] divides them into three categories:

1. Comparative equilibrium measurements of reactions in solvents
2. Comparative spectroscopic measurements based on solvatochromic effects of solvents
3. Comparative measurements of reaction rates in various solvents

Table 2, taken from Reichardt's review, describes representative approaches from the above three categories. Especially interesting are the parameter's donor number (DN), acceptor number (AN), and $E_T 30$, which appears in Table 3.

Figure 1, taken from Refs. 12, 13, and 14 presents both the compound whose solvatochromic effect defines the polarity scale $E_T 30$ and the typical absorption in the UV-VIS of this compound in three different solvents of different polarities (in decreasing order): ethanol, acetonitrile and 1-4-dioxane. It is clearly seen in this figure that as the solvent is less polar its absorption in the visible appears at a lower wave number. The scale itself is defined by the following equation:

$$E_T 30 \quad (\text{kcal/mole}) = hc\overline{v}L \tag{1}$$

where $h$ is Planck's constant, $c$ is the speed of light, $\overline{v}$ is the absorption's wave number, and $L$ is Avogadro's number. The importance of this solvent parameter lies in the ease of its measuring and the widely successful experience obtained to date in correlating this parameter to reaction rate coefficients and equilibrium constants in many solvent systems. The $E_T 30$ values for different solvents are listed in Table 3 [12,15].

The donor number (DN), introduced by Gutmann and Mayer [15–19], represents a measure of the donor properties of solvents and is defined as the numerical value of the heat of adduct formation between the donor molecule and the reference acceptor $SbCl_5$ in dilute 1,2-dichloroethane solution:

$$SbCl_5 + \text{solvent} \rightarrow [SbCl_5 * \text{solvent}] \text{ in 1,2-dichloromethane} \tag{2}$$

Donor numbers of various solvents are listed in Table 3. The values range between 0 and about 60 kcal/mole. Polar solvents such as nitromethane are very weak donors. Solvents such as acetonitrile and propylenecarbonate are moderately strong donors. Ketone, ester, and ether solvents show medium donor properties. Amides and sulfones are strongly basic solvents with donor numbers 25 to 30, and aliphatic amines are very strong donors.

The characterization of the *acceptor* properties of solvents is a more difficult problem. The definition of an analogous thermochemical quantity is not possible because many solvents contain atoms with lone pair electrons and thus may undergo adduct formation with strong reference acceptors.

The problem was solved by Mayer [15–20] with the use of triethylphosphine $Et_3P$ oxide ($Et_3P^{\delta+}$---$O^{\delta}$) as a basic reference donor in a solvent acting as an acceptor.

Triethylphosphine oxide contains a highly basic oxygen atom, which is easily accessible to solvent electrophilic attack. This causes a polarization of the $P{=}O$ bond and a downfield shift of the $^{31}P$ NMR signal. The observed chemical shifts ($\delta$) referred to the reference solvent *n*-hexane and extrapolated to infinite dilution may be taken as a measure of the acceptor properties of the solvents. Hence, the acceptor number (AN) is defined as follows:

$$AN = \frac{\delta(Et_3PO \text{ in solvent S}) - \delta(Et_3PO \text{ in hexane})}{\delta(SbCl_5 \cdot Et_3PO \text{ in } C_2H_4Cl_2)} \qquad (3)$$

Acceptor numbers of various solvents are also listed in Table 3. The values range from zero, for the reference solvent *n*-hexane, to about 130, for trifluoromethane sulfonic acid. For instance, the acceptor number of aliphatic alcohols varies between 27 and 41 (methanol). Within the group of dipolar aprotic solvents there are considerable differences in acceptor properties. Solvents such as propylenecarbonate, tetramethylene-sulfone, acetonitrile, dimethylsulfoxide, or nitromethane are stronger acceptors than solvents such as acetone, *N*-methylpyrrolidone, or dimethylacetamide. The acceptor strengths of amine solvents vary considerably with the degree of substitution. For instance, triethylamine has no acceptor properties.

The importance of solvent parameters such as DN and AN and the advantage of their use over physical-electrostatic parameters was further demonstrated by Mayer et al. [21], who studied correlations between the DN and AN of solvents and redox potentials and their chemical equilibrium and ion pair equilibria. According to the Born theory, redox potentials should depend linearly on the reciprocal of the solvent's dielectric constant. Plotting $E_{1/2}$ values of a redox such as Cd/$Cd^{2+}$ versus $1/\varepsilon$ of the solvents in which it is measured results in a very scattered picture. In contrast, it has been clearly shown by Mayer et al. [15] that redox potentials of metals (e.g., Zn/$Zn^{2+}$, Cd/$Cd^{2+}$, Eu/$Eu^{2+}$) can be nicely

**Table 2** Typical Approaches for Defining Solvent Polarity

| Symbol (name) | Physical quantity measured | Solvent-dependent standard process |
|---|---|---|
| *From equilibrium measurements* | | |
| $L$ (desmotropic constant) | Equilibrium constant | Keto-enol tautomerism equilibrium of ethyl acetoacetate at ca. 20°C. |
| $-\Delta G°$ | Free energy of the reaction in standard state | NH/OH tautomerism equilibrium of Schiff bases of pyridoxal-5'-phosphate at 25°C. |
| DN (donor number) | Reaction enthalpy $-\Delta H_{\text{EPD-SbCl5}}$ | 1:1 adduct formation between antimonyl(V) chloride as standard EOA and EPD solvents in 1,2-dichloroethane at 25°C. |
| *From measurements of reaction rates* | | |
| $Y$ | Relative rate constant $k_1$ | $S_N1$ Solvolysis of *tert*-butyl chloride at 25°C. |
| $X$ | Relative rate constant $k_2$ | $S_E2$ reaction of tetramethyltin with bromine at 20°C. |
| $\Omega$ | *Endo-exo* product ratio | Diels-Alder $[_\pi4 + _\pi2]$ cycloaddition of cyclopentadiene to methyl acrylate at 30°C. |
| *From spectroscopic measurements* | | |
| $Z$ | Molar transition energy | Charge transfer absorption of 1-ethyl-4-methoxycarbonylpyridinium at 25°C. |
| $E_T, E_T(30)$ | Molar transition energy | $\pi-\pi^*$ absorption of pyridiniophenolate at 25°C. |
| $Z^R$ | Molar transition energy | $\pi-\pi^*$ absorption of a positively solvatochromic undecamethinemerocyanine dye at 25°C. |

| | | |
|---|---|---|
| $Z^B$ | Molar transition energy | π-π* absorption of a negatively solvatochromic nonamethinemerocyanine dye at 25°C. |
| $\phi(F)$ | Wave number difference | π-π* absorption of saturated aliphatic ketones. |
| $E_K$ | Molar transition energy | d-π* absorption of tetracarbonyl [N-(2-pyridylmethylene) benzyl-amino]molybdenum(o). |
| $E_T^{SO}$ | Molar transition energy | π-π* absorption of N,N-dimethylthiobenzamide S-oxide. |
| $\pi^*$ | Absorption wave number | π-π* absorption of several compounds particularly nitrosubstituted arenes (e.g., 4-nitroanisole (1-methoxy-4-nitrobenzene). |
| $S$ | Equilibrium constant, rate constant, molar transition energy | Mixed parameter, calculated from various solvent dependent processes. |
| $G$ | Relative wave number difference | IR stretching vibration absorption of X=O and X—H · · · B groups (X=C,S,N,O, or P, B = solvent) in the gas phase and in solution. |
| $B$ | Wave number difference | IR stretching vibration absorption of O—D and O—H groups in CH$_3$OD or C$_6$H$_5$OH in the gas phase and in solution. |
| $P$ | Relative $^{19}$F NMR chemical shift | $^{19}$F NMR absorption of 1-fluoro-4-nitrosobenzene. |
| $AN$ (acceptor number) | Relative $^{31}$P NMR chemical shift | $^{31}$P NMR absorption of triethylphosphane oxide. |

*Source:* From Ref. 12 with permission from Willey-VCH Verlag GmbH, 1998.

**Table 3**  $E_T 30$ Values, Donor Numbers (DN), and Acceptor Numbers (AN) of Various Solvents

| Solvent | $E_T 30$* | DN[†] | AN[†] |
|---|---|---|---|
| Acetic acid | 51.2 | — | 52.9 |
| Acetone (AC) | 42.2 | 17.0 | 12.5 |
| Benzene | 34.5 | 0.1 | 8.2 |
| 1,2-Dichloroethane (DCE) | 39.4 | 0.0 | 16.7 |
| Diethylamine | 35.4 | — | 9.4 |
| Diethyl ether | 34.6 | 19.2 | 3.9 |
| N,N-Dimethylformamide (DMF) | 43.8 | 26.6 | 16.0 |
| Dimethyl sulfoxide (DMSO) | 45 | 29.8 | 19.3 |
| Ethanol (ET) | 51.9 | — | 37.9 |
| Formamide | 56.6 | 24 | 39.8 |
| N-Hexane | | 0.0 | 0.0 |
| Methyl acetate | | 16.5 | 10.7 |
| N-Methyl-2-pyrrolidone (NMP) | | 27.3 | 13.3 |
| Propylene carbonate | 46.6 | 15.1 | 18.3 |
| Propionitrile | 43.6 | 16.1 | — |
| Thionyl chloride | | 0.4 | — |
| Water | 63.1 | 16.4 | 54.8 |

*Source*: *From Ref. 12. Copyright Willey-VCH Verlag GmbH 1998. [†]From Ref. 15. Reproduced by permission of The Electrochemical Society, Inc., 1998.

correlated to the donor numbers of the solvents, and that the $E_{1/2}$ values measured behaved linearly versus the solvents' DN. Similarly, $E_{1/2}$ values for reductions of quinones to semiquinones are linear (with a very high correlation coefficient) versus the acceptor number of nonaqueous solvents.

Another important example is the correlation that could be found between $\Delta G_r$ values of chemical reactions in different solvents and their DN and AN values according to the following equation [15]:

$$\Delta G_r^S - \Delta G_r^R = A(DN^S - DN^R) + B(AN^S - AN^R) + C\Delta E \qquad (4)$$

where $\Delta G_r^S$ and $\Delta G_r^R$ denote the standard free energies of the reaction of any chemical reaction in a solvent S and a reference solvent R, and where $DN^S$, $DN^R$, $AN^S$, and $AN^R$ denote the corresponding donor and acceptor numbers, respectively. The third term, $\Delta E$, which is related to the free energy of hole formation, is empirical and may usually be neglected. The constants $A$ and $B$ are related to the donor and acceptor properties of the reactants.

The application of such a correlation is straightforward. The values $A$ and $B$ can be found for each specific reaction by experimental equilibria measurements in various solvents. These empirical constants can then be used to calculate $\Delta G_r$ values in a variety of solvents whose DN and AN are known. As shown by

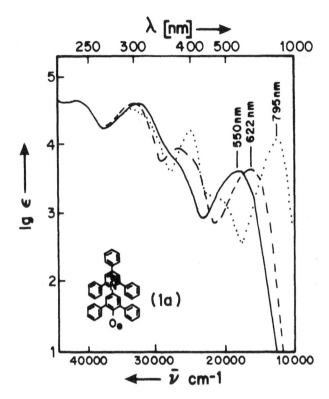

**Figure 1** UV-VIS absorption spectrum of 2,6-diphenyl-4-(2,4,6-triphenyl-1-pyridino) phenolate (*la*) in ethanol (solid line), acetonitrile (dashed line), and 1,4-dioxolane (dotted line) at 25°C [12]. (With copyrights from WILLEY-VCH Verlag GmbH, 1998.)

Mayer [22], the above correlations indeed work well and are quite useful for predicting values such as the free energy of salt solutions and complex formation in various solvents. Another typical example of the importance of the use of DN and AN as solvent parameters, instead of properties such as the dielectric constant, would be ion pair association constants in isodielectric solvents. For instance, as shown by Mayer [15], association constants of various perchlorates isocyanates, and halides (alkali metal, ammonium, and tetraalkyl ammonium cations) are very different in isodielectric solvents such as nitromethane (DN = 2.7), acetonitrile (DN = 14.1), and DMF (DN = 26.6), whose dielectric constant is around 26 at room temperature.

The association constants of the above salts are the highest in nitromethane (low DN), lower in acetonitrile (medium DN), and negligible in DMF (high DN). These results are well understood in light of the meaning of the donor number, as they reflect the ability of a solvent to interact with cations, acids, and other species having electron deficiency: hence, the higher the donor number, the higher the interaction of solvent molecules with cations, and hence, the lower the degree of anion-cation association.

In conclusion, extensive work on solvent properties has revealed that simple physical properties, such as the dielectric constant or dipole moment, are inadequate measures for solvent polarity (which can correlate well with the influence of solvents on thermodynamic and kinetic reaction parameters in them). Better solvent parameters, which correlate well with the impact of the solvent chosen on electrochemical and chemical reactions, are donor and acceptor numbers or parameters based on solvatochromic effects, because these reflect not only pure electrostatic effects but rather the entire electronic properties of a solvent.

## IV. CORRELATIONS BETWEEN THERMODYNAMIC PROPERTIES OF SOLUTES IN DIFFERENT SOLVENTS

The possibility of predicting thermodynamic properties of redox couples and solutes in different solvents is very important. It should be very useful to develop procedures of "transferring" thermodynamic data such as redox potentials from solvent to solvent. In fact, the correlation found between kinetic and thermodynamic parameters of reactions in solutions, and solvent parameters such as DN, AN, dielectric constant, etc., indicates that it may be quite feasible to draw empirical formulas which predict, for instance, redox potentials in some solvents, based on well-established data obtained experimentally with other solvents. Thus, it may be possible to define transfer parameters ($\Delta G_t^\circ$, $\Delta H_t^\circ$, $\Delta S_t^\circ$, etc.) reflecting the difference between aqueous and polar aprotic solutions in the thermodynamic properties of solutes.

As discussed by Coetzee [23] for soluble electrolytes, the transfer free energy ($\Delta G_t^\circ$) can be obtained by vapor pressure measurements or, more conveniently, by solubility measurements, *provided* that the same solid phase is in equilibrium with the saturated solutions in the two solvents. For example, for a salt of the type MX which is completely dissociated in both saturated solutions, the following relationship holds for $\Delta G_t^\circ$ at 298 $K$ from the reference solvent, R, to another solvent, S:

$$\Delta G_t^\circ = 1.36\log(K_{SP}^R/K_{SP}^S) \text{ kcal mol}^{-1} \tag{5}$$

where $K_{sp}$ is the solubility product of the electrolyte.

A second method for the measurement of the transfer free energies of electrolytes involves a comparison of the potentials of corresponding galvanic cells in different solvents.

As Coetzee [23] has concluded, there is no fundamental problem (although there may be practical difficulties) in measuring the transfer free energy of an uncharged solute or of an electrolyte as a whole. In his review [23], Coetzee describes the reference systems based on ferrocene and $Ph_4AsPh_4B$ and the $Ag/Ag^+$ couple. He further explains [23] that transfer free energies can also be expressed in terms of shifts in standard potentials, which would be particularly appropriate for battery technology. For example, for a couple of

$$M^{z+} + ze \rightleftharpoons M^0 \text{ at } 298 \ K,$$

$$\Delta E_0 = \left(\frac{0.043}{z}\right) \Delta G_t^\circ = \left(\frac{0.059}{z}\right) \log \gamma_t \quad (6)$$

where $E_0$ is in volts and $\Delta G_t^\circ$ is in kcal $mol^{-1}$. For instance, the standard reduction potential of the $Ag^+/Ag^0$ couple is 0.32 V more negative in dimethyl sulfoxide and 0.21 V more positive in propylene carbonate than it is in water [23].

Table 4 presents the $\Delta G_t^\circ$, $\Delta H_t^\circ$, and $\Delta S_t^\circ$ of transfer of monovalent cations from water to six different aprotic solvents [23–25]. Some useful consequences of the data in Table 4 may be summarized as follows [23]:

(a) Compared with many nonaqueous solvents studied, water is a poor solvent in terms of enthalpy of solvation, but a good solvent for most ions in terms of entropy of solvation, undoubtedly owing to the large entropy gain associated with the disruption of the unique, highly ordered water structure by the hydrated metal ions. The nonaqueous solvents have much less ordered structure. A solute may act either as a net structure maker or a net structure breaker, depending mainly on its size and charge. Usually, singly charged inorganic ions are net structure breakers, and the degree of structure breaking increases with an increased size of the ion. Small, multicharged inorganic ions, such as nickel or copper (II) ions, are net structure makers. Large, uncharged solutes and even ions containing large, essentially nonpolar moieties are net structure makers.

(b) As expected, since a Lewis basicity scale depends on the nature of the reference acceptor, the order of donor strengths of solvents is not the same for all cations. However, toward most multicharged cations, the order of donor strengths is DMSO > W > AN > SL, PC. Toward the single-charge metal cations listed, the order remains largely the same.

(c) Expected differences in the redox potentials of $M^+/M$ couples can be predicted. For instance, the free energy of transfer of the lithium ion varies from $-3.5$ (in DMSO) to $+5.7$ (in PC) kcal $mol^{-1}$. Hence, the standard reduction potential of the $Li^+/Li^0$ couple varies from 0.15 V more negative (in DMSO) to 0.24 V more positive (in PC) than its value in water.

(d) Small, "hard" anions, such as chloride ions, and also large anions with

**Table 4** Standard Free Energies, Enthalpies, and Entropies of Transfer of Singly Charged Ions from Water to Dipolar Aprotic Solvents (kcal mol$^{-1}$)

| Ion | $\Delta G_t^\circ$ | $\Delta H_t^\circ$ | $-T\Delta S_t^\circ$ | $\Delta G_t^\circ$ | $\Delta H_t^\circ$ | $-T\Delta S_t^\circ$ | $\Delta G_t^\circ$ | $\Delta H_t^\circ$ | $-T\Delta S_t^\circ$ |
|---|---|---|---|---|---|---|---|---|---|
| | Dimethylformamide | | | Dimethyl sulfoxide | | | Sulfolane | | |
| Li$^+$ | $-2.3$ | $-7.7$ | $+5.4$ | $-3.5$ | $-6.3$ | $+2.8$ | — | — | — |
| Na$^+$ | $-2.5$ | $-7.9$ | $+5.4$ | $-3.3$ | $-6.6$ | $+3.3$ | $-0.7$ | $-3.6$ | $+2.9$ |
| K$^+$ | $-2.3$ | $-9.4$ | $+7.1$ | $-2.9$ | $-8.3$ | $+5.4$ | $-1.0$ | $-6.0$ | $+5.0$ |
| Rb$^+$ | $-2.4$ | $-9.0$ | $+6.6$ | $-2.6$ | $-8.0$ | $+5.4$ | $-2.1$ | $-6.4$ | $+4.3$ |
| Cs$^+$ | $-2.2$ | $-8.8$ | $+6.6$ | $-3.0$ | $-7.7$ | $+4.7$ | $-2.4$ | $-5.9$ | $+3.5$ |
| Ag$^+$ | $-4.1$ | $-9.2$ | $+5.1$ | $-8.0$ | $-13.1$ | $+5.1$ | $-0.9$ | $-3.2$ | $+2.3$ |
| Ph$_4$As$^+$ | $-9.1$ | $-4.7$ | $-4.4$ | $-8.8$ | $-2.8$ | $-6.0$ | $-8.5$ | $-2.5$ | $-6.0$ |
| Cl$^-$ | $+11.0$ | $+5.1$ | $+5.9$ | $+9.2$ | $+4.5$ | $+4.7$ | $+12.6$ | $+6.2$ | $+6.4$ |
| Br$^-$ | $+7.2$ | $+0.8$ | $+6.4$ | $+6.1$ | $+0.8$ | $+5.3$ | $+9.5$ | $+2.9$ | $+6.6$ |
| I$^-$ | $+4.5$ | $-3.3$ | $+7.8$ | $+2.2$ | $-3.2$ | $+5.4$ | $+4.9$ | $+2.1$ | $+7.0$ |
| SCN$^-$ | $+3.9$ | $-2.4$ | $+6.3$ | $+2.0$ | — | — | — | — | — |
| ClO$_4^-$ | $+0.1$ | $-5.4$ | $+5.5$ | $-0.3$ | $-4.6$ | $+4.3$ | — | — | — |
| PH$_4$B$^-$ | $-9.1$ | $-4.7$ | $-4.4$ | $-8.8$ | $-2.8$ | $-6.0$ | $-8.5$ | $-2.5$ | $-6.0$ |
| | Acetonitrile | | | Propylene carbonate | | | N-Methylpyrrolidone | | |
| Li$^+$ | — | — | — | $+5.7$ | $+0.9$ | $+4.8$ | — | — | — |
| Na$^+$ | $+3.3$ | $-3.1$ | $+6.4$ | $+3.6$ | $-1.6$ | $+5.2$ | $-3.9$ | $-9.4$ | $+5.5$ |
| K$^+$ | $+1.9$ | $-5.4$ | $+7.3$ | $+1.4$ | $-5.0$ | $+6.4$ | $-3.3$ | $-10.5$ | $+7.2$ |
| Rb$^+$ | $+1.6$ | $-5.5$ | $+7.1$ | $-0.7$ | $-5.6$ | $+4.9$ | — | — | — |
| Cs$^+$ | — | — | — | $-2.9$ | $-6.2$ | $+3.3$ | — | — | — |
| Ag$^+$ | $-5.2$ | $-12.6$ | $+7.4$ | $+3.8$ | $-3.0$ | $+6.8$ | — | — | — |
| Ph$_4$As$^+$ | $-7.8$ | $-2.5$ | $-5.3$ | $-8.5$ | $-3.6$ | $-4.9$ | $-9.5$ | $-4.2$ | $-5.3$ |
| Cl$^-$ | $+8.7$ | — | — | $+9.0$ | $+6.7$ | $+2.3$ | $+13.2$ | $+5.8$ | $+7.4$ |
| Br$^-$ | $+7.6$ | $+2.0$ | $+5.6$ | $+7.1$ | $+4.2$ | $+2.9$ | $+9.7$ | $+2.3$ | $+7.4$ |
| I$^-$ | $+4.5$ | $-1.7$ | $+6.2$ | $+4.2$ | $-0.2$ | $+4.4$ | $+5.8$ | $-0.4$ | $+6.2$ |
| SCN$^-$ | $+3.0$ | — | — | — | — | — | — | — | — |
| ClO$_4^-$ | $+1.1$ | — | — | — | — | — | — | — | — |
| PH$_4$B$^-$ | $-7.8$ | $-2.5$ | $-5.3$ | $-8.5$ | $-3.6$ | $-4.9$ | $-9.5$ | $-4.2$ | $-5.3$ |
| | Dimethylformamide | | | Dimethyl sulfoxide | | | Acetonitrile | | |
| Ba$^{2+}$ | $-5$ | $-20$ | $+15$ | $-6$ | $-19$ | $+13$ | $+14$ | $-2$ | $+16$ |
| Zn$^{2+}$ | $-7$ | $-15$ | $+8$ | $-12$ | $-15$ | $+3$ | $+16$ | $+5$ | $+11$ |
| Cd$^2$ | $-8$ | $-15$ | $+7$ | $-13$ | $-17$ | $+4$ | $+10$ | $+2$ | $+8$ |
| Mn$^{2+}$ | — | — | — | $-12$ | $-15$ | $+3$ | $+16$ | $+5$ | $+11$ |
| Ni$^{2+}$ | — | — | — | $-23$ | $-26$ | $+3$ | $0$ | $-11$ | $+11$ |

*Source*: From Refs. 23–25. Reproduced by permission of the The Electrochemical Society, Inc., 1998.

localized charges, such as benzoate ions (not listed), require hydrogen bonding for stabilization. In aprotic solvents, therefore, such anions have high activities (and, hence, reactivities). Some consequences have already been indicated: in aprotic solvents, such anions tend to form stable complexes (e.g., ion pairs), and the solubilities of their salts with "unreactive" cations, such as the alkali metal cations, are low.

(e)  Large, "soft" anions, such as iodide, thiocyanante, and perchlorate ions, derive considerable stabilization from dispersion interactions [23] in polarizable solvents. Therefore, salts of such anions are usually soluble in aprotic solvents.

(f)  Solvation of very large cations and anions with a reasonably shielded charge is much stronger in nonaqueous solvents than in water, owing mainly to the large positive entropy of transfer.

## V.  CONDUCTIVITY OF LIQUID ELECTROLYTE SOLUTIONS BASED ON POLAR APROTIC SOLVENTS

Electrical conductivity is a critical issue in nonaqueous electrochemistry, since the use of nonaqueous solvents, which are usually less polar than water, means worse electrolyte dissolution, worse charge separation, and, hence, worse electrical conductivity compared with aqueous solutions. In this section, a short course on electrical conductivity in liquid solutions is given, followed by several useful tables summarizing representative data on solution conductivity and conductivity parameters.

The elementary condition for obtaining conductivity in solutions is obviously the dissolution of electrolyte and considerable charge separation. The most common and important conduction mechanism relates to the motion of separated ions in the liquid medium, retarded by friction (solvent-solute and ion-ion interactions). The ions thus reach a constant drift speed ($S$) under the electric field:

$$S = \frac{ZeE}{f} \tag{7}$$

where $f = 6\pi\eta a$ (Stock's law [26]), $E$ = electric field, $Z$ = ion valence, $\eta$ = solvent viscosity, and $a$ = ion radius. This drift speed is related to the ion mobility $U$:

$$S = UE \quad \text{and thus} \quad U = \frac{Ze}{6\pi\eta a} \tag{8}$$

The specific conductivity of an ion is $k_i = Z_i U_i \upsilon_i C_i F$, and thus the molar conductivity of a single ion is

$$\lambda_i = \frac{k_i}{C_i \upsilon_i} = Z_i U_i F \tag{9}$$

Hence, the molar conductivity of a $1:1$ electrolyte is

$$\Lambda_m = \frac{\kappa}{C} = (Z_+ U_+ \upsilon_+ + Z_- U_- \upsilon_-)F$$

where $\kappa$ is the electrolyte's specific conductivity and $\upsilon_i$ is the stoichiometric number of the specific ion in the salt. The mobility relates to the diffusion coefficient $D$ by the Einstein equation [27]:

$$D = \frac{URT}{ZF} \tag{10}$$

Therefore,

$$\Lambda_m = \frac{F^2}{RT}(\upsilon_+ Z_+^{\,2} D_+ + \upsilon_- Z_-^{\,2} D_-) \tag{11}$$

and thus, for a single ion,

$$\lambda_i = \frac{Z_i^2 F^2 D_i}{RT} \tag{12}$$

We distinguish two cases: strong electrolytes, where a complete charge separation is obtained as the salt dissolves and the solvent molecules stabilize its ions separately; and weak electrolytes, where the ions of the dissolved salt are only partially separated in solution. Thus, the following equilibrium exists:

$$M_x A_y \text{ (solution)} \rightleftharpoons XM^{y+} + YA^{x-} \tag{13}$$

In the case of strong electrolyte systems, the molar conductivity $\Lambda_m$ depends on concentration, according to Kohlrausch's law [28]:

$$\Lambda_m = \Lambda_m^0 - \text{const.}C^{1/2} \quad [28] \tag{14}$$

where $\Lambda_m^0$ is the molar conductivity at infinite dilution (i.e., extrapolation of $\Lambda_m$ to the limit of zero concentration). This behavior reflects a weak dependence of the molar conductivity on the electrolyte concentration due to the fact that conductivity depends on the mobility of the ions in solutions. The higher the salt concentration, the greater is the importance of ion-counterion interactions in solutions which increase the friction for free ion mobility under the electrical field. The Kohlrausch constant can be calculated based on a prediction of ionic interactions by the Debye-Hückel-Onsager theory [29]:

$$\text{Const.} = A + B\Lambda_m^0 \tag{15}$$

where $A$ and $B$ are solution constants calculated from solvent properties (viscosity $\eta$ and dielectric constant $\varepsilon$), salt properties (ionic strength and ion charge), and temperature. For instance, for $1:1$ electrolytes,

$$A = \frac{Z^2 eF^2}{3\pi\eta}\left(\frac{2}{\varepsilon RT}\right)^{1/2}, \quad B = \frac{0.586Z^3 eF}{24\pi\varepsilon RT}\left(\frac{2}{\varepsilon RT}\right)^{1/2} \tag{16}$$

where $Z$ is the ion valence, $e$ is the charge of an electron, and $F$ is the Faraday number. It can be stated that the various ions migrate independently. Therefore, $\Lambda_m$ and $\Lambda_m^0$ may be taken as the sum of the contributions of all the ions. Hence,

$$\Lambda_m = \sum_i \upsilon_i \lambda_i \tag{17}$$

and thus, a transference number of ion may be defined as follows:

$$t_i = \frac{\upsilon_i \lambda_i}{\sum\limits_i \upsilon_i \lambda_i} = \frac{Z_i \upsilon_i U_i}{\sum\limits_i Z_i \upsilon_i U_i} \tag{18}$$

For weak $1:1$ electrolytes, the dissociation constant ($K_d$) reflects the degree of ionization of the salt $\alpha$. Ignoring the activity coefficient,

$$K_d = \frac{\alpha^2 C}{1 - \alpha} \tag{19}$$

and

$$\alpha = \frac{K_d}{2C}\left\{\left(1 + \frac{4C}{K_d}\right)^{1/2} - 1\right\} \tag{20}$$

The actual molar conductivity $\Lambda_m$ relates to the hypothetical molar conductivity of a fully ionized electrolyte $\Lambda_m'$ by $\Lambda_m = \alpha\Lambda_m'$, which can be approximated for low electrolyte concentration to $\Lambda_m = \alpha\Lambda_m^0$. The relation $1/\alpha = 1 + \alpha C/K_d$ leads to the well-known Ostwald dilution law [30]:

$$\frac{1}{\Lambda_m} = \frac{1}{\Lambda_m^0} + \frac{\Lambda_m C}{K_d \Lambda_m^0} \tag{21}$$

In solvents of low polarity the formation of triple ions and ion quadrupoles with an increasing concentration of ionic salts has been observed [31]. The conductance versus concentration is a function with a minimum. The increase in conductivity versus concentration (after the minimum) is due to the formation of triple ions by coulombic interaction between ion pairs and ionic species. The equation describing this equilibrium is [32]

$$\Lambda_m = \sqrt{\frac{K_d}{C}}\,\Lambda_m^0 + \sqrt{\frac{K_d C}{k}}\,\Lambda_{m3}^0 \tag{22}$$

where $K_d$ and $k$ are the dissociation constants for ion pair and triple ion formation, and $\Lambda_m^0$ and $\Lambda_{m3}^0$ are the molar conductivities at infinite dilution for the single and triple ions, respectively. When plots of $\Lambda C^{1/2}$ versus $C^0$ in these systems yield straight lines, it indicates that such interactions do predominate.

As predicted by Fuoss and Shedlovsky, in more complicated cases, when the activity coefficient must be taken into account, the following expression may be valid for the relationship between the molar conductivity, the concentration, and the activity coefficient [33]:

$$\frac{1}{\Lambda_m S(z)} = \frac{1}{\Lambda_m^0} + \frac{CS(z)f^2}{K_d(\Lambda_m^0)^2} \tag{23}$$

where $S(z)$ is a function of solvent properties and solution conductance, and $f$ is the activity coefficient.

As determined by Venkatassety [34], the conductivity of solutions of Li-AsF$_6$ in THF and PC, which are very important for Li battery systems, does indeed behave according to Eq. (22), and thus $\Lambda_3^0$ and $\Lambda^0$ values for these solutions may be obtained by the appropriate plots $\Lambda C^{1/2}$ versus $C^0$, etc.

If the association of ions to ion pairs is solely due to electrostatic forces, then there should be a correlation between the association constant $K_A$ and the dielectric constant of the solvent. The relation proposed by Bjerrum [35] has been found to describe satisfactorily ion association in solvents of low dielectric constants [36]. In the case of solvents of moderate to high dielectric constants, the electrostatic theory of association leads to the equation [34,37]

$$K_A = \left(\frac{4\pi N a_f^3}{3000}\right) \exp\left(\frac{e^2}{a\varepsilon T}\right) \tag{24}$$

where $a_f$ is equal to $R^+ + R^-$ (the ions radii) and the other terms have the usual connotations. A plot of $\log K_A/a_f^3$ versus $1/\varepsilon$ should give a straight line, and the ion size parameter can be obtained from the slope. For instance, the plot of $K_A$ versus $1/\varepsilon$ for LiAsF$_6$ in solvents such as PC, DMSO, dimethylsulfide, and nitromethane yields a straight line from which the ion size parameter $a$ was obtained [34] and has been found to be in good agreement with the crystallographic ionic radii 4.44 Å.

As reported by Venkatassety [34], evaluation of the conductivity of dipolar aprotic solutions must take into account, in addition to ion pair association and triple ion formation, the possibility of strong ion-solvent interactions and the pronounced effect of solvent viscosity on the conductivity. A typical example is PC solution of Li salts, where the $\Lambda_0$ values calculated (based on conductivity measurements) were found to be very low in spite of the high polarity of this solvent and the expected high degree of dissociation of the electrolytes, due to the high viscosity of this solvent.

An important approach to forming conductive nonaqueous solutions is the use of solvent mixtures. There are many important highly polar aprotic solvents whose polarity leads to strong, intermolecular interactions, and thus to high viscosity. Hence, the advantage of the high electrolyte dissociation due to the high polarity is balanced by the high viscosity, which leads to the relatively low conductivity, since the latter property depends inversely on the solvent viscosity, as shown above [Eqs. (6), (7), (13), (14), (15)]. Combining these solvents with a low dielectric constant and low viscosity solvents may form mixtures of higher conductivity, compared with the single solvent systems. Typical examples are mixtures of alkyl carbonates and ethers that, together with Li salts, form highly conductive electrolyte solutions for lithium batteries. In most cases, the conductivity of these systems versus the molar ratio of the polar and viscous solvent and the less polar and nonviscous solvent has a maximum around equimolar concentrations of the solvents. Typical examples are presented in Figures 2–4

The dependence of the conductivity on the salt concentration may also be complicated. As the salt concentration increases, the total amount of charge carriers in the solution increases. However, a high salt concentration leads to high viscosity and strong ion-ion interactions, which increase the friction for ion mobility under the electric field. Thus, in many cases the dependence of the conductivity of polar aprotic solutions on the salt concentration is a function with a maximum. Typical examples are presented in Figures 5 and 6.

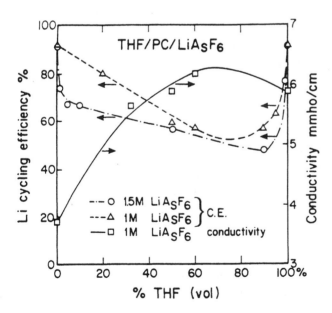

**Figure 2** Conductivity and Li cycling efficiency obtained with PC-THF/LiAsF$_6$ solutions [50]. (Reproduced by permission of The Electrochemical Society, Inc., 1998.)

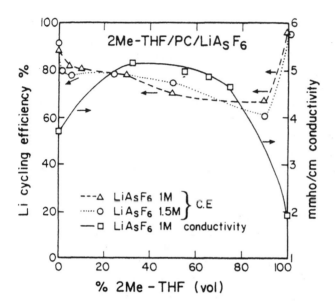

**Figure 3**  Conductivity and Li cycling efficiency obtained with PC-2Me-THF/LiAsF₆ solutions [50]. (Reproduced by permission of The Electrochemical Society, Inc., 1998.)

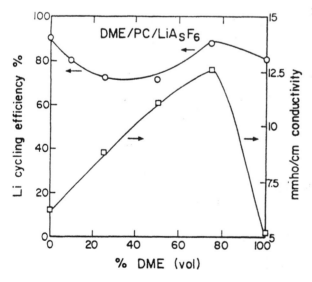

**Figure 4**  Conductivity and Li cycling efficiency obtained with PC-DME/LiAsF₆ solutions [50]. (Reproduced by permission of The Electrochemical Society, Inc., 1998.)

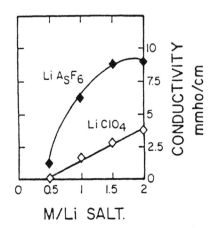

**Figure 5**  Conductivity obtained with DN/LiAsF$_6$ and DN/LiClO$_4$ solutions.

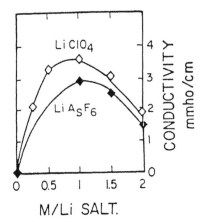

**Figure 6**  Conductivity obtained with PC/LiAsF$_6$ and PC/LiClO$_4$ solutions.

**Table 5**  Typical Conductance Parameters for Lithium Salts in Different Solvents

| Solvent | Methyl formate | Dimethyl sulfoxide | Propylene carbonate | Thionyl chloride |
|---|---|---|---|---|
| Electrolyte | $LiBF_4$, $LiClO_4$ | $LiPF_6$, $LiBF_4$ | LiBr | $LiAlCl_4$ |
| $\Lambda_0$ | 9.1, 16.6 | 56.8, 52.9 | 26.5 | 16.5 |
| $K_d$ | $2.4 \times 10^{-4}$, $2.64 \times 10^{-4}$ | $4.6 \times 10^{-3}$, $13.9 \times 10^{-2}$ | $4.7 \times 10^{-2}$ | $1.6 \times 10^{-3}$ |
| $k/\Lambda_{03}$ | $4.9 \times 10^{-3}$, $1.8 \times 10^{-3}$ | $1.7 \times 10^{-5}$, $1.2 \times 10^{-5}$ | $7.3 \times 10^{-4}$ | $1.2 \times 10^{-3}$ |

*Source:* From Ref. 34. Reproduced by permission of The Electrochemical Society, Inc., 1998.

**Table 6** Conductance Parameters for LiAsF$_6$ in Different Solutions

| | Dielectric constant | Viscosity (millipoise) | $\Lambda_0$ ($\Omega^{-1}\cdot$cm$^2\cdot$equiv$^{-1}$) | Dissociation constant, $K_d$ | Ion size parameter Bjerrum (Å) |
|---|---|---|---|---|---|
| Methyl formate (MF) | 8.5 | 3.298 | 71.5 | $3.13 \times 10^{-4}$ | 7.8 |
| Tetrahydrofuran (THF) | 7.39 | 4.6 | 28.9 | $2.0 \times 10^{-4}$ | 10.5 |
| Propylene carbonate (PC) | 64.4 | 25.3 | 22.4 | $1.5 \times 10^{-2}$ | 0.7 |
| Dimethyl sulfoxide (DMSO) | 46.6 | 19.6 | 49.0 | $7.0 \times 10^{-3}$ | 1.1 |
| Dimethyl sulfite (DMSI) | 22.5 | 7.71 | 118.1 | $5.96 \times 10^{-4}$ | 1.9 |
| Nitromethane (NM) | 35.94 | 6.27 | 128.20 | $7.16 \times 10^{-3}$ | 1.5 |

*Source:* From Ref. 34. Reproduced by permission of The Electrochemical Society, Inc., 1998.

**Table 7** Selected Data on the Conductivity of Electrolyte Solutions Based on Polar Aprotic Systems

| Solvent system | Salt (1 $M$) | Specific conductivity ($m\Omega^{-1} \cdot cm^{-1}$) | Refs. |
|---|---|---|---|
| PC | LiAsF$_6$ | 5.28 (20°) | 38 |
| PC | LiClO$_4$ | 6.0 (RT) | 40 |
| EC | LiAsF$_6$ | 6.97 (20°) | 38 |
| EC:PC 1:1 | LiAsF$_6$ | 5.94 (20°) | 38 |
| EC:PC 1:1 | LiN(SO$_2$CF$_3$)$_2$ | 5.12 (20°) | 38 |
| EC:PC 1:1 | LiSO$_3$CF$_3$ | 2.22 (RT) | 43 |
| EC:PC 1:1 | LiPF$_6$ | 6.56 (20°) | 38 |
| EC:PC 1:1 | LiBF$_4$ | 4.25 (20°) | 38 |
| THF | LiAsF$_6$ | 12.87 (20°) | 38 |
| THF | LiClO$_4$ | 3.3 (RT) | 40 |
| 2Me-THF | LiAsF$_6$ | 2.73 (20°) | 38 |
| 2Me-THF | LiClO$_4$ | 0.88 (RT) | 40 |
| 2Me-THF:EC 1:1 | LiAsF$_6$ | 10.14 (20°) | 38 |
| 2Me-THF:PC 1:1 | LiClO$_4$ | 7.4 (RT) | 40 |
| THF:EC 1:1 | LiClO$_4$ | 13.5 (RT) | 45 |
| THF:EC 1:1 | LiAsF$_6$ (1.5M) | 13.7 (RT) | 45 |
| 2Me-THF:EC:PC 6:1:1 | LiN(SO$_2$CF$_3$)$_2$ | 7.06 (RT) | 43 |
| DME:EC 1:1 | LiAsF$_6$ | 14.52 (RT) | 43 |
| γ-Butyrolactone | LiAsF$_6$ | 10.62 (20°) | 38 |
| 1,3-Dioxolane | LiClO$_4$ | 1.0 (RT) | 40 |
| 1,3-Dioxolane | LiN(SO$_2$CF$_3$)$_2$ | 4.2 (RT) | 44 |
| Ethylmonoglyme: sulfolane 1:1 | LiN(SO$_2$CF$_3$)$_2$ | 3.59 (RT) | 43 |

*Source*: From Refs. 38–44.

Extensive studies on conductivity and related parameters of nonaqueous electrolyte solutions have been carried out in connection with lithium batteries, due to the importance of lowering their internal resistance, power losses, and the consequent heat dissipation, during operation. Tables 5–8 present some typical data on the conductivity of nonaqueous electrolyte solutions [38–44].

**Table 8** Summary of Bibliographic Sources on Conductivity of Nonaqueous Electrolyte Solutions

| Subject | References |
|---|---|
| Conductivity of Li salt solutions | 38, 39, 44 |
| Conductivity of mixtures | 38, 39, 40, 41, 43 |
| Conductivity of tetraalkyl ammonium salts | 43 |

# VI. ELECTROCHEMICAL WINDOWS OF POLAR APROTIC ELECTROLYTE SOLUTIONS AND THEIR STABILITY

In order to review briefly the electrochemical stability of nonaqueous electrolyte solutions, as expressed by their electrochemical windows, classification of the electrode types used is very important. The relevant electrodes may be divided into the following categories.

## A. Noble Metal Electrodes

Noble metal electrodes include metals whose redox couple $M/M^{z+}$ is not involved in direct electrochemical reactions in all nonaqueous systems of interest. Typical examples that are the most important practically are gold and platinum. It should be emphasized, however, that there are some electrochemical reactions which are specific to these metals, such as underpotential deposition of lithium (which depends on the host metal) [45]. Metal oxide/hydroxide formation can occur, but, in any event, these are surface reactions on a small scale (submonolayer $\rightarrow$ a few monolayers at the most [6]).

## B. Reactive Metal Electrodes

Reactive electrodes refer mostly to metals from the alkaline (e.g., lithium, sodium) and the alkaline earth (e.g., calcium, magnesium) groups. These metals may react spontaneously with most of the nonaqueous polar solvents, salt anions containing elements in a high oxidation state (e.g., $ClO_4^-$, $AsF_6^-$, $PF_6^-$, $SO_3CF_3^-$) and atmospheric components ($O_2$, $CO_2$, $H_2O$, $N_2$). Note that all the polar solvents have groups that may contain C—O, C—S, C—N, C—Cl, C—F, S—O, S—Cl, etc. These bonds can be attacked by active metals to form ionic species, and thus the electrode-solution reactions may produce reduction products that are more stable thermodynamically than the mother solution components. Consequently, active metals in nonaqueous systems are always covered by surface films [46].

When introduced to the solutions, active metals are usually already covered by "native" films (formed by reactions with atmospheric species), and then these initial layers are substituted by surface species formed by the reduction of solution components [47]. In most of these cases, the open circuit potentials of these metals reflect the potential of the $M/MX/M^{z+}$ half-cell, where MX refers to the metal salts/oxide/hydroxide/carbonates which comprise the surface films. The potential of this half-cell may be close to that of the $M/M^{z+}$ couple [48].

At potentials positive to the OCV, the electrochemical window may be very limited because dissolution of the active metal takes place either via migration of $M^{z+}$ ions through the surface films or by the breakdown of the surface films, exposing the active metal to the solution and enabling a "direct" dissolution process. At potentials lower than the OCV, the situation is more complicated. Since there is no possibility of a breakdown mechanism of the surface films, the possibility of obtaining active metal deposition depends either on the feasibility of migration of the $M^{Z+}$ ion through the surface films or electron tunneling though them to form metal deposits outside the surface layer [49]. Consequently, in the case of alkaline earth metals (Ca, Mg, etc.), whose ions are bivalent, and whose migration through thin layers of their salt is difficult (if not impossible), there may be no possibility of metal deposition [50]. In such cases, the cathodic limit of the electrochemical window may be determined by reduction reactions of solutions components. In the case of alkaline metals such as lithium, the electrochemical window is very narrow because Li dissolution and deposition take place above and below the OCV at very low overpotentials.

## C.  Nonactive/slightly Reactive Metal Electrodes

Nonactive/slightly reactive electrode materials include metals whose reactivity toward the solution components is much lower compared with active metals, and thus there are no spontaneous reactions between them and the solution species. On the other hand, they are not noble, and hence their anodic dissolution may be the positive limit of the electrochemical windows of many nonaqueous solutions. Typical examples are mercury, silver, nickel, copper, etc. It is possible to add to this list both aluminum and iron, which by themselves may react spontaneously with nonaqueous solvent molecules or salt anions containing atoms of high oxidation states. However, they are not reactive due to passivation of the metal which, indeed, results from the formation of stable, thin anodic films that protect the metal at a wide range of potentials, and thus the electrochemical window is determined by the electroreactions of the solution components [51,52].

A typical example is aluminum, which is used as a current collector for cathodes in lithium batteries [53]. The stability of aluminum in many Li salt solutions at potentials as high as 4.5 V versus $Li/Li^+$ is due to the formation of highly insoluble Al-halides on the aluminum surface which remain stable and thus protect this active metal from corrosion [53]. In any event, in the evaluation of metals as electrode materials in nonaqueous systems, each case needs to be dealt with separately because the level of passivity and stability of most of the transition metals in polar aprotic systems depends on the solution composition.

## D.  Intercalation Electrodes

Intercalation electrodes may include either carbon electrodes or transition metal oxide and sulfide electrodes, which intercalate with solution components such as metal cations. The most important example is intercalation of lithium ions with many types of carbonaceous materials at low potentials (0–1.5 V versus $Li/Li^+$)

[54] and with transition metal oxides and sulfides at high potentials (2.5–4.5 V versus Li/Li$^+$) [55]. These electrochemical intercalation reactions are the basis for rechargeable Li ion battery systems and are usually the limiting electrochemical reactions in the Li salt solutions in which they are used.

## E. Redox Electrodes

Redox electrode materials mostly includes electronically conducting polymers such as polypyrrole, polythiophene, or polyaniline that may undergo reversible oxidation-reduction reactions accompanied by doping of the polymer with the counterion that balances the charge [56,57]. These processes are the anodic limit for these electrodes in many nonaqueous systems. In some cases, such as conductive polythiophene or polyparaphenylene electrodes, it is also possible to dope the polymer at low potentials and thus charge it negatively, when this negative doping is balanced by the entrance of cations into the polymer's matrix [58]. The electrochemical window of these systems may be determined solely by the redox activity of the electrode materials, which may leave quite a wide range of potentials between the two processes in which the electrode is electrochemically inert.

Besides the effect of the electrode materials discussed above, each nonaqueous solution has its own inherent electrochemical stability which relates to the possible oxidation and reduction processes of the solvent, the salts, and contaminants that may be unavoidably present in polar aprotic solutions. These may include trace water, oxygen, CO, $CO_2$ protic precursor of the solvent, peroxides, etc. All of these substances, even in trace amounts, may influence the stability of these systems and, hence, their electrochemical windows. Possible electroreactions of a variety of solvents, salts, and additives are described and discussed in detail in Chapter 3. However, these reactions may depend very strongly on the cation of the electrolyte. The type of cation present determines both the thermodynamics and kinetics of the reduction processes in polar aprotic systems [59]. In addition, the solubility product of solvent/salt anion/contaminant reduction products that are anions or anion radicals, with the cation, determine the possibility of surface film formation, electrode passivation, etc. For instance, as discussed in Chapter 4, the reduction of solvents such as ethers, esters, and alkyl carbonates differs considerably in Li or in tetraalkyl ammonium salt solutions [6]. In the presence of the former cation, the above solvents are reduced to insoluble Li salts that passivate the electrodes due to the formation of stable surface layers. However, when the cation is TBA, all the reduction products of the above solvents are soluble.

In conclusion, determination of the electrochemical window of a nonaqueous electrolyte solution depends on the electrode used, the solvent, the electrolyte's cation and anion, and the possible presence of atmospheric contaminants in solution. Chapter 6, which discusses the behavior of active metal electrodes, and Chapter 4, which describes the electrochemical behavior of nonaqueous systems with nonactive metal electrodes, deal with the stability of a wide variety of electrochemical systems of interest.

## 7. EXAMPLES FOR THE USE OF NONAQUEOUS SYSTEMS IN MODERN ELECTROCHEMISTRY

### A. Batteries

One of the most important challenges in modern electrochemistry relates to the development of novel, high energy density batteries. Naturally, the high energy density demands that attention be focused on the use of highly reactive electrode materials. These include anodes of alkaline and alkaline earth metals (Li, Na, Mg, Ca, etc.) and their alloys which possess low redox potentials, as well as high voltage cathodes of redox couples or intercalation compounds (e.g., $M_{1x}M_2O_y$, where $M_1$ and $M_2$ are active metals and transition metals, respectively) [55]. The most important requirement for electrolyte solutions for such batteries is a large electrochemical window [60]; i.e., these solutions have to be stable and safe for use in contact with both highly reactive materials (reductive agents) and oxidizing redox couples or transition metal oxides in which the transition metal is at a high oxidation state (e.g., $MnO_2$, $V_2O_5$). Such a demand is only relevant to nonaqueous systems. Due to the importance of this field in modern electrochemistry, a chapter in this book is devoted to nonaqueous batteries. However, a few typical examples are briefly reviewed in this introductory section.

The most important nonaqueous batteries are lithium and lithium ion batteries. In the former, the anode is lithium metal or lithium alloy. Both primary [61] and secondary [62] Li batteries have been developed and commercialized. Table 9 summarizes some information on several types of lithium batteries, their relevant cathode materials, voltage, classification, and electrolyte solution used [63–75]. As presented in this table, most of the Li battery systems contain electrolyte solutions which consist of salts from the $LiAsF_6$, $LiBF_4$, $LiSO_3CF_3$, $LiClO_4$, $LiPF_6$, and $LiN(SO_2CF_3)_2$ list and three major classes of solvents: alkyl carbonates (PC, EC, DMC, DEC), ethers (THF, 2-Me-THF, DME), and esters (MF and $\gamma$-BL). These salts are characterized by large, monovalent anions whose relatively low charge density enables high solubility and good ion separation in the above solvents.

As shown in Table 9, another class of Li batteries includes inorganic solutions in which the solvent may be $SO_2$ or $SOCl_2$ and in which the electrolyte is $LiAlCl_4$. In these batteries the solvents are also the cathode active material. Another type of Li battery, which is very important because of properties such as high cycle life, rechargeability, and improved safety, is the lithium ion battery.

**Table 9** Summary of Information on Several Types of Li Battery Systems (Li metal anodes)

| System (cathode)[a] | Classification | Potential (V) | Status (1998) | Electrolyte solution[c] | Ref. |
|---|---|---|---|---|---|
| Li-CuO | Primary | 1.5 | Commercial | $LiClO_4$/DN | 63 |
| Li-FeS$_2$ | Primary | 1.5 | Commercial | $LiCF_3SO_3$/DME + 1,3-Dioxolane-3 methyl-2-oxazolidone | 64 |
| Li-MnO$_2$ | Primary | 3.0 | Commercial | $LiClO_4$/PC + DME | 65 |
| Li-(CF)$_n$ | Primary | 2.6 | Commercial | $LiAsF_6$/dimethyl sulfite or $LiBF_4$/BL + THF or $LiBF_4$/PC + DN | 66 |
| Li-VO$_x$ | Primary | 3.3 | Commercial | $LiAsF_6$ 2M + $LiBF_4$ 0.4M/MF | 67 |
| Li-SO$_2$ | Primary | 3.0 | Commercial | LiBr/acetonitrile | 68 |
| Li-SOCl$_2$ | Primary | 3.6 | Commercial | $SOCl_2$ (some with $LiAlCl_4$ additives) | 69 |
| Li-MnO$_2$ | Secondary | 3.0 | Commercial | $LiAsF_6$/DN | 70 |
| Li-MoS$_2$ | Secondary | 1.7 | Was commercial and was withdrawn[b] | $LiAsF_6$ 1M/EC:PC 1:1 | 71 |
| Li-S | Secondary | 2.2 | Prototype | $LiClO_4$ 1M/THF + toluene + polysulfite | 72 |
| Li-VO$_x$ | Secondary | 2.3 | R&D | $LiAsF_6$ 1.4M/2Me-THF | 73 |
| Li-SO$_2$ | Secondary | 3.1 | R&D | $LiAlCl_4$*$xSO_2$ | 74 |

[a] The systems are classified by the cathodes used.

[b] This battery system was abandoned due to some safety hazards and accidents.

[c] See structure formulae of all the abbreviated solvents appearing in this table in Scheme 1.

These batteries consist of carbonaceous anodes which may insert lithium reversibly up to a stochiometry of $LiC_3$ [54,76] (true for disordered, hard carbons [77] and lithiated transition metal oxide cathodes of high redox potential [55] (>4 V versus $Li/Li^+$, such as $Li_xMn_2O_4$ (spinel) [78], $Li_xNiO_2$ [79], and $Li_x CoO_2$ [80]). The most common electrolyte solutions of these batteries include a mixture of alkyl carbonate solvents (EC-DMC are commonly used) [81], and salts from the $LiPF_6$ (most important), $LiAsF_6$, $LiN(SO_2CF_3)_2$, and $LiC(SO_2CF_3)_3$ list [82].

Several of these battery systems have already been commercialized and are now available on the market. However, it should be emphasized that, in addition to Li batteries comprised of the liquid electrolyte systems mentioned above, there are increasing efforts and extensive work toward the development of solid state Li batteries based on ionically conducting polymers [83], membranes [84], and ceramic materials [85]. In addition to Li batteries, there have been attempts to develop nonaqueous calcium batteries [86]. The most important Ca battery system studied contains $SOCl_2$ as the solvent and the cathode active material [87]. Another interesting development in this field relates to magnesium batteries. There are several positive signs that it is possible to dissolve and plate magnesium reversibly and also to reversibly insert magnesium ions into a few transition metal oxides such as $VO_x$ [88] and $CoO_x$ [89].

## B.  Supercapacitors

Another important development in the field of electrochemical power sources is that of supercapacitors. The electrical double layer in the usual interfaces between conducting electrodes and aqueous solutions has a differential capacitance ranging from 10–30 $\mu F/cm^2$ [90]. The availability of very porous, high surface area, electrically conducting carbonaceous materials (thousands of square meters per gram) enables electrode-electrolyte systems to be obtained, which can accumulate electric charges at a capacity of 100–150 F/gr [91]. Combining two electrodes of this type and operating them within a potential window which is sufficiently narrow to prevent electrochemical redox reactions of the electrolyte solutions/ electrode materials forms a highly reversible power source. These power sources are termed supercapacitors and may reach an energy density of the same order of magnitude as some practical rechargeable batteries, such as lead acid and NiCd. Their major advantages over conventional battery systems are their very high reversibility and cycle life (millions of charge-discharge cycles can be reached) [92] and very good safety features. These advantages are due to the fact that their operation involves no redox reactions but interfacial electrostatic interactions only.

The theoretical (maximal) energy stored by a capacitor depends on its capacitance and the voltage applied according to the formula $E_c = \frac{1}{2}CV^2$. Hence, the higher the possible potential window in which such a device can be operated, without any interference by solution redox reactions, the higher the energy density that can be reached. Since many of the nonaqueous systems mentioned in the previous sections may have electrochemical windows as wide as 4–5 V, they are excellent candidates as electrolyte solutions for supercapacitors of very high energy density that may compete with that of some battery systems.

Typical examples are combinations of polar aprotic solvents such as alkyl carbonates, esters, and acetonitrile with tetralkyl ammonium salts (a typical anion that can be used is $BF_4^-$ [93]), all of which can be operated within a potential window wider than 3 V (compared with 1.2 V for aqueous solutions) in supercapacitors. Table 10 summarizes data of several types of supercapacitors that are currently being developed [94–97]. Typical current uses of these devices are as back-up power sources for computers, electric toys, solar watches, VCRs, cameras, etc. Several reviews are available in the literature on the trends in this field [91–98], the expected power and energy densities of these devices in their current state of the art and in possible future developments, and comparisons with state-of-the art battery systems. It is evident from these reports that the choice of the appropriate porous electrode materials with suitable current collectors, as well as the use of nonaqueous systems of optimized composition which may enable the application of higher voltages than those currently applied (<3 V [94–97]), will lead to the development of supercapacitors possessing energy density not too far from that of some rechargeable batteries which are in current use today.

## C. Electroorganic Synthesis

The use of nonaqueous systems increases the applicability of electrochemistry for the synthesis and study of a large variety of compounds that cannot be synthesized in aqueous solutions for three major reasons:

1. The solubility of precursors in water
2. The relatively narrow electrochemical window of the aqueous solutions
3. The reactivity of the reagents, intermediates, or products with water

In many cases, only the first point is important, and thus a great deal of work in this field has been carried out with protic nonaqueous solvents such as alcohols (MeOH, EtOH, IPrOH) and acids (e.g., acetic acid) [99–105]. However, there are many reactions for which the second and third points are also relevant. Hence, there are many reports on the use of solvents such as acetonitrile, methylene chloride, THF, DMSO, DMF, and electrolytes from the $R_4N^+X^-$ family (R = alkyl, X = $Cl^-$, $BF_4^-$, etc.) in unique organic synthesis. Several typical examples are summarized in Table 11 [106–115].

**Table 10** Examples of Nonaqueous Supercapacitors

| Electrodes | Solution components[a] | Potential applied | Configuration | Energy density | Cycle life | Reference J. Power Sources, 60 (1996) |
|---|---|---|---|---|---|---|
| Aluminum $Al_2O_3$ | EC, PC, BL $(CH_3CH_2)_4N^+$ $RCOO^-$ | >1.5 V | Lab apparatus | Still irrelevant | — | Page 179 |
| Polyacene | PC $(CH_3CH_2)_4N^+BF_4^-$ | 2.5 V | Coin cells Jelly-rolled cells | — | >1,000,000 | Page 207 |
| Activated carbon | PC $(CH_3CH_2)_4N^+BF_4^-$ | 2.5 V | Jelly-rolled cells | — | — | Page 213 |
| Activated carbon | PC $R_4P^+BF_4^-$ or $R_4 N^+BF_4^-$ | 3 V | Coin cells | 10 W-h/L | >10,000 | Page 239 |

[a] See Scheme 1 for solvents' structure formulae.

With the advantage of nonaqueous electrochemistry, which has opened the door for many new reaction paths of unique selectivity and product distribution, this area of research has become very popular, and an immense amount of work is currently being performed. Among the many books, review articles, and proceedings volumes from international symposia and special issues in electrochemistry journals, the following examples are of interest.

1.   Special Volume on Electrochemical Organic Synthesis, ed. S. Toris, *Electrochimica Acta, 142* (1997), vol. 13–14.
2.   Organic Electrochemistry, eds. H. Lund and M. M. Baizer, Marcel Dekker, New York (1991).
3.   H. J. Schaefer, in Comprehensive Organic Synthesis, eds. B. M. Trost, I. Fleming, and G. Pattender, vol. 3, Pergamon Press, Oxford (1991).
4.   Techniques of Electroorganic Synthesis, Parts I–III, ed. N. L. Weinberg, Wiley, New York (1975).

## D.   The Use of Nonaqueous Electrochemical Systems Which Comprise Nonpolar Solvents

One of the most important uses of electrochemistry in industrial processes is, obviously, the plating of metals. Since the most common use of these processes relates to surface finishing, corrosion protection, and surface electrical conductivity, the relevant metals are usually noble or nonactive, such as gold, silver, copper, nickel, tin, and lead. Electroplating of all these metals requires the use of aqueous solutions. However, there are applications in which deposition of active metals such as aluminum is required, and where only nonaqueous solutions can be used due to the high reactivity of bare aluminum to any protic solvent. Several commercial processes for electrodeposition of aluminum based on nonaqueous systems (e.g., etheral solutions) have been developed [116–122]. Interesting solutions for electrodeposition of aluminum were studied and developed by Gileadi et al. [123–125]. The uniqueness of these systems lies in the use of nonpolar solvents such as aromatic hydrocarbons, and, thus, the conducting mechanisms in these solutions are quite different from the conventional ones discussed in Section IV.

As shown by Gileadi et al. [123–125], it is possible to form solutions based on aromatic hydrocarbons such as ethyl benzene, toluene, benzene, and mesitylene. The electrolyte comprises $Al_2Br_6$ and KBr [126]. By using a nearly molar concentration of these species, it is possible to obtain specific conductivity of the same order of magnitude as that measured in *polar* aprotic solutions (around 5 $m\Omega\cdot cm^{-1}$) at ambient temperatures. It was found that these electrolytes do not dissolve in the above solvents to form ions or ion pairs, but rather form clusters which are charged aggregates of $Al_2Br_7^-K^+$. The conductivity mechanism is thus

**Table 11** Typical Examples of the Use of Nonaqueous Electrochemistry in Organic Synthesis Which Emphasize Its Unique Advantages for This Field

| Basic reaction | Electrolyte system | Remarks | Refs. |
|---|---|---|---|
| $CCl_3CF_3 + RCHO \xrightarrow{2e^-, H^+} RCHCl_2CF_3 + Cl^-$ <br> $R\text{-SiR}'_3 \xrightarrow{-e^-, F^-}$ [R–SiR'3–F] $\xrightarrow{}$ F-SiR'3 + R· $\xrightarrow{H·}$ R-R, R-H | DMF, CTMS | CTMS = chlorotrimethylsilane | 106 |
|  | $Et_4NBF_4/CH_3CN$ | R = alkyl, Ph, allyl <br> 2.3V vs. Ag/AgNO$_3$ | 107 |
| $Ph_3Sb(OCOCH_3)_2 \xrightarrow{-2e^-, 2F^-} Ph_3Sb(F)_2 + 2CH_3COO·$ | $Et_3N*3HF/CH_3CN$ |  | 108 |
| (imidazole/fluoroaryl structure, CHO) | $DMSO + 0.1\ M\ Et_4NBF_4$ | $E = 1.60V$ vs. SCE | 109 |
| $Ph(H)C{=}C(H)H + CO_2 \rightarrow Ph(H)C{=}C(CO_2^-)H + \tfrac{1}{2}H_2$ | $0.1\ M\ Bu_4NBF_4/DMF$ | cathode = Pt, anode = Mg | 110 |

| Reaction | Conditions | | Ref. |
|---|---|---|---|
| (CH₃)₂CNO₂ + MeOOC-C≡C-COOMe ⟶ ... | $CH_2Cl_2/Bu_4NBF_4$ | −0.45V/Hg | 111 |
| (EtO)₂PH + cyclohexane-CHO ⟶ ... | DMF/Et₄NOTs | | 112 |
| cyclopentanone ⟶ lactone | $CF_3COOH—CH_2Cl_2$ + Et₃N | | 113 |
| cyclohexanone-CH₂CH₂COOH ⟶ bicyclic alcohol | $Bu_3P/CH_3SO_3H$ Et₄NCl in CH₃CN | cathode = Sn, anode = carbon, 0°C | 114 |
| dibromocyclopropane + Me₃SiCl ⟶ ... | THF/TBAP | | 115 |

a non-Stokesian one (i.e., it does not depend on ion migration under conditions of viscous flow) and resembles the conductivity of $H^+$ and $OH^-$ in aqueous solutions. Hence, it may be concluded that the surprisingly high conductivity of these nonpolar systems is due to intercluster transport of $Br^-$ by a "jumping" type transfer. It has been proven that these systems can be used for aluminum plating without the complications related to the highly reactive aluminum deposits [123–126], and it can be assumed that a similar approach may be used for plating other active metals, such as magnesium. It should be noted that similar behavior and conductivity mechanism have been found by Gileadi et al. in other unique systems based on liquid bromine [127] and on liquid and solid iodine [128], where $Br^-$ and $I^-$ are the hopping ions.

## E.  Electrochemistry in Inorganic Nonaqueous Systems

The examples of the practical use of nonaqueous electrochemical systems reviewed briefly in the last four sections related mostly to solutions based on organic solvents. However, there are several interesting uses of nonaqueous electrochemical solutions based on inorganic solvents. In fact, two inorganic solvents, $SOCl_2$ and $SO_2$, were mentioned in connection with lithium batteries. Lithium/ $SO_2$/$LiAlCl_4$ and lithium/$SOCl_2$/$LiAlCl_4$ are already being used as commercial primary Li batteries and have the unique feature of the solvent also being the major cathode material [129]. However, in addition to batteries, there are other interesting uses of inorganic solvents in electrochemistry. Bard et al. reported on the study of unique electrochemical reactions in liquid ammonia [130] and $SO_2$ [131]. Especially important are the studies of electrochemical reactions under supercritical and near-critical conditions [132]. It is clear from this pioneering work that electrochemistry in liquid $NH_3$ and $SO_2$, especially under near-critical and supercritical conditions, opens the door for unique paths of electrosynthesis and selectivity. It should be emphasized that a major driving force for the use of electrochemical methods in synthesis is creating new possibilities in terms of product distribution and selectivity, which could not otherwise be obtained. This considerably enriches the spectrum of compounds that can be obtained and facilitates the synthesis of new products at high selectivity. Hence, the study of the systems by Bard et al. is a further step in this direction. Several examples of unique reactions studied in $NH_3$ and $SO_2$ by this group are

1.  Reduction of $C_{60}$ in liquid $NH_3$ [133]
2.  Photohole emission from platinum in liquid $SO_2$ [134]
3.  Oxidation of the $Ni(2,2'$ bipyridine$)_3^{2+}$ complex to $Ni(2,2'$ bipyridine$)_3^{4+}$, which possess thermochronic properties [135]
4.  Reduction of halonitrobenzenes in liquid $NH_3$ to form stable radical anions [136]

## 8. NONAQEOUS SYSTEMS OTHER THAN ELECTROLYTE SOLUTIONS

### A. Electrically Conducting Polymers

There are two types of electrically conducting polymers: electron conducting and ion conducting. The first type includes polymers of species possessing double and triple bonds or aromatic rings which, upon polymerization, form conjugated π systems in which electrons are mobile along the polymeric chains. Typical examples are polyacetylene, polypyrrole, polypheniline, polythiophene, and polyaniline. However, this group is beyond the scope of this book because these electronically conducting polymers must be considered not as electrolyte systems but as electrode materials. There are indeed important applications of these polymers as electrode materials for rechargeable batteries [137], biosensors [138], and sensors [139]. The second group, ionically conducting polymers, can indeed be considered as nonaqueous electrolyte systems which provide solid matrices in which the electric current flows via ion migration, as in liquid electrolyte solutions. Consequently, a chapter in this book has been dedicated to these systems. However, within the framework of this introductory section, some classification and overview of these systems appear below. Basically, conducting polymeric matrices contain two key elements: polymers possessing the appropriate functional groups that can interact with ions, and electrolytes which are, in many cases, the same as those used in liquid solutions. They may be divided into two classes:

1. Systems composed of polymers which have functional groups that can solvate ions, wherein the charge separation required for ionic motion is obtained by dispersion and dissolution of electrolytes in the polymeric matrix.
2. Systems in which the polymers' functional groups are not sufficient to create charge separation between electrolyte ions. Thus, special additives are required to promote sufficient charge separation, which enables the ions in the solid matrix to respond to an electric field. Hence, in these systems, the major role of the polymer is mostly to maintain a solid, stable matrix, whereas the ion migration within the matrix under an electrical field is feasible because of the additives.

Typical examples of the first group are polyethylene oxide [140–144] and polypropylene oxide [145–146] mixed with lithium salts. The oxygen atoms of these polymers interact with the Li ions and solvate them, and thus the Li salt may be uniformly dissolved in these polymers.

Typical examples for the second group are polyacrylonitrile [147–149] or polyvinylidene difluoride (PVDF) with Li salts. In these cases, a reasonable solubilization of the Li salts and the required charge separation are maintained by the addition of solvents such as PC and/or EC in amounts sufficient to maintain electrolyte solubility and, thus, ionic conductivity, but which are still small enough to keep the matrix in a solid/rigid state. Typical applications of the above systems are rechargeable lithium batteries. Replacement of the liquid electrolyte solutions in these batteries by rigid, ionically conducting polymeric membranes has been proven to reduce the interfacial reactivity of lithium and lithiated compounds (used as anodes) in the battery systems and thus increase their cycle life [150] and their stability [151]. A key point in the use of the ionically conducting polymers is their conductivity. Many of these systems show an activation controlled behavior, i.e., an exponential dependence of their specific conductivity in $1/T$ [152]. At certain low temperatures, some of these systems may form a crystalline structure which further decreases their conductivity. Hence, considerable efforts are directed toward increasing the conductivity of these systems in order to make them compatible with liquid systems. Common directions in this respect are

1.  Applications which require elevated temperatures (e.g., Li batteries for electric vehicles [153]) at which a reasonable conductivity is reached.
2.  The use of copolymerization which adds functional groups which prevent the crystallization of these polymers at low temperature and thus maintains an amorphous structure within an extended temperature range [154].
3.  Further use of additives which solubilize the salts in the solid matrix and separate between the electrolyte's ions [155–159]. Table 12 summarizes some typical examples of these systems.

As demonstrated in Chapter 7, this field is very vital and innovative. A large variety of systems are currently being studied, and several promising battery systems based on these conducting polymer systems are being developed and commercialized.

## B.  Molten Salts

Molten salt electrolyte systems comprise salts in the liquid state (molten), which form ionic phases and are highly ionically conductive, thus reaching specific conductivities comparable to those of room temperature concentrated aqueous solutions ($0.1 < \lambda < 10 \ \Omega^{-1} \cdot cm^{-1}$) [160]. These systems can be divided into two classes:

1.  High temperature molten salts
2.  Ambient temperature molten salts [161]

**Table 12** Typical Examples of Ionically Conducting Polymeric Systems

| System | Monomer | Common additives | Specific conductivity at certain temperatures | Applications | Ref. |
|---|---|---|---|---|---|
| PEO | (epoxide) | $\gamma$-LiAlO$_2$ | $(PEO)_8LiClO_4$ + 10–30 w/o $\gamma$-LiAlO$_2$, 5 × $10^{-4}$ mS·cm$^{-1}$ at 25°C | Lithium batteries, electrochromic devices, sensors | 155 |
| PEO | (epoxide) | MEEP | 45 w/o PEO-$(LiBF_4)_{0.13}$, 55 w/o MEEP 2.4 × $10^{-3}$ mS·cm$^{-1}$ at 25°C | | 156 |
| PAN | —CH$_2$CH— \| CN | PC,EC,BL | 21 m/o PAN, 38 m/o EC, 33 m/o PC, 8 m/o LiClO$_4$ 1.7 mS·cm$^{-1}$ at 20°C | | 157 |
| PVDF | —CH$_2$CF$_2$— | PC,EC | PVDF/LiClO$_4$/PC in mol ratio 49.7/21.3/29.0 1 × $10^{-2}$ mS·cm$^{-1}$ at 25°C | | 158 |
| SBR–NBR | Styrene–butadiene rubber acrylonitrile–butadiene rubber | BL,DME | SNR/NBR (1:1), 1M LiClO$_4$, BL 49 w/o >1 mS·cm$^{-1}$ at 25°C | | 159 |

The first type includes salts of metallic cations and anions which may be halides, carbonates, etc., and thus melt at high temperatures (several hundred degrees Celsius and above). The melting temperature of these systems may be reduced by the use of eutectic compositions of salt mixtures [162]. Classical uses of these systems include the manufacture of pure metals (e.g., aluminum, alkaline, and alkaline earth metals) [163] and fuel cells (e.g., molten carbonate fuel cells) [164]. Due to their extensive industrial importance, Chapter 9 in this book is devoted to these systems and describes their unique properties and utilities. The second class of molten salts includes substituted nitrogen containing aromatic compounds which can be charged positively and can thus form complexes with anions such as halides, $AlCl_4^-$, and even $AsF_6^-$ and $PF_6^-$. Typical examples are pyridine and imidazole. There are a number of publications on room temperature electrochemical systems based on $n$-alkylpyridinium $AlCl_4^-$ [165] and 1-alkyl-3alkyl-imidazolium $AlCl_4^-$ [166] (e.g., $n$-butylpyridinium X [167] and 1-methyl-3-ethyl-imidazolium X [168], where X=anion, usually $Cl^-$ or $AlCl_4$).

Extensive work has been devoted to these systems over the past two decades. Electrochemical windows were determined and found to depend on the acidity, basicity, or neutrality of the melts. For instance, an electrochemical window of more than 4 V was proven for neutral 1-methyl-3-ethyl-imidazolium $AlCl_4^-$ [169]. The RT conductivity of these melts is around 0.01 $\Omega^{-1} \cdot cm^{-1}$, which is quite comparable to that of electrolyte solutions based on polar aprotic solvents [170]. These systems have advantages over the molten salts of the first class in several respects and over electrolyte solutions of polar aprotic solvents. Their relatively high RT conductivity is a major advantage over high temperature molten salts. They dissolve inorganic salts better than organic solvents do and may be less reactive than they are, especially at low potentials [169]. Consequently, a major potential application of these systems lies in the processing of active metals which cannot be deposited in aqueous or protic systems due to reactivity problems and which cannot be treated in polar aprotic systems due to the solubility limitations of their salts. Indeed, extensive work has been devoted to the study of lithium [171], sodium [172], aluminum [173], and palladium [174] deposition in substituted imidazolium based molten salts. Lithium intercalation into carbon was also studied in these systems. These studies obviously pave the way for a possible application of the RT molten salts for rechargeable batteries with active metal anodes.

In usual polar aprotic solvents, all active metals (e.g., Li, Mg, Ca, and Al) are covered with surface films due to the reduction of solution components by the active metal and the consequent precipitation of insoluble species, but in the above molten salts these corrosion/passivation phenomena may be much less pronounced. Thus, deposition of divalent metals such as magnesium and calcium, which is not possible in usual polar aprotic systems because of the surface film barrier, may be feasible in electrolyte systems based on these molten salts.

## C. Miscellaneous Nonaqueous Electrolyte Systems

The nonaqueous electrolyte systems mentioned in this section are solid state systems which may be very important and promising for some types of batteries and fuel cells applications.

### 1. Ion Conductive Alumina

β-Alumina containing lithium or sodium ions may serve as a solid electrolyte system for high temperature lithium and sodium batteries [175]. For instance, extensive work has been devoted to the development of sodium sulfur batteries that operate at $T > 350°C$ [176]. The basic cell reaction, which is fully reversible, is $2Na + xS \rightleftharpoons Na_2S_x$ (polysulfide). The common electrolyte systems for these batteries are $β-Al_2O_3$ which contain $Na_2O$ [176]. At the above elevated temperatures, the specific conductivity of these systems is about $0.1-1 \ \Omega^{-1} \cdot cm^{-1}$ [177]. Fabrication of these materials in rigid structure and their corrosion protection against sodium are the most important issues in their application as Na-S battery electrolyte systems. Similar application of lithiated β-alumina were studied for lithium-sulfur batteries. The major uses for such high temperature batteries should be electric vehicles and load leveling [178]. It should be noted that there are similar developments for high temperature fuel cells; i.e., several types of oxides can conduct at elevated temperature oxygen ions and $H_3O^+$ ions. These oxides may serve as electrolyte systems for fuel cells in which the cathodic reaction is oxygen reduction and the anodic reaction is hydrogen oxidation [179]. In addition to alumina, other oxides such as zirconia [180] and ceria [181] are also currently investigated in this connection.

### 2. Composite Solid Electrolyte Systems for Li Batteries

Ionically conducting polymers and their relevance to lithium batteries were mentioned in a previous section. However, there are several developments which contain both ionically conducting materials and other supporting agents which improve both the bulk conductivity of these materials and the properties of the anode (Li)/electrolyte interface in terms of resistivity, passivity, reversibility, and corrosion protection. A typical example is a composite electrolyte system comprised of polyethylene oxide, lithium salt, and $Al_2O_3$ particles dispersed in the polymeric matrices, as demonstrated by Peled et al. [182]. By adding alumina particles, a new conduction mechanism is available, which involved surface conductivity of ions on and among the particles. This enhances considerably the overall conductivity of the composite electrolyte system. There are also a number of other reports that demonstrate the potential of these solid electrolyte systems [183].

## 3. Solid Electrolyte Based on Lithium Salts

A unique approach in nonaqueous electrochemistry which may be applicable to
several fields, especially for batteries, was recently presented by Koch et al. (private communication). They showed that it is possible to use solid matrices based
on lithium salts contaminated with organic solvents as electrolyte systems. These
systems demonstrate several advantages over liquid systems based on the same
solvents and salts as solutions. Their electrochemical windows are larger, especially in the anodic direction (oxidation reactions), and it appears that their reactivity toward active electrodes (e.g., Li, Li—C) is much lower than that of the
liquid electrolyte systems.

## 4. Solid Electrolytes for Thin Film Technology

The wide interest in thin batteries has led to research and development of solid
electrolytes which are mechanically stable, chemically inert with active electrode materials (e.g., low voltage Li and Li/C anodes or high voltage transition
metal oxides), and which are electronic insulators and ionic conductors as thin
films. Table 13 presents some typical compositions, deposition methods, and
specific conductivity [184] of examples of these materials. Several materials
in the table have already found use in novel developments of thin film, rechargeable Li batteries. Such microbatteries, all of which are solid state, can be
integrated into electronic circuits, operate microcomputers, and, in general, are
highly attractive as rechargeable power sources for microelectronic devices
[184].

**Table 13** Typical Solid Electrolytes Suggested for All Solid State Li
Batteries

| Electrolyte | Deposition method | $\sigma_i \, (\Omega^{-1}\cdot cm^{-1})$ |
|---|---|---|
| $Li_2O\text{—}B_2O_3$ | Thermal evaporation | $10^{-8}\text{–}10^{-10}$ |
| $Li_2O\text{—}B_2O_3\text{—}LiI$ | RF sputtering | $10^{-6}$ |
| $Li_2O\text{—}B_2O_3\text{—}P_2O_5$ | Thermal evaporation | $10^{-10}$ |
| $Li_2O\text{—}SiO_2\text{—}ZrO_2$ | Magnetron sputtering | $10^{-6}$ |
| $Li_{3.6}SiO_{0.6}P_{0.4}O_4$ | RF-sputtering | $5 \times 10^{-6}$ |
| $Li_2S\text{—}B_2S_3$ | Thermal evaporation | $10^{-3}$ |
| $Li_2S\text{—}SiS_2$ | Thermal evaporation | $10^{-5}$ |
| $Li_2O\text{—}P_2O_5\text{—}Nb_2O_5$ | Thermal evaporation | $10^{-8}$ |
| $Li_2O\text{—}P_2O_5\text{—}Nb_2O_5$ | RF sputtering | $2 \times 10^{-7}$ |

*Source*: Reprinted from Ref. 184 with permission from Elsevier Science, 1998.

## REFERENCES

1. G. E. Blomgren, in *Lithium Batteries*, J. P. Gabano, ed., Academic Press, London, 1983, Table 1, p. 17.
2. L. A. Dominey, in *Lithium Batteries, New Materials, Developments and Perspectives*, G. Pistoia, ed., Elsevier, New York, 1994, Table 8.1, p. 160.
3. *Handbook of Chemistry and Physics*, 78th ed., D. R. Lide, editor-in-chief, CRC Press, Boca Raton, FL, 1997–1998.
4. T. C. Waddington, *Non-Aqueous Solvents*, Nelson, London, 1973.
5. J. Barthel and H. Gores, in *Chemistry of Nonaqueous Solutions*, G. Mamontov and A. I. Popov, eds., VCH, New York, 1994, p. 1.
6. D. Aurbach, M. Daroux, P. Faguy, and E. Yeager, *J. Electroanal. Chem., 297*, 225 (1991).
7. F. A. Cotton and G. Wilkinson, *Advanced Inorganic Chemistry—A Comprehensive Text*. 2nd ed., Interscience, New York, 1966, p. 240.
8. D. Aurbach, O. Youngman, and P. Dan, *Electrochim. Acta, 35*, 639 (1990).
9. E. Goren, O. Chusid (Youngman), and D. Aurbach, *J. Electrochem. Soc., 138*, L6 (1991).
10. W. R. Fawcett, *Russ. Elektrokhim.* (Engl. ed.) *19*, 1044 (1983).
11. M. Garreau and S. Fouache, in *Proceedings of the Symposium on Lithium Batteries*, A. N. Dey, ed., The Electrochemical Society, Inc., softbound series PV 87-1, 1987, p. 240.
12. C. Reichardt, *Angew. Chem. Int. Ed. Engl., 18*, 98 (1979).
13. K. Dimroth, C. Reichardt, T. Siepmann, and F. Bohlmann, *Justus Liebigs Ann. Chem., 661*, 1 (1963); K. Dimrot and C. Reichardt, *ibid., 727*, 93 (1969); C. Reichardt, *ibid., 752*, 64 (1971).
14. K. Dimrot and C. Reichardt, *Z. Anal. Chem., 215*, 344 (1966); Z. B. Maksimpvic, C. Reichardt, and A. Spiric, *ibid., 270*, 100 (1974).
15. U. Mayer, in *Proceedings of the Workshop on Lithium Nonaqueous Battery Electrochemistry*, E. B. Yeager, B. Schumm, Jr., G. Blomgren, D. R. Blankenship, V. Leger, and J. Akridge, eds., The Electrochemical Society, Inc., softbound series PV 80-7, 1980, p. 13.
16. U. Mayer, *Pure Appl. Chem., 51*, 1697 (1979).
17. U. Mayer, *Monatsh. Chem., 109*, 421 (1978).
18. V. Gutmann and E. Wychera, *Inorg. Nucl. Chem. Lett., 2*, 257 (1966).
19. V. Gutmann, *The Donor-Acceptor Approach to Molecular Interactions*, Plenum Press, New York, 1978.
20. U. Mayer, V. Gutmann, and W. Gerber, *Monatsh. Chem., 106*, 1235 (1975).
21. U. Mayer, *Monatsh. Chem., 108*, 1479 (1977); U. Mayer, A. Kotocova, V. Gutmann, and W. Gerber, *J. Electroanal. Chem., 100*, 875 (1979).
22. U. Mayer, *Monatsh. Chem., 109*, 775 (1978).
23. J. F. Coetzee, in *Proceedings of the Workshop on Lithium Nonaqueous Battery Electrochemistry*, E. B. Yeager, B. Schumm, Jr., G. Blomgren, D. R. Blankenship, V. Leger, and J. Akridge, eds., The Electrochemical Society, Inc., softbound series PV 80-7, 1980, p. 70.

24. B. G. Cox, G. R. Hedwig, A. J. Parker, and D. W. Watts, *Austral. J. Chem., 27*, 477 (1974).
25. G. R. Hedwig, D. A. Owensby, and A. J. Parker, *J. Am. Chem. Soc., 97*, 3888 (1975); I. M. Kolthoff and M. K. Chantooni, Jr., *J. Phys. Chem., 76*, 2024 (1972); J. f. Coetzee and E. J. Subak, Jr., *Rev. Chim. Minérale, 15*, 40 (1978).
26. P. W. Atkins, *Physical Chemistry*, 4th ed. Oxford University Press, Oxford, Melbourne, and Tokyo, 1990, p. 754, Eq. 6.
27. P. W. Atkins, *Physical Chemistry*, 4th ed., Oxford University Press, Oxford, Melbourne, and Tokyo, 1990, p. 764, Eq. 16.
28. S. H. Maron and J. B. Lando, *Fundamentals of Physical Chemistry*, Macmillan, New York; Collier Macmillan, London, 1974, Ch. 13, Eq. 40.
29. P. W. Atkins, *Physical Chemistry*, 4th ed., Oxford University Press, Oxford, Melbourne, and Tokyo, 1990, p. 760, Eq. 13.
30. V. Fried, U. Blaks, and H. F. Hameka, *Physics and Chemistry*, Collier Macmillan International Editions, New York and London, 1977. Ch. 24, Eqs. 24–30.
31. C. A. Kraus and R. M. Fuoss, *J. Am. Chem. Soc., 57*, 1 (1935).
32. R. M. Fuoss and C. A. Kraus, *J. Am. Chem. Soc., 55*, 2387 (1933).
33. R. M. Fuoss and T. Shedlovsky, *J. Am. Chem. Soc., 71*, 1496 (1949).
34. H. V. Venkatasetty, in *Proceedings of the Workshop on Lithium Nonaqueous Battery Electrochemistry*, E. B. Yeager, B. Schumm, Jr., G. Blomgren, D. R. Blankenship, V. Leger, and J. Akridge, eds., The Electrochemical Society, Inc., softbound series PV 80-7, 1980, p. 46.
35. N. Bjerrum, *Kgl. Danske Vidensk. Selskab., 7*, 9 (1926).
36. R. M. Fuoss and C. A. Kraus, *J. Am. Soc., 55*, 1019 (1933).
37. R. M. Fuoss, *J. Am. Soc., 80*, 5059 (1958).
38. J. T. Dudley, D. P. Wilkinson, G. Thomas, R. LeVae, S. Woo, H. Blom, C. Horvath, M. W. Juzkow, B. Denis, P. Aghakian, and J. R. Dahn, *J. Power Sources, 35*, 59 (1991).
39. Y. Matsuda, *J. Power Sources, 20*, 19 (1987).
40. S. Tobishima and T. Okada, *Electrochim. Acta, 30*, 1715 (1985).
41. Y. Matsuda and M. Morita, *J. Power Sources, 20*, 273 (1987).
42. H. V. Vankatasetty, in *Lithium Battery Technology*, H. V. Vankatasetty, ed., The Electrochemical Society, Inc., Pennington, NJ, 1984, p. 1.
43. M. Salomon and E. J. Plichta, *Electrochim. Acta, 1*, 113 (1985).
44. L. A. Dominey, *Novel Stable, Non-complexing Anions for Rechargeable Lithium Batteries*, Covalent Associates, Inc., Woburn, MA; NSF Grant ISI8660048, Prepared for: National Science Foundation, Washington, DC, 1987.
45. D. M. Kolb, M. Przasnyski, and H. Gerischer, *J. Electroanal. Chem., 54*, 25 (1974).
46. E. Peled and E. Gileadi, *Plating, 62*, 342 (1975); E. Peled and H. Straze, *J. Electrochem. Soc., 124*, 1030 (1977).
47. K. Kanamura, H. Tamura, and Z. I. Takehara, *J. Electronal. Chem., 333*, 127 (1992).
48. M. Odziemkowski and D. E. Irish, *J. Electrochem. Soc., 139*, 11 (1992); M. Odziemkowski and D. E. Irish, *J. Electrochem. Soc., 140*, 6 (1993).
49. E. Peled, in *Lithium Batteries*, J. P. Gabano, ed., Academic Press, London, 1983, p. 43.

50. D. Aurbach, R. Skaletsky, and Y. Gofer, *J. Electrochem. Soc.*, *138*, 3529 (1991).
51. J. Kruger, D. A. Shifler, J. F. Scanlon, and P. J. Morita, *Russian J. Electrochem.*, *31*, 1087 (1995).
52. R. G. Kelly and P. J. Moran, *Corros. Sci.*, *30*, 495 (1990).
53. M. Wohlfahrt-Mehrens, A. Butz, R. Oesten, G. Arnold, R. P. Hemmer and R. A. Huggins, *J. Power Sources*, *68*, 582 (1997).
54. R. Yazami, in *Lithium Batteries. New Materials, Developments and Perspectives*, G. Pistoia, ed., Elsevier, Amsterdam, 1994, p. 49.
55. T. Ohzuku, in *Lithium Batteries. New Materials, Developments and Perspectives*, G. Pistoia, ed., Elsevier, Amsterdam, 1994, p. 239.
56. P. Novak, K. Muller, K. S. V. Santhanam, and O. Hass, *Chem. Rev.*, *97*, 207 (1997); M. Gazard, in *Handbook of Conducting Polymers*, Vol. 1, T. A. Skotheim, ed., Marcel Dekker, New York, 1988, Ch. 19.
57. K. Nishio, M. Fujimoto, N. Yoshinaga, N. Furukawa, O. Ando, H. Ono, and T. Suzuki, *J. Power Sources*, *34*, 153 (1991).
58. T. Yamamoto, T. Saneckika, and A. Yamamoto, *J. Polym. Sci. Lett. Ed.*, *18*, 9 (1980).
59. D. Aurbach, A. Zaban, Y. Ein-eli, I. Weissman, O. Chusid, B. Markovsky, M. Levi, E. Levi, A. Schechter, and E. Granot, *J. Power Sources*, *68*, 91 (1997).
60. L. A. Dominey, in *Lithium Batteries, New Materials, Developments and Perspectives*, G. Pistoia, ed., Elsevier, Amsterdam, 1994, p. 137.
61. D. Linden, in *Handbook of Batteries*, 2nd ed., D. Linden, ed., McGraw-Hill, Inc., New York, 1995, p. 14.1.
62. S. Hossain, in *Handbook of Batteries*, 2nd ed., D. Linden, ed., McGraw-Hill, New York, 1995, p. 36.1.
63. R. Bates and Y. Jumel, in *Lithium Batteries*, J. P. Gabano, ed., Academic Press, London, 1984, p. 73.; T. Iijima, Y. Toyoguchi, J. Nishimura, and H. Ogawa, *J. Power Sources*, *5*, 1 (1980).
64. M. B. Clark, in *Lithium Batteries*, J. P. Gabano, ed., Academic Press, London, 1984, p. 115.
65. H. Ikeda, in *Lithium Batteries*, J. P. Gabano, ed., Academic Press, London, 1984, p. 169.
66. M. Fukuda and T. Iijima, in *Lithium Batteries*, J. P. Gabano, ed., Academic Press, London, 1984, p. 211.
67. C. R. Walk, in *Lithium Batteries*, J. P. Gabano, ed., Academic Press, London, 1984, p. 265.
68. C. R. Walk, in *Lithium Batteries*, J. P. Gabano, ed., Academic Press, London, 1984, p. 281; D. Linden and B. Mcdonald, *J. Power Sources*, *5*, 35 (1980).
69. C. R. Schlaikjer, in *Lithium Batteries*, J. P. Gabano, ed., Academic Press, London, 1984, p. 303.
70. P. Dan, E. Mengeritski, Y. Geronov, D. Aurbach, and I. Weissman, *J. Power Sources*, *54*, 143 (1995).
71. J. A. R. Stilb, *J. Power Sources*, *26*, 233 (1989).
72. H. Yamin and E. Peled, in *Proceedings of the Symposium on Lithium Batteries*, A. N. Dey, ed., The Electrochemical Society, Inc., softbound series PV 84-1, 1984, p. 301.

73.  K. M. Abraham, J. L. Goldman, and M. D. Demosey, *J. Electrochem. Soc., 128,* 2493 (1981); K. M. Abraham, J. L. Goldman, and G. L. Holleck, in *Proceedings of the Symposium on Lithium Batteries,* H. V. Vankatasetty, ed., The Electrochemical Society, Inc., softbound series PV 81-4, 1981, p. 271.
74.  A. N. Dey, H. C. Kuo, P. Pilliero, and M. Kalianidis, *J. Electrochem. Soc., 135,* 2115 (1988).
75.  J. O. Besenhard and G. Eichinger, *J. Electroanal. Chem., 68,* 1 (1976); G. Eichinger and J. O. Besenhard, *J. Electroanal. Chem., 72,* 1 (1976).
76.  T. Zheny, Y. Liu, E. W. Fuller, S. Tseny, U. Von Sacken, and J. R. Dahn, *J. Electrochem. Soc., 142,* 2581 (1995).
77.  K. Sato, M. Noguchi, A. Demachi, N. Oki, and M. Endo, *Science, 264,* 556 (1994).
78.  R. J. Gummow, A. de Kock, and M. M. Thackeray, *Solid State Ionics, 69,* 59 (1994).
79.  L. Xie, W. Ebner, D. Fouchard, and S. Megahed, in *Proceedings of the Symposium on Rechargeable Lithium and Lithium-Ion Batteries,* S. Megahed, B. M. Barnett, and L. Xie, eds., The Electrochemical Society, Inc. softbound series PV 94-28, 1995, p. 263.
80.  K. Mizushima, P. C. Jones, P. J. Wiseman, and J. B. Goodenough, *Mat. Res. Bull., 17,* 783 (1980).
81.  D. Guyomard and J. M. Tarascon, *J. Electrochem. Soc., 140,* 11, 3071 (1993).
82.  D. Aurbach, B. Markovsky, A. Schechter, Y. Ein-Eli, and H. Cohen, *J. Electrochem. Soc., 143,* 3809 (1996).
83.  M. Gauthier, D. Fauteux, G. Vassort, A. Belanger, M. Duval, R. Ricoux, J. M. Chabagbo, D. Muller, P. Rigaud, M. B. Armand, and D. Deroo, *J. Electrochem. Soc., 132,* 1333 (1985).
84.  M. Alamgir and K. M. Abraham, in *Lithium Batteries, New Materials, Developments and Perspectives,* G. Pistoia, ed., Elsevier, Amsterdam, 1994, p. 93.
85.  L. W. Zhang, M. Kobayashi, and S. Goto, *Solid State Ionics, 18–19,* 741 (1986).
86.  W. L. Wade, Jr., C. W. Walker, Jr., and M. Binder, *J. Power Sources, 28,* 295 (1989).
87.  E. Peled, E. Elster, R. Tulman, and J. Kimel, *J. Power Sources, 14,* 93 (1985).
88.  P. Novak, *Z. Phys. Chem., 185,* 51 (1994).
89.  T. D. Gregory, R. J. Hoffman, and R. C. Winterton, *J. Electrochem. Soc., 137,* 775 (1990).
90.  W. Lorenz, F. Möckler, and W. Müller, *Z. Phys. Chem., N. F. Frankfurt 25,* 145 (1960); A. Hamelin, *J. Electroanal. Chem., 138,* 395 (1982).
91.  A. Nishino, *J. Power Sources, 60,* 137 (1996).
92.  T. C. Murphy and W. E. Kramer, in *Proceedings of the Fourth International Seminar on Double Layer Capacitor and Similar Energy Storage Devices,* Florida Educational Seminars, December 1994.
93.  A. B. McEwen, S. F. McDevitt, and V. R. Koch, *J. Electrochem. Soc., 144,* L84 (1997).
94.  M. Morita and Y. Matsuda, *J. Power Sources, 60,* 179 (1996).
95.  S. Yata, E. Okamoto, H. Satake, H. Kubota, M. Fujii, T. Taguchi, and H. Kinoshita, *J. Power Sources, 60,* 207 (1996).
96.  A. Yoshida, S. Nonaka, I. Aoki, and A. Nishino, *J. Power Sources, 60,* 213 (1996).

97. T. Morimoto, K. Hiratsuka, Y. Sanada, and K. Kurihara, *J. Power Sources, 60,* 239 (1996).
98. J. P. Zheng, J. Huang, and T. R. Jow, *J. Electrochem. Soc., 144,* 2026 (1997).
99. Y. N. Ogibin, A. I. Ilovaisky, and G. I. Nikishin, *Electrochim. Acta, 42,* 1933 (1997).
100. K. C. Möller and H. J. Schäfer, *Electrochim. Acta, 42,* 1971 (1997).
101. M. Sugawara, K. Mori, and J. I. Yoshida, *Electrochim. Acta, 42,* 1995 (1997).
102. F. Bergamini, A. Citterio, N. Gatti, M. Nicolini, R. Santi, and R. Sebastiano, *J. Chem. Res.,* 364 (1993).
103. A. Bryan and J. Grimshaw, *Electrochim. Acta, 42,* 2101 (1997).
104. T. Inokuchi, H. Kawafuchi, K. Aoki, A. Yoshida, and S. Torii, *Bull. Chem. Soc. Jpn., 67,* 595 (1994).
105. M. N. Elinson, S. K. Feducovich, T. L. Lizunova, and G. I. Nikishin, in *Novel Trends in Electroorganic Synthesis,* S. Torii, ed., Kodansha, Tokyo, 1995, p. 47.
106. T. Shono, N. Kise, and H. Oka, *Tetrahedron Lett., 32,* 6567 (1991).
107. I. Y. Alyev, In. Rozhkov, and I. L. Knunyants, *Tetrahedron Lett.,* 2469 (1976).
108. T. Fuchigami and M. Miyazaki, *Electrochim. Acta, 42,* 1979 (1997).
109. M. Médebielle, J. Pinson, and J. M. Savéant, *Electrochim. Acta, 42,* 2049 (1997).
110. H. Kamekawa, H. Senboku, and M. Tokuda, *Electrochim. Acta, 42,* 2117 (1997).
111. G. Montero, G. Quintannila, and F. Barba, *Electrochim. Acta, 42,* 2177 (1997).
112. M. Kimura, T. Kurata, T. Yamashita, H. Kawai, and Y. Sawaki, *Electrochim. Acta, 42,* 2225 (1997).
113. K. Fujimoto, N. Yamashita, Y. Tokuda, Y. Matsubara, H. Maekawa, T. Mizuno, and I. Nishiguchi, *Electrochim. Acta, 42,* 2265 (1997).
114. H. Ohmori, T. Maki, and H. Maeda, in *Novel Trends in Electroorganic Synthesis,* S. Torrii, ed., Kodansha, Tokyo, p. 345.
115. A. J. Fry and J. Touster, *Electrochim. Acta, 42,* 2057 (1997).
116. A. Brenner, in *Advances in Electrochemistry and Electrochemical Engineering,* Vol. 5, C. W. Tobias, ed., Wiley-Interscience, New York, 1967.
117. R. Suchentrunk, in *68th Annual Technical Conference of the American Electroplating Society,* Vol. 1, 1981, paper I.
118. J. C. Beach, L. D. McGraw, and C. L. Faust, *Plating, 55,* 936 (1968).
119. M. Yoshio, H. Nakamura, H. Nogouchi, and M. Nagamatsu, in *Seventh International Conference on Nonaqueous Solutions,* Vol. 1, Regensburg, 1980, p. E3.
120. M. Yoshio, N. Ishibashi, H. Naki, and T. Peiyama, *J. Inorg. Nuclear Chem., 34,* 2439 (1972).
121. S. Birkle, in *Proceedings of the Symposium on Electrodeposition Technology, Theory and Practice,* L. T. Romankiw and D. R. Turner, eds., The Electrochemical Society, Inc., softbound series PV 87-17, 1987, p. 369.
122. J. Eckert and H. Koelling, East German Patent 143,088 (1980); *Chem. Abstr., 94*: 164816n.
123. M. Elam, E. Peled, and E. Gileadi, *J. Electrochem. Soc., 131,* 2058 (1984).
124. E. Peled, M. Elam, and E. Gileadi, *J. App. Electrochem., 11,* 463 (1981).
125. E. Peled and E. Gileadi, *J. Electrochem. Soc., 123,* 15 (1976).
126. E. Peled, M. Brand, and E. Gileadi, *J. Electrochem. Soc., 128,* 1697 (1981).
127. I. Rubinstein, M. Bixon, and E. Gileadi, *J. Phys. Chem., 84,* 715 (1980).

128. A. Brestovisky, E. Kirowa-Eisner, and E, Gileadi, *Electrochimica Acta, 31*, 1553 (1986).
129. C. R. Schilaikjer, in *Lithium Batteries*, J. P. Gabano, ed., Academic Press, London, 1984, p. 304.
130. F. A. Uribe, P. R. Sharp, and A. J. Bard, *J. Electroanal. Chem. 152*, 173 (1983).
131. C. R. Cabrera, E. Garcia, and A. J. Bard, *J. Electroanal. Chem., 260*, 457 (1989).
132. R. M. Crooks, F. F. Fan and, A. J. Bard, *J. Am. Chem. Soc., 106*, 6851 (1984).
133. F. Zhou, C. Jehoulet, and A. J. Bard, *J. Am. Chem. Soc., 114*, 11004 (1992).
134. E. Garcia and A. J. Bard, *Chem. Phys. Lett., 120*, 437 (1985).
135. J. B. Chlistunoff and A. J. Bard, *Inorg. Chem., 31*, 4582 (1992).
136. T. Teherani and A. J. Bard, *Acta Chem. Scand. B, 37*, 413 (1983).
137. P. Novak, K. Muller, K. S. V. Santhanam, and O. Haas, *Chem. Rev., 97*, 207 (1997).
138. T. Osaka, *Electrochim. Acta, 42*, 3015 (1997).
139. W. E. Morf, *The Principles of Ion Selective Electrodes and of Membrane Transport*, Elsevier, Amsterdam, 1981.
140. A. Killis, J. F. LeNest, and H. Cheradame, *Macromol. Chem. Rapid Commun., 1*, 595 (1980).
141. K. M. Abraham, *Electrochim. Acta, 38*, 1233 (1993).
142. S. Chintapalli and R. Frech, *Electrochim. Acta, 40*, 2093 (1995).
143. D. Martin-Vosshage and B. V. R. Chowdari, *Electrochim. Acta, 40*, 2109 (1995).
144. P. G. Bruce, *Electrochim. Acta, 40*, 2077 (1995).
145. M Watanabe and N. Ogata, in *Polymer Electrolyte Reviews 1*, J. R. MacCallum and C. A. Q. Vincent, eds., Elsevier, London, 1987, p. 39.
146. R. Frech and J. P. Manning, *Electrochim. Acta, 37*, 1499 (1992).
147. Z. Bashir, S. P. Church, and D. M. Price, *Acta Polym., 44*, 221 (1993).
148. M. Watanabe, M. Kanba, K. Nagaoka, and I. Shinohara, *J. Polym. Sci. Polym. Phys. Ed., 21*, 939 (1983).
149. S. Reich and I. Michaeli, *J. Polym. Sci. Polym. Phys. Ed., 13*, 9 (1975).
150. M. C. Borghin, M. Mastragostino, and A. Zanelli, *J. Power Sources, 68*, 52 (1997).
151. G. Eichinger and M. Fabian, Annual Meeting of the International Society of Electrochemistry, Xiamen, China, 1995. Extended Abstracts 1, Abst. 5-01.
152. J. Fan and P. S. Fedkiw, *J. Electrochem. Soc., 144*, 399 (1997).
153. G. Ardel, D. Golodnitsky, and E. Peled, in *Proceedings of the Symposium on Batteries for Portable Applications and Electric Vehicles*, C. F. Holmes and A. R. Landgrebe, eds., The Electrochemical Society, Inc., softbound series PV 97-18, 1997, p. 272.
154. C. Booth, C. V. Nicholas, and D. J. Wilson, in *Polymer Electrolyte Reviews 2*, J. R. MacCallum and C. A. Vincent, eds., Elsevier, New York, 1989, p. 229; F. M. Gray, *Solid State Ionics, 40/41*, 637 (1990); M. Watanabe, H. Nagasaka, and N. Ogata, *J. Phys. Chem., 99*, 12294 (1995).
155. F. Capuano, F. Croce, and B. Scrosati, *J. Electrochem. Soc., 138*, 1918 (1991).
156. K. M. Abraham, M. Alamgir, and R. K. Reynolds, *J. Electrochem. Soc., 136*, 3576, (1989).
157. K. M. Abraham and M. Alamgir, *J. Electrochem. Soc., 137*, 1657 (1990).

158. H. Ohno, H. Matsuda, K. Mizoguchi, and E. Tsuchida, *Polym. Bull., 7*, 271 (1982).
159. M. Matsumoto, T. Ichino, J. S. Rutt, and S. Nishi, *J. Electrochem. Soc., 140*, L151 (1993).
160. G. J. Janz, F. W. Dampier, G. R. Lakshminoraganan, P. K. Lorenz, and R. P. T. Tomkins, *Molten Salts: Vol. 1, Electrical Conductance, Density and Viscosity Data*, United States Department of Commerce, National Standard Reference Data Series, Natural Bureau of Standards 15, Washington, 1968.
161. F. H. Hurley and T. P. Wier, *J. Electrochem. Soc., 98*, 203, 207 (1951).
162. H. A. Hjuler, S. von Winbush, R. W. Berg, and N. J. Bjerrum, *J. Electrochem. Soc., 136*, 901 (1989).
163. K. Grjotheim, C. Krohn, M. Malinovsky, K. Matiasovsky, and J. Thonstad, *Aluminium Electrolysis, Fundamentals of the Hall-Heroult Process*, Aluminium-Verlag, Düsseldorf, 1982.
164. A. J. Appleby and F. R. Foulkes, *Fuel Cell Handbook*, Van Nostrand Reinhold, New York, 1989, p. 15.
165. H. L. Hjuler, S. Von Winbush, R. W. Berg, and N. J. Bjerrum, *J. Electrochem. Soc., 136*, 901 (1989).
166. R. C. Carlin and J. S. Wilkes, in *Chemistry of Nonaqueous Solutions*, G. Mamantov and A. I. Popov, eds., VCH, New York, 1994, p. 277.
167. C. L. Hussey, in *Chemistry of Nonaqueous Solutions*, G. Mamantov and A. I. Popov, eds., VCH, New York, 1994, p. 227.
168. S. Pye, J. Winnick, and P. A. Kohl, *J. Electrochem. Soc., 144*, 1933 (1997).
169. C. Scordilis-Kelley, J. Fuller, R. T. Carlin, and J. S. Wilkes, *J. Electrochem. Soc., 139*, 694 (1992).
170. J. S. Wilkes, J. A. Levisky, R. A. Wilson, and C. L. Hussey, *Inorg. Chem., 21*, 1263 (1982).
171. J. Fuller, R. T. Carlin, and R. A. Osteryoung, *J. Electrochem. Soc., 143*, L145 (1996); J. Fuller, R. A. Osteryoung, and R. T. Carlin, *J. Electrochem. Soc., 142*, 3632 (1995); C. Scordilis-Kelley and R. T. Carlin, *J. Electrochem. Soc., 140*, 1606 (1993).
172. G. E. Gray, J. Winnick, and P. A. Kohl, *J. Electrochem. Soc., 143*, 3821 (1996).
173. Q. Liao, W. R. Pitner, G. Stewart, C. L. Hussey, and G. R. Stafford, *J. Electrochem. Soc., 144*, 936 (1997).
174. H. C. De Long, J. S. Wilkes, and R. T. Carlin, *J. Electrochem. Soc., 141*, 1000 (1994).
175. P. S. Nicholson, in *Proceedings of the Symposium on Sodium-Sulfur Batteries*, A. Landgrebe, R. D. Weaver, and R. K. Sen, eds., The Electrochemical Society, Inc., softbound series PV 87-5, 1987, p. 80.
176. R. K. Sen and A. Landgrebe, in *Proceedings of the Symposium on Sodium-Sulfur Batteries*, A. Landgrebe, R. D. Weaver, and R. K. Sen, eds., The Electrochemical Society, Inc., softbound series PV 87-5, 1987, p. 1.
177. I. Bloom, G. H. Kureca, J. Bradley, P. A. Nelson, and M. F. Roche, in *Proceedings of the Symposium on Sodium-Sulfur Batteries*, A. Landgrebe, R. D. Weaver, and R. K. Sen, eds., The Electrochemical Society, Inc., softbound series PV 87-5, 1987, p. 125.
178. H. Birnbreier, W. Fischer, and G. N. Benninger, in *Proceedings of the Symposium*

*on Sodium-Sulfur Batteries*, A. Landgrebe, R. D. Weaver, and R. K. Sen, eds., The Electrochemical Society, Inc., softbound series PV 87-5, 1987, p. 49.

179. A. J. Appleby, *Proceedings of the Second Symposium on Electrode Materials and Processes for Energy Conversion and Storage*, S. Srinivasan, S. Wagner, and H. Wroblowa, eds., The Electrochemical Society, Inc., softbound series PV 87-12, 1987, p. 129.

180. K. West, in *Lithium Batteries. New Materials, Developments and Perspectives*, G. Pistoia, ed., Elsevier, Amsterdam, 1994, p. 323.

181. M. Godickemeier and L. J. Gauckler, *J. Electrochem. Soc., 145*, 2 (1998).

182. E. Peled, D. Golodnitsky, G. Ardel, and A. Peled, in *Proceedings of the Symposium on Rechargeable Lithium and Lithium-Ion Batteries*, S. Megahed, B. Barnett, and L. Xie, eds., The Electrochemical Society, Inc., softbound series PV 94-28, 1994, p. 389.

183. B. K. Choi, Y. W. Kim, and K. H. Shin, *J. Power Sources, 68*, 357 (1997).

184. C. Julien, in *Lithium Batteries. New Materials, Developments and Perspectives*, G. Pistoia, ed., Elsevier, Amsterdam, 1994, p. 167.

# 2

# Physical and Chemical Properties of Nonaqueous Electrolyte Solutions

**George E. Blomgren**
*Energizer, Westlake, Ohio*

## I. INTRODUCTION

The electrolyte phase in electrochemical studies is the bridge between the anode and cathode (or working and counterelectrodes) in the system. Faraday understood this role of electrolyte in his electrolysis experiments in the early part of the nineteenth century. However, it remained for the latter part of that century for Arrhenius [1] to advance his theory of electrolytic dissociation to reveal the details of electrolyte behavior. The essential property of the electrolyte solution is to possess negligible electronic conductivity, and have adequate ionic conductivity to support the current needed for the study without introducing large ohmic corrections to the experiment. To this end, it is necessary for the solvent to have some polarity in order to dissolve electrolyte salts and cause dissociation into free ions. The solvent polarity gives rise to a dielectric constant or dielectric permittivity ($\epsilon$) greater than about 2 or 3, a typical value for nonpolar liquids. Some nonpolar solvents (dipole moment equal to 0) have high polarizability, however, which gives them some capability of dissolving large salts or highly polarizable salts to form solutions with modest conductivity. In dilute solutions of electrolyte salts (of the order of milli- to micromolar concentration), the ions are completely dissociated in high permittivity solvents, while in low permittivity solvents ($\epsilon > 4$) at least some ions are dissociated. The nature of ion association is of great importance to the electrochemistry of solutions and is discussed in some detail later in the chapter. Most of the solvents of interest in this book are aprotic. That is, they do not have active or mobile hydrogen atoms or ions which may be donated to other molecules or ions and do not undergo autodissociation into ions. These types of solvents have been well studied for some time in electrochemistry and in battery work [2,3]. Ion-ion-solvent interactions have been studied for many years by classical conductivity methods. More recently, spectroscopic methods and molecular theory and modeling methods have been applied in attempts to understand these subtle interactions. These aspects will form the bulk of the final section of the chapter. This chapter discusses only organic and inorganic solvent based electrolytes. Other chapters discuss molten salt and polymer electrolytes.

## II. NONAQUEOUS SOLVENTS USED IN MODERN ELECTROCHEMISTRY

While most of the solvents of interest here are aprotic, in the sense defined in the introduction, they may be further classified with regard to their solvent type. Kolthoff [4] has categorized these solvents in a useful way. Only the classifications useful for nonaqueous studies have been included below. That is, the protic solvents (amphiprotic solvents in Kohltoff's terminology) are not included in this chapter. The classes and definitions are as follows:

- Protophillic H-bond donor solvents: solvents such as amides, amines or and other compounds with at least one N—H bond, which may be shared or donated. These solvents also have a highly basic character in the Bronsted sense; i.e., they have a likelihood of accepting a free proton or a proton from a proton donor molecule (protophillic). These solvents also show high electron donor and acceptor properties (basic and acidic in the Lewis sense).
- Aprotic protophillic solvents: all nitrogen bases without protons, e.g., pyridine, and other conjugated amines, and fully alkylated amines or amides. The solvents function as Bronsted bases as well as Lewis bases and may have high permittivity.
- Aprotic protophobic solvents: like the above, but do not act as proton or Bronsted bases. This group includes ketones, nitriles and esters.
- Low permittivity electron donor solvents: saturated cyclic and linear ethers which are good Lewis bases but have dielectric constants less than about 10.
- Low polarity solvents of high polarizability: these solvents have small or zero dipole moments and are generally poor Lewis bases. They may act as weak proton acceptors, however.
- Inert solvents: alkanes and fluoroalkanes constitute the main members of this class, which is quite large, but does not play a great role in

electrochemistry because of the difficulty of dissolving salts to a sufficiently high degree and the very high resistivity of the solutions.

• Aprotic inorganic solvents: these solvents are used widely as reagents in organic chemistry as halogenating agents and for other purposes. Their electrochemical properties have led to considerable use in batteries, in particular those solvents which are able to form passivating, electronically nonconducting films such as those listed in Table 1.

Barthel and Gores [5] have tabulated the physical properties of many solvents, and Table 1 gives their compilation with a number of additions which the author has found useful in nonaqueous electrochemistry listed according to the above categories. The table includes common acronyms and the most important physical properties which bear on electrochemical experiments, i.e., melting and boiling points, dielectric constant, viscosity, density, and dipole moment. The liquid range defined by the melting and boiling points of the solvent is essential for choosing experimental equipment and the likelihood of technical applications resulting from the study. The dielectric constant is the most important property influencing the dissociation of ions to form conductive electrolytes. However, the viscosity is closely related to the mobility of ions in the solution. Actually, the viscosity of the solvent and the inverse of the mobility of ions are both related to some degree to the cohesive energy of the liquid, especially at low concentration of electrolyte. This is the relationship which gives rise to Walden's rule (see, for example, Ref. 6), which states that the product of the viscosity of the solvent and the equivalent conductance at infinite dilution is approximately a constant for a given solvent and various electrolyte salts:

$$\Lambda_0 \eta_0 = \text{constant} \tag{1}$$

The cohesive energy of the solvent also is related to the boiling point, so there is a correlation of boiling point ot solvent viscosity as well. A further relationship of the equivalent conductance at infinte dilution is that it is composed of the individual ionic conductances at infinite dilution:

$$\Lambda_0 = \lambda_0^+ + \lambda_0^- \tag{2}$$

where $\lambda_0^+$ and $\lambda_0^-$ are the individual ionic conductances determined from a variety of salts at infinite dilution, e.g., NaCl, NaBr, KCl and KBr. These four measurements are sufficient to determine the conductances of Na, K, Cl and Br and then these values may be used to determine other ion conductances and check for self consistency. A more accurate method involves the determination of the transference number for one solution which then gives the partition of individual ion conductances directly [6]. These relationships have received a lot of attention over many years and have given a level of understanding of the individual ion moving at infinite dilution. For practical electrochemical studies and for applica-

**Table 1**  Physical Properties of Solvents at 25°C (unless noted)

| Solvent | Acronym | Melting point (°C) | Boiling point (°C) | Dielectric permittivity | Viscosity (cP) | Density (g/cm³) | Dipole moment (D) |
|---|---|---|---|---|---|---|---|
| 1. Protophilic H-bond donor solvents | | | | | | | |
| Ammonia | | −77.7 | −33.4 | 23.9@−33°C | 0.1660@0°C | 0.7601@−33°C | 1.82 |
| 2-Aminoethanol | | 10.53 | 170.95 | 37.72 | 19.346 | 1.01159 | 2.27 |
| Aniline | | −6.0 | 184.4 | 6.98 | 3.770 | 1.0175 | 1.51 |
| Cyclohexylamine | | −17.7 | 134.8 | 4.73 | 1.097 | 0.8622 | 1.26 |
| 1,2-Diaminoethane | | 11.3 | 117.26 | 12.9 | 1.54 | 0.8922 | 1.90 |
| Formamide | FA | 2.55 | 218 | 109.5 | 3.30 | 1.129 | 3.37 |
| N-methylacetamide | NMA | 30.56 | 206 | 171.7 @35°C | 3.38 @35°C | 0.94553 | 1.50 |
| N-methylformamide | NMF | −5.4 | 180–185 | 186.9 | 1.65 | 0.998244 | 3.86 |
| Morpholine | | −3.1 | 128.94 | 7.42 | 1.792 | 0.99547 | 1.50 |
| Pyrrole | | −23.41 | 129.76 | 8.13 | 1.233 | 0.9656 | 1.80 |
| 2-Pyrrolidinone | | 25 | 245 | 33 | 13.3 | 1.107 | 3.55 |
| 2. Aprotic protophilic solvents | | | | | | | |
| N,N-diethylacetamide | DEA | — | 184 | 30.4 | 1.226 | 0.904 | 3.69 |
| N,N-diethylformamide | DEF | — | 178.3 | 28.4 | 1.139 | 0.9017 | — |
| N,N-dimethylacetamide | DMA | −20 | 166.1 | 37.78 | 0.927 | 0.9350 | 3.72 |
| N,N-dimethylformamide | DMF | −60.44 | 153.0 | 36.71 | 0.794 | 0.9439 | 3.86 |
| Dimethyl sulfoxide | DMSO | 18.54 | 189.0 | 46.5 | 1.99 | 1.0955 | 3.9 |
| Hexamethylphosphoric triamide | HMPA, HMPT | 7.2 | 233 | 29.6 | 3.22 | 1.0201 | 5.54 |
| 3-Methyl-2-oxazolidinone[a] | 3-Me-2-Ox | 15.9 | — | 77.5 | 2.450 | 1.1702 | ~5 |
| N-methyl-2-pyrrolidinone | NMP | −24.4 | 202 | 32.0 | 1.663 | 1.0286 | 4.09 |
| 3-Methyl sydnone | | 36 | — | 144.0 @40°C | 5.50 @40°C | 1.3085 @40°C | 7.3 @40°C |
| Pyridine | Py | −41.55 | 115.256 | 12.4 @21°C | 0.884 | 0.97824 | 2.37 |
| Quinoline | Qu | −14.9 | 237.1 | 9.00 | 3.37 | 1.0898 | 2.18 |

| Solvent | Abbr. | | | | | | |
|---|---|---|---|---|---|---|---|
| Sulfolane | TMS | 28.45 | 287.3 | 43.3 @30°C | 1.262 @30°C | 4.81 | 4.81 |
| Tetramethylurea | TMU | −1.2 | 175.2 | 23.5 | 1.398 | 0.9619 | 3.47 |
| Triethylamine | TEA | −114.7 | 89.5 | 2.42 | 0.363 | 0.723 | 0.66 |
| 3. Aprotic protophobic solvents | | | | | | | |
| Acetic anhydride | AA | −73.1 | 140.0 | 20.7 @19°C | 0.971 @15°C | 1.0686 @30°C | 2.82 |
| Acetone | | −94.7 | 56.29 | 20.56 | 0.303 | 0.7843 | 2.69 |
| Acetonitrile | AN | −48.835 | 81.60 | 35.95 | 0.341 | 0.7767 | 3.44 |
| Acetophenone | | 19.62 | 202.0 | 17.48 | 1.511 @30°C | 1.02382 | 2.96 |
| Benzonitrile | BN | −12.75 | 191.10 | 25.20 | 1.237 | 1.0006 | 4.05 |
| 2-Butanone | MEK | −86.69 | 79.64 | 18.04 | 0.368 | 0.7996 | 2.76 |
| γ-Butyrolactone | GBL | −43.53 | 204 | 39.1 | 1.7315 | 1.1242 | 4.12 |
| Diethyl carbonate | DEC | −43.0 | 126.8 | 2.820 @20°C | 0.748 | 0.9693 | 0.90 |
| Diethyl sulfite | DES | — | 157 | 15.6 @20°C | 0.839 | 1.0829 @20°C | 2.96 |
| Dimethyl carbonate[b] | DMC | 3 | 90 | 3.12 | 0.585 | — | — |
| Ethyl acetate | EA | −84.0 | 71.1 | 6.02 | 0.426 | 0.8946 | 1.88 |
| Ethyl formate | EF | −79.6 | 54.31 | 7.16 | 0.358 | 0.9153 | 1.94 |
| Ethylene carbonate | EC | 36.5 | 238 | 90.36 @40°C | 1.9 @40°C | 1.321 @40°C | 4.87 |
| Ethylene glycol sulfite[c] | EGS | −11 | 173 | 39.6 | 2.056 | 1.4158 | — |
| Methyl acetate | MA | −98.05 | 56.868 | 6.68 | 0.364 | 0.9279 | 1.61 |
| Methyl formate | MF | −99.0 | 31.75 | 8.5 @ 20°C | 0.328 | 0.9664 | 1.77 |
| Nitromethane | NM | −28.55 | 101.20 | 38.0 | 0.62 | 1.131 | 3.56 |
| 3-Pentanone | | −38.97 | 101.99 | 17.00 @20°C | 0.442 | 0.80945 | 2.82 |
| Propionitrile | PN | −92.78 | 97.35 | 28.86 @20°C | 0.389 @30°C | 0.77682 | 3.57 |
| Propylene carbonate | PC | −54.53 | 242 | 64.95 | 2.51 | 1.1996 | 4.98 |
| Trimethyl phosphate | TMP | −46.0 | 197.2 | 22.3 | 2.03 | 1.0695 @20°C | 3.02 |
| 4. Low permittivity electron donor solvents | | | | | | | |
| Anisole | | −37.5 | 153.60 | 4.331 | 1.52 @15°C | 0.98932 | 1.245 |
| Diethyl ether | DEE | −116.3 | 34.55 | 4.23 | 0.242 | 0.70760 | 1.15 |
| Diglyme | DG | −64 | 163 | 7.23 | 1.06 | 0.9440 | 1.97 |

**Table 1**  Continued

| Solvent | Acronym | Melting point (°C) | Boiling point (°C) | Dielectric permittivity | Viscosity (cP) | Density (g/cm³) | Dipole moment (D) |
|---|---|---|---|---|---|---|---|
| Diisopropyl ether | | −85.5 | 68.3 | 3.88 | 0.379 | 0.7182 | 1.22 |
| 1,2-Dimethoxyethane | DME | −58 | 84.50 | 7.075 | 0.407 | 0.8612 | 1.71 |
| 1,4-Dioxane | DX | 11.80 | 101.32 | 2.21 | 1.20 | 1.0280 | 0.45 |
| 1,3-Dioxolane | DIOX | −97.22 | 75.6 | 7.13[c] | 0.589[c] | 1.0647 @20°C | 1.47 |
| 2-Methyltetrahydrofuran | Me-THF | −137.2 | 79.9 | 6.97 | 0.467[b] | 0.848[c] | — |
| o-Dimethoxybenzene[d] | Veratrole | 22.5 | 206 | 4.45 @20°C | | 1.0810 | 1.3 |
| Tetrahydrofuran | THF | −108.5 | 65.965 | 7.43 | 0.459 | 0.8819 | 1.75 |
| Tetrahydropyran[b] | THP | −45 | 88 | 5.61 | 0.764 | | |
| Tetraglyme[e] | | | | | | | |
| Triglyme[e] | | −45 | 216 | 7.5 | | | |
| 5. Low polarity solvents of high polarizability | | | | | | | |
| Benzene | | 5.533 | 80.07 | 2.274 | 0.6028 | 0.87360 | 0.00 |
| Carbon tetrachloride | | −23.0 | 76.8 | 2.22793 | 0.905 | 1.5844 | 0.00 |
| Chloroform | | −63.6 | 61.2 | 4.7218 | 0.540 | 1.4799 | 1.15 |
| 1,1-Dichloroethane | 1,1-DCE | −97.0[e] | 57.28[f] | 9.90[f] | 0.464[d] | 1.1757[d] | 2.06[f] |
| 1,2-Dichloroethane[d] | 1,2-DCE | −35.5 | 83.5 | 10.42 @20°C | 0.779 | 1.2351 @20°C | 1.8 |
| Hexafluorobenzene | | 5.10 | 80.255 | 2.029 | 0.860 | 1.60732 | 0.33 |
| Methylene chloride | | −95.14 | 39.75 | 8.93 | 0.413 | 1.31678 | 1.14 |

|  |  |  |  |  |  |  |
|---|---|---|---|---|---|---|
| Toluene | −95.0 | 110.6 | 2.379 | 0.552 | 0.8623 | 0.31 |
| Trichloroethylene | −86.4 | 87.19 | 3.42 @16°C | 0.532 | 1.4514 @30°C | 0.8[d] |
| 6. Inert solvents |  |  |  |  |  |  |
| Cyclohexane | 6.544 | 80.725 | 2.01714 | 0.898 | 0.77374 | 0.00 |
| n-Heptane | −90.582 | 98.424 | 1.9246 @20°C | 0.3967 | 0.67946 | 0.0 |
| n-Hexane | −95.5 | 68.7 | 2.0231 | 0.299 | 0.6548 | 0.09 |
| Perflouro-n-heptane |  |  | 1.765 |  |  |  |
| Perfluoro-n-hexane |  |  | 1.76 |  |  |  |
| Perfluoro-n-octane |  |  | 1.85 |  |  |  |
| 7. Aprotic inorganic solvents |  |  |  |  |  |  |
| Phosphoryl chloride[c] | 1 | 108 | 13.9 @22°C | 1.15 @20°C | 1.460 @20°C | 2.40 |
| Sulfur dioxide[c] | −75.46 | −10.02 | 14.1 @20°C | 0.291 | 1.3695 | 1.60 |
| mSulfur monochloride[g] | −76.5 | 137.1 | 4.9 @ 22°C | — | 1.6828 | 1.60 |
| Sulfuryl chloride | −54.1 | 69.4 | 11.5 | 0.674 | 1.657 | — |
| Thionyl chloride[c] | −104.5 | 75.8 | 9.25 @20°C | 0.603 @20°C | 1.629 @20°C | 1.38 |

[a] H. L. Huffman, Jr., Thesis, University of Kentucky, Ph.D., Chemistry, Physical.
[b] Dominey [18].
[c] Blomgren [2].
[d] D. R. Lide, *Handbook of Organic Solvents*, CRC Press, Boca Raton, 1995.
[e] Reichardt [15].
[f] Popovych and Tomkins [33].
[g] R. C. Paul and G. Singh, in *The Chemistry of Nonaqueous Solvents*, Vol. Vb, J. J. Lagowski, ed., Academic Press, New York, 1978, Chap. 4.
*Source:* From Ref. 5 unless noted otherwise.

tions in electrodeposition or in batteries, however, these relationships are of little value. The properties of more concentrated solutions will be discussed later.

The temperature dependencies of the solvent parameters are also of considerable importance in determining electrolyte behavior. The viscosity and dielectric permittivity of solvents invariably decrease with increasing temperature. For fully dissociated dilute solutions, i.e., solutions for which no ion pairing occurs, the conductivity thus increases, according to Walden's rule, as the temperature is increased. Since the dielectric constant also decreases steadily, however, it is often the case that ion pairing sets in at elevated temperature and the conductivity may decrease, depending on the relative growth of nonconducting ion pair species compared to the increased mobility of free ions due to decreased viscosity. Data are widely scattered in the literature, but the Lagowski series of books (Ref. 6 and succeeding volumes) gives values for specific solvents, and Barthel and Gores [5] give general guidance on the use of empirical equations to describe the temperature dependence of viscosity, density (or molar volume) and dielectric constant for pure solvents and for solvent mixtures. For example, a rather complete compilation of properties for propylene carbonate, an important solvent, is given by Lee [7], and the table of dielectric constants is reproduced in Table 2. The inclusion of

**Table 2** Dielectric Permittivity ($\varepsilon$) and Kirkwood $g$ Factor of Propylene Carbonate at Various Temperatures

| Temperature (°C) | $\epsilon$ (Debyes) | $g$ |
|---|---|---|
| −60 | 89.3 | 1.04 |
| −50 | 86.1 | |
| −40 | 83.0 | |
| −30 | 80.0 | |
| −20 | 76.9 | |
| −10 | 73.9 | |
| 0 | 71.0 | 0.99 |
| 10 | 68.4 | |
| 20 | 66.1 | 1.17 |
| 25 | 64.92 | |
| 30 | 63.7 | |
| 35 | 62.6 | 1.01 |
| 40 | 61.42 | 1.02 |
| 45 | 60.27 | |
| 50 | 59.17 | |
| 55 | 58.05 | |
| 60 | 56.89 | 1.02 |

*Source*: Adapted from Ref. 7.

the Kirkwood $g$ factor calls attention to the fact that the value of unity as observed for propylene carbonate indicates that solvent self-association does not occur. This is typical of the aprotic solvents discussed here. It is not the case for the protophilic hydrogen bonding solvents of the first category. These solvents are generally highly associated and have $g$ factors which deviate greatly from unity. Thus, for example, the extremely high dielectric constants of formamide, $N$-methyl formamide and $N$-methyl acetamide (Table 1) are due to the strong association of these hydrogen bonded solvents. Table 3 gives values of $d \ln \epsilon / d \ln T$, from Swarc [8], and values for PC derived from the data in Table 2. This term can be interpreted as the electrostatic part of the ion pair dissociation enthalpy. The fact that it is negative and greater than unity for many solvents is responsible in part for the exothermicity of ion pair dissociation reactions (see Ref. 8 for details).

In many electrochemical studies, and particularly in practical applications, the pure solvents are not adequate for the task. The liquid range may not be suitable, the dielectric constant may be too low, or the viscosity may be too high. Therefore, solvent mixtures are frequently employed in experiments and applications. Fortunately, several mixing rules exist which can be employed to estimate the properties of solvent mixtures. These mixing rules are generally derived from ideal or nearly ideal solution theory and are of course not exact for real solutions. However, in many cases, particularly for dipolar aprotic solvents, the mixing rules predict the properties of the mixtures to within a few percent. For protic solvents and for some protophilic solvents which are especially prone to hydrogen bond formation, the mixing rules do not work well and the experimentalist is well advised to measure the properties of solvent mixtures to be used. Table 4 summarizes the rules which have been gathered by the author [9]. For interpretation of nonideal solvent mixtures, the reader is referred to the extensive work of Hildebrand [10].

**Table 3**  Temperature Dependence of Dielectric Permittivity for Selected Solvents

| Solvent | Temperature range (°C) | $d \ln \epsilon / d \ln T$ | $D$ at 0°C |
|---|---|---|---|
| Diethyl ether | −70 to +25 | −1.33 | 4.88 |
| Dimethoxyethane | −70 to +25 | −1.28 | 8.00 |
| 2-Methyltetrahydrofuran | −70 to +25 | −1.125 | 6.92 |
| Propylene carbonate[a] | −60 to +60 | −1.018 | 71.0 |
| Tetrahydrofuran | −70 to +25 | −1.16 | 8.23 |
| Tetrahydropyran | −40 to +25 | −0.97 | 6.12 |

[a] Data from Lee [7].
*Source*: Adapted from Ref. 8.

**Table 4**  Mixing Rules for Nonaqueous Solvent Properties

---

Molar volume of mixture ($V_m$, cm³/mole)

  $V_m = x_1V_1 + x_2V_2 + \cdots$     where $x_i$ is the mole fraction of component $i$ and $V_i$ is the molar volume of component $i$

Average molecular weight of mixture ($M_m$, g/mole)

  $M_m = x_1M_1 + x_2M_2 + \cdots$     where $M_i$ is the molecular weight of component $i$

Density of mixture ($\rho_m$, g/cm³)

  $\rho_m = M_m/V_m$

Viscosity of mixture ($\eta_m$, Poise)

  $\eta_m = x_1 \log \eta_1 + x_2 \log \eta_2 + \cdots$     where $\eta$ is the viscosity of component $i$

Dielectric permittivity of mixture ($\varepsilon_m$, Debyes)

  $\varepsilon_m = y_1\varepsilon_1 + y_2\varepsilon_2 + \cdots$     where $_i$ is the volume fraction of component $i$ ($=x_iV_i/V_m$) and $\varepsilon_i$ is the dielectric permittivity of component $i$

---

*Source*: From Ref. 9.

## III.  EMPIRICAL PARAMETERS

Empirically measured parameters are additional solvent properties, which have been developed through the efforts of physical chemists and physical organic chemists in somewhat different, but to some extent related, directions. They have been based largely on the Lewis acid base concept, which was defined by G. N. Lewis. The concept originally involved the theory of chemical bonding which stated that a chemical bond must involve a shared electron pair. Thus, an atom in a molecule or ion which had an incomplete octet in the early theory, or a vacant orbital in quantum mechanical terms, would act as an electron pair acceptor (an acid) from an atom in a molecule or ion which had a complete octet or a lone pair of electrons (a base). Further developments have included the concepts of partial electron transfer and a continuum of bonding from the purely electrostatic bonds of ion-ion interactions to the purely covalent bonds of atoms and molecules. The development of the concept has been extensively described (see Ref. 11 for details).

Many physical chemists have embraced the concepts of donicity (donor numbers, DN) and acceptor numbers (AN) as developed by Gutmann and his co-workers [12]. The DN is measured by the heat of reaction of the donor solvent and antimony pentachloride in a 1:1 ratio as a dilute solution in 1,2-dichloroethane. It is taken to be a measure of the strength of the Lewis base. The AN is measured as the relative shift of the [31]P NMR peak in triethylphosphine oxide dissolved in the sample solvent. Hexane is given the value of zero on the scale, and antimony pentachloride is given the value of 100. The AN is taken to be a measure of the strength of the Lewis acid. The applications of the concepts have been largely qualitative, but a large number of observations have been correlated, at least in a consistent series relative to these parameters. Thus, for example, in comparing solvent interactions with relatively weak Lewis acids (electron acceptors), such as lithium or other alkali or alkaline earth cations, or strong Lewis acids, such as aluminum or boron halides, solvents with large DNs invariably are found to coordinate preferentially with the Lewis acid. Since neither property can be rationalized in any direct way as a molecular quantity, it is surprising that so many other measurements correlate with the two series (Ref. 11 gives many examples).

An elaboration of the Lewis acid base concept is given by Pearson [13,14]. This is called the hard-soft acid base (HSAB) concept and is based upon the fact that polarizability of the acid or base can also play an important role in the degree of interaction between acids and bases. Thus, a highly polarizable molecule or ion acts as a soft acid or base and tends to interact more strongly with a corresponding soft base or acid. Likewise, a less polarizable molecule or ion acts as a hard acid or base and tends to interact more strongly with a strong base or acid. The HSAB theory reconciles many of the conflicts found with the Gutmann theory, as shown by Jensen [11]. However, it is notable that most of the solvents commonly used in nonaqueous electrochemistry may be considered as hard bases or acids, while many of the cations employed are hard acids, so the effects of this refinement may be small in many experiments. Whenever a soft acid or base is used, however, the multiparameter HSAB approach of Pearson is desirable. The connection of this type of equation to molecular orbital theory was also discussed [11].

Physical organic chemists have tended to examine parameters based on shifts in the absorption peaks in the spectra of various dyes or indicator molecules. The $\alpha$ and $\beta$ scales of Taft and Kamlet, the $E_T(30)$ scale of Dimroth and Reichardt, the $\pi^*$ scale of Taft and co-workers and the $Z$ value of Kosower are all examples of this type of parameter. The definitions and measurement means for these parameters, as well as important references, are shown in Table 5. An alternative definition of the Dimroth-Reichardt parameter is the dimensionless, $E_T^N$, which is now preferred by some organic chemists (for a discussion see Ref. 15). The $Z$ value is important in that it led to the scale of Dimroth and Reichardt, which overcomes many of the limitations of the earlier scale. Several workers have shown that relationships exist, with good correlation coefficients, between similar parameters. Thus, DN is linearly related to $\beta$, both parameters being designed to measure the donor properties (or Lewis basicity) of solvent molecules. Also, $E_T(30)$ is related to $\alpha$ as well as to AN: all three parameters purport to measure the electron acceptor properties (or Lewis acidity) of solvent molecules. It has been found that different solvent types have different coefficients in linear relationships between $\pi^*$ and the dipole moment. The Taft and Dimroth-Reichardt parameters, in particular, have been found to correlate with free energies and

**Table 5**  Empirical Acid-Base Parameters for Nonaqueous Solvents

| Solvent | DN[a] | $E_T(30)$[b] | $(E_T^N)$[c] | AN[d] | $\alpha$[e] | $\beta$[f] | $(\pi^*)$[g] | Z[h] |
|---|---|---|---|---|---|---|---|---|
| **1. Protophilic H-bond donor solvents** | | | | | | | | |
| Ammonia | 59 | — | — | | | | | |
| 2-Aminoethanol | | 51.8 | 0.651 | 33.7 | | | | |
| Aniline | 35 | 44.3 | 0.420 | | | | | |
| Cyclohexylamine | 55 | 42.0 | 0.349 | 20.9 | | | | |
| 1,2-Diaminoethane | | | | | | | | |
| Formamide | 24 | 56.6 | 0.799 | 39.8 | 0.71 | | 0.97 | 83.3 |
| N-methylacetamide | | 52.0 | 0.657 | | | | | 77.9 |
| N-methylformamide | | 54.1 | 0.722 | 32.1 | | | | |
| Morpholine | | 41.0 | 0.318 | | | | | |
| Pyrrole | | 48.3 | 0.543 | | | | | |
| 2-Pyrrolidinone | | | | | | | | |
| **2. Aprotic prophilic solvents** | | | | | | | | |
| N,N-diethylacetamide | 32.3 | 41.1 | 0.330 | 13.6 | 0.0 | 0.78 | | |
| N,N-diethylformamide | 30.9 | | | | | | | |
| N,N-dimethylacetamide | 27.8 | 43.7 | 0.401 | | 0.0 | 0.76 | | 66.9 |
| N,N-dimethylformamide | 26.6 | 44.0 | 0.404 | | 0.0 | 0.69 | | 68.4 |
| Dimethyl sulfoxide | 29.8 | 45.1 | 0.444 | 19.3 | 0.0 | 0.76 | 1.00 | 71.1 |
| Hexamethylphosphoric triamide | 38.8 | 40.9 | 0.315 | 10.6 | 0.0 | 1.05 | 0.87 | 62.8 |
| 3-Methyl-2-oxazolidinone | | | | | | | | |
| N-methyl-2-pyrrolidinone | 27.3 | 42.2 | 0.355 | 13.3 | 0.0 | 0.77 | 0.92 | |
| 3-Methyl sydnone | | | | | | | | |
| Pyridine | 33.1 | 40.5 | 0.302 | 14.2 | 0.0 | 0.64 | 0.87 | 64.0 |
| Quinoline | 32 | 39.4 | 0.269 | | 0.0 | 0.80 | | |
| Sulfolane | 14.8 | 44.0 | 0.410 | 19.2 | 0.0 | 0.70 | | 77.5 |
| Tetramethylurea | 29.6 | 41.0 | 0.318 | | 0.0 | 0.0 | 0.71 | |
| Triethylamine | 61.0 | 32.1 | 0.043 | 1.4 | 0.0 | 0.71 | 0.14 | |

| Solvent |  |  |  |  |  |  |  |  |
|---|---|---|---|---|---|---|---|---|
| **3. Aprotic protophobic solvents** |  |  |  |  |  |  |  |  |
| Acetic anhydride | 10.5 | 43.9 | 0.407 |  | 0.0 |  | 0.76 |  |
| Acetone | 17.0 | 42.2 | 0.355 | 12.5 | 0.08 | 0.48 | 0.71 | 65.5 |
| Acetonitrile | 41.1 | 45.6 | 0.460 | 18.9 | 0.19 | 0.31 | 0.75 | 71.3 |
| Acetophenone |  | 40.6 | 0.306 |  |  | 0.49 | 0.90 |  |
| Benzonitrile | 11.9 | 41.5 | 0.333 | 16.0 | 0.0 | 0.41 | 0.90 | 65.0 |
| 2-Butanone |  | 41.3 | 0.327 |  | 0.6 | 0.48 | 0.67 |  |
| γ-butyrolactone |  | 44.3 | 0.420 | 17.3 | 0.0 | 0.49 | 0.87 |  |
| Diethyl carbonate | 16.0 | 37.0 | 0.194 |  | 0.0 | 0.40 | 0.45 |  |
| Diethyl sulfite |  |  |  |  |  |  |  |  |
| Dimethylcarbonate |  | 41.1 | 0.321 | 9.3 | 0.0 | 0.0 | 0.38 | 59.4 |
| Ethyl acetate | 17.1 | 38.1 | 0.228 |  | 0.0 | 0.45 | 0.55 |  |
| Ethyl formate | 16.4 | 40.9 | 0.315 |  |  | 0.36 | 0.61 |  |
| Ethylene carbonate | 15.3 | 48.6 @40°C | 0.552 |  |  |  |  |  |
| Ethylene glycol sulfite | 16.5 | 50.0 | 0.596 |  |  | 0.42 |  |  |
| Methyl acetate |  | 40.0 | 0.287 | 10.7 | 0.0 | 0.37 | 0.60 |  |
| Methyl formate |  | 45.0 | 0.441 |  | 0.0 | 0.45 | 0.62 |  |
| Nitromethane | 2.7 | 46.3 | 0.481 | 20.5 | 0.22 | 0.37 | 0.85 | 71.2 |
| 3-Pentanone |  | 39.3 | 0.265 |  |  | 0.40 | 0.72 |  |
| Propionitrile | 16.1 | 43.7 | 0.401 | 18.3 | 0.0 |  | 0.71 |  |
| Propylene carbonate | 15.1 | 46.6 | 0.491 | 16.3 | 0.0 |  | 0.83 |  |
| Trimethyl phosphate | 23.0 | 43.6 | 0.398 |  |  |  |  | 72.4 |
| **4. Low permittivity electron donor solvents** |  |  |  |  |  |  |  |  |
| Anisole |  | 37.1 | 0.198 |  | 0.0 | 0.22 | 0.73 |  |
| Diethyl ether | 19.2 | 34.5 | 0.117 | 3.9 | 0.0 | 0.57 | 0.27 |  |
| Diglyme |  | 38.6 | 0.244 | 9.9 | 0.0 | 0.49 | 0.27 |  |
| Diisopropyl ether |  |  |  |  |  |  |  |  |
| 1,2-Dimethoxyethane | 24 | 38.2 | 0.231 | 10.2 | 0.0 | 0.41 | 0.53 | 59.1 |
| 1,4-Dioxane | 14.8 | 36.0 | 0.164 | 10.8 | 0.0 | 0.37 | 0.55 |  |
| 1,3-Dioxolane |  | 43.1 | 0.383 |  |  |  |  |  |

**Table 5**  Continued

| Solvent | DN[a] | $E_T(30)$[b] | $(E_T^N)$[c] | AN[d] | $\alpha$[e] | $\beta$[f] | $(\pi^*)$[g] | $Z$[h] |
|---|---|---|---|---|---|---|---|---|
| 2-Methyltetrahydrofuran | | 36.5 | 0.179 | | | | | |
| o-Dimethoxybenzene | | 38.4 | 0.238 | | | | | 55.3 |
| Tetrahydrofuran | 20.0 | 37.4 | 0.207 | 8.0 | 0.0 | 0.55 | 0.58 | 58.8 |
| Tetrahydropyran | | 36.6 | 0.182 | | | | | |
| Tetraglyme | | | | | | | | |
| Triglyme | | 38.9 | 0.253 | | | | | 61.3 |
| 5. Low polarity solvents of high polarizability | | | | | | | | |
| Benzene | 0.1 | 34.4 | 0.111 | 8.2 | 0.0 | 0.10 | 0.59 | 54 |
| Carbon tetrachloride | 0 | 32.4 | 0.052 | 8.6 | 0.0 | 0.0 | 0.28 | |
| Chloroform | 4 | 39.1 | 0.259 | 23.1 | 0.44 | 0.0 | 0.58 | 63.2 |
| 1,1-Dichlorethane | | 39.4 | 0.269 | | | | | 62.1 |
| 1,2-Dichloroethane | 0 | 41.3 | 0.327 | 16.7 | | | | 63.4 |
| Hexafluorobenzene | | 34.2 | 0.108 | | | | | |
| Methylene chloride | 1 | 40.7 | 0.309 | 20.4 | 0.30 | 0.0 | 0.82 | 64.7 |
| Toluene | 0.1 | 33.9 | 0.099 | 0.0 | 0.0 | 0.11 | 0.54 | |
| Trichloroethylene | | 35.9 | 0.160 | 0.0 | 0.0 | 0.0 | 0.53 | |
| 6. Iner solvents | | | | | | | | |
| Cyclohexane | | 30.9 | 0.006 | | 0.0 | 0.0 | 0.0 | |
| n-Heptane | 0 | 31.1 | 0.012 | | 0.0 | 0.0 | −0.08 | |
| n-Hexane | 0 | 31.0 | 0.009 | 0.0 | 0.0 | 0.0 | −0.08 | |
| Perflouoro-n-heptane | | | | | 0.0 | 0.0 | −0.39 | |

| | | | |
|---|---|---|---|
| Perfluoro-*n*-hexane | 0.0 | 0.0 | −0.40 |
| Perfluoro-*n*-octane | 0.0 | 0.0 | −0.41 |
| 7. Aprotic inorganic solvents | | | |
| Phosphoryl chloride | 11.7 | | |
| Sulfur dichloride | | | |
| Sulfur dioxide | | | |
| Sulfur monochloride | | | |
| Sulfuryl chloride | 0.1 | | |
| Thionly chloride | 0.4 | | |

[a] DN is measured calorimetrically by the enthalpy of reaction of the donor solvent in a 1:1 ratio with $SbCl_5$ as a dilute solution in 1,2-dichloroethane. Compilation from Ref. 15, pp. 20–21, a Lewis base scale. Marcus [44] has given a normalized scale.

[b] $E_T(30)$ is the spectrochemical shift of the $\pi \to \pi^*$ transition of a dye (number 30 in a list) in the investigated solvent measured in kcal/mol, a Lewis acid scale. Values are taken from Ref. 15, pp. 365–371.

[c] $E_T^N$ is a normalized $E_T(30)$ scale based on water and tetramethylsilane as extreme values. See Ref. 15, pp. 364–371, for a compilation and discussion.

[d] AN is measured by the $^{31}P$ NMR shift values of triethylphosphine oxide phosphorus in a 1:1 complex with the test solvent measured against the 1:1 complex with $SbCl_5$ test solvent using *n*-hexane as a neutral solvent. This is a normalized scale of Lewis acidity. Values are taken from Ref. 15, pp. 23–24.

[e] $\alpha$ is a scale based on solvatochromic shifts of the test hydrogen bond acceptor solvent in a hydrogen bond donor solvent. Several comparison solvents are used in the scale, which is similar to a Lewis acidity scale. Values are from Ref. 5, pp. 18–21.

[f] $\beta$ is the coordinate to $\alpha$, using hydrogen bond acceptor solvents in reference hydrogen bond donor solvents. It is similar to a Lewis basicity scale. Values are also from Ref. 5, pp. 18–21.

[g] $\pi^*$ is a measured by a solvatochromic shift of the test solvent in a substitued nitrophenol. It is a measure of the polarity/polarizability of the test solvent. Values are also from Ref. 5, pp. 18–21.

[h] Z is measured by the solvatochromatic shift of a substituted pyridinium iodide dye due to the test solvent. It is a rough measure of the solvent polarity. Values are from Ref. 15, pp. 361–362.

enthalpies of transfer from water to organic solvents. Further relationships have been established for other spectroscopic and kinetic measurements in electrolyte solutions. Still other empirical parameters have been investigated, such as the four parameter equation of Drago and co-workers [15,16]. For further information, the reader is referred to Jensen [11]. These tools are especially powerful in correlating information from mechanistic studies of organic reactions and have provided physical organic chemists with much impetus to carry on such research, although the implications for electrochemical kinetics have not been investigated nearly as much.

The topic of interactions between Lewis acids and bases could benefit from systematic ab initio quantum chemical calculations of gas phase (two molecule) studies, for which there is a substantial body of experimental data available for comparison. Similar computations could be carried out in the presence of a dielectric medium. In addition, assemblages of molecules, for example a test acid in the presence of many solvent molecules, could be carried out with semiempirical quantum mechanics using, for example, a commercial package. This type of neutral molecule interaction study could then be enlarged in scope to determine the effects of ion-molecule interactions by way of quantum mechanical computations in a dielectric medium in solutions of low ionic strength. This approach could bring considerable order and a more convincing picture of Lewis acid base theory than the mixed spectroscopic (molecular) parameters in interactive media and the purely macroscopic (thermodynamic and kinetic) parameters in different and varied media or perturbation theory applied to the semiempirical molecular orbital or valence bond approach [11 and references therein].

## IV.  ELECTROLYTE SALTS

The choice of salt for nonaqueous electrochemistry is primarily dependent on the application. For lithium or lithium ion batteries, the salt is normally a lithium salt. For electroanalytical studies, the choice may be an alkali, alkaline earth or quaternary ammonium salt. The need for high purity lithium salts in battery applications has led to the development of battery grade salts which are of general utility for all electrochemical studies. While lithium salts are in general not inexpensive, they are available in substantial quantities and have well-controlled impurity levels, especially those which might be electrochemically active. Table 6 gives a summary of the salts which are available and have had some utility in lithium battery work. The most important salts at present are lithium trifluoromethanesulfonate ($LiCF_3SO_3$), which is mainly used in primary batteries with solid cathodes like manganese dioxide and iron disulfide; lithium tetrachloroaluminate ($LiAlCl_4$), used in thionyl chloride liquid cathode batteries; lithium bromide (LiBr), used in sulfur dioxide batteries; lithium perchlorate ($LiClO_4$), used

**Table 6** Salts Used in Lithium Battery Electrochemistry

| Salt | Evaluation |
|---|---|
| LiClO$_4$ | Widely used in small cells, safety hazard when nearly dry in presence of organic material |
| LiBr | Used in some sulfur dioxide batteries, but generally solubility is too low and potential window is too small |
| LiI | Used in solid state batteries where the voltage window is small. Even smaller potential window than LiBr. |
| LiAsF$_6$ | Used commonly in lithium metal batteries because of good cycle life predicated on beneficial film formed on lithium and high solution conductivity. |
| LiPF$_6$ | The most common salt used in lithium ion batteries. Low toxicity is combined with good conductivity and wide electrochemical window. The salt has thermal stability problems and reacts with glass, giving electrolyte storage problems. |
| LiAlCl$_4$ | The most common salt used in oxyhalide batteries, giving best stability and high conductivity. |
| LiBF$_4$ | Similar to LiPF$_6$ in electrochemical window, but somewhat more difficult to purify. It has poor cycling efficiency with lithium metal. |
| LiCF$_3$SO$_3$ | Used extensively in primary lithium batteries, but has lower conductivity than several other salts. Is more environmentally stable than the fluoride salts above and is less expensive. |
| LiN(SO$_2$CF$_3$)$_2$ | Electrochemically stable with a wide window and high conductivity solutions, but causes corrosion of aluminum in high voltage cathodes, which has limited use in lithium ion cells. |
| LiC(SO$_2$CF$_3$)$_3$ | Very stable thermally and electrochemically and high conductivity solutions. Still quite expensive because of limited production. |

in some miniature batteries, lithium hexafluoroarsenate (LiAsF$_6$), used in rechargeable lithium metal batteries; and lithium hexafluorophosphate (LiPF$_6$) and lithium tetrafluoroborate (LiBF$_4$), used in rechargeable lithium ion batteries, the latter salt also used in lithium primary cells with CF$_x$ cathodes. Salts which are under study for batteries under development include lithium bis(trifluoromethylsulfonyl)imide [LiN(SO$_2$CF$_3$)$_2$], and lithium tris(trifluoromethylsulfonyl)methide [LiC(SO$_2$CF$_3$)$_3$]. An excellent review of the physical and electrochemical properties of these and other salts which have been considered for lithium batteries is given by Dominey [18].

The use of quaternary ammonium salts has a long history in both aqueous and nonaqueous electrochemistry. In fact, quaternary ammonium salts of tetraphenyl borides are important in the theory of individual ion conductivities, since it is presumed that such large ions will not undergo ion pairing, even in low dielectric media, and because of similar size and charge will have similar mobilities. Of course, they are used widely in electroanalytical studies as background salts. They also have technical importance as they are used as the electrolyte salt in supercapacitors and for electrolysis of organic species to form conductive polymers.


## V.  SOLUTION CONDUCTIVITY

The measurement of solution conductivity is a long-established field of physical electrochemistry and, as discussed in the introduction, has been used to develop the ideas of solution structure. The interest in this section is in the practical values displayed by concentrated solutions and in the values of dilute solutions for the light it sheds upon the ion-ion association. These topics will be taken up in turn. For guidance in measuring electrical conductance, see Evans and Matesich [19] or Coetzee and Ritchie [20]. For guidance in measuring the related transference numbers, see Kay [21] or Spiro [22].

The specific conductance, measured in Siemens per centimeter (S/cm), has been studied for many nonaqueous electrolytes as a function of concentration (often measured in molality). The use of molality for the concentration unit is experimentally convenient, since it is easier in dry box manipulations to weigh the solvent than it is to measure the volume of solution, and the measurement is also independent of the temperature. The connection to the more usual chemical property of molarity is through the density of the solution at the temperature in question. Unfortunately, solution densities are frequently not measured by experimentalists, so the more theoretically meaningful quantity, the equivalent conductance, is not accessible from these measurements. This is not a significant handicap, since the theoretical development of transport properties of concentrated electrolyte solutions is not very advanced at present. However, as theory develops, it would be quite useful to have densities available for testing equivalent conductances. As the field has developed, a number of empirical and semiempirical equations have been advanced as ways to bring some order to the data. These equations have generally related the specific conductance to the molality of the solution, and take advantage of the fact that a maximum in the specific conductance is always observed when plotted against the molality if the solubility of the salt is sufficiently high. The Casteel-Amis equation [23] is a good example of this type:

$$\frac{\kappa}{\kappa_{max}} = \left(\frac{m}{\mu}\right)^a \exp\left\{ b(m - \mu)^2 - \frac{a}{\mu}(m - \mu) \right\} \tag{3}$$

where $\kappa$ is the specific conductance, $\kappa_{max}$ is the maximum value of the specific conductance, $\mu$ is the value of the molal concentration at which the maximum value of conductance occurs, and $a$ and $b$ are empirical parameters. Generally, a least square fitting procedure is used to determine $\kappa_{max}$, $\mu$, $a$ and $b$. This is a useful way of determining the values of conductance and molality at the maximum, since the curve is often quite broad and measurements are not often taken at close intervals near the maximum.

Another approach has been developed by Bruno and Della Monica [24–26]. This work takes the Vogel-Tamman-Fulcher (VTF) equation, which has been used to rationalize transport properties in molten salts and glassy electrolytes, and modifies it for nonaqueous solutions. The work follows the development of Angell and co-workers [27,28], who carried out a similar development for aqueous solutions. The expression used is

$$\kappa(m,T) = AT^{-1/2} \exp\left\{\frac{-B}{R[T - T_0(m)]}\right\} \qquad (4)$$

where $A$ and $B$ are constants which are characteristic of the solvent, and $T_0$ is a function of the molality. In most cases the preexponential temperature term has little effect on the goodness of fit for the parameters, so it is often omitted. In the VTF theory, $T_0$ plays the role of an ideal glass transition temperature (in another version, $T_0$ is the temperature at which the configurational entropy of the system goes to zero), but in this application it is better regarded as an empirical parameter. Bruno and Della Monica used a linear dependence of $T_0$ upon the concentration, which gave an equivalent expression to Eq. (4), where the concentration replaces the temperature.

Barthel and co-workers [29] found that a polynomial expression for this quantity gave a better fit for several solvents:

$$T_0(m) = T_0(0) + am + bm^2 \qquad (5)$$

The Casteel-Amis and the VTF equations give reasonable fits for moderate to high dielectric solvents and are of use in correlating the data. The Casteel-Amis equation generally fits better at low to intermediate concentrations and the VTF at moderate to high concentrations. These treatments completely miss an important behavior in low dielectric solvents or in solutions with highly associating salts, however. That unusual behavior is a minimum in the equivalent conductance as a function of concentration. This behavior is reflected in the specific conductance as a point of inflection. This minimum has given rise to considerable discussion in attempts to understand the origin and will be discussed later in this section.

Figures 1 and 2 give typical plots for solvents of low and high dielectric permittivity from Venkatasetty [30]. Figure 3 shows the results of an unusual series of measurements from the work of Day and co-workers [31] in which an

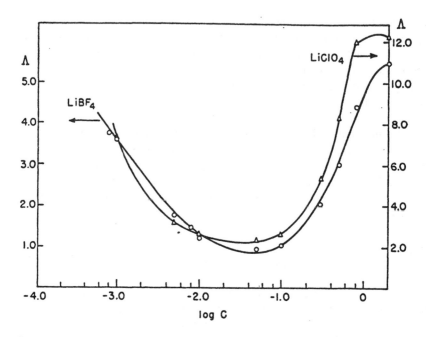

**Figure 1**   Equivalent conductance vs. logarithm of concentration of LiClO$_4$ and LiBF$_4$ in methyl formate illustrating conductivity minimum with low dielectric permittivity solvent. (From Venkatasetty [30], reprinted by permission of the Electrochemical Society.)

inert solvent, benzene, is used with two tetraalkyl aluminate salts. In these latter cases, the entire concentration range from dilute solution to molten salt is plotted. Clearly, the molten salt conductivity is in a monotonic relationship with the highest concentration region of the solution, even though a minimum and a maximum are observed in both cases. Also, the range of conductance is over many orders of magnitude from the minimum to the maximum in a very low dielectric permittivity solvent like benzene.

Another equation which has received use is the Wishaw-Stokes equation, which is discussed in Robinson and Stokes [32]. This equation is based on the extended Onsager equation and is corrected for viscosity in a Walden rule sort of term:

$$\Lambda = \frac{\eta_0}{\eta}\{\Lambda_0 - B_2\sqrt{c}\ (1 + Ba\sqrt{c})^{-1}\}\ \{1 - B_1F\sqrt{c}(1 + Ba\sqrt{c})^{-1}\} \qquad (6)$$

where $B_1$, $B_2$ and $B$ are calculated constants involving the dielectric constant and temperature (see Robinson and Stokes [32] for definitions), $\eta_0$ and $\eta$ are the

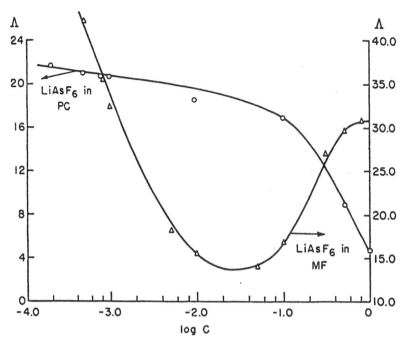

**Figure 2** Equivalent conductance vs. logarithm of concentration for LiAsF$_6$ in methyl formate and in propylene carbonate illustrating difference between solvents of high and low dielectric permittivity. (From Venkatasetty [3], reprinted by permission of the Electrochemical Society.)

solvent viscosity and the solution viscosity at the temperature and concentration concerned, $\Lambda_0$ is the equivalent conductance at infinite dilution and $a$ is the distance of closest approach of unlike ions from the Debye-Huckel theory and is usually an adjusted parameter. This equation is most successful for intermediate concentrations. Table 7 gives a compilation of the maximum specific conductance for LiClO$_4$ in various solvents as well as the concentration at which the maximum occurs. Note that the low dielectric permittivity solvents have maxima at high concentrations, while the moderate to high permittivity solvents have maxima at much lower concentrations. Furthermore, there is no clear reason for the magnitude of the conductivity maximum, except for a rough correlation with viscosity. Understanding the reasons for observations like these are clear goals for theoretical studies.

One of the imporant facets of these nonaqueous solutions is the ion association which occurs frequently. The subject of ion association is a complex one,

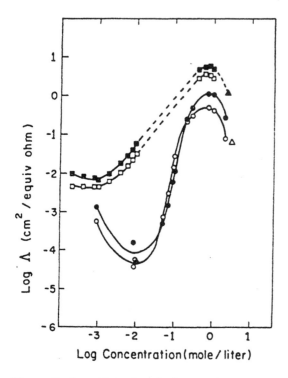

**Figure 3** Logarithm of equivalent conductance vs. logarithm of concentration for NaAlBu$_4$ at 25 (○) and 50°C (●) in benzene and Bu$_4$NAlBu$_4$ at 25 (□) and 50°C (■). Broken lines are extrapolations because of two phase behavior in the Bu$_4$NAlBu$_4$ system. Equivalent conductances of molten salts are given by filled and open triangles for Bu$_4$NAlBu$_4$ and NaAlBu$_4$, respectively. (From Imhof et al. [31], reprinted by permission of The Journal of Solution Chemistry.)

**Table 7** Maximum Specific Conductance ($\kappa_{max}$) and Concentration of Maximum Concentration ($\mu$) for LiClO$_4$ in Various Solvents at 25°C

| Solvent | Dielectric Permittivity | $\kappa_{max}$ | $\mu$ | Concentration units |
|---------|------------------------|----------------|-------|---------------------|
| DIOX | 7.13 | 11.1 | 2.9 | $M$ |
| DMF | 36.71 | 22.2 | 1.16 | $m$ |
| DMSO | 46.5 | 10 | 1.5 | $M$ |
| EF | 7.165 | 16 | 3.0 | $m$ |
| MF | 8.5 | 32 | 2.8 | $M$ |
| NMP | 32.0 | 11.2 | 0.85 | $M$ |
| PC | 64.95 | 5.420 | 0.6616 | $m$ |

*Source*: Adapted from Barthel [5].

which has occupied the attention of solution chemists for many decades. Arrhenius and Ostwald recognized the phenomenon in the last century through the importance of weakly dissociated electrolytes in aqueous solutions. The topic was heavily developed through acid-base theory by Bronsted and Lowry, but the general concept which is applicable to all solutes in all solvents is the Lewis acid-base theory referred to above. An excellent discussion on the effects of the theory is given in Jensen [11], but the key relationships are through the following expressions. For the case of strong donor solvents:

$$A:B \text{ (electrolyte ion-ion pair)} + S \text{ (donor solvent)} \rightarrow A^+ S_n + B^- \quad (7)$$

where each of the species is soluble and the ion pair A:B may be solvated. For the case of strong acceptor solvents:

$$A:B \text{ (electrolyte ion-ion pair)} + S' \text{ (acceptor solvent)} \rightarrow A^+ + B^- S_n \quad (8)$$

where again the ion pair may be solvated. The anion in Eq. (7) is shown as unsolvated, as is the cation in (8), but for somewhat amphoteric solvents they may be solvated as well. In addition to the Lewis acid-base characteristics of solvents and ions, an important part of the association energy is due to the dielectric permittivity of the solvent, since a high dielectric permittivity medium can shield the ion-ion attraction to such an extent that the association energy is greatly weakened.

The best-developed way to measure the association of ions is through the measurement of electrical conductance of dilute solutions. As mentioned, this realization occurred in the nineteenth century to Arrhenius and Ostwald. An elaborate development of conductance equations suitable to a range of ion concentrations of millimolar and lower by many authors (see Refs. 5, 33 and 34 for critical reviews) has made the determination of association constants common. Unfortunately, in dealing with solutions this dilute, the presence of impurities becomes very difficult to control and experimenters should exercise due caution, since this has been the source of many incorrect results. For example, 20 ppm water corresponds to 1 m$M$ water in PC solution, so the effect of even small contaminants can be profound, especially if they upset the acid-base chemistry of association. The interpretation of these conductance measurements leads, by least squares analysis of the measurements, to a determination of the equivalent conductance at infinite dilution, $\Lambda_0$, the association constant for a positively and negatively charged ion pair, $K_A$, and a distance of close approach, $d$, using a conductance equation of choice. One alternative is to choose the Bjerrum parameter for the distance, which is defined by

$$q = \frac{|z_1 z_2| e^2}{2\epsilon kT} \quad (9)$$

where $\epsilon$ is the dielectric permittivity, $z_1$ and $z_2$ are the cation and anion charges, and $e$ is the electronic charge ($k$ and $T$ have their usual definitions). The Bjerrum distance $q$ is somewhat arbitrary and has some problems at very large and very small dielectric constants, but was chosen by Bjerrum to represent a distance at which it is reasonable that short-range forces would dominate and that interionic effects, defined by the Poisson-Boltzmann equation in the Debye-Huckel theory, would not be important (Fuoss and Accascina [34] give a critical account). In Bjerrum's theory, the only attractive force considered in the space between the distance $q$ and the hard sphere distance $d$ was the Coulomb force, while repulsion was modeled as that of hard spheres at the distance $d$. In two elaborations of the Bjerrum theory, phase space was again divided into multiple regions. In the version of Barthel (see Ref. 5 for an extensive discussion), the potential in the space between the hard sphere region and the region of short-range force interaction (a distance $R$) is given by an ion-ion Coulomb attractive term and a short-range potential. In the region beyond the distance $R$, the normal Poisson-Boltzmann solution of Debye and Huckel is applied. The short-range potential is analyzed as a potential of mean force between the cation and anion (averaged over all positions of the other ions), and, like the Hildebrand theory of nonelectrolytes referred to above, is taken to be a constant, $W_{+-}$, independent of distance.

In the second version of Bjerrum's theory [35,36], the ion pair formation is regarded as a solvent displacement reaction as in Eq. (7). An approximation to the short-range potential for this reaction is taken as a square well potential added to the Coulomb potential of the two ions of a width equal to the ion-ion radius sum (the hard sphere distance, $d$) plus a distance parameter. A barrier to the displacement reaction is taken as a plateau potential of width equal to the difference between the ion-solvent-ion distance at closest contact and the beginning of the potential well. The remaining space is taken up to the Bjerrum distance in which the potential is defined by the Coulomb interaction as in the earlier theory. The barrier width and the well depth were fitted to a linear behavior with the series of alkali ions, and a good agreement was found over seven orders of magnitude in comparison to various experimental measurements. An important aspect of the theory is the proportionality of the well depth to the difference in the donor numbers between solvent and anion, which is consistent with the Lewis acid-base theory for solvation energies and the conceptual framework of the association as a solvent displacement reaction. The advantage of the Barthel model is that the association constant can be expressed in terms of the single parameter, $W_{+-}$, at a given temperature. Both approaches may be refined. It would be interesting, for example, to carry out semiempirical quantum calculations of the reaction coordinate for the solvent displacement reaction in the presence of a dielectric continuum. This would be a good test of the thermodynamics of the association reaction as well as the kinetics.

Many spectroscopic measurements have been carried out on solutions

which have been analyzed by conductance measurements to display ion association. The vibrational studies have been reviewed by Popovych and Tomkins [33],
Gill [37], Irish [38], and Edgell [39]. These methods have established the existence on ion pairs in an unequivocal way and have been used to measure approximate ion pair association constants which are in good agreement with those obtained by conductance measurements. A difficulty arises in the comparison
because spectroscopic measurements determine the concentration of pairs, while
the conductance measurements determine the activity of pairs. The sensitivity of
the spectroscopic methods is such that the best value of a concentration based
constant is determined at higher concentrations in the region that activity coefficients show large deviations from unity. This could be rectified if advanced statistical mechanical methods were applied to the determination of activity coefficients of free ions, but such studies have not as yet been carried forward. Electron
spin resonance (ESR) studies have also been carried out to show the presence
of ion pairs with free radical ions and have been reviewed by Sharp and Symons
[40]. Nuclear magnetic resonance (NMR) studies have also been carried out to
verify the existence of ion pairs, and this field has been reviewed by de Boer
and Sommerdijk [41]. The importance of the various spectroscopic measurements
is the internal self-consistency of the observation of ion pairs and the wide range
of lifetimes sampled by the various techniques. Typically, the range is from about
$10^{-4}$ s for NMR to $10^{-12}$ s for IR and Raman techniques with ESR lifetimes in
between. This means that ion pairs exist and are stable for the period of a great
many diffusion steps, which is the basis for the conductance measurements.

The postulation of higher associates has been controversial for a number
of years since their introduction by Fuoss and Kraus in the 1930s [42]. The postu-
late is that, since the conductance in low dielectric constant media increases after
reaching a minimum, a new conducting species must be responsible for the
change in direction of what should be a monotonic decreasing function. Positive
and negative charge triple ions were assigned this function. Fuoss and Accascina
[34] describe the rationale and the statistical mechanical treatment used to analyze
conductance data of this type. In more recent years, however, the realization
has occurred that the conductance does not necessarily have to be a monotonic
decreasing function. In fact the association to ion pairs may be present in the
mid-range of concentrations, but as the concentration increases toward a molten-
salt-like behavior, the alternation of charge and equal distances of separation of
anions and cations typical of the molten salt may also apply to concentrated
solutions. Thus, solutions of this type may be dissociated at very low concentrations, form increasing fractions of ion pairs as the concentration increases due
to the association constant and finally show decreasing fractions of ion pairs as
concentrations increase toward the molten salt due to activity effects (in actuality
due to the long-range nature of the Coulomb force). The spectroscopic measurements are not very accurate in this type of study, since the estimated concentration

of triple ions is much less than that of ion pairs. That higher associates may occur, such as ion quadruples consisting of two positive and two negative ions, seems to be better established in some special cases of strongly associating anions such as thiocyanate. UV and visible spectroscopy have given some evidence of the transitory existence of such species, but the lifetimes of the measuring technique are so short that they are not pertinent to diffusion/conductance lifetimes. Some semiquantitative calculations of activity coefficients based on an approximate statistical mechanical treatment (the mean spherical approximation) suggest that the idea of association of ions to form ion pairs in the dilute range, followed by redissociation of ion pairs in the concentrated range due to activity coefficient variation, is a valid picture of ion association [9,43]. This field is still a developing one and will be the subject of much interest in the years to come.

## REFERENCES

1. S. Arrhenius, *Z. Phys. Chem.*, 1887, 1, 613.
2. G. E. Blomgren, in *Lithium Batteries*, J.-P. Gabano, ed., Academic Press, New York, 1983, p. 13.
3. H. V. Venkatasetty, in *Lithium Battery Technology*, H. V. Venkatasetty, ed., The Electrochemical Society, Pennington, NJ, 1984, p. 1.
4. I. M. Kolthoff, in *Nonaqueous Electrochemistry*, F. C. Marchon, ed., Butterworths, London, 1971.
5. J. Barthel and H.-J. Gores, in *Chemistry of Nonaqueous Solutions*, G. Mamontov and A. I. Popov, eds., VCH, New York, 1994, p. 1.
6. E. Price, in *The Chemistry of Nonaqueous Solvents*, Vol. 1, J. J. Lagowski, ed., Academic Press, New York, 1966, p. 67.
7. W. H. Lee, in *The Chemistry of Nonaqueous Solvents*, Vol. IV, J. J. Lagowski, ed., Academic Press, New York, 1976, pp. 188–189.
8. M. Swarc, in *Ions and Ion Pairs in Organic Reactions*, Vol. 1, M. Swarc, ed., Wiley-Interscience, New York, 1972, Chap. 1.
9. G. E. Blomgren, *J. Power Sources*, 1985, 14, 39.
10. J. H. Hildebrand and R. L. Scott, *Regular Solutions*, Prentice-Hall, Englewood Cliffs, NJ, 1962.
11. W. B. Jensen, *The Lewis Acid-Base Concepts*, Wiley, New York, 1980.
12. V. Gutmann, *Coordination Chemistry in Non-aqueous Systems*, Springer-Verlag, Vienna, 1968.
13. R. G. Pearson, *J. Am. Chem. Soc.*, 1963, 85, 3533.
14. R. G. Pearson, *Chem. Brit.*, 1967, 3, 103.
15. C. Reichardt, *Solvents and Solvent Effects in Organic Chemistry*, 2nd ed., VCH, New York, 1988, Chap. 7.
16. R. S. Drago and B. Wayland, *J. Am. Chem. Soc.*, 1965, 87, 3571.
17. A. P. Marks and R. S. Drago, *J. Am. Chem. Soc.*, 1975, 97, 3324; *Inorg. Chem.*, 1976, 15, 1800.

18. L. A. Dominey, in *Lithium Batteries, New Materials, Developments and Perspectives*, G. Pistoia, ed., Elsevier, New York, 1994, p. 137.
19. D. F. Evans and M. A. Matesich, in *Techniques of Electrochemistry*, Vol. II, E. Yeager and A. J. Salkind, Jr., eds., Wiley-Interscience, New York, 1973, Chap. 1.
20. J. F. Coetzee and C. D. Ritchie, Eds., *Solute-Solvent Interactions*, Vol. II, Marcel Dekker, New York, 1976.
21. R. L. Kay, in *Techniques of Electrochemistry*, Vol. II, E. Yeager and A. J. Salkind, Jr., eds., Wiley-Interscience, New York, 1973, Chap. 2.
22. M. Spiro, in *Physical Chemistry of Organic Solvent Systems*, A. K. Covington and T. Dickinson, eds., Plenum Press, London, 1973, Chap. 5.
23. J. F. Casteel and E. A. Amis, *J. Chem. Eng. Data*, 1972, 17, 55.
24. P. Bruno and M. Della Monica, *J. Phys. Chem.*, 1972, 76, 3034.
25. P. Bruno and M. Della Monica, *Electrochim. Acta*, 1975, 20, 179.
26. P. Bruno, C. Gatti and M. Della Monica, *Electrochim. Acta*, 1975, 20, 533.
27. C. A. Angell and R. D. Bressel, *J. Phys. Chem.*, 1972, 76, 3244.
28. C. A. Angell and J. C. Tucker, *J. Phys. Chem.*, 1974, 78, 278.
29. J. Barthel, H. J. Gores, P. Carlier, F. Feuerlein and M. Utz, *Ber. Bunsenges. Phys. Chem.*, 1983, 87, 436.
30. H. V. Venkatasetty, *J. Electrochem. Soc.*, 1975, 122, 245.
31. J. Imhof, T. D. Westmoreland and M. C. Day, *J. Solution Chem.*, 1974, 3, 83.
32. R. A. Robinson and R. H. Stokes, *Electrolyte Solutions*, Academic Press, New York, 1955, pp. 152–155.
33. O. Popovych and R. P. T. Tomkins, *Nonaqueous Solution Chemistry*, Wiley, New York, 1981.
34. R. M. Fuoss and F. Accascina, *Electrolytic Conductance*, Interscience, New York, 1959, pp. 213 et. seq.
35. G. E. Blomgren, Proc. Vol. 80-4, The Electrochemical Society, Pennington, NJ, 1980, pp. 368–377.
36. G. E. Blomgren, Proc. Vol. 80-7, The Electrochemical Society, Pennington, NJ, 1980, pp. 35–45.
37. B. Gill, in *Chemistry of Nonaqueous Solutions*, G. Mamontov and A. I. Popov, eds., VCH, New York, 1994, Chap. 2.
38. D. Irish, in *Ionic Interactions*, Vol. II, S. Petrucci, ed., Academic Press, New York, 1971, Chap. 9.
39. W. F. Edgell, in *Ions and Ion Pairs in Organic Reactions*, Vol. I, M. Swarc, ed., Wiley, New York, 1972, Chap. 4.
40. J. H. Sharp and M. C. R. Symons, in *Ions and Ion Pairs in Organic Reactions*, Vol. I, M. Swarc, ed., Wiley, New York, 1972, Chap. 5.
41. E. de Boer and J. L. Sommerdijk, in *Ions and Ion Pairs in Organic Reactions*, Vol. I, M. Swarc, ed., Wiley, New York, 1972, Chaps. 7, 8.
42. R. M. Fuoss and C. A. Kraus, *J. Am. Chem. Soc.*, 1933, 55, 2387.
43. W. Ebeling and M. Grigo, *J. Sol. Chem.*, 1982, 11, 151.
44. Y. Marcus, *J. Sol. Chem.*, 1983, 12, 135.

# 3

# Experimental Considerations Regarding the Use and Study of Nonaqueous Electrochemical Systems

**Doron Aurbach and Arie Zaban**
*Bar-Ilan University, Ramat-Gan, Israel*

## I. INTRODUCTION

The study of nonaqueous electrochemical systems, especially those which consist of active electrodes (e.g., active metals, hygroscopic composites and lithiated carbons), requires special experimental considerations. Ignoring the high sensitivity of such systems to atmospheric and other contaminants, even at a ppm level, renders the results obtained meaningless. The following experimental aspects have to be considered during research and/or the use of nonaqueous systems.

    a.  Purity standards for nonaqueous solvents and salts, how they are attained, and what the optimized purification methods and solution preparations are

    b.  The working environment—how to maintain an inert, contaminant-free atmosphere

    c.  How to monitor contaminants in the atmosphere and in the solutions

    d.  How to keep the solutions dry

    e.  Basic aspects of glove box operation; how to organize it for electrochemical studies

    f.  Cells for electrochemical studies, construction materials, reference electrodes

g.  Electrode preparation, active and nonactive metals, composite electrodes
h.  Application of electrochemical techniques, transients, microelectrodes, micropolarization-impedance spectroscopy, PITT and standard techniques (e.g., LSV, chronopotentiometry)
i.  How to use spectroscopic tools for the studies of nonaqueous electrochemical systems; techniques and transfer systems
j.  Safety considerations

In this chapter, the above issues are dealt with point by point. The emphasis is on experimental aspects related to liquid electrolyte solutions, since these are the most commonly used and have the broadest applications.

Other nonaqueous systems, which were mentioned in the first chapter, such as ionically conducting polymers, molten salts and solid electrolytes, have uses that are more specific. Hence, experimental aspects that are related to polymer based systems and molten salts are mentioned in the chapters that deal with them.

## II.  PREPARATION AND PRESERVATION OF HIGHLY PURE NONAQUEOUS ELECTROLYTE SOLUTIONS

### A.  Sources of Solvents and Salts

Since the advantage of using nonaqueous systems in electrochemistry lies in their wide electrochemical windows and low reactivity toward active electrodes, it is crucial to minimize atmospheric contaminants such as $O_2$, $H_2O$, $N_2$, $CO_2$, as well as possible protic contaminants such as alcoholic and acidic precursors of these solvents. In aprotic media, these contaminants may be electrochemically active on electrode surfaces, even at the ppm level. In particular, when the electrolytes comprise metallic cations (e.g., Li, Mg, Na), the reduction of all the above-mentioned atmospheric contaminants at low potentials may form surface films as the insoluble products precipitate on the electrode surfaces. In such cases, the metal-solution interface becomes much more complicated than their original design. Electron transfer, for instance, takes place through electrode-solution rate limiting interphase. Hence, the commonly distributed solvents and salts for usual R & D in chemistry, even in an analytical grade, may not be sufficient for use as received in electrochemical systems.

The excellent development of nonaqueous batteries (mostly lithium, metal based and Li ion ''rocking chair'' type batteries) created a market for the ''Li battery grade'' materials which are manufactured at high purity standards. These materials are manufactured by companies whose expertise lies in the production and distribution of the highly pure solvents and salts. Currently, a great part of

**Table 1**  Partial List of Companies Which Distribute Highly Pure Solvents and Salts and Their Products

| Company | Country | Typical products (Li battery grade) |
|---------|---------|-------------------------------------|
| Merck | Germany | DN, DMC, EC, prepared electrolyte solutions |
| Burdick and Jackson | USA | THF, PC, AN |
| Grant Chemicals | USA | DN, glymes (polyethers) |
| Tomiyama | Japan | All solvents and salts for Li batteries |
| Mitsubishi | Japan | All solvents and salts for Li batteries |
| Lithco | USA | $LiAsF_6$ |
| 3M | USA | $LiN(SO_2CF_3)_2$, $LiSO_3CF_3$ |
| Hashimoto | Japan | $LiPF_6$, $LiBF_4$ |

the solvents and salts mentioned at the beginning of Chapter 1 and which appear in Scheme 1 are available as Li battery grade materials. Thus they can be used as received, provided that the appropriate working atmosphere is maintained (as discussed later in this chapter). It should be emphasized that the proper handling and distribution of highly pure nonaqueous solvents and salts are key factors in the possibility of using them as received; they should be packed and stored in appropriate containers under highly pure argon or nitrogen. For instance, salts that are composed of anions, such as $PF_6^-$ or $BF_4^-$ and their solutions *cannot* be stored in glass because they are usually contaminated with HF, which reacts with the hydroxy groups of the glass. Other common contaminants in these salt solutions, such as $PF_5$ or $BF_3$, can also react with the hydroxy groups of the glass to form P—F—O and B—F—O compounds and liberate HF, which further attacks the glass.

Table 1 lists several companies and the typical highly pure products they offer. In any event, users need to be aware that from time to time certain batches may be below standard. Thus, even when using the Li battery grade materials, routine checks have to be applied to ensure quality. Other solvents and salts which may be needed should be purchased only in the highest quality available and used only after intensive purification that will be described later.

## B.  Purification of Solvents and Salts

### 1.  Solvent Distillation

In spite of the availability of Li battery grade solvents and the large number of applications for which they can be used as received, anyone working intensively with nonaqueous systems should be able to purify solvents for the following reasons:

1.  Many solvents can be spoiled during prolonged storage, which requires redistillation before use. For example, ethers exposed to oxygen develop peroxide contamination, and cyclic carbonates may decompose to derivatives of their mother glycols and $CO_2$.
2.  Some solvents, such as 1,3-dioxolane, may contain stabilizers that can be electrochemically active. These stabilizers should be eliminated before the solvents are used.
3.  Many important solvents are available only at normal analytical grade as their highest purity level. Most of these solvents *are not* pure enough for many applications in electrochemistry.

The following guidelines for distillation are suggested:

1.  Distillation should be carried out in a fractionating glass system, under highly pure argon (or vacuum and back-filled argon, depending on the relevant vapor pressures).
2.  All the parts, including valves and taps, should be made from glass and Teflon, since metallic or plastic parts may introduce contamination. The parts should be connected with viton O rings.
3.  The distillation system should be capable of collecting the first fraction separately and then the main fractions in closed vessels, which can be introduced, hermetically sealed, into a glove box via the vacuum chamber.
4.  For each solvent, the appropriate contaminant absorber should be chosen, as well as the distillation temperature at which it is stable.

A scheme for such a system is shown in Figure 1. These systems can be built by simple glass blowing. Alternatively, it is possible to use commercially available distillation systems such as the spinning band systems.

In a typical batch distillation, the lower bulb of the system is loaded with the solvent and a suitable absorber in the glove box, after which it is connected to the system under flowing argon. The system is purged with argon, the appropriate pressure is adjusted, and the lower bulb is heated to boiling point. The first fraction is taken out, and total reflux is applied. Then the middle fraction, the pure material, is collected in the terminal bulb. As the distillation stops, the terminal bulb is closed hermetically and introduced into the glove box.

Table 2 summarizes distillation conditions (pressure, temperature and contaminant absorbing materials for several important solvents). In brief, the best contaminant absorber for ether solvents is benzophenone radical anion, which is formed by the introduction of alkali metal chips into a solution of benzophenone in the ether. The metal is chosen, depending on the difference between its melting

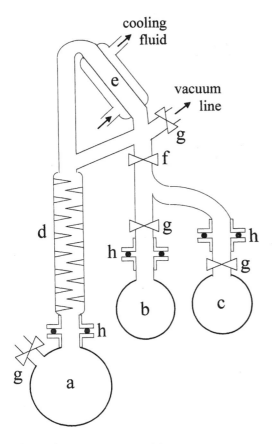

**Figure 1** Scheme for a distillation system. These systems can be built by simple glass blowing: (a) source solvent; (b) first fraction bulb; (c) collection bulb; (d) fractionating column; (e) condenser; (f) reflux ratio valve; (g) regulator Teflon valve; (h) O ring connector.

point [$T_m$(metal)] and the boiling point of the ether [$T_b$(ether)]. $T_m$(metal) should be higher than $T_b$(ether), since it is very dangerous to melt alkali metal in boiling ethers.

It should be noted that if the ethereal solution is contaminated with protic substances at too high a level, the alkali metal might be passivated with the reduction products of the contaminants. Consequently, the reaction between the active metal and the benzophenone does not take place, and no absorption of contaminants occurs. When the benzophenone does react with the alkali metal and the

**Table 2** Distillation Conditions for Several Important Nonaqueous Solvents

| Solvent | Common impurities | Recommended absorber | Normal boiling point | Recommended distillation parameters | |
|---|---|---|---|---|---|
| | | | | Boiling point | Pressure (mmHg) |
| PC | Propylene glycol, $CO_2$, $H_2O$ | $CaH_2$ | 240 | <80°C | 5–10 |
| EC | Ethylene glycol, $CO_2$, $H_2O$ | $CaH_2$ | 243 | <80°C | 5–10 |
| DMC | $H_2O$, $CH_3OH$ | $CaH_2$ | 90 | Normal $T_b$ | Atmospheric |
| DEC | $H_2O$, $CH_3CH_2OH$ | $CaH_2$ | 127 | Normal $T_b$ | Atmospheric |
| DME | $H_2O$, ethylene glycol, methanol | Na, B.P.[b] | 85 | Normal $T_b$ | Atmospheric |
| Ethylglyme | Ethanol, $H_2O$ | Li, B.P.[b] | 121 | Normal $T_b$ | Atmospheric |
| Diglyme | Methanol, $H_2O$ | Li, B.P.[b] | 162 | Normal $T_b$ | Atmospheric |
| THF | $H_2O$, alcohols | Na, B.P.[b] | 66 | Normal $T_b$ | Atmospheric |
| 2Me THF | $H_2O$, alcohols | Na, B.P.[b] | 79 | Normal $T_b$ | Atmospheric |
| DN | $H_2O$, alcohols | Li, B.P.[b] | 74 | Normal $T_b$ | Atmospheric |
| AN | $H_2O$ | $CaH_2$ | 81 | Normal $T_b$ | Atmospheric |
| BL | $H_2O$, acid | $CaH_2$ | 204 | <80°C | 5–10 |
| MF | $H_2O$, methanol, acid | $P_2O_5$ | 32 | Normal $T_b$ | 5–10 |
| DMF | Methanol, $H_2O$ | $CaH_2$ | 153 | <80°C | 5–10 |
| DMSO | $H_2O$ | $CaH_2$ | 189 | <80°C | 5–10 |

[a] In an analytical grade material.
[b] B.P. = benzophenone.

absorber-benzophenone radical anion is formed, these solutions turn blue. The blue color indicates that the contamination level is sufficiently low to allow massive formation of the absorber radical anion. During distillation of these blue solutions, all the common atmospheric and protic contaminants are reduced to nonvolatile products (alkali metal salts), which remain in the bulb.

For the alkyl carbonates and esters such as butyrolactone, $CaH_2$ can be used as the contaminant absorber. EC, PC and BL have to be distilled in vacuum due to their high boiling points. For esters such as methyl formate, $P_2O_5$ can be used as a contaminant absorber (highly efficient for absorbing protic species).

## 2. Salt Purification

Purification of salts may be too complicated for most electrochemistry groups. Fortunately, all salts of interest can now be purchased at a very high level of purification, which allows using them as received. However, it is always necessary to dry these salts further. Commonly used salts which are composed of metallic or tetraalkyl ammonium cations and anions, such as $AsF_6^-$, $ClO_4^-$, $BF_4^-$, $PF_6^-$, $SO_3CF_3^-$ $N(SO_2CF_3)_2^-$, $C(SO_2CF_3)_3^-$, are extremely sensitive to heating. Hence, their drying should be based on prolonged evacuation (several days) at high vacuum ($p < 10^{-3}$ torr). The temperature should be kept below 80°C. The exceptions are $LiClO_4$, which can be heated up to 150°C, and metal halides, which can be heated even further ($T > 200°C$). It should be emphasized that drying has to be carried out in vessels that can be closed under vacuum, to prevent exposure of the salts to atmospheric contamination during the transfer to the glove box. Salts of $PF_6^-$ and $BF_4^-$ should never be treated in glass.

## 3. Drying Solvents

The traditional drying agents for nonaqueous solvents include anhydrous salts (e.g., $MgSO_4$, $CaCl_2$), molecular sieve, alumina ($Al_2O_3$), silica gel, and alkali metals (Li, Na) for ethers. Of all the above, only molecular sieves should be used (4 Å pore size, activated by evacuation at $T > 250°C$). Inorganic salts may contaminate polar solvents with ions. Alkali metals are readily passivated in ethers and may be dangerous to use. Alumina was also found to be inadequate as a drying agent because it decomposes solvents such as alkyl carbonates [1] and ethers [2]. Molecular sieves, which may keep solvents at a water contamination level below 10 ppm, are suitable for drying if the following precautions are taken:

1. The molecular sieve produces dust, which precipitates in the bottle. Solvents kept over a molecular sieve should be removed carefully by decantation to prevent the dust at the bottom of the bottle from floating.

2. Solutions containing salts should not be kept over a molecular sieve due to possible cationic exchange between this drying agent and the electrolyte.

## C.  Preparation and Storage of Electrolyte Solutions

Most of the nonaqueous solvents are highly hygroscopic. The presence of oxygen also affects the purity of nonaqueous solutions, as solvent oxidation may occur during prolonged exposure to oxygen. Hence, solutions should be prepared and stored under an inert atmosphere, using a glove box (see glove box section for details).

Most of the nonaqueous solutions can be stored in glass containers, exceptions being those that contain $PF_6^-$ and $BF_4^-$. These solutions should be stored in Teflon containers because they are, unavoidably, always contaminated by HF that attacks glass. Another important point relates to the method by which the salt and the solvent are mixed. The enthalpy of dissolution of many salt-solvent combinations may be very high (in absolute values). Therefore, upon mixing, when a large amount of heat is liberated, solvents such as ethers, esters and alkyl carbonates may decompose. Thus, the salts should be added slowly to the solvents (and not inversely) while good agitation is maintained. In many cases, it is also necessary to cool the solutions during mixing. For instance, the uncontrolled addition of $LiAsF_6$ to ethers such as THF or 2Me-THF leads to their partial decomposition.

## D.  Monitoring the Purity of Solvents and Salts

### 1.  Solvents

The standard tools such as FTIR, NMR, MS and UV-Vis are *not* sufficient for the requirements of nonaqueous electrochemistry, that is, detection of impurities in the solvents at a ppm level. The best of these methods, the NMR, can detect impurities only at a subpromil level, which is too high. As discussed in the chapter dealing with the electrochemical windows of nonaqueous systems, a few tens of ppm of contaminants can considerably affect the electrochemical behavior. Therefore, GC or HPLC are the only appropriate methods available today [3–5]. However, it is important to note that there are solvents that can partially decompose in the injection port or in the column of the GC. The relevant solvents are those whose normal boiling points are high, and thus their GC analysis requires high temperatures. For example, PC, EC, BL and SL are typically problematic for GC analysis.

The following column compositions are recommended for the analysis of various solvents (J&W Scientific names).

| Solvent | Column |
|---------|--------|
| AN | DB-Wax |
| BL | DB-1 |
| DEC | DB-1; DB-23 |
| DMC | DB-1; DB-23 |
| DN | DB-624 |
| EC | DB-1; DB-23 |
| MA | DB-1 |
| MF | DB-1 |
| PC | DB-1; DB-23 |
| THF | DB-624 |
| 2Me-THF | DB-624 |

An immediate test for the purity of ethers is the appearance of a blue color when they come into contact with alkali metal and benzophene.

## 2. Salts

Purity tests for salts are much more problematic than are purity tests for solvents. The only way to follow the salt purity rigorously is by careful element analysis by a combination of methods, which include atomic absorption, ICP (in solutions), and, to some extent, MS. Water contamination can be detected relatively easily by IR spectroscopy. All the above-mentioned salts have characteristic IR spectra in which the hydration water has specific bands.

Figure 2 presents several typical FTIR spectra of commonly used salts (in KBr pellets). Peak assignments, including the location of the hydration water, may be found in Ref. 6.

## 3. Water Contamination

The water contamination level in solutions is usually measured by Karl Fisher (KF) titration. The water detection is based on the mechanism described below. In brief, the reaction includes two steps:

$$R'OH + SO_2 + RN \rightarrow [RNH^+]SO_3R' \tag{1}$$

$$H_2O + I_2 + RNHSO_3R' + 2RN \rightarrow [RNH]SO_4R' + 2RNH^{(+)} + 2I^- \tag{2}$$

RN is a base (e.g., pyridine), R'R are alkyl groups, and $I_2$ is generated separately in a different compartment. Hence, in the presence of water, $RNHSO_3R'$ is oxidized by the iodine. The amount of water is determined by a coulombic titration of the $I^-$ formed, back to iodine:

$$2I^- + 2e^- \rightarrow I_2 \tag{3}$$

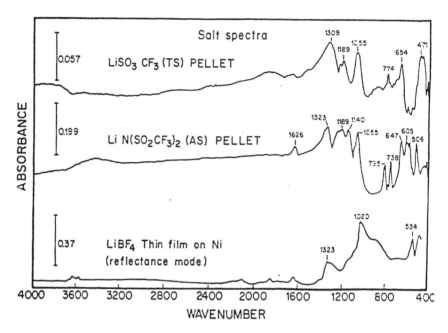

**Figure 2**   Typical FTIR spectra of commonly used salts. LiSO$_3$CF$_3$ and LiN(SO$_2$CF$_3$)$_2$ in KBr pellets and LiBF$_4$ as a thin layer on nickel mirror, reflectance mode [6]. (Reprinted with copyright from Elsevier Science LTD.)

There are several types of automated KF titrators available from leading companies that supply electrochemical equipment (Metrohm, for example). It should be noted that the mother solutions of these instruments are highly sensitive to side reactions with components of the nonaqueous solution. Hence, the users have to consult the suppliers of the KF mother solutions to ensure that they are compatible with the composition of the studied solution.

## III.   GLOVE BOX OPERATION

### A.   Introductory Remarks

Nonaqueous systems are highly sensitive to atmospheric components, e.g., $H_2O$, $O_2$, $CO_2$, and even $N_2$. This sensitivity is caused by the wide potential window at which these systems operate, in which all the above contaminants are electrochemically active. Elimination of atmospheric contaminants is therefore a first and mandatory condition for carrying out our work with nonaqueous systems. Several modes of operation may be applicable.

## 1. Use of Vacuum Systems

There are several reports in the literature on the performance of electrochemical studies in vacuum systems. These reports include studies by Yeager [7] and Bard et al. [8–10]. Of special importance are recent reports on electrochemical measurements in ultrahigh vacuum systems by Scherson et al. [11–13].

## 2. Use of Dry Rooms

The Li battery industry uses dry rooms in which the humidity is maintained at around 1–2%. For certain applications, especially for mass production, dry rooms can provide an appropriate atmosphere. However, for most of the rigorous studies with nonaqueous solutions, dry rooms are definitely inadequate.

## 3. Glove Boxes

Glove box operation is the most common way to enable a comfortable and continuous study under inert atmosphere. Therefore, special attention is given in this chapter to the various aspects of glove box operation. We aim at sharing our experience with the readers and providing appropriate guidelines for newcomers to this field in order to save their time and prevent them from making mistakes.

## B. Basic Features of a Good Glove Box System

The basic working station is a box constructed of metal or plastic with an appropriate transparent window onto which replaceable rubber gloves are mounted. This configuration enables all areas within the box to be reached and viewed. The best materials for constructing the box are aluminum or stainless steel. Aluminum is preferred for most of the applications since it is lighter, easier to handle, and is probably cheaper. Polycarbonate (lexane) is the best material for the window.

The entire system should contain the following elements:

1. Vacuum chambers attached to the box through which items are introduced into the glove box. The vacuum chamber should be equipped with the appropriate hermetically closed doors, valves and a vacuum pump. The chamber should maintain a vacuum of $10^{-2}$–$10^{-3}$ torr in order to ensure a low level of contamination upon its being opened to the glove box.
2. $H_2O$ and $O_2$ absorbers (molecular sieves and other active materials) through which the glove box atmosphere is continuously circulated.

The absorbers column should be attached to the glove box by the appropriate pipes, valves and blowing system. It should include a regeneration capability.

3. $H_2O$ and $O_2$ analyzers which are capable of monitoring these contaminants at the ppm level.

4. A trapping system for vapor of organic solvents and $CO_2$. These systems are based on massive circulation of the glove box atmosphere through a cooled trap outside the glove box. Such a system should include an appropriate blower, valves and a connection to a vacuum line to enable emptying of the trap.

5. Electrically controlled connections to the inert gas supply and the vacuum pump line which are capable of maintaining any desirable pressure inside the box. This control system should include pedal switches to allow operation by foot while the hands are engaged inside the glove box.

6. A separate set of vacuum and inert gas lines should be introduced into the box for operations such as drying, evacuation and gas purging.

7. High voltage electrical outlets for operation of electrical instruments inside the box.

8. Low impedance connectors for electrochemical measurements taken with equipment that is located outside the glove box. High impedance or bad shielding of these electrical lines may introduce artifacts.

9. (Optional) $N_2$ absorber for special cases in which nitrogen is also reactive, e.g., when the electrochemistry of active metals such as Na or Li is investigated.

Figure 3 shows a typical layout of a glove box system, which includes the above capabilities. Excellent glove boxes are commercially available from several manufacturers, such as VAC (USA) and M. Brown, (Germany). Figure 4 is a glove box system made by M. Brown. State-of-the-art glove boxes include a microprocessor control unit that monitors the pressure, $H_2O$ and $O_2$ contamination level, and the circulation/regeneration life cycle of the $H_2O$ and $O_2$ absorbers. Commercially available glove box systems can maintain a contamination level of $O_2$ and $H_2O$ below 1 ppm.

Especially attractive are glove boxes that can be evacuated. The advantage of these systems is that they enable replacement of a contaminated atmosphere by a new, highly pure one within a few minutes. However, such systems require special operation that takes into account the effect of the evacuation on the glove box contents. The evacuated glove box requires strong construction and a system of glove covers which can be evacuated simultaneously with the box. Such systems are available commercially.

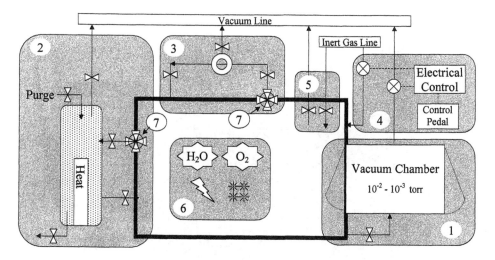

**Figure 3** Schematic of a glove box that includes the basic capabilities: (1) Vacuum chambers equipped with doors, valves and a vacuum pump. (2) $H_2O$ and $O_2$ absorbers through which the glove box atmosphere is continuously circulated. The absorber column should include a regeneration capability. (3) A trapping system for vapor of organic solvents and $CO_2$ based on massive circulation of the glove box atmosphere through a cooled trap. (4) Electrical control of the inert gas supply and the vacuum pump line including pedal switches for foot operation. (5) A separate set of vacuum and inert gas lines for various operations. (6) $H_2O$ and $O_2$ analyzers, high voltage electrical outlets and low impedance connectors for electrochemical measurements. (7) Air blowers.

Research groups which have good machine shop services should not be discouraged from building their own glove boxes. A well-organized machine shop, with facilities for aluminum welding under an argon atmosphere, can make all the required parts with one exception, the $H_2O$ and $O_2$ absorption device. The manufacture of an $H_2O$ and $O_2$ absorption device with the appropriate circulation and regeneration units requires professional skill. Therefore, it is strongly recommended that people wishing to build their own glove boxes should purchase this unit from a commercial company and incorporate it into their homemade system.

Developing the capability to make ''homemade'' glove boxes may be very important for special needs, such as spectroelectrochemical studies. For instance, FTIR spectroscopy, AFM or STM measurements of active metal electrodes are much more accurate and reliable when done inside a glove box. To enable such measurements, the instruments should be installed in glove boxes that are specially designed for them. We ourselves built special glove boxes for a FTIR

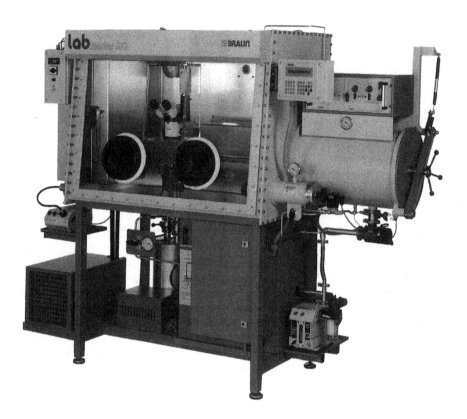

**Figure 4** The M. Brown ''Lab master 130'' glove box system. A typical example of state-of-the-art glove boxes that include a microprocessor control unit that monitors the pressure, $H_2O$ and $O_2$ contamination level and the circulation/regeneration life cycle of the $H_2O$ and $O_2$ absorbers.

spectrometer and for an AFM-STM system, the latter being designed so that it can be evacuated.

## C.  Guidelines for Glove Box Operation

Work with sensitive materials, such as nonaqueous electrochemical systems, utilizing glove boxes is well established and common now. In fact, purchasing a commercial system is usually accompanied by reasonable support and service from the manufacturer, which enables a good start-up. However, efficient and useful glove box operation also requires extensive experience, and therefore in the next section we will provide glove box users with a list of guidelines based on our own experience.

1.  The best commercially available glove boxes can provide an inert atmosphere containing $H_2O$ and $O_2$ at the ppm level. The lowest level of each contaminant which can be achieved is 1 ppm. However, this level can be achieved only when highly pure inert gas is used (below 1 ppm of $H_2O$ and $O_2$ content). Commonly used gases are Ar, He and $N_2$, although $N_2$ cannot be considered inert for any active metals of the alkali metal group, Li, Na and K. Argon may be better than helium because it is heavier, and thus its diffusion rate from the glove box is slower.

2.  It should be emphasized that a noble gas environment containing atmospheric contaminants at the ppm level *cannot* be considered inert for sensitive electrodes such as active metals. The inertness of a glove box atmosphere can be sufficient for preparation and work with nonaqueous solutions at low $H_2O$, $O_2$, $CO_2$ and $N_2$ contamination levels. Work with active metal electrodes, especially when spectroelectrochemical studies are important, requires special electrode surface preparation, as described later. As a rule, a glove box may be sufficient for maintaining a low contamination level in bulk materials, but *not* on active surfaces.

3.  It is essential to have $H_2O$ and $O_2$ monitors in the glove box. For instance, there are standard $H_2O$ and $O_2$ detectors made by the American companies Teledine and Ondyne. The sensors of these detectors are placed inside the glove box and are usually based on electrochemical devices. For example, the $O_2$ analyzer is a micro fuel cell whose anode may be Pb and whose cathode may be an oxygen reduction catalytic electrode with a concentrated KOH solution as the electrolyte. Each of these detectors requires four wires, two for the signal and two for a thermistor placed in the cell. The thermistor is important because, like any electrochemical process, these processes are highly temperature sensitive. It should be noted that the lifetime of an oxygen detector is only a few months. It is extremely sensitive to shocks and high vacuum. The $H_2O$ analyzer is much less sensitive and has a longer lifetime. Of the two, the most important analyzer is the one for $O_2$. Due to the high concentration of $O_2$ in the air, its response is the most immediate and indicative in case of leaks in the system.

4.  In the usual glove box equipped with only $H_2O$ and $O_2$ absorbers, the concentration of $CO_2$ and $N_2$ may increase during normal operation with no control. Both gases are electroactive, particularly on active metal and noble metal electrodes polarized to low potentials. When elimination of $N_2$ is required (e.g., in the case of spectroscopic studies of active metal electrodes), the glove box system has to include a commercial $N_2$ absorber. Companies such as VAC pro-

duce modular $N_2$ absorbers which can be connected to any existing glove box system. $CO_2$ can be removed by applying liquid $N_2$ to the vapor trapping system mentioned below.

5. Working with volatile solvents requires the current removal of their vapor from the glove box atmosphere. Volatile solvents are absorbed by plastic parts of the glove box, including the gloves, window, seals, O rings, etc., which causes a deterioration in the quality of these parts and their ability to insulate the glove box from the outside atmosphere. Therefore, when working with volatile materials, the glove box must include a solvent trap, which can be purchased or made in house. The easiest and cheapest system should contain a standard glass cold trap equipped with a dewar for liquid nitrogen. The trap should be connected to the box with sufficiently wide pipes ($d > 2$ cm), through ball valves that can isolate the trap when needed. The system should be connected to a vacuum line for removal of the trapped contaminants. The glove box atmosphere may be circulated through the trap, using a regular air blower. By knowing the box volume and the flow rate of the atmosphere through the system, it is easy to calculate how much time is required to clean the atmosphere, provided that the efficiency of the liquid $N_2$ trap is absolute. The operation also removes the $CO_2$ contamination completely.

6. The glove box atmosphere may contain vapors that are hazardous. When the vacuum chamber, which is filled with the glove box atmosphere, is opened outside, the user may be exposed to dangerous materials. It is therefore recommended that the vacuum chamber be evacuated and back-filled with fresh air before it is opened.

7. A critical point in the maintenance of a highly pure glove box atmosphere is the quality of the vacuum chamber through which all substances are introduced into the glove box after being evacuated. When this chamber is filled with inert gas and opened to the glove box, atmospheric contaminants can be introduced in a concentration that depends on the vacuum reached in the chamber. The volume of a typical vacuum chamber is about one-tenth that of the glove box. In order to prevent contamination above the ppm level when the vacuum chamber is opened to the glove box, it should be able to reach a vacuum below $10^{-1}$ torr. To further minimize the contamination level, one can evacuate the chamber more than once, filling it with inert gas between evacuations. Another option is to use a small chamber when only small items are transferred into the glove box. Such minichambers, in addition to standard ones, are offered by glove box manufacturers.

8. A key issue in a glove box operation is *order*. The box must be equipped with shelves and should contain an appropriate set of tools. It is usually useful to have in the glove box a pair of long tweezers, paper tissues, hermetically sealed containers for disposal of used materials and a balance.

9. When the glove box is used for conducting electrochemical measure-

ments, one must take into account the long cables that connect the cell in the glove box to the outside instrumentation. Therefore, all wires must be properly screened. It is very important that the electrometer (high impedance) of the reference electrode be inside the glove box so that the connection between it and the reference electrode in the cell is as short as possible.

10.  It *is not* recommended that the user work directly with rubber gloves when touching materials and tools. Doing so shortens the lifetime of the gloves and, more importantly, introduces contamination into the experiments. Therefore, gloves should be internally lined with disposable polyethylene gloves that provide a clean surface. The polyethylene gloves should be replaced frequently.

11.  Rubber gloves are punctured easily and, thus, have to be checked frequently. It is always a good idea to keep the glove box under positive pressure to ensure that, if there is a leak, the gas will flow outside from the box and not vice versa. This can also be used as a standard test for leaks, provided that the room temperature and, thus, the gas volume do not change. When the gloves are punctured, it is necessary to replace or fix them. For this, the glove box has to be equipped with an internal glove port cover which can isolate the gloves from the inside of the glove box. With this port, the damaged glove can be removed while the box remains isolated.

Another part that may require replacement is the window. The usual transparent plastics used for these windows—polycarbonates—are readily attacked by a number of organic solvents. Thus, during intensive work with organic solvents, the window may lose its transparency. Replacement of the window requires careful evacuation of all the sensitive materials from the glove box and its full exposure to air.

12.  Putting a glove box system to work requires efficient replacement of its air atmosphere by the inert gas. Using the $H_2O/O_2$ absorbers to clean the atmosphere is not sufficient in cases when nitrogen and $CO_2$ should also be eliminated. Since the glove boxes are usually equipped with $H_2O/O_2$ detectors, they can be used to examine the concentration of all the other gases existing in the air. The glove box should be washed with the inert gas until the concentrations of $H_2O$ and $O_2$, and thus the other gases, are sufficiently low. In order to save expensive gas in this step, it is recommended that balloons be used. As recommended in the previous section, a glove box system should include additional auxiliary inert gas and vacuum lines that can be used to fill the entire volume of the glove box with balloons filled with the inert gas. The remaining free volume can then be washed several times by a few gas filling/evacuation cycles, after which the $H_2O/O_2$ absorbers can be put to work. The balloons should be evacuated, using the vacuum line inside the glove box.

13.  The regeneration of the $H_2O/O_2$ absorber is performed by a cycle which includes three steps: heating, washing with a mixture of the inert gas with hydrogen (5–10%), and evacuation of the absorber. The exhaust gas that escapes

during the regeneration process is highly contaminated; thus it should be extracted via a hood. The regeneration process produces a large amount of water which is trapped by the vacuum pump's oil. Therefore, after a regeneration cycle, the oil in the relevant vacuum pump has to be replaced.

14.   The active material in the $H_2O/O_2$ absorber has to be replaced from time to time after prolonged and intensive use, and users must be prepared for that. The producers of these devices have to provide the users with this option and instruct them on how to do it correctly.

15.   When the system is not working, e.g., during vacation time, it is recommended that it be left with a positive pressure and that all electric valves (solenoids) be turned off. The vacuum chamber should be left evacuated but disconnected from the vacuum pump. Under these conditions, nothing unexpected can happen when no one is around.

## IV.   ELECTROCHEMICAL MEASUREMENTS

### A.   Cell Configurations

There are several typical studies related to nonaqueous electrochemical systems which require specific cell configurations:

1.   Quantitative electrolysis
2.   Quick voltammetric analysis of metal salt solutions
3.   Quick voltammetric analysis of tetraalkyl ammonium salt solutions
4.   Testing of electrode materials for batteries
    a. Li anodes
    b. intercalation anodes (e.g., lithiated carbons)
    c. transition metal oxide cathodes
    d. composite electrodes of redox species (e.g., electrically conducting polymers)
5.   Fine electroanalytical measurements, e.g., impedance spectroscopy and fast transients
6.   Preparation of electrodes for in situ spectroscopic studies
7.   Cells for in situ spectroscopic measurements

There are several key points which have to be considered in the preparation and construction of cells for the above measurements.

1.   The solvents are usually involved in the counterreaction which takes place on the counterelectrode. Since many of these solvents are not as viscous as water, the diffusion of undesired species from the counterreaction's compartment may interfere severely with the studied reaction.

Hence, the catholyte and the anolyte have to be well separated and isolated.

2.  The choice of the appropriate reference electrode and its separation from the rest of the cell may be critical.
3.  Nonaqueous solutions are usually much more resistive than aqueous solutions, therefore, the location of the reference electrode is very important.
4.  It is important that the cell's geometry should lead to a uniform current distribution. This is especially important in the performance of fine electroanalytical measurements and electrode surface preparation of spectroscopic studies.
5.  Aqueous solutions are usually saturated with air, and therefore the usual electrochemical cell for aqueous solutions includes an inert gas bubbler which de-aerates the solution. Many nonaqueous solvents are too volatile to be de-aerated by gas bubbling. Hence, the conception is that the solutions should be introduced into the cell when they are sufficiently pure and degassed (due to the appropriate distillation and other pretreatments described in the previous sections).
6.  Many plastic materials are not compatible with nonaqueous solutions. The plastics themselves or the plasticizers may dissolve into the solution and contaminate it. In addition, active metals (including their alloys or low potential insertion compounds such as Li—C) are incompatible with glass and any plastic materials which contain F or Cl, such as Teflon and PVC. Hence, for nonaqueous systems which do not contain active metal electrodes, glass or Teflon may be the most desirable cell construction materials. Cell parts that must be in contact with active metals should, however, be made of polyethylene or polypropylene.

Figure 5 shows a simple cell for a quick study of the voltammetric behavior of a solution containing an electrolyte with metal ion [14]. The working electrode may be noble metal wire embedded in glass, and the counter- and reference electrodes may be the metal of the electrolyte's cation, provided that this metal is apparently stable in the solution. In such a case, the counter- and reference electrodes are the $M/M^{z+}$ couples. The electrodes may be of a cylinder, ring or wire form. In a case in which a wire is used, the edge of the wire should also be embedded in a glass bulb to avoid the nonuniform current distribution due to edge effects. To achieve a short distance between the working and reference electrodes, a luggin capillary, as shown in the figure, should be used. Since the counterreactions in such a study are either metal deposition or dissolution, they should not interfere with the processes occurring on the working electrodes.

In many cases, it is impossible to use solutions containing metal ions be-

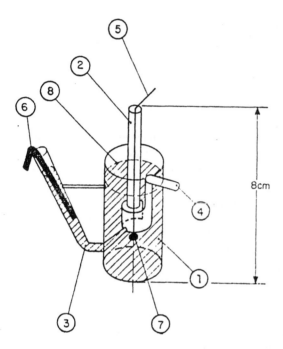

**Figure 5** Simple cell for a quick electrochemical study: (1) glass body, (2) working electrode (metal wire sealed with glass), (3) luggin capillary, (4) counter electrode (cylinder geometry), (5) working electrode metal, (6) reference electrode, (7) glass sealing to the edge of the working electrode, (8) solution level [14]. (Reprinted with copyright from The Electrochemical Society Inc.)

cause the reaction products precipitate on the electrode as insoluble metal salts. In such cases, the electrolyte's cation has to be tetraalkyl ammonium, and it is thus impossible to use simple counter- and reference electrodes based on a M/$M^{z+}$ couple in direct contact with the studied solution. A two-compartment cell in which the working electrode is separated by a glass frit from the other two electrodes will enable the use of such electrodes, depending on the duration of the study.

Figure 6 presents a scheme of an electrolysis cell for the isolation of reduction and oxidation products of nonaqueous solutions [15]. The electrolyte of the W.E. solution must be an alkyl ammonium salt because the reduction products of most of the commonly used solvents in the presence of metal cations precipitate as insoluble metal salts. The counter- and reference electrode compartments are separated from the working electrode compartment by two frits each. The separating units have pipes which enable the sampling of their solutions in order

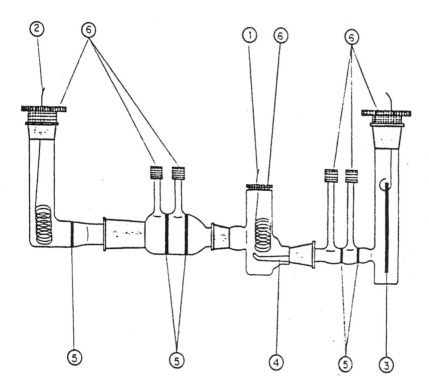

**Figure 6** Four separated compartments cell for isolation of electrolysis products of non-aqueous solutions: (1) working electrode, (2) counterelectrode, (3) reference electrode, (4) luggin capillary, (5) glass frits, (6) Teflon caps [15]. (Reprinted with copyright from Elsevier Science Ltd.)

to monitor possible diffusion of undesirable species. If prolonged electrolysis needs to be conducted, it is recommended that the solution of the counterelectrode be replaced from time to time.

Figure 7 shows a cell which provides parallel plate geometry for the study of the behavior of active and noble metal electrodes versus active metal C.E., in cases in which the $M/M^{z+}$ couple (M = active metal) does not interfere with the reaction of the counterelectrode [16]. Such a cell is useful for half-cell cycling efficiency testing of active metal anodes, such as Li electrodes for rechargeable batteries, and for preparation of electrode surfaces for ex situ observation by SEM and spectroscopic techniques. It should be emphasized that the cells in Figures 5–7 are designed to operate in glove boxes under an inert atmosphere.

The study of active metal electrodes such as Li and magnesium, which are

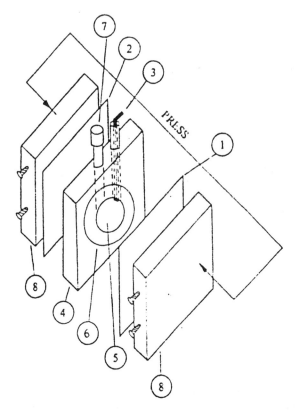

**Figure 7** One compartment parallel plate geometry cell: (1) working electrode, (2) counterelectrode, (3) reference electrode, (4) polyethylene or Teflon profile, (5) cylindrical space for solution, (6) O ring to avoid leaks, (7) glass tubes with Teflon caps for solution filling, (8) stainless steel bars that press the cell components together and serve as electrical contacts [16]. (Reprinted with copyright from Elsevier Science Ltd.)

always covered with surface films, may require in situ electrode surface preparation in solution. Figure 8 describes a cell and the procedure for the in situ preparation of Li electrodes in solutions for a study by impedance spectroscopy and fast transient techniques [17]. As shown in this figure, Li rods embedded in polyethylene are sheared in solution by a stainless steel rod, and, thus, two similar electrodes with a smooth, fresh area are formed. These two electrodes are then separated by the R.E. compartment (containing Li wire R.E.) so that a parallel plate geometry is maintained.

As described in the first chapter, one of the most important uses of nonaqueous solutions is in the field of lithium and Li ion batteries. Within the framework

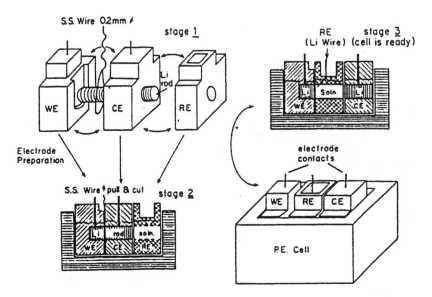

**Figure 8** The cell and the procedure for the in situ preparation of active metal electrodes such as Li and magnesium. The metal rod is pressed through the C.E. into the W.E. and a SS wire is wrapped around it. The three parts are placed in the polyethylene bath, which is filled with solution. Once in solution the SS wire is pulled to cut the rod and prepare fresh surfaces in solution. The last stage includes rearrangement of the electrodes in the bath [17]. (Reprinted with copyright from Elsevier Science Ltd.)

of the R&D of these batteries, it is important to use standard cells to test the battery materials for electrodes, as well as prototypes of batteries. The basic preferred cell configuration for testing battery electrodes, as well as prototype battery systems, should have a parallel plate geometry in which the electrodes are held together with a separator film which is soaked with the electrolyte solution between them. The film may be composed of porous polypropylene or woven glass cloth. This configuration leads to highly uniform current distribution, low effect of contaminants in the electrolyte solutions (high electrode area to solution volume ratio), and high energy density. It is possible to use standard, commercial button cells, a spirally wound configuration ("jelly-rolled"), or any other packaging method which holds these three elements in the parallel plate configuration in good and uniform electrochemical contact. A simple scheme of such a cell is described in Figure 9 and in Refs. 18 and 19. It should be emphasized that the above cell configuration is the best for the study of intercalation processes into solid compounds, as well as doping processes, repeated redox activity of electrically conducting polymers, etc. In the study of systems based on ionically con-

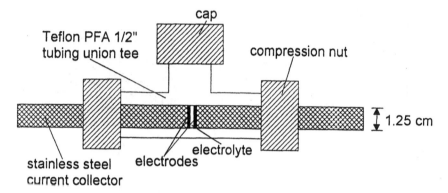

**Figure 9** A simple T cell for tests in battery configuration. The two electrodes and the separator film are held in a parallel plate configuration [18]. (Reprinted with copyright from The Electrochemical Society Inc.)

ducting polymers as the electrolyte systems, the conducting polymeric membrane replaces the separator film in the above cell configuration.

## B. Reference Electrodes

### 1. Calomel Electrode [aqueous Pt|Hg|Hg Cl$_2$|saturated KCl (aq)]

In principal, there are applications in which standard, commercial calomel electrodes can be used as reference electrodes in nonaqueous solutions. A schematic of such an electrode is shown in Figure 10. The saturated KCl solution is separated from the rest of the cell by a fine glass frit which allows only a very slow diffusion of water through it. Hence, when such an electrode is placed in a compartment with a luggin capillary (as shown in Figure 10), it can definitely be used in short experiments. During prolonged experiments, however, pronounced junction potential difference may develop in the interface between the wet glass frit and the solution. Hence, it is important to check the stability of the electrode in the specific solution. This can be done by potentiometric measurements versus stable reference electrodes such as Li/Li$^+$.

### 2. Ag|AgX|X$^-$ Electrode

The Ag|AgX|X$^-$ electrode, which is widely used in aqueous solutions, may also be used in nonaqueous systems. For instance, when the solutions contain perchlorate salts, the Ag|AgClO$_4$|ClO$_4$$^-$ system can provide a stable reference elec-

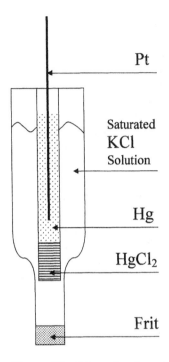

**Figure 10**  Schematic of standard, commercial calomel electrode.

trode. When using such electrodes it is recommended that their potential and stability be checked frequently versus the well-established ferrocene bis($\eta$-cyclopentadienyl)iron(II)/ferricenium couple [20].

## 3.  M|M$^{Z+}$ Electrode

Active metals such as lithium and sodium can be used as stable reference electrodes in nonaqueous solutions in which they are apparently stable. To a limited extent this may be true for the Mg/Mg$^{2+}$, Ca/Ca$^{2+}$ and Al/Al$^{3+}$ couples as well (though they must be checked separately for each specific solution). It is important to note that in most of the commonly used nonaqueous systems, the above active metals are thermodynamically unstable and react readily with the solvent, the salt anions and the unavoidably present atmospheric contaminants. However, the active metals are apparently stable in many systems because the above reaction products, which are usually insoluble (metal salts), precipitate as protective passivating surface films. These films prevent further corrosion of the active metals in solutions [21]. Hence, the active metal covered by the surface films may form a stable electrode of the $M|film|M_{(sol.)}^{z+}$ type, which can be regarded as a reversible R.E.

A typical and very useful example is the lithium reference electrode (Li/Li$^+$), which is simply a strip of lithium placed close enough to the working electrodes. (Note that the symbol Li/Li$^+$ is just an abbreviation of the Li|film|Li$^+$, which is the true system). It should be emphasized that the critical condition for the use of such a simple R.E. is its perfect passivation and stability in solutions. Lithium R.E. can be used in ethers, cyclic alkyl carbonates, dimethyl carbonates, γ-butyrolactone, methyl formate and thionyl chloride. It cannot be used in acetonitrile, open chain alkyl carbonates (except for DMC) and methyl acetate. In applications where it is impossible to use solutions with metal cation based electrolytes and where tetraalkyl ammonium salts are preferred, it is still possible to use the $M|M^{z+}$-R.E. in spite of the absence of metal cations. In the case of Li/Li$^+$ for example, this can be done by using cells in which the Li strip is placed in a salt solution in a compartment separated from the rest of the cell by a fine glass frit, as shown in Figure 6. Such an arrangement may be quite useful in providing highly stable R.E. with an omission of contamination of the solution near the W.E. with Li ions which may promote the formation of insoluble surface films on the electrodes [22].

### 4. Ferrocene/Ferricenium Derivatives

The ferrocene bis(η-cyclopentadienyl)iron(II)/ferricenium couple has been found to have a very stable potential in many nonaqueous solvents. It can be used as a reference system for many applications in which a separate compartment for the reference electrode is possible because it is difficult to totally separate the ions from the measured electrolyte. The recommended use and potential of this system are described in Ref. 20.

## C. Electrode Preparation

### 1. General

In this section we comment on the preparation of a few types of electrodes, with emphasis on specific issues which relate to nonaqueous systems. These electrodes include active and nonactive metals and composite electrodes.

### 2. Nonactive Metal Electrodes

For many applications, it is necessary to use metal disks embedded in electrical insulating materials, e.g., rotating disk electrodes. During preparation of these electrodes the following facts need to be taken into account.

1. Many common adhesives dissolve in nonaqueous solutions and thus cannot be used.

2. Many plastic materials contain plasticizers that can dissolve into organic solutions and contaminate them.

3. Many nonaqueous solutions have a low viscosity. Therefore, in the preparation of metal embedded in a plastic material, the machining has to be sufficiently good to prevent there being a space between the metal and the insulator into which the nonviscous solution can percolate.

4. At first glance, Teflon may be an ideal plastic material compatible with nonactive metals in nonaqueous systems. However, it should be noted that all plastics which contain halogens (F, Cl) are reactive with active metals such as Li. Hence, they should also be regarded as being *electroactive* in contact with metals *at low potentials* in the presence of alkali metal salts (e.g., Li and Na salts).

## 3. Active Metal Electrodes

Alkaline, alkaline earth metals and aluminum are naturally covered with anodic films. The removal of these native films, even in the best glove box atmosphere, exposes the fresh metal to reactive atmospheric contaminants at a high enough concentration and quickly cover the metal with new surface films. As discussed above, even the glove box atmosphere of an inert gas containing atmospheric components at the ppm level should be considered as being quite reactive to active metals such as lithium. Therefore, anyone intending to study the intrinsic behavior of active metal electrodes in solution must prepare a fresh electrode surface *in solution*.

There are reports on the preparation of lithium electrodes in which Li wires [17,23–25] and Li rods have been embedded in glass or plastic materials, and fresh Li surfaces have been prepared in solutions by shearing off the edges of the metal [17,23,24]. An example of such an approach was noted in the discussion on the construction of various cells. It should be mentioned that the plastics that can be used with metals such as lithium are highly pure (low plasticizer content) polyethylene and propylene.

There are studies in which the fact that active metal electrodes are covered with surface films is not so important, e.g., when these metals are used as counterelectrodes, or when they are studied as practical anodes in batteries. However, even in these cases, the "native active" metals as received may be covered with two thick films. It is therefore, necessary to remove the "initial native" film covering the active metal under an inert atmosphere. The passivating films of lithium and calcium can be scraped off with a stainless steel knife. In the case of harder active metals such as magnesium and aluminum, an abrasive cloth or a frazer apparatus may be needed. After removal of the native films from these metals, the new surface films formed in a glove box atmosphere are much thinner than the native ones. Indeed, active metals such as Li, Ca and Mg whose surfaces were freshly prepared in a glove box atmosphere containing atmospheric contaminants at the ppm level remain shiny, reflective, and apparently stable for prolonged storage periods (months).

## 4. Composite Electrode Preparation

Many applications require the preparation of composite electrodes in which the active materials are in a powder form which may be nonconductive. Hence, the electrode must include a rigid current collector, a binder and some electrically conducting additive, in addition to the active substance. Such electrodes are important for electrocatalysis and as cathodes for batteries. For instance, many cathode materials for rechargeable Li and Li ion batteries are lithiated transition metal oxides, which appear as a nonconductive powder.

It would be helpful to describe a general procedure for the preparation of typical composite electrodes. The current collector could be a metal net or foil. The choice of the metal depends on the potential window which the electrodes are operated. For low potentials ($<2$ V versus Li/Li$^+$), nickel or copper can be used, while for high potentials ($>2.5$ V versus Li/Li$^+$), aluminum may be a good choice. While aluminum is an active metal, it passivates well at high potentials in nonaqueous solutions, especially when the salt anion contains fluorine [e.g., $AsF_6^-$, $PF_6^-$, $BF_4^-$, $C(SO_2CF_3)_3^-$, etc.]. A layer of highly passivating $AlF_3$ is probably formed, preventing the corrosion of aluminum at high potentials. The conductive additive generally used is carbon black of a high surface area. The binder should be a polymer which is insoluble in the relevant solvent, but which still can be cast and processed. A good choice may be polyvinylidene difluoride (PVDF). This polymer is insoluble in a variety of solvents, including ethers, esters and alkyl carbonates, but is soluble in *n*-methyl pyrrolidone. A reasonable cathode composition for an active material which is nonconductive can be 80% of active mass, 15% carbon black and 5% binder (PVDF powder), by weight. This mixture should be milled together with a mortar and pestle (small amount) or a ball mill (low energy) in order to achieve homogeneity. A slurry is then prepared by adding *n*-methyl pyrrolidone and is then spread on the current collector at the desired thickness. The wet electrode should be pressed between filter papers in order to obtain a uniform and homogeneous active mass and should then be dried in an oven. It is possible to use spraying techniques as well to spread the slurry on foils. In some cases, it is possible to use electrically conducting polymers as a conductive additive. For instance, cathodes for rechargeable lithium batteries in which the active mass is redox polymers, such as polydisulfides [(—SRS—)$_n$], operate much better when polyaniline is used instead of carbon black as the conductive additive [26]. It should be noted that sintering processes by pressure could also be used for composite electrode preparation. For instance, there are commercial products such as Teflonized carbon black which can be mixed with active mass such as lithiated transition metal oxide powders (used as cathodes for Li batteries). The mixture can be pelletized in a dye (pressure) to form disk-shaped electrodes. These porous electrodes are usually used in cells that have a parallel plate geometry. The electrical contact to the electrode is maintained by attaching a metallic current collector to the back outer side of this disk. The electrolyte solution percolates into the pores of the cathode from the inner side of the disk and thus does not interfere with the electrical contact on its back. It should be noted that binders which contain species with carbon-halogen bonds, such as Teflon and PVDF, may be reactive at low potentials in solutions containing alkali metal salts, such as lithium salts. For instance, it has been shown that the C—F bond in PVDF is reduced at low potential in the presence of Li salts to form LiF and Li carbide groups [27].

## V. ELECTROCHEMICAL MEASUREMENTS

The application of electroanalytical techniques is a broad subject that is well covered in many comprehensive books, for example [28–33]. Hence, a detailed description of the commonly used electroanalytical techniques is not necessary. However, we include a few comments on the application of some electroanalytical techniques for nonaqueous systems.

1. Nonaqueous systems are generally more resistive than aqueous ones, and thus the *IR* drop in potentiostatic operation may be pronounced. The *IR* drop should be taken into account especially in fine measurements, such as mechanistic research, and potentiostats with *IR* drop compensation devices should be used.

2. For many studies, it is important to conduct electrochemical measurements within glove boxes. In such cases, long cables that connect the instruments outside the glove box to the electrochemical cell inside the glove box are needed. Long cables are problematic for sensitive measurements. It is therefore recommended that all the cables used be as short as possible and screened (e.g., coaxial cables, in which the screening envelope is grounded). In addition, a high impedance electrometer device, which connects the reference electrode and the potentiostat, should be placed in the glove box.

3. For sensitive measurements such as impedance spectroscopy and fast transients, it is not advisable to perform the measurement in the glove box. It is possible to perform the measurements near the measuring equipment, using a transfer system that maintains the inert atmosphere. Such a transfer device may be a hermetically closed metal box which contains the appropriate electrical connections, such as BNC plugs. Thus, the electrochemical cell should be assembled

in the glove box, sealed in the transfer system, and placed outside the glove box near the measuring system. Grounding of the metal transfer box provides Faraday cage shielding conditions.

## VI. BRIEF REVIEW OF THE APPLICATION OF SPECTROSCOPIC TOOLS TO THE STUDY OF NONAQUEOUS SYSTEMS

### A. Introduction

The application of spectral and spectroelectrochemical tools for the study of electrochemical systems, in general, and nonaqueous electrochemical systems, in particular, is very important. These measurements result in a better understanding of reaction mechanisms, surface phenomena, and the correlation among surface chemistry, interfacial electrical properties, morphology, three-dimensional structure, and electrochemical behavior. In view of this, Chapter 5 is devoted to the use of spectroscopic methods, especially in situ methods, for the study of nonaqueous electrochemical systems.

We briefly review some experimental aspects which are unique for the study of nonaqueous systems by spectroscopic tools. We include some comments on the appropriate choice of spectroelectrochemical tools for different nonaqueous systems. We also describe the use of transfer methods and spectroelectrochemical cells that should be used in order to protect the sensitive nonaqueous electrochemical systems from contamination, and enable the performance of reliable measurements.

### B. Comments on the Choice of Appropriate Spectroscopic Tools for Different Electrodes and Nonaqueous Systems

Various spectroscopic techniques can provide surface analysis of electrodes; the most important and commonly used are discussed.

#### 1. Fast Fourier Transformed Infrared Spectroscopy (FTIR)

The application of FTIR in chemistry, its unique features, and the relevant instrumentation are well documented [34,35]. In brief, an FTIR spectrometer is based on a Michelson interferometer that provides a spectrum in the time domain which is Fourier-transformed by a computer to a spectrum in the frequency domain. The sample can be scanned repeatedly, and the accumulated spectra can be averaged, thus producing a representative IR spectrum of a very high signal to noise ratio. This enables the measurement of samples containing a very low concentration of the active materials. FTIR is a nondestructive method that can be used for the study of surfaces.

The application of FTIR spectroscopy in electrochemistry is based on the use of specific accessories for each purpose. The simplest mode of operation for the analysis of bulk liquid (thin layer cells) or solids (pelletized with KBr powder) is the transmission mode. It is possible to analyze surface species on electrodes in this mode using grid-type electrodes. For the study of thin layers adsorbed on reflective electrodes, an external reflectance mode should be used. Of special importance is the application of grazing angle reflectance in which the incident IR beam hits the surface at an angle >80° [36–38]. Improved sensitivity is achieved by filtering the incident beam with a polarizer, which allows only P polarized light to reach the surface (or hit the detector). Commercial accessories which provide the appropriate beam alignment and polarization are available (Harrick, Spectratech, etc.). A typical accessory for grazing angle reflectance spectroscopy is presented in Figure 11. An internal reflectance mode may also be useful for the study of electrode surfaces, especially in cases of nonreflective surfaces. The ATR mode is particularly important [39–42]. Both external and internal reflectance modes can be used for in situ studies of electrodes in solutions

**Figure 11**   The Spectra-Tech FT-80 grazing angle accessory. A typical accessory for grazing angle reflectance spectroscopy. A polarizer that transmits only P polarized incident light improves the sensitivity of the measurement.

**Figure 12**  The Spectra-Tech Collector diffuse reflectance accessory. A typical accessory for the study of particles and powders by diffuse reflectance FTIR.

under potential control [43,44]. For the study of powders, e.g., surface species on powdered active electrode materials, the diffuse reflectance mode is the appropriate tool. Figure 12 shows a typical commercial accessory for the study of particles and powders by the diffuse reflectance mode [45].

## 2.  Raman Spectroscopy

Raman Spectroscopy provides information comparable to that obtained by FTIR. The sample is illuminated by a laser beam (visible light), and the light dispersed from the sample (Raman effect) [46] is analyzed. The frequency differences between the light dispersed and that of the initial laser beam reflect the various functional groups of surface and solution species. This method can also be used in situ for the study of electrode surfaces in solutions under potential control [34]. However, it should be noted that the laser beam which heats the electrode surface may be destructive to the surface species. Except for unique phenomena (SERS) in which species adsorbed to metallic surfaces (Ag, Au) provide very strong signals [47], the signal to noise ratio of Raman spectra from surface species on electrodes that are measured in situ is low. Since FTIR and Raman provide similar information, the former method is usually preferable for electrode surface studies (especially in situ), unless the adsorbed species of a specific system have functional groups which are active in the Raman and not in the IR. Raman spectroscopy may be useful for analyzing solvent-solute interactions in studies of solution phases (possible shifts of $Raman_{peaks}$ [48]). There is also a unique application of surface sensitive Raman spectroscopy in the field of intercalation processes. For instance, intercalation of Li into graphite changes the typical Raman graphite peaks [49].

## 3. Ultraviolet, Visible Light (UV-Vis)

As is widely known, bulk species which have chromophores that absorb in the UV-Vis can be analyzed quantitatively and qualitatively by this spectroscopy. The study of electrodes or species adsorbed as thin layers by UV-Vis is more difficult, due to sensitivity problems and the availability of the appropriate chromophores [50]. Another use of this type of analysis is electroreflectance [51,52]. Adsorption of species on reflective electrode surfaces changes their reflectivity. Thus, this method can indicate electroadsorption processes very sensitively, in situ, although it does not provide specific information on the structure and composition of surface layers.

## 4. Extended X-ray Absorption Fine Structure (EXAFS), X-ray Absorption Near-Edge Structure (XANES)

The Extended X-ray Absorption Fine Structure (EXAFS) and X-ray Absorption Near-Edge Structure (XANES) methods provide unique information on the composition of surface species and their structure. They can be used as an in situ tool. However, EXAFS and XANES require a synchrotron radiation source (X-ray) [53].

## 5. X-ray Photoelectron Spectroscopy (XPS)

The XPS technique is based on the analysis of the energy of electrons emitted due to irradiation of surfaces by an X-ray beam. This energy reflects very specifically the elements present on the surface, as well as their oxidation states. This method requires a vacuum system which provides a background vacuum of $10^{-9}$–$10^{-10}$ mmHg. This is an ex situ technique, and its application for the study of sensitive electrodes requires special transfer arrangements, as described in the next section. Any modern XPS system includes the option of depth profiling of the surface studied by sputtering the surface with argon ions, followed by XPS analysis. The information thus obtained is highly specific, both qualitatively and quantitatively, and a completed comprehensive element analysis is provided. However, it should be noted that this method *may be destructive* to surfaces. The sputtering, as well as the X-ray beam, may change the oxidation states of elements and induce surface reactions. In view of this, the user has to check carefully, by repeated measurements, that the sample investigated remains stable under the conditions of the analysis [52,54–57].

## 6. Auger Electron Spectroscopy (AES)

Auger electron spectroscopy is somewhat similar to XPS and involves the analysis of Auger electrons emitted from surfaces due to irradiation with an X-ray beam. This technique is less important than XPS as a qualitative tool, but is very useful for quantitative analysis of elements on the surface, including quantitative depth profiling by measurements which follow the sputtering of the measured surface (argon ions) [58–62]. AES also requires ultrahigh vacuum (UHV $< 10^{-9}$ mmHg). It is therefore, applicable only for ex situ measurements requiring the use of special transfer methods in cases of sensitive surfaces.

## 7. Energy Dispersive Analysis of X-rays (EDAX)

The EDAX technique involves an analysis of the X-ray radiation emitted from surfaces which are studied by scanning electron microscopy (SEM) [63]. The surface studied by SEM is hit by the electron beam, emitting X-rays of a limited penetration depth which are specific to the elements present on the surface. This method provides qualitative and quantitative element analysis of electrode surfaces. Its disadvantage compared with XPS is that it cannot provide information on the oxidation states of elements and is not equipped for depth profiling. The advantages are that the instrumentation is much cheaper than XPS systems and the vacuum required is only $10^{-6}$ mmHg. Since the electron beam of the SEM can be efficiently focused to provide a submicronic resolution, the accompanying element analysis by EDAX provides the same spatial resolution, which is higher than that usually achieved by XPS or AES. EDAX is, of course, an ex situ technique. It requires a special transfer method when sensitive samples are measured.

## 7. Secondary Ion Mass Spectrometry (SIMS)

Secondary ion mass spectrometry is based on surface bombardment by argon ions in UHV, followed by mass spectrometry of the charged species which are sputtered from the sample's surface. It provides specific information on surface species, high spatial resolution and depth profiling [64]. The SIMS requires an ultrahigh vacuum ($>10^{-11}$) and is an ex situ method which requires the use of a special transfer method for sensitive samples.

It should be noted that surface analysis by SIMS is not always straightforward. What is measured, in fact, is the distribution of charged particles by their mass/charge values. It is often very difficult to relate these charged particles to their original precursors, which are the actual surface layers that cover the electrode.

## 8.  Electrochemical Quartz Crystal Microbalance (EQCM)

The EQCM method is based on the piezoelectric effect of thin quartz crystals (5–10 μm thick). Two electrodes are deposited on two sides of the quartz crystal, and the resonant frequency of the crystal under an alternating electric field is measured. The resonant frequency depends linearly on the mass accumulating on any of the electrodes used as the working electrode in the electrochemical system studied (possible resolution of nanograms per cm²) [65]. By recording the mass and the charge in an electrochemical process in which adsorption and/ or precipitation of species occur, it is possible to estimate the equivalent weight of surface species formed under different experimental conditions, e.g., potential, concentration, temperature. Hence, this in situ method can serve as an attractive electroanalytical tool for the in situ study of adsorption processes. Its use for nonaqueous systems requires the development of special cells, as described in the next section. In spite of the linear relationship between mass accumulation and frequency changes in EQCM, other factors influence the frequency response of the QC, such as electrode roughness, the solvent viscosity, and temperature. In addition, the calculation of the equivalent weight of adsorbed species by EQCM may be accurate only if all of the charge measured is involved in mass accumulation on the electrode surfaces, and only in the absence of secondary, nonelectrochemical reactions of the surface species precipitating on the electrode. Hence, the results measured by this method should be analyzed carefully. This method has proven itself to be extremely useful for the study of many precipitation processes in electrochemical systems [55,66–68].

## B.  Morphological Studies

### 1.  Atomic Force Microscopy (AFM)

The relatively novel method of atomic force microscopy can be used both ex situ and in situ for the study of the surface morphology of electrodes. It is based on a thin and sensitive cantilever to which a sharp microscopic tip is attached. The tip is raster-scanned along the studied surface, changing its deflection as a result of topographic changes. The deflection is measured by a laser beam which is reflected from the back of the cantilever to a detector that measures the position of the laser beam. The changes in the cantilever, as a function of the tip position with respect to the sample plane, are translated by sophisticated software into a 3D picture of the surface topography [69].

The application of this technique for the study of electrodes in nonaqueous systems which are highly sensitive to atmospheric contaminants, and which may be volatile, is difficult and requires the design of a special cell and transfer method, as described in the next section. It should be noted that there are several variations in the application of AFM in electrochemistry. These include a noncontact mode, in which the tip is not in direct contact with the surfaces [70], friction forces between the tip and surface species (lateral forces) [71], and magnetic force microscopy, in which a magnetic tip senses magnetic surface species [72]. The major advantage of this technique is its possible application as an in situ tool for electrode surface morphology measurements. Its disadvantages are the possibility that the tip will interfere with the original surface morphology, and the experimental difficulties in applying it to sensitive and reactive systems.

## 2.  Scanning Tunneling Microscopy (STM)

The scanning tunneling microscope is also a tool for surface morphological studies that is widely used in situ [73]. It is based on the analysis of a tunneling current between a very sharp microscopic tip and the electrode surface caused by a bias potential applied between the two. This method is well established for the study of electrochemical systems [58,59,74–76]. Its advantage over AFM is that it is technically much simpler to use for in situ studies of electrochemical systems, and it obtains better resolution. However, the application of STM to nonaqueous systems may be complicated by the following factors:

1.  When the electrode surfaces are covered by surface species, which are electrically insulating, there is no possibility for a tunneling current to flow between the tip and the electrode.
2.  When the electrochemical systems are too reactive, it may be difficult to apply a bias potential in which the system is electrochemically inert.
3.  Isolation from atmospheric contaminants and volatility of the solvents may present problems in the application of STM to nonaqueous electrochemical systems.

## 3.  Scanning Electron Microscopy (SEM)

Scanning electron microscopy is the traditional method for ex situ morphological studies of electrode surfaces. Basically, this is the simplest tool for morphological studies of the electrode. It should be noted, however, that in cases in which the electrodes are covered with electrically insulating films, the resolution of this method is limited, due to static charging of the surface. SEM application for the study of sensitive electrodes requires the use of transfer systems, as described in the next section.

## D. Bulk Analysis and Three-Dimensional, Structural Studies

### 1. X-ray Diffractometry (XRD)

X-ray diffractometry is widely used for the characterization of electrode materials for the battery field, electrocatalysis, etc. Both areas require the development of new materials whose three-dimensional structure is critical for their electrochemical activity. In brief, XRD is based on a monochromatic X-ray beam that hits the sample and is reflected from it at a variety of scattering angles. Since the X-rays are reflected by the atoms in the sample's lattice, and since the wavelength is of the same order of magnitude as interatom distances in the solid state, interference among the reflected X-rays occurs, leading to typical, unique diffraction patterns for each specific material [77]. A completed analysis of lattice structures can be obtained from judicious treatments of the XRD patterns. It should be noted that, as the structure becomes more complicated, the XRD data has to be more precise. This depends on factors such as the power of the X-ray beams, the diffractometer structure and precision, measurement time, the amount of material measured and its homogeneity. There are several software packages available for computerized analysis and modeling of XRD data which provide precise sample structure fitting [78]. The most common way of analyzing samples by XRD is in a powdered form. The active mass is pressed into a metallic holder and is thus measured in a pelletized from. However, it is possible to measure composite electrodes directly by XRD.

Many practical electrodes are prepared from powdered active mass and some conductive additive, such as carbonaceous materials, that are bonded to a metallic current collector with a polymeric binder such as PVDF (described in the previous section). Such electrodes can be measured directly. Is it very useful to measure such electrodes in their pristine from, before any electrochemical treatment, and then as a function of their electrochemical history. For quantitative analysis (phase composition, evaluation of concentration of constituents in mixtures, etc.) it is important to use internal standards in the sample. Fortunately, several components of composite electrodes, which are, in any event, contained in the sample measured, can be used as internal standards. These include the current collector (Cu, Al), the conductive additive, such as graphite, or the binder, such as Teflon.

It should be noted that XRD can also be applied as an in situ technique. It requires the use of specific cells with windows which do not absorb the X-ray beam. For instance, polyethylene and mylar films are suitable. A typical cell for in situ XRD measurements of composite electrodes, e.g., lithiated graphite, is presented in the next section. There are already reports on the use of in situ XRD measurements for the study of composite electrodes in nonaqueous systems [52,57,61,79–81] and the study of surface layers on electrodes (e.g., a lithium electrode in an aprotic medium) [82].

## 2.  NMR, ESR Spectroscopy

There are reports on the use of both NMR and ESR for the study of electrode materials [83–85] and bulk products of electrochemical processes [86,87]. For instance, $^7$Li NMR may be found to be very useful for the study of Li intercalation into carbonaceous materials [88]. A major advantage of these techniques is that they are applied in situ. The electrochemical cell is, in fact, an NMR tube in which the studied electrode is mounted so that it can be placed within the magnet's cavity. While NMR provides information on the environment of the element studied, within the electrode measured, ESR provides information on the formation and stability of radical ions when formed during the course of an electrochemical process.

## 3.  Mossbauer Spectroscopy

Mossbauer spectroscopy may be important and useful when applied to electrodes which contain ferromagnetic components. It is basically an in situ tool which provides valuable information on possible orientation and oxidation states of ferromagnetic species in the electrodes as a function of the electrochemical process and the potential applied. For example, electrodes for oxygen reduction may be highly catalytic when containing macrocycles with transition metal cations such as $Fe^{2+}$, $Ni^{2+}$, $Co^{2+}$ [89,90]. A typical apparatus for this technique is described in Ref. 91.

## E.  Comments on the Purchase of Major Spectroelectrochemical Tools

Acquiring equipment for the spectroscopic techniques described above requires difficult decisions based on a careful optimization of needs and costs. It is therefore very important to have at hand comprehensive information on the performance and prices of instruments from different companies. This information is, in fact, available to all users, who should refer to special articles published each year on this subject in the *Journal of Analytical Chemistry*.

## F. Cells and Transfer Methods for the Application of the Preceding Techniques with Sensitive Electrodes and Nonaqueous Systems

### 1. FTIR Spectroscopy

For intensive studies of sensitive samples by FTIR spectroscopy, it is highly recommended to place the FTIR spectrometer in a glove box. It is not necessary to purchase an expensive glove box system for this purpose. The authors can provide schemes and instructions for the buildup of a simple and cheap glove box. The atmosphere of the FTIR spectrometer must be cleaned of $H_2O$ and $CO_2$, which are strong IR absorbers and thus mask the spectra. In addition, both contaminants should be considered reactive for nonaqueous systems, especially when active electrodes are involved (e.g., Li, Ca, Mg, Li-carbon, $Li_xMO_y$, M = transition metal). Hence, a FTIR spectrometer usually requires a continuous supply of $CO_2$ and $H_2O$ free purging gas. There are commercial systems which use compressed air, filter it, and absorb $H_2O$ and $CO_2$ in regeneratable systems which utilize columns of 3 Å and 10 Å molecular sieves (Balson Ltd., UK, for example). Such systems need an inlet pressure of about 6 atmospheres (gauge) and can provide purified air at a flow rate of $>50$ SCFH and a dew point $< -70°C$ ($H_2O$ contamination at a sub-ppm level).

A standard size glove box purged continuously with such air at this flow rate can maintain an atmosphere whose $H_2O$ or $CO_2$ content is much below the detection limit of any research grade FTIR spectrometer. The operation of an FTIR spectrometer under such conditions provides excellent isolation from $CO_2$ and $H_2O$ and eliminates the time wasted in waiting for the stabilization of the spectrometer atmosphere after each measurement (as is the case in the normal operation of a purged instrument).

The experimental aspects of ex situ FTIR spectroscopic studies of sensitive electrodes (e.g., Li and Ca surfaces) using a purged spectrometer (not in a glove box) have been described in detail in Refs. 36–38, 92, and 93. Briefly, it is possible to analyze thin surface films on active metal, using a reflectance mode, while the active surface is in contact with a KBr or NaCl polished window. In Refs. 37 and 38, a possible organization of the measurement chamber of an FTIR spectrometer for such measurements is described. It should be noted, however, that the performance of such measurements when the spectrometer is placed in the glove box is much more elegant and easier.

The experimental aspects of the performance of in situ FTIR measurements are described in Refs. 43 and 44. Figure 13 shows a typical cell for in situ external reflectance mode (e.g., SNIFTIRS type measurements) [94,95]. The experimental aspects of its use are described in Ref. 96. Figures 14 and 15 show cells for in situ internal reflectance modes; multiple internal reflectance ATR and single internal

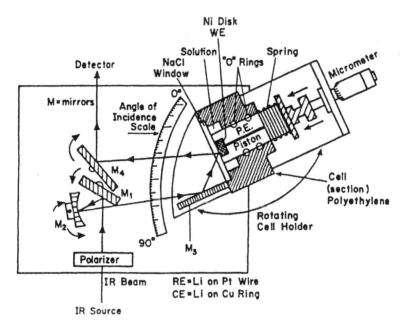

**Figure 13** A cell for in situ FTIR external reflectance mode of thin layers (SNIFTIRS type measurements) [96]. (Reprinted with copyright from The Electrochemical Society Inc.)

reflectance modes, respectively [44]. The former approach requires crystals which have a high refractive index (>2). The common materials which have such a high refractive index and are transparent to the IR in the 500–4000 cm$^{-1}$ range, which is the most useful optical window for the characterization of most of functional groups, are KRS-5, ZnSe and germanium (R.I. = 2.37, 2.4 and 4, respectively). The use of the ATR mode requires the facilities of thin metal film deposition under UHV.

It should be noted that the ATR crystals of these materials are very expensive. All of the above materials are reactive with nonaqueous systems at low potentials. Hence, a single experiment may be extremely expensive because the crystal surfaces may be damaged during these experiments. While the crystals may be polished and regenerated, it appears that after repolishing, the quality of the surface finishing is not always satisfactory for the ATR mode. Thus, only part of the in situ FTIR spectroelectrochemical ATR experiments in nonaqueous solutions may be successful. Another approach for in situ single internal mode is described in detail in Ref. 43 and Figure 14. This method is easier, much cheaper, and thus may be more adequate for the in situ studies of electrode sur-

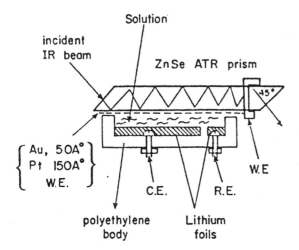

**Figure 14** Spectroelectrochemical cell for in situ multiple internal reflectance mode ATR. The working electrode is a thin Pt layer deposited on the ZnSe paralleloid. The polyethylene body which contains the electrolyte and the other electrodes is pressed against the prism to form good sealing [44]. (Reprinted with copyright from The Electrochemical Society Inc.)

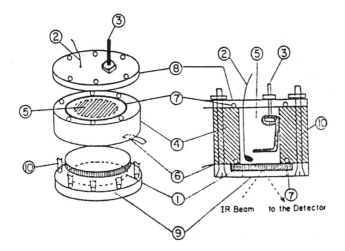

**Figure 15** A cell for in situ single internal reflectance spectroscopy: (1) working electrode—an NaCl optical window covered by a thin Pt deposited layer, (2) reference electrode, (3) counterelectrode, (4) polyethylene cell body, (5) space for solution, (6) electrical contact to the working electrode—a thin nickel foil, (7) O ring, (8) polyethylene cover, (9) brass holder for the optical window, (10) bolts that hold the cell [44]. (Reprinted with copyright from The Electrochemical Society Inc.)

faces in nonaqueous systems. Its major advantage is the possibility of using KBr or NaCl, which are completely inert for most of the nonaqueous systems and are relatively cheap and easy to handle and polish. The IR beam is reflected from the interface between the optical window and the thin metallic cover which serves as the working electrode. A thin metallic cover of a thickness of hundreds of Ångstroms on polished KBr is rigid and sufficiently integrated to serve as an adequate working electrode. It is also partially transparent to the IR beam, and thus the beam reflected from the thin metallic film has sufficient penetration depth, so that the absorption of the environment close to the metal solution interface influences quite pronouncedly the reflected IR beam. The only problem related to the use of the ATR or SIR modes for spectroelectrochemistry is the adhesion of the thin metallic coating on the IR transparent crystal. Phenomena such as surface film formation on the coating may induce stress, causing it to peel off the crystal.

The above cells may be mounted and loaded with solution in a glove box, and are then transferred to the FTIR spectrometer, to which a potentiostat/galvanostat is attached. The external reflectance and ATR modes are performed, using standard accessories from Harrick or Spectra-Tech for these modes. The SIR mode is performed using the horizontal FT-80 or FT-85 grazing angle reflectance accessory from Spectra-Tech, equipped with a polarizer (standard equipment, see Figure 11).

## 2. Cells for Raman Spectroscopy

Figure 16 shows a scheme of a suitable cell for in situ electrode surface studies by Raman spectroscopy. The working electrode is embedded in an insulating piston made of a plastic material such as Teflon or polyethylene. The optical window is made of quartz adhered to the glass cell by an epoxy-based adhesive. The laser beam that hits this surface is reflected, and the light that is dispersed perpendicular to the reflected beam is analyzed. Solution masking may be a severe problem, and therefore the cell operates at a thin layer configuration adjusted by the micrometer, as shown in the figure. The cell should be loaded with solution in a glove box before the transfer to the spectrometer. It maintains the appropriate insulation, and thus the measurements can be carried out in a normal atmosphere.

## 3. Cells for AFM, EQCM and in situ XRD

Figure 17 shows a scheme of a cell apparatus for in situ AFM studies of nonaqueous systems and active metal electrodes such as lithium. It is based on the Topometrix's commercial cell (Discoverer 2010 AFM system). The cell is hermetically sealed, and the solution is contained in a latex bag which maintains reasonable isolation from atmospheric contaminants. The compatibility of the latex with the solvents under study is, of course, a key issue that needs to be

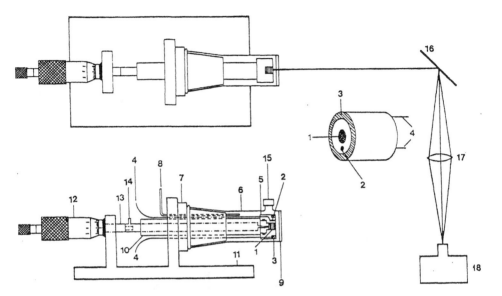

**Figure 16**  A cell for in situ electrode surface studies by Raman spectroscopy. The working electrode (1) is embedded in an insulating piston that can be moved back and forth for the measurement and the electrochemical process: (2) reference electrode, (3) counterelectrode (4) electrical contacts to the reference and counter electrodes, (6) glass cell, (7) Teflon cell holder, (8) Teflon tube for argon, (9) glass optical window, (10) Teflon piston, (11) base, (12) micrometer, (13) micrometer shaft, (14) electrical contacts to the working electrode, (15) solution entry (via septum), (16) mirror, (17) focusing lens, (18) detector.

checked. However, it is possible to replace the latex with polyethylene, which is compatible with most of the nonaqueous systems that may be of interest. The use of the above cell and the performance of the measurements are described in detail in Refs. 97 and 98.

It should be emphasized that the plastic bag, which is a necessary part of the cell in Figure 17, is *not* totally impermeable to atmospheric contaminants. Thus, during prolonged measurements, the solution may be contaminated as a result of the slow diffusion of atmospheric components through the latex or the polyethylene bag. In order to perform prolonged in situ AFM studies of very sensitive electrodes and solution, we built a special glove box in which the AFM system is placed for measurements under highly pure argon. The glove box can be fully evacuated so that its atmosphere is replaced before each set of measurements. To prevent vibration, the glove box is hung on springs and is provided with accessories and connections which enable its disconnection from the feeding pipes when measurements are being taken. Figure 18 show pictures of this apparatus and the cell.

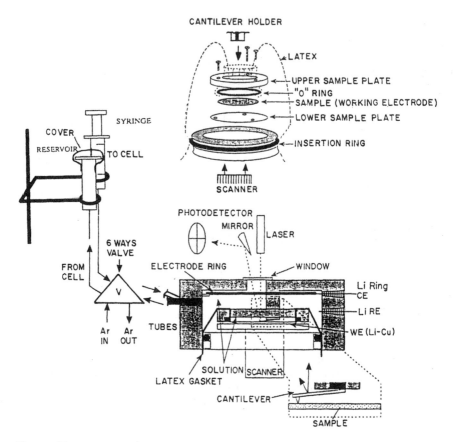

**Figure 17** Illustration of the cell for in situ AFM studies of nonaqueous systems and active metal electrodes. The scheme shows the solution filling arrangement and the way in which the latex gasket is attached to the cell components. The cell is based on Topometrix's commercial cell (Discoverer 2010 AFM system) [97] (Reprinted with copyright from The Electrochemical Society Inc.)

The solutions and cells are prepared in a normal glove box and are transferred to the AFM glove box using the special hermetically sealed transfer system (Figure 19). This mode of operation provides complete isolation of the systems studied from atmospheric contamination, and yet allows for excellent shock and vibration protection. There are other approaches for the performance of in situ AFM measurements under controlled atmosphere, such as those presented by molecular imaging [99].

Figures 20 and 21 show cells for EQCM measurements (electrodes and

**(a)**

**(b)**

**Figure 18** A glove box for long in situ AFM studies: (a) Before each set of measurements the glove box is fully evacuated to replace its atmosphere by inert one. (b) During measurements the glove box is hung on springs disconnected from the feeding pipes to prevent vibration.

**Figure 19**  A sealed transfer system for AFM cells that are prepared in a normal glove box and then transferred to the AFM glove box shown in Figure 17.

equipment from Elchema, Potsdam, NY, USA). Their operation is described in detail in Refs. 100 and 101. Figure 22 shows a cell for in situ XRD measurements. The cell operation is described in detail in Ref. 102. The basic principle is to use a thin polyethylene or mylar window to isolate the back of the cell from the environment. Since the working electrode is hit by the X-ray beam from the back, the electrode must be sufficiently porous to enable solution penetration for electrochemical reaction at the back. This may indeed be true for composite electrodes, such as lithiated graphite, and transition metal oxide cathodes for batteries, which are usually porous.

## 4.  Transfer Methods for Techniques Performed ex situ under High or Ultrahigh Vacuum

SEM, SIMS, XPS and AES measurements are performed in HV or UHV systems. Hence, the study of sensitive electrodes requires special procedures for the transfer of the samples from the glove box atmosphere in which they are prepared to the vacuum system. Here an innovation is required for choosing the most suitable transfer mode for the specific measuring system used. In principle, the best mode of operation is to load the samples onto a substrate which fits the measuring

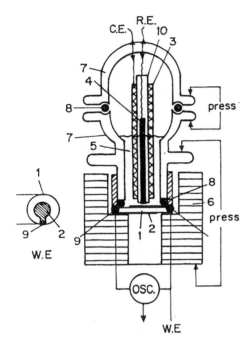

Figure 20 A cell for EQCM measurements: (1) quartz crystal, (2) gold electrodes deposited on both sides of the quartz crystal, (3) counterelectrode, (4) reference electrode, (5) solution, (6) polyethylene body, (7) glass cell parts, (8) O rings, (9) electrical contacts for the working electrode, (10) glass tube [100]. (Reprinted with copyright from The Electrochemical Society Inc.)

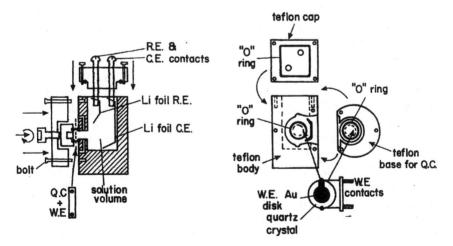

Figure 21 A cell for EQCM measurements [101]. (Reprinted with copyright from The Electrochemical Society Inc.)

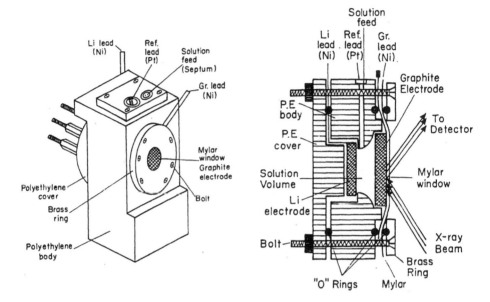

**Figure 22** A cell for in situ XRD measurements, isometric and section views [102]. (Reprinted with copyright from The Electrochemical Society Inc.)

system and which can be hermetically sealed under argon. The apparatus has to fit both the sample's entry port of the instrument and the manipulator that introduces the sample from the inlet port into the measurement chamber. Usually, there are no commercial transfer systems available, and each research group has to design and build its own system. Figure 23 shows a scheme of a transfer system from a glove box to a SEM system (JEOL 840). The operation of this system is described in Ref. 103.

Another approach may be the use of a glove bag (polyethylene). The operation of the glove bag is based on the following procedure. First, the glove bag is attached to the inlet port of the instrument, forming a hermetically isolated space that includes this port, the samples in closed vessels and the necessary tools. Then the atmosphere inside the bag is replaced by an inert atmosphere by repeated evacuation and purging of highly pure gas. Using this approach with the appropriate regulators, it is possible to maintain a highly pure atmosphere inside the glove bag. Finally, the inlet port may be opened and the samples can be introduced into the UHV system inside the glove bag (under a highly pure, inert atmosphere). As described in Ref. 104, we could successfully measure highly sensitive lithium samples by XPS, using the above transfer procedure.

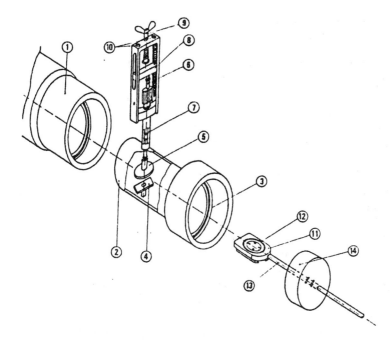

**Figure 23** Scheme of a transfer system for air sensitive samples from a glove box to a SEM system: (1) SEM inlet, (2) system body, (3) O rings, (4) fixed tray, (5) brass disk that seals the samples on the tray, (6) brass shaft, (7) brass cylinder with two rubber O rings, (8) bridge attached to the edge of the shaft, (9) bolt by which the bridge is pressed down, (10) two bolts by which the bridge is raised up to release the tray when evacuated, (11) sample tray and its O rings (12), (13) manipulator, (14) cover through which the manipulator is moved with two O rings [103]. (Reprinted with copyright from The Electrochemical Society Inc.)

## VII. COMMENTS ON SAFETY CONSIDERATIONS

1. Any work with nonaqueous electrochemical systems may involve exposure to extremely dangerous materials.

- The solvents are volatile, and they may be highly flammable and even explosive.
- Some salts may be carcinogenic (e.g., $LiAsF_6$).
- Active metals may be highly flammable, especially in high surface area form.
- Many composite materials used as catalysts or cathodes for batteries

are powdered transition metal compounds which may be carcinogenic (e.g., Ni and Co compounds).

Indeed, the glove box operation, previously described in detail, may provide adequate isolation for the users. However, the transfer of equipment and materials out of and into the glove box through the vacuum chamber may expose the user to the glove box atmosphere, which may contain dangerous carcinogenic and flammable dust. Thus, when using the glove box's vacuum chamber, the following recommendations should be considered: the vacuum chamber should be pumped and purged with air before opening it, and masks with an active filter should be used in cases in which the vacuum chamber contains dangerous dust, or in cases in which it is impossible to pump it (e.g., because it may contain volatile liquid waste).

2.   When a glove box system develops too deep a vacuum or too high a pressure, the window can explode. Hence, a glove box system has to be fully protected from unexpected changes in the pressure limits which the system can withstand. This should be done by a safety control for the solenoid valves of the purging gas and vacuum inlets. For glove boxes which have automatic pressure control, it is recommended that when the system is not in use, e.g., overnight, the gas and vacuum inlets be left closed.

3.   All the outlets of the vacuum pumps and the regeneration lines should be connected by solid piping to a ventilation system such as a hood.

4.   Waste materials taken out of the glove box should first be separated under the inert atmosphere:

- Active metals should be removed in closed vessels and neutralized immediately with a mixture of ethanol (water free) and propanol in a hood, under nitrogen purging.
- Waste paper should be kept separately.
- Waste solutions should be removed in hermetically sealed vessels and treated according to the laws governing waste regulation.

5.   It is recommended that all materials which form dust be processed in a hood and not on regular working tables in the open air. This is particularly important for the preparation of composite electrodes.

6.   Distillation of solvents may be extremely dangerous because many nonaqueous solvents are volatile and flammable. Hence, the distillation systems have to be explosion proof and equipped with a safety valve which can liberate pressure which develops in the system. It should be noted that ether solvents may contain peroxides that can cause an explosion upon heating. In fact, any ether solvent that is exposed to air is suspected of containing dangerous peroxides.

The following procedure is recommended for checking and neutralizing the peroxides in ethers before their distillation:

Testing

    a.  Mix the following three solutions (made fresh on testing day).

1.4 g of ammonium iron(II) sulfate hexahydrate in 100 ml of water
10 ml of 0.5 $M$ $H_2SO_4$
75 mg ammonium thiocyanate in 10 ml water

    b.  Mix 5 ml of the testing solution with 5 ml of the tested solvent and shake vigorously. Appearance of brown, yellow or red color in the water phase indicates the existence of peroxides.

Neutralization

    a.  For each liter of solvent, dissolve 12 g $FeSO_4$ in 30 ml water.
    b.  Stir the neutralization solution with the solvent for 2 h.
    c.  After neutralization, dry and clean the solvent by the procedures known for this solvent.

    7.  FTIR, RAMAN and AFM involve the use of laser beams. Special care is required to properly protect the user's eyes.
    8.  XPS and XRD involve the use of X-rays; hence special care is required to protect the user from exposure. Usually, modern instruments provide built-in protection, but it is still recommended that this protection be double-checked.

## REFERENCES

1. Aurbach, D.; Gofer, Y.; Ben-Zion, M.; Aped, P. *J. Electroanal. Chem.* **1992**, *339*, 451.
2. Aurbach, D.; Zaban, A.; Gofer, Y.; Abramson, O.; Ben-Zion, M. *J. Electrochem. Soc.* **1995**, *142*, 687.
3. Coetzee, J. F., Ed. *Recommended Methods for Purification of Solvents and Tests for Impurities*; Pergamon Press: New York, 1982.
4. Mori, S.; Asahina, H.; Suzuki, H.; Yonei, A.; Yokoto, K. *J. Power Sources* **1997**, *68*, 59.
5. Kerr, J. B. *J. Electrochem. Soc.* **1985**, *132*, 2839.
6. Aurbach, D.; Weissman, I.; Zaban, A.; Youngman, O. C. *Electrochim. Acta* **1994**, *39*, 51–71.
7. Yeager, E. B. Workshop on Lithium Nonaqueous Battery Electrochemistry, 1980, Cleveland, Ohio.
8. Chlistunoff, J. B.; Bard, A. J. *Inorg. Chem.* **1993**, *32*, 3521.
9. Smith, W. H.; Bard, A. J. *J. Am. Chem. Soc.* **1975**, *97*, 5203.
10. Smith, W. H.; Bard, A. J. *J. Electroanal. Chem.* **1997**, *76*, 19.
11. Zhuang, G.; Wang, K.; Chottiner, G.; Barbour, R.; Lou, Y.; Bae, I. T.; Tryk, D. A.; Scherson, D. A. *J. Power Sources* **1995**, *54*, 20.

12. Zhuang, G.; Chottiner, G.; Swang, E.; Tolmachev, Y. V.; Bae, I. T.; Scherson, D. A. *Rev. Sci. Instrum.* **1994**, *65*, 2494.
13. Zhuang, G.; Chottiner, G. S.; Scherson, D. A. *J. Phys. Chem.* **1995**, *99*, 7009.
14. Aurbach, D. *J. Electrochem. Soc.* **1989**, *136*, 906.
15. Aurbach, D.; Gottlieb, H. *Electrochim. Acta* **1989**, *34*, 141.
16. Aurbach, D.; Youngman, O.; Gofer, Y.; Meitav, A. *Electrochim. Acta* **1990**, *35*, 625.
17. Zaban, A.; Aurbach, D. *J. Electroanal. Chem.* **1993**, *348*, 155.
18. Koch, V. R.; Nanjundian, C.; Battista Appetecchi, G.; Scrosati, B. *J. Electrochem. Soc.* **1995**, *142*, L116.
19. Ein-Eli, Y.; Thomas, S. R.; Chadha, R.; Blakley, T. J.; Koch, V. R. *J. Electrochem. Soc.* **1997**, *144*, 823.
20. Gritzner, G.; Kuta, J. *Pure Appl. Chem.* **1982**, *54*, 1527.
21. Peled, E. In *Lithium Batteries*; Gabano, J. P., Ed.; Academic Press: New York, 1983; Chap. 3.
22. Aurbach, D.; Daroux, M. L.; Faguy, P.; Yeager, E. *J. Electroanal. Chem.* **1991**, *297*, 225.
23. Aurbach, D.; Zaban, A. *J. Electroanal. Chem.* **1994**, *367*, 15.
24. Aurbach, D.; Zaban, A. *J. Electroanal. Chem.* **1994**, *365*, 41.
25. Odziemkowski, M.; Krell, M.; Irish, D. E. *J. Electrochem. Soc.* **1992**, *139*, 3063.
26. Oyama, N.; Tatzuma, T.; Sotomura, T. *J. Power Sources* **1997**, *68*, 135.
27. Aurbach, D.; Markovsky, B.; Schechter, A.; Ein-Eli, Y.; Cohen, H. *J. Electrochem. Soc.* **1996**, *143*, 3809.
28. Bard, A. J.; Faulkner, L. R. *Electrochemical Methods: Fundamentals and Applications*; Wiley: New York, 1980.
29. Gileadi, E. *Electrode Kinetics for Chemists, Chemical Engineers, and Materials Scientists*; VCH: New York, 1993.
30. Gileadi, E.; Kirowa-Eisner, E.; Penciner., J. *Interfatial Electrochemistry. An Experimental Approach*; Addison-Wesley: London, 1975.
31. Christensen, P. A.; Hammet, A. *Techniques and Mechanisms in Electrochemistry*; Blackie: New York, 1994.
32. Kissinger, P. T.; Heineman, W. R. *Laboratory Techniques in Electroanalytical Chemistry*; Marcel Dekker: New York, 1984.
33. Bard, A. J.; Rubinshtain, I., Eds. *Electroanalytical Chemistry. A Series of Advances*; Marcel Dekker: New York, 1966–1996.
34. Schrader, B., Ed. *Infrared and Raman Spectroscopy Methods and Applications*; VCH: New York, 1995.
35. Roeges, N. P. G. *A Guide to the Complete Interpretation of Infrared Spectra of Organic Structures*; Wiley: New York, 1994.
36. Aurbach, D.; Daroux, M. L.; Faguy, P.; Yeager, E. *J. Electrochem. Soc.* **1987**, *134*, 1611.
37. Aurbach, D. *J. Electrochem. Soc.* **1989**, *136*, 1610.
38. Aurbach, D. *J. Electrochem. Soc.* **1989**, *136*, 1606.
39. Bae, I. T.; Sandifer, M.; Lee, Y. W.; Tryk, D. A.; Sukenik, C. N.; Scherson, D. A. *Anal. Chem.* **1995**, *67*, 4508.

40. Pham, M.-C.; Adami, F.; Lacaze, P.-C. *J. Electroanal. Chem.* **1989**, *265*, 247.
41. Johnson, B. W.; Pettinger, B.; Doblhofer, K. *Ber. Bunsenges. Phys. Chem.* **1993**, *97*, 412.
42. Patty, D. B.; Harris, J. M. *Langmuir* **1990**, *6*, 209.
43. Goren, E.; Chusid, O.; Aurbach, D. *J. Electrochem. Soc.* **1991**, *138*, L6.
44. Chusid (Youngman), O.; Aurbach, D. *J. Electrochem. Soc.* **1993**, *140*, L1.
45. Maulhardt, H.; Kunath, D. *Talanta* **1982**, *29*, 237.
46. Duyne, R. P. In *Chemical and Biochemical Applications of SERS*; Moore, C. B., Ed.; Academic Press: New York, 1979; Vol. 4, Chap. 4.
47. Moskevitz, M.; Dillela, D. P. In *Surface Enhanced Raman Scattering*; Furtak, T. E., Chang, R. K., Eds.; Plenum Press: New York, 1982; pp. 264–267.
48. Venkatasetty, H. V. The Workshop on Lithium Nonaqueous Battery Electrochemistry, **1980**, Cleveland, Ohio.
49. Kostecki, R.; Tran, T.; Song, X.; Kinoshita, K.; Mclarnon, F. *J. Electrochem. Soc.* **1997**, *144*, 3111.
50. Zaban, A.; Ferrere, S.; Gregg, B. A. *J. Phys. Chem. B* **1997**, *102*, 452–460.
51. Tolmachev, Y. V.; Wang, Z. H.; Hu, Y. N.; Bae, I. T.; Scherson, D. A. *Anal. Chem.* **1998**, *70*, 1149.
52. Ernst, B.; Bensaddik, A.; Hilaire, L.; Chaumette, P.; Kiennemann, A. *Catal. Today* **1998**, *39*, 329.
53. Hu, Y. N.; Bae, I. T.; Mo, Y. B.; Antonio, M. R.; Scherson, D. A. *Can. J. Chem.* **1997**, *75*, 1721.
54. Alonso, C.; Pascual, M. J.; Salomon, A. B.; Abruna, H. D.; Gutierrez, A.; Lopez, M. F.; GarciaAlonso, M. C.; Escudero, M. L. *J. Electroanal. Chem.* **1997**, *435*, 241–254.
55. Herlem, G.; Goux, C.; Fahys, B.; Dominati, F.; Goncalves, A. M.; Mathieu, C.; Sutter, E.; Trokourey, A.; Penneau, J. F. *J. Electroanal. Chem.* **1997**, *435*, 259–265.
56. Zamborini, F. P.; Campbell, J. K.; Crooks, R. M. *Langmuir* **1998**, *14*, 640.
57. Pham, M. T.; Zyganow, I.; Matz, W.; Reuther, H.; Oswald, S.; Richter, E.; Wieser, E. *Thin Solid Films* **1997**, *310*, 251.
58. Zei, M. S.; Ertl, G. *Z. Phys. Chem. Int. J. Res. Phys. Chem. Chem. Phys.* **1997**, *202*, 5–19.
59. Huang, B. M.; Lister, T. E.; Stickney, J. L. *Surf. Sci.* **1997**, *392*, 27–43.
60. Conway, B. E. *Prog. Surf. Sci.* **1995**, *49*, 331.
61. Fernandez, A. M. *Adv. Mater. Opt. Electron* **1998**, *8*, 1.
62. Carlson, T. A. *Photoelectron and Auger Spectroscopy*; Plenum Press: New York, 1975.
63. Goodhew, P. J.; Humphreys, F. J. *Electron Microscopy and Analysis*; Taylor & Francis: London, 1988.
64. Gouerec, P.; Savy, M.; Riga, J. *Electrochim. Acta* **1998**, *434*, 743.
65. Buttry, D. A. In *Electroanalytical Chemistry. A Series of Advances*; Bard, A. J., Ed.; Marcel Dekker: New York, 1991; Vol. 17.
66. Jusys, Z.; Stalnionis, G. *J. Electroanal. Chem.* **1997**, *431*, 141–144.
67. Chen, S. L.; Wu, B. L.; Cha, C. S. *J. Electroanal. Chem.* **1997**, *431*, 243–247.
68. Nomura, T.; Iijima, M. *Anal. Chim. Acta* **1981**, *131*, 97.

69. Sarid, D. *Scanning Force Microscopy with Application to Electric, Magnetic and Atomic Forces*; Oxford University Press: New York, 1994; Vol. 5.
70. Hamers, R. J. *J. Phys. Chem.* **1996**, *100*, 13103.
71. Hugstad, G.; Gladfelter, W. L.; Weberg, E. B.; Weberg, R. T.; Weatherill, T. D. *Langmuir* **1994**, *10*, 4295.
72. Sidles, J. A.; Garbini, J. L.; Bruland, K. J.; Rugar, D.; Zuger, O.; Hoen, S.; Yannoni, C. S. *Rev. Mod. Phys.* **1995**, *67*, 249.
73. Stroscio, J. A.; Kaiser, W. J., Eds. *Scanning Tunneling Microscopy*; Academic Press: New York, 1993.
74. Nanjo, H.; Newman, R. C.; Sanada, N. *Appl. Surf. Sci.* **1997**, *121*, 253–256.
75. Lebreton, C.; Wang, Z. Z. *Surf. Sci.* **1997**, *382*, 193–200.
76. Ge, M.; Zhong, B.; Clemperer, W. G.; Gewirth, A. A. *J. Am. Chem. Soc.* **1996**, *118*, 5812.
77. Cheetham, A. K.; Day, P., Eds. *Solid State Chemistry Techniques*; Oxford Science: New York, 1988.
78. Young, R. A. *J. Appl. Cryst.* **1995**, *28*, 366.
79. Luo, J. L.; Cui, N. *J. Alloys Compounds* **1998**, *264*, 299.
80. Rivera, M.; Seetharaman, R.; Girdhar, D.; Wirtz, M.; Zhang, X. J.; Wang, X. Q.; White, S. *Biochem. Usa* **1998**, *37*, 1485.
81. Tsumura, T.; Inagaki, M. *Solid State Ionics* **1997**, *104*, 183.
82. Nazri, G.; Muller, R. H. *J. Electrochem. Soc.* **1985**, *132*, 1385.
83. Kranke, P.; Wahren, M.; Findeisen, M. *Isot. Environ. Health Stud.* **1997**, *33*, 245.
84. Sanders, J. K. M.; Hunter, B. K. *Modern NMR Spectroscopy: A Guide for Chemists*; Oxford University Press: London, 1993.
85. MacOmber, R. S. *A Complete Introduction to Modern NMR Spectroscopy*; Wiley: New York, 1997.
86. Raper, E. S.; Creighton, J. R.; Clegg, W.; Cucurull-Sanchez, L.; Hill, M. N. S.; Akrivos, P. D. *Inorg. Chim. Acta* **1998**, *271*, 57.
87. Demadis, K. D.; Bakir, M.; Klesczewski, B. G.; Williams, D. S.; White, P. S.; Meyer, T. J. *Inorg. Chim. Acta* **1998**, *270*, 511.
88. Tkami, N.; Satoh, A.; Oguchi, M.; Sasaki, H.; Ohsaki, T. *J. Power Sources* **1997**, *68*, 283.
89. Prakash, J.; Tryk, D.; Aldred, W.; Yeager, E. In *Electrochemistry in Transition*; Murphy, O. J., Srinivasan, S., Conway, B. E., Eds.; Plenum Press: New York, 1992; pp. 93–106.
90. Carbonio, R. E.; Tryk, D.; Yeager, E. B. In *Proceedings of the Symposium on Electrode Materials and Processes for Energy Conversion and Storage*; Srinivasan, S., Wagner, S., Wroblova, H., Eds.; Electrochemical Soc.: Pennington, NJ, 1987; Vol. PV 87-12; p. 238.
91. Scherson, D. A. In *Spectroelectrochemistry: Theory and Practice*; Gale, R., Ed.; Plenum Press: New York, 1988.
92. Aurbach, D.; Ein-Eli, Y.; Zaban, A. *J. Electrochem. Soc.* **1994**, *141*, L1.
93. Aurbach, D.; Skaletsky, R.; Gofer, Y. *J. Electrochem. Soc.* **1991**, *138*, 3536.
94. Christensen, P. A.; Hamnett, A.; Trevellick, P. R. *J. Electroanal. Chem.* **1988**, *242*, 23.
95. Seki, H.; Kunimatzu, K.; Golden, G. *Appl. Spectrosc.* **1985**, *39*, 437.

96. Aurbach, D.; Chusid, O. *J. Electrochem. Soc.* **1993**, *140*, L155.
97. Aurbach, D.; Cohen, Y. *J. Electrochem. Soc.* **1996**, *143*, 3525.
98. Aurbach, D.; Ein-Eli, Y. *J. Electrochem. Soc.* **1997**, *144*, 3355.
99. Tao, N. J. *Phys. Rev. Lett.* **1996**, *76*, 4066.
100. Aurbach, D.; Zaban, A. *J. Electroanal. Chem.* **1995**, *393*, 43.
101. Aurbach, D.; Moshkovits, M. *J. Electrochem. Soc.* **1998**, *145*, 2629.
102. Aurbach, D.; Ein-Eli, Y. *J. Electrochem. Soc.* **1995**, *142*, 1746.
103. Aurbach, D.; Gofer, Y.; Langzam, Y. *J. Electrochem. Soc.* **1989**, *136*, 3198.
104. Aurbach, D.; Weissman, I.; Schecter, A.; Cohen, H. *Langmuir* **1996**, *12*, 3991.

# 4

# The Electrochemical Window of Nonaqueous Electrolyte Solutions

**Doron Aurbach and Yosef Gofer**
*Bar-Ilan University, Ramat-Gan, Israel*

## I. INTRODUCTORY REMARKS

This chapter deals with the electrochemical behavior of nonaqueous electrolyte solutions in contact with nonactive electrodes, which are defined as electrodes in which there is no spontaneous chemical reaction between the solution and the electrode material under open circuit conditions. This term serves to differentiate these electrodes from active metals such as lithium, calcium, magnesium, etc., which react spontaneously with the solution under open circuit conditions.

A nonactive electrode may include noble metals such as gold, silver, and platinum, the so-called *sp*-metals such as In, Ga, Cd, Bi, as well as transition (or *d*) metals such as nickel or cobalt. Carbon electrodes and semiconductors such as indium tin oxide [1], diamond [2], and conducting polymers may fall into the category of nonactive electrodes in appropriate solutions, as do composite materials that contain metal oxides or chalcogenides. The behavior of active electrodes in nonaqueous solution is discussed separately in the next chapter.

An electrochemical stability window is defined as the potential range in which an electrode can be polarized in a solution without the passage of substantial Faradaic currents. This definition is only a practical one, as it is impossible to define a precise value for the term ''substantial.''

In general, the electrochemical stability window for a solution-electrode system is limited by the electrochemical stability of the salt or the solvent or by the dissolution or degradation of the electrode.

The electrochemical stability window of electrolyte-solution systems, as well as the electrochemical behavior of the systems within the window, depends on the

1. Nature of the solvent
2. Nature of the salt
3. Nature of the electrode material
4. Presence of contaminants

Although the stability window of any electrochemical system depends on the inherent electrochemical stability of its individual components, it is practically impossible to predict this a priori from the possible interactions among the known components of the electrochemical system. Hence, an electrochemical window reported for any electrochemical system is unique to its specific components, including solvents, salts, electrode materials, cleanliness of the system, and its preparation procedure.

For instance, the reduction potential of many solvents depends on the salt used and, in particular, on the cation. The reduction potentials of alkyl carbonates and esters in the presence of tetraalkyl ammonium salts (TAA) are usually much lower than in the presence of alkaline ions ($Li^+$, $Na^+$, etc.). Similar effects were observed with the reduction potential of some common contaminants (e.g., $H_2O$, $O_2$, $CO_2$). Moreover, the reduction products of many alkyl carbonates and esters are soluble in the presence of tetraalkyl ammonium salts, while in the presence of lithium ions, film formation occurs, leading to passivation of the electrode [3].

The presence of contaminants in the solution, such as $H_2O$, $CO_2$, $O_2$, and alcohols may limit the electrochemical window of nonaqueous systems. These reactive contaminants are commonly present in nonaqueous solutions and may be reduced at higher potentials (or oxidized at lower potentials), compared with the reduction (or oxidation) potentials of the other components of the solutions.

In some cases, the electrode material is the limiting factor of the electrochemical stability window. In a metal salt solution, underpotential deposition (UPD) may occur. In some examples, such as gold or platinum electrodes in the presence of lithium ions, the UPD appears at potentials that are substantially higher than the bulk metal deposition [4–6]. In addition, some metals may possess catalytic activity for specific reduction or oxidation processes [7–12]. Many nonactive metals (distinguished from the noble metals), including Ni, Cu, and Ag, which are commonly used as electrode materials, may dissolve at certain potentials that are much lower than the oxidation potentials of the solvent or the salt. In addition, some electrode materials may be catalytic to certain oxidation or reduction processes of the solution components, and thus we can see differences in the stability limits of nonaqueous systems depending on the type of electrode used.

In this chapter we review the data accumulated over the years for various electrochemical reduction and oxidation reactions of nonaqueous systems, with special attention to the impact of the above factors on the practical electrochemical windows of commonly used nonaqueous electrolyte solutions.

The first comprehensive review of electrochemical windows of nonaqueous systems, by Mann, appeared 20 years ago in the series *Electroanalytical Chemistry* [13], edited by A. J. Bard. In general, the picture provided by Mann is quite reliable. However, over the years, a vast amount of work has been done with nonaqueous systems, particularly within the framework of basic studies related to high energy density batteries and the application of novel spectroelectrochemical tools. The accumulated data provide a more precise picture of the various reactions that limit the electrochemical window of commonly used systems.

In the following sections we present data that was published from the time of Mann's review to date, on the behavior of a variety of nonaqueous systems with the so-called nonactive electrodes defined in Section I. The chapter is divided into two major sections: reduction and oxidation reactions.

## II. REDUCTION REACTIONS OF TETRAALKYL AMMONIUM–BASED SALT SOLUTIONS

### A. General

Tetraalkyl ammonium (TAA) salts are characterized by very low reduction potentials, along with good solubility in many organic solvents. Thus, nonaqueous solutions composed of such salts (e.g., tetrabutyl ammonium perchlorate and organic solvents such as ethers, esters, and alkyl carbonates) can be electrolyzed using noble metal electrodes. In contrast to lithium salt solutions, in TAA-based solutions there is no precipitation of insoluble products on the electrode, which leads to its passivation. Therefore, it is possible to isolate and identify the electrolysis products and thus outline precise reduction mechanisms for the various systems.

### B. Ether Solutions

Ether solutions based on TAA salts are not reduced on noble metal electrodes. The major cathodic reaction of these solutions involves the cation reduction to trialkyl amine, alkane, and alkene (which are the stable disproportion products of the alkyl radical formed by the electron transfer to the cation) [3]. Electrolysis of ethers such as THF or DME containing TBAP, formed in the catholyte tributyl amine, butane and butene, were unambiguously identified by NMR and GCMS analysis [3]. In the presence of water (several hundred ppm and more), the electrolysis products were found to be tributyl amine and butene (butane was not detected) [3]. The potential of this reduction reaction is higher than that of the dry solution, and it is clear that the initial electroactive species in this case is the

a.  $(CH_3CH_2CH_2CH_2)_4N^+ + e^- \xrightarrow{\text{THF, DME}} (CH_3CH_2CH_2CH_2)_3N + CH_3(CH_2)_3\bullet$

b.  $2\ CH_3CH_2CH_2CH_2\bullet \xrightarrow{\text{disproportionation}} CH_3CH_2CH_2CH_3 + CH_3CH_2CH = CH_2$

c.  $H_2O + e^- \rightarrow OH^- + \tfrac{1}{2}H_2 \uparrow$

d.  $OH^- + CH_3CH_2CH_2CH_2N^+R_4 \rightarrow H_2O + CH_3CH_2CH{=}CH_2 + (CH_3CH_2CH_2CH_2)_3N$

**Scheme 1**   Reduction patterns for TBA salt solutions in ethers.

water. Scheme 1 describes reaction mechanisms for TBA reduction in dry and wet ethereal solutions, based on the above product distribution. The reduction potential of TBAP on gold electrodes was found to be 0.V versus $Li/Li^+$ (depending on the salt concentration).

Figure 1 shows a typical voltammogram of dry and wet 0.2 $M$ TBAP solution in ether, obtained with a gold electrode, together with voltammograms related to other solvents [3]. The non-Faradaic currents which characterize these voltammograms over a wide potential range, as shown in Figure 1, reflect a double layer capacitance of the order of 10–20 $\mu F/cm^2$, which is anticipated and typical for a bare metal-solution interface (in contrast to cases for electrodes covered by surface films, as shown further on). The larger the amount of water present, the higher is the reduction onset potential [14]. This simply reflects water reduction to $OH^-$, which further attacks the TBA cation and deprotonates one of the alkyl group, as shown in Scheme 1. Note that the smaller the cation's R group, the less sterically hindered is the $N^+$ center, thus its reduction is easier and the reduction potential is higher.

## C.  Ester Solutions

We have studied the electrolysis of $\gamma$-butyrolactone (BL) and methyl formate (MF) in TBAP solutions. A typical voltammogram of $\gamma$-BL/TBAP with a gold electrode is also shown in Figure 1. Butyrate ($CH_3CH_2CH_2COO^-$) and a cyclic $\beta$-keto ester were identified as the major electrolysis products. The latter is a product of a nucleophilic attack of $\gamma$-BL anion (in the $\alpha$ position) on the carbonyl center of another molecule [3]. The FTIR spectra of this product, as well as its lithiated derivative, are shown in Figure 2. The basic reduction mechanisms of $\gamma$-BL, based on the above product analysis, as well as on other arguments [3], are presented in Scheme 2.

Water, as a contaminant, is reduced to $OH^-$, which further attacks the $\gamma$-BL molecules. $\gamma$-Hydroxy butyrate is thus the final reduction product in wet

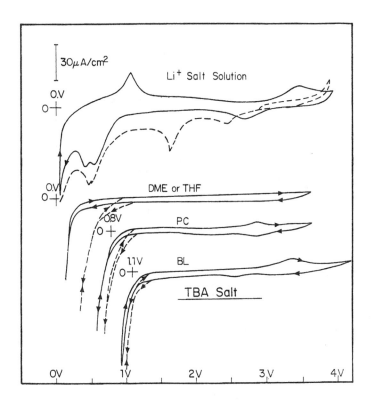

**Figure 1** Steady state cyclic voltammograms obtained from the studied four systems. The three lower ones were obtained from 0.2 $M$ TBAP solutions. The upper one was obtained from 0.2 $M$ LiAsF$_6$ or 0.2 $M$ LiClO$_4$ solution in BL. The potential scale is referred to Li/Li$^+$ 0.1 $M$ reference electrode. Sweep rate was 20 mV s$^{-1}$. The current scale is common. Solid lines = dry solutions ($\approx$15–30 ppm H$_2$O); dashed lines = water-contaminated solutions (0.01 M). The potential scan in all cases started from open circuit potential 2.7–3 V in these systems [3]. (With copyrights from Elsevier Science Ltd.)

γ-BL/TBAP solutions, as also presented in Scheme 2. As expected, the cathodic limit of the wet solutions shifts positively, as the water content is higher.

Electrolysis of MF/TBAP solution produces methoxide (CH$_3$O$^-$), formate anion (HCO$_2$$^-$), and CO [15]. The voltammograms of MF/TBAP solutions with gold electrodes are quite similar to those of the γ-BL solutions in Figure 1. Scheme 3 describes the reduction pattern of MF based on the above product distribution. In wet MF/TBAP solutions, both formate and methoxide are formed, and it is clear that, as in the case of wet γ-BL solutions, H$_2$O is predominantly

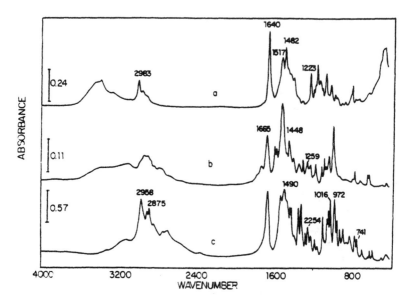

**Figure 2**  FTIR spectra obtained from BL electrolysis product and reference compounds (KBr pellets). (a) Spectrum of tetraethylammonium ethylacetoacetate anion [3]. (With copyrights from Elsevier Science Ltd.)

$$\underset{(CH_3CH_2)_4N^+CH_3CH_2O\overset{\overset{O}{\|}}{C}CH^-\overset{\overset{O}{\|}}{C}CH_3}{}$$

(b) Spectrum of the product of the reaction between BL and *t*-butoxide (potassium) in *t*-butanol. (c) Spectrum of the product of BL electrolysis in THF-TBAP medium.

reduced to OH⁻, followed by hydrolysis of the ester to its two components (see Scheme 3).

## D.  Alkyl Carbonate Solutions

A typical steady state voltammogram in propylene carbonate (PC) TBAP solution with a gold electrode is shown in Figure 1. Electrolysis of PC/TBAP solution, followed by precipitation as a lithium salt with LiClO₄, produced a single major product whose FTIR spectrum is shown in Figure 3. Similar electrolysis and treatment of ethylene carbonate (EC) TBAP solutions also produced a single product whose FTIR spectrum is shown in Figure 4 [16]. The NMR spectrum of the PC reduction product, precipitated as a Li salt and, dissolved in D₂O, showed a triplet, a multiplet, and a doubleton at 1.5, 3.4, and 3.6 ppm, respectively [3],

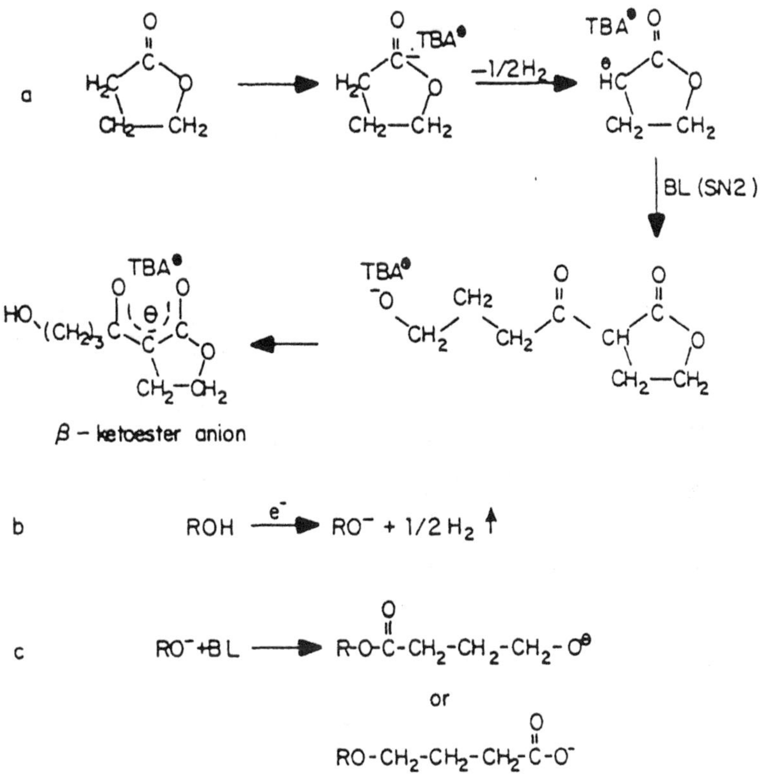

**Scheme 2** Reduction of TBA salt solutions in BL.

$$HCOOCH_3 + e^- \rightarrow HCOOCH_3^{\bullet -}$$

$$HCOOH_3^{\bullet -} \rightarrow HCOO^- + CH_3 \bullet$$

$$HCOO^- + Li^+ \text{ or } Ca^{2+} \rightarrow$$
$$HCOOLi \downarrow \text{ or } (HCOO)_2Ca \downarrow$$

$$CH_3 \bullet \xrightarrow{\quad H\bullet \text{ or } CH_3\bullet \quad} CH_4 \uparrow \text{ or } C_2H_6 \uparrow$$

**Scheme 3** Reduction patterns for MF.

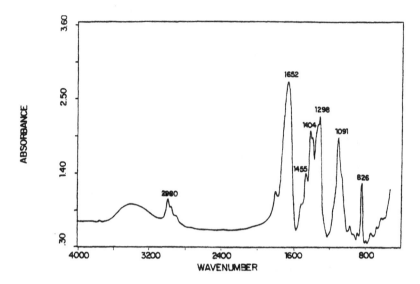

**Figure 3**  FTIR spectrum obtained from KBr pelletized major product of PC electrolysis, precipitated as lithium salt [3]. (With copyrights from Elsevier Science Ltd.)

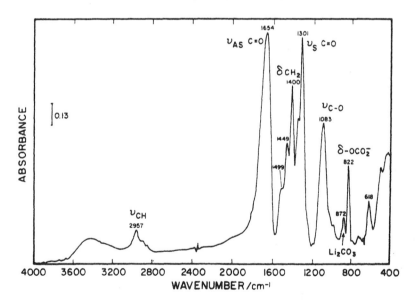

**Figure 4**  FTIR spectrum of the electrolysis product of EC in THF + 0.5 *M* TBAP on gold, isolated as Li salt (pelletized with KBr) [16]. (With copyrights from Elsevier Science Ltd.)

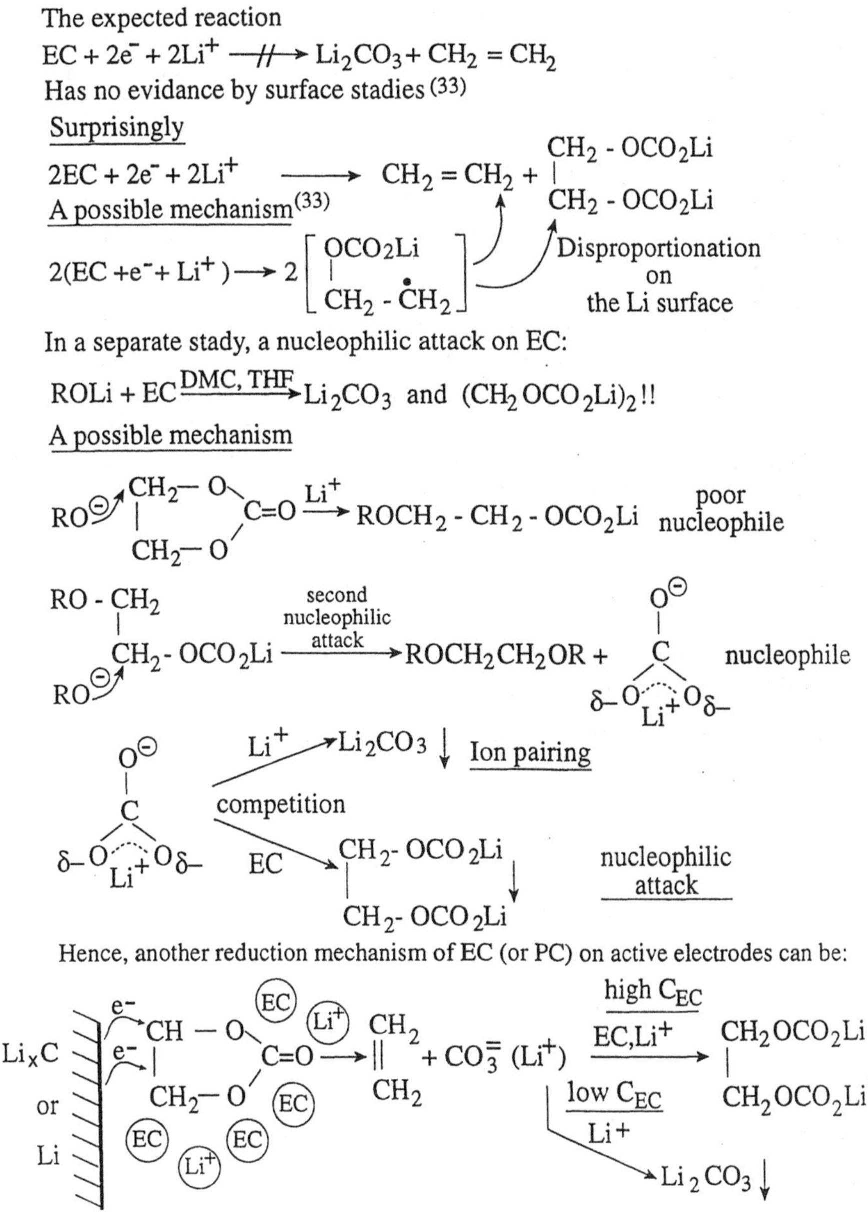

The expected reaction

$$EC + 2e^- + 2Li^+ \xrightarrow{\;/\!/\;} Li_2CO_3 + CH_2 = CH_2$$

Has no evidence by surface stadies [33]

<u>Surprisingly</u>

$$2EC + 2e^- + 2Li^+ \longrightarrow CH_2 = CH_2 + \begin{array}{l} CH_2\text{-}OCO_2Li \\ | \\ CH_2\text{-}OCO_2Li \end{array}$$

A possible mechanism[33]

$$2(EC + e^- + Li^+) \longrightarrow 2 \begin{bmatrix} OCO_2Li \\ | \\ CH_2\text{-}\overset{\bullet}{C}H_2 \end{bmatrix}$$

Disproportionation on the Li surface

In a separate stady, a nucleophilic attack on EC:

$$ROLi + EC \xrightarrow{DMC,\ THF} Li_2CO_3 \text{ and } (CH_2OCO_2Li)_2 !!$$

A possible mechanism

$$RO^{\ominus}\!\!\begin{array}{c} CH_2\!-\!O \\ | \quad\;\; \diagdown \\ CH_2\!-\!O \end{array}\!\!C{=}O \xrightarrow{Li^+} ROCH_2\text{-}CH_2\text{-}OCO_2Li \quad \begin{array}{c} poor \\ nucleophile \end{array}$$

$$\begin{array}{c} RO\text{-}CH_2 \\ | \\ CH_2\text{-}OCO_2Li \\ RO^{\ominus}\!\nearrow \end{array} \xrightarrow[\text{attack}]{\begin{array}{c}\text{second}\\\text{nucleophilic}\end{array}} ROCH_2CH_2OR + \begin{array}{c} O^{\ominus} \\ | \\ C \\ \delta\text{-}O\overset{\cdots}{\underset{Li^+}{}}O\delta\text{-} \end{array} \quad nucleophile$$

$$\begin{array}{c} O^{\ominus} \\ | \\ C \\ \delta\text{-}O\overset{\cdots}{\underset{Li^+}{}}O\delta\text{-} \end{array} \begin{array}{c} \xrightarrow{Li^+} Li_2CO_3 \downarrow \;\; \text{Ion pairing} \\[4pt] \xrightarrow{\text{competition}} \\[4pt] \xrightarrow{EC} \begin{array}{c} CH_2\text{-}OCO_2Li \\ | \\ CH_2\text{-}OCO_2Li \end{array} \downarrow \;\; \begin{array}{c}\text{nucleophilic}\\\underline{\text{attack}}\end{array} \end{array}$$

Hence, another reduction mechanism of EC (or PC) on active electrodes can be:

$$Li_xC \text{ or } Li \quad \begin{array}{c} e^- \\ e^- \end{array}\!\!\!\begin{array}{c} \text{\small(EC)} \\ CH\!-\!O \\ | \quad\;\; \diagdown \\ CH_2\!-\!O \\ \text{\small(EC)} \end{array}\!\!C{=}O \xrightarrow[\text{\small(EC)} \; \text{\small(Li}^+\text{)}]{\text{\small(Li}^+\text{)}} \begin{array}{c} CH_2 \\ \| \\ CH_2 \end{array} + CO_3^{=} (Li^+) \xrightarrow[\begin{array}{c}\underline{low\ C_{EC}}\\ Li^+\end{array}]{\begin{array}{c}high\ C_{EC}\\ \overline{EC,Li^+}\end{array}} \begin{array}{c} CH_2OCO_2Li \\ | \\ CH_2OCO_2Li \end{array}\!\!\downarrow$$

$$\searrow Li_2CO_3 \downarrow$$

**Scheme 4**   EC (PC) reduction mechanisms.

while the EC reduction product showed a singlet at 3.5 ppm [16]. The spectra in Figures 3 and 4 are typical of $ROCO_2Li$ species [17], and thus both NMR and FTIR spectra prove that the major reduction products of EC and PC in dry TBAP solutions are alkylene dicarbonates (propylene and ethylene for PC and EC, respectively). Hence, Scheme 4 suggests comprehensive mechanisms for PC and EC reduction in light of the above product distribution. Based on recent studies [18], we cannot exclude the possibility that EC and PC undergo two-electron reduction, which produces unpaired $CO_3^{2-}$, which further nucleophilically attacks another solvent molecule, thus forming the final product, namely, alkylene dicarbonate dianion.

In wet PC or EC solutions with TBAP, the inorganic carbonate anion $CO_3^{2-}$ becomes a major reduction product. FTIR spectra of the major reduction product of the wet PC/TBAP solution, precipitated as a Li salt, is characteristic of $Li_2CO_3$ (two peaks around 1500 and 1480 cm$^{-1}$ and a sharp peak around 880 cm$^{-1}$). It should be emphasized, however, that FTIR measurements conducted during electrolysis proved that the $ROCO_2^-$ species are present as intermediates on the reaction path [3]. Thus, in the case of wet EC or PC/TBAP solutions, the solvents are reduced to $ROCO_2Li$ species, as described in Scheme 4, and these solvents reduction products further react with water, as shown in Scheme 5.

### E.  Interactions between TBA Cations and Some Electrode Materials at Low Potentials

It was found by Kariv-Miller et al. [19–21] that under certain conditions TBA$^+$ cations may be reduced to form alloy-type compounds with metals such as Hg,

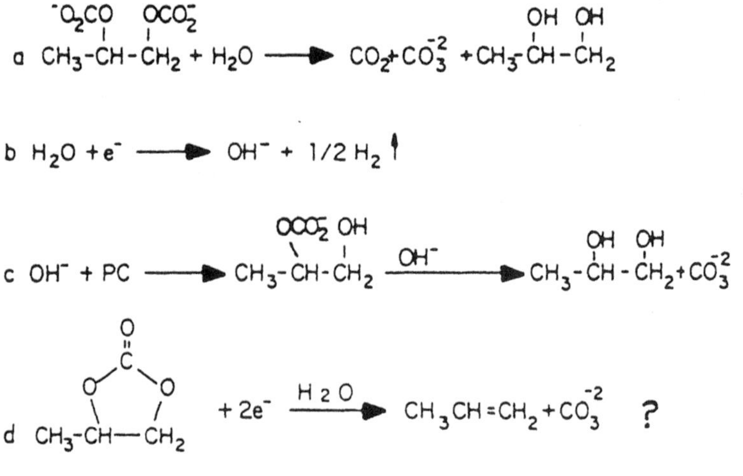

**Scheme 5**  Reduction reactions of wet PC solutions.

Pb and Sn. In these cases, this "alloying" process determines the cathodic limit of the electrochemical window.

## F. Oxygen Electrochemistry in Nonaqueous Tetrabutylammonium Salt Solutions

Oxygen electrochemistry in several nonaqueous solutions containing TBA salts as the electrolyte was studied by several groups [22–24]. We have also investi-

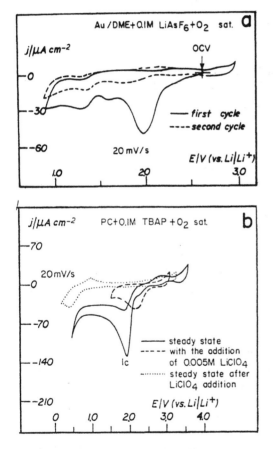

**Figure 5** Effect of $O_2$ on the cyclic voltammetry of Au electrode in DME and PC. (a) DME + 0.1 $M$ LiAsF$_6$ saturated with full-stop $O_2$•20 mV/s. (—) First cycle with a new electrode. (---) Second cycle, same electrode. (b) PC + 0.1 $M$ TBAP. 20 mV/s. (—) Steady state behavior. (---) Next cycle after an addition of 0.005 $M$ LiClO$_4$ [12]. (With copyrights from Elsevier Science Ltd.)

gated oxygen reduction in ethereal and alkyl carbonate solutions [12]. In nonelectrophilic solvents, such as ethers or acetonitrile, the voltammetric behavior of oxygen containing TBA solutions is characterized by two peaks around 2 V versus Li/Li$^+$. These correspond to $O_2$ reduction to the superoxide $O_2^\pm$ radical anion and its reoxidation to $O_2$. These reactions are kinetically very sluggish, as reflected by peak separation that can reach 200 mV [23–24] (as compared with about 60 mV for a typical reversible one-electron redox process at 25°C [25]). In the presence of electrophiles such as water in the solution, or in an electrophilic solvent such as PC, the superoxide formed continues to react nucleophilically, and thus only the cathodic peak is observed (unless the sweep rate applied is faster than its following reaction). Typical voltammetric behavior of oxygen in the two types of solvents (as described above) is shown in Figure 5.

## III.  VOLTAMMETRIC BEHAVIOR OF LITHIUM SALT SOLUTIONS IN ETHERS, ESTERS AND ALKYL CARBONATES

### A.  General Remarks

The presence of Li ions in nonaqueous solutions, such as those dealt with in the previous section, completely changes the behavior of these systems in the following respects.

a.  The onset of solvent reduction processes is observed at higher potentials than when the cations are tetraalkylammonium [3,8,12].

b.  Reduction of most of the relevant solvents, as well as salt anions and common contaminants, such as $H_2O$, $O_2$ and $CO_2$, precipitates surface films, which comprise insoluble Li salts. Hence, the electrodes become covered with surface films whose onset for formation may be at potentials as high as 2 V versus Li/Li$^+$ [3,7–12,26].

c.  These surface films "passivate" the electrode, and thus further reduction of most of the solution species is inhibited. Therefore, in contrast to TAA-based solutions, there is no bulk solution reduction. At this stage, the cathodic potential limit becomes Li deposition at 0. V Li/Li$^+$. This process is *not* inhibited by the surface films because most of these surface films are good Li ion conductors (as described by the solid electrolyte interphase model for Li electrodes in these solutions [27]).

d.  With certain electrode materials, after the formation of the "passivation" layer, Li may be deposited as surface monolayers and/or form alloys at potentials higher than those of Li bulk deposition. Hence, the voltammograms of Li salt solutions may be characterized by pronounced Li UPD, Li UPD stripping and/or Li alloy formation and Li alloy decomposition peaks. As described in the next section, this is the case for many commonly used electrode materials, including gold, platinum, silver, and copper [5,6,12].

When carbon electrodes are used, Li may be inserted/intercalated reversibly into the carbon at potentials as high as 1 V versus Li/Li$^+$ (after the formation of surface films). In the case of disordered carbons, insertion may occur at even higher potentials. In the case of graphite (as described in the next section), the onset for lithium intercalation is around 0.3 V versus (Li/Li$^+$). With glassy carbon, there is no considerable lithium insertion, and hence this electrode behavior depends solely on the solvent and anion used and their cathodic stability [28].

## B. Steady State Voltammetric Behavior of Ethers, Esters, and Carbonate Solutions Containing Lithium Ions with Noble Metal Electrodes (Ag, Au, Pt) [7–12,14]

This section deals only with solvents whose reduction products are insoluble in the presence of lithium ions. The list includes open chain ethers such as diethyl ether, dimethoxy ethane, and other polyethers of the glyme family: cyclic ethers such as THF, 2Me-THF, and 1,4-dioxane; cyclic ketals such as 1,3-dioxolane and 1,3-dioxane, esters such as γ-butyrolactone and methyl formate; and alkyl carbonates such as PC, EC, DMC, and ethylmethyl carbonate. This list excludes the esters, ethyl and methyl acetates, and diethyl carbonate, whose reduction products are soluble in them (in spite of the presence of Li ions). Solutions of solvents such as acetonitrile and dimethyl formamide are also not included in this section for the same reasons. Figure 6 presents typical steady state voltammograms obtained with gold, platinum, and silver electrodes in Li salt solutions in which solvent reduction products are formed and precipitate at potentials above that of lithium metal deposition. These voltammograms are typical of the above-mentioned solvent groups and are characterized by the following features:

a. The voltammograms are strongly dependent on the metal used. At potentials below 0.7 V, Li UPD peaks appear. With silver, one UPD peak appears around 0.4 V versus Li/Li$^+$, with a corresponding anodic stripping peak around 1.05 V. Two Li UPD peaks appear with gold electrode (around 0.6 and 0.4 V) and sometimes merge into one peak (depending occasionally on the gold surface used). The corresponding Li UPD stripping peak appears at 1.1 V (Li/Li$^+$). With polycrystalline platinum, two Li UPD peaks appear at 0.7 and 0.3 V (Li/Li$^+$), with corresponding anodic stripping peaks around 1.1 and 1.3 V (Li/Li$^+$), respectively. It should be emphasized that the sharpness and resolution

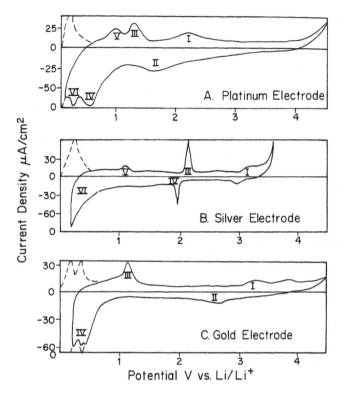

**Figure 6**  Effect of electrode metal on cyclic voltammetry of 0.2 $M$ LiAsF$_6$/DME electrolyte. Sweep rate = 20 mV/s. Dashed lines show lithium bulk deposition and dissolution processes.

of these peaks depends strongly on the purity of these solutions, as discussed below. These potentials of the stripping peaks of Li UPD are in general agreement with the correlation between $\Delta E_{upd}$ and $\Delta\phi$ (work functions) of the corresponding metals suggested by Gerisher et al. [5,6].

b.  The cathodic limit of these voltammograms is Li-alloying process that occurs with all three metals at potentials below 0.2 V (Li/Li$^+$). This alloying process is the fastest and the most pronounced with gold. However, this also depends on the solution composition, as discussed below.

c.  Below 0. V (Li/Li$^+$), pronounced bulk lithium metal deposition occurs.

d.  On the anodic side, when the metals are gold or platinum, solvent oxidation limits the electrochemical window. In general, the ethers are oxidized around 4.5 V (Li/Li$^+$); alkyl carbonates and esters are oxidized between 4.5 and 5 V (Li/Li$^+$). 1,3-Dioxolane is oxidatively polymerized at potentials above 3.5 V (Li/Li$^+$).

e.  In the case of a silver electrode, the anodic limit of the electrochemical window is silver dissolution occurring around 3.7 V (Li/Li$^+$). Another unique feature of the voltammograms of these Li salt solutions with silver electrodes is the presence of two peaks appearing around 2 V (Li/Li$^+$), which correspond to a reversible redox couple. It should be emphasized that these peaks do not depend on the solution composition. We tentatively attribute these peaks to an intercalation process of lithium into the oxide layer that covers the silver.

f.  In the potential range between the above-described peaks that correspond to the Li UPD processes and the above-described anodic limits of the electrochemical window of these systems, the voltammograms are nearly featureless and reflect mostly non-Faradaic, capacitive currents. It should be noted that these capacitive currents reflect quite a high interfacial capacity of the order of several hundreds of $\mu F/cm^2$. This interfacial capacity is at least one order of magnitude higher than typical metal-solution double layer capacitance (which is obtained in the voltammograms measured with TAA salt solutions).

When the solutions are contaminated with water (even in a level of a few tens of ppm), the following features are seen in the *steady state* voltammograms:

a.  With a platinum electrode, a water reduction peak appears around 1.6 V (Li/Li$^+$) and has a corresponding anodic peak at around 2.2 V. We explain this behavior, which is unique to platinum, as being related to the catalytic properties of this metal and its high affinity to hydrogen adsorption. Hence, trace water is reduced on the platinum to LiOH and hydrogen, which remain partly adsorbed on the metal, and is oxidized to H$^+$ in the reverse anodic sweep. This H$^+$ reacts further with surface LiOH to form H$_2$O and Li$^+$.

b.  With a gold electrode, two corresponding peaks are observed: a cathodic one at potentials around 2.8 V versus Li/Li$^+$, and an anodic one at around 3.5 V. These peaks appear only after LiOH is formed on the surface at low potentials ($<$1.5 V versus Li/Li$^+$), and thus they may reflect the following redox couples [3]:

$$Au + 3LiOH \rightleftharpoons Au(OH)_3 + 3Li^+ + 3e^- \tag{1}$$

$$Au + 3LiOH \rightleftharpoons Au(OH)_3ads + 3Li^+ + 3e^- \tag{2}$$

To conclude this section, it should be emphasized that the steady state voltammograms described above are quite different from the first scan of the cyclic voltammetry of these systems. During the first polarization of these electrodes to low potentials, pronounced reduction processes of solution components are observed. As a result of these processes, a stable precipitate forms on the electrodes as insoluble films, and hence the above steady state voltammetric behavior reflects electrochemistry which is surface film controlled. The outer, solution side of these films is probably porous, leading to the high interfacial capacity which is reflected by the relatively high non-Faradaic currents which characterize these voltammograms. The next section describes in detail the initial voltammetric behavior of these systems and the surface film formation on the electrodes.

## C.  The First Cycle in the Voltammetric Behavior of Ethers, Esters, and Alkyl Carbonate Solutions Containing Lithium Salts, and the Role of Atmospheric Contaminants ($O_2$, $H_2O$, $CO_2$)

The open circuit voltage (OCV) of many nonactive electrodes in nonaqueous Li salt solutions is around 2.5–3 V (Li/Li$^+$). When the potential is initially scanned anodically from the OCV up to the solvent's oxidation potentials, the currents measured are capacitive, non-Faradaic and reflect a double layer capacitance of the order of 10–20 $\mu F/cm^2$. This reflects a bare electrode surface, free of films. The initial cathodic scanning to low potentials yields voltammograms that are characterized by an irreversible wave of increasing cathodic current as the potential is scanned toward lower potentials. The onset of this wave is at around 2 V (Li/Li$^+$). Irreversible $O_2$ and $H_2O$ peaks, as well as reversible Li UPD peaks, are superimposed on this wave. It should be emphasized that even highly pure solutions treated in a glove box atmosphere contain unavoidable traces of $O_2$ and $H_2O$, which affect their initial voltammograms. Figures 7–9 present typical initial voltammograms measured with different solutions, as indicated [3,7–12,14,29,30]. The oxygen reduction peak is clearly seen around 2 V, and the water reduction peak appears around 1.5–1.1 V (Li/Li$^+$). However, a rigorous drying of these solutions and bubbling highly pure argon through them leads to the disappearance of these peaks from the voltammograms obtained with noble metal electrodes in these solutions.

In the absence of trace $H_2O$ and $O_2$ in esters, ethers and alkyl carbonate solutions, the initial voltammograms obtained with noble metals in these solutions are featureless and show the irreversible reduction wave with an onset at 1.5 V (Li/Li$^+$), which corresponds to salt anion and solvent reduction. In addition, in the absence of $H_2O$ and $O_2$, the peaks related to Li UPD shown in Figure 6 and the peaks related to the Au/Au(OH)$_3$ [or Au/Au(OH)$_{ads}$] couple described in the previous section do not appear. This is demonstrated in Figure 10.

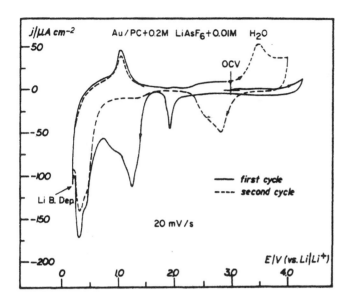

**Figure 7** Effect of $H_2O$ on cyclic voltammetry on Au in PC + 0.2 $M$ LiAsF$_6$ + 0.01 $M$ $H_2O$. Sweep rate 20 mV/s. (—) First cycle with a fresh electrode. (---) Second cycle [12]. (With copyrights from Elsevier Science Ltd.)

Another interesting result that relates to the presence of these contaminants is the onset of massive Li-metal alloy formation. As shown in Ref. 30, in uncontaminated solutions it is impossible to distinguish between the onset of Li-Au alloy formation and Li bulk deposition, while in oxygen-containing solutions a pronounced Li-Au alloying takes place at the onset of 0.2 V (Li/Li$^+$). The strong influence of the cations on the voltammetric behavior of these systems was demonstrated in the following experiments. Steady state voltammograms were obtained with gold and platinum electrodes in TBAP solutions in THF and PC, after which LiClO$_4$ was introduced. Although the concentration of the Li$^+$ was only 0.005 $M$, the voltammetric behavior changed drastically and resembled those shown in Figures 7–9 (first scan voltammograms) and 6 (steady state voltammograms). These experiments further show how strongly the surface chemistry of these systems depends on the nature of the cation. The major conclusions from the above data are summarized below:

    a.    When Li cations are present, the reduction of solvents and salt anions occurs at a potential as high as 1.5 V (Li/Li$^+$). These reduction processes are all irreversible and result in the precipitation of films that passivate these electrodes. This onset potential of the surface film for-

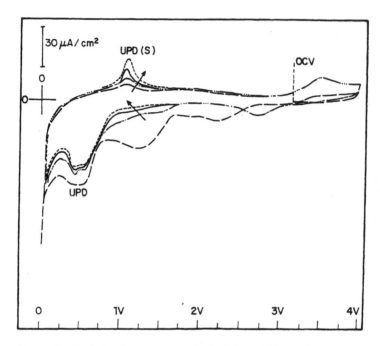

**Figure 8** Typical voltammograms obtained from lithium salt solutions of BL with gold electrodes (0.5 *M* LiAsF$_6$, 20–30 ppm H$_2$O); – – – first cycle, –····– second cycle, ——— third cycle, and ---- steady state cycle. Sweep rate = 20 mV/s [30]. (With copyright from The Electrochemical Society Inc.) [30] (With copyrights from Elsevier Science Ltd.)

mation could also be probed by UPD of metals such as Cd on noble metals in solvents such as PC [31].

b.  Hence, once noble metal electrodes are polarized to low potentials in Li salt solutions, they can no longer be considered to be bare since they become covered with stable surface films.

c.  As shown later in the chapter, the stability of these surface films depends on the nature of the solvent, the purity of the solutions, and the ratio between the solution volume and the electrode surface. The higher this ratio, the more pronounced the possible dissolution of the surface species initially formed.

d.  The presence of H$_2$O and O$_2$ traces in the solutions plays an important role in the surface chemistry developed [7–12,14]. The specific surface species formed in each solution will be discussed in the next section. However, at this stage one can state that since Li$_2$O or LiOH are formed (by O$_2$ and H$_2$O reduction, respectively) at higher potentials than those

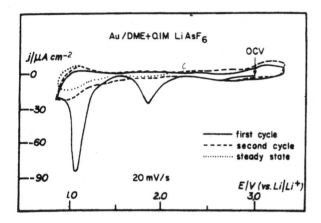

**Figure 9** Cyclic voltammetry, gold in DME + 0.1 $M$ LiAsF$_6$. Sweep rate-20 mV/s. (—) First cycle, no additives (---). (⋯) Second cycle, steady state, no difference if solution is saturated with O$_2$, once steady state is achieved [12]. (With copyrights from Elsevier Science Ltd.)

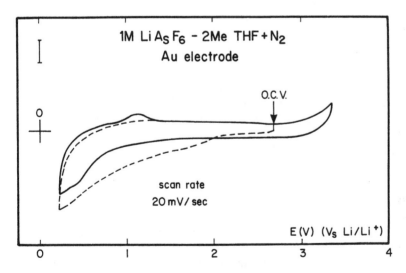

**Figure 10** Voltammogram obtained from argon-saturated 0.5 $M$ LiAsF$_6$ + 2Me-THF solution with a polycrystalline gold electrode. Sweep rate = 20 mV/s. Uncontaminated solutions: (---) first cycle; (—) second cycle [29]. (With copyrights from Elsevier Science Ltd.)

for the solvent and salt anion reduction potentials, they attenuate other reduction processes and dominate the nature of the surface film. This is clearly proven by the influence of the presence of the contaminants on processes such as Li UPD and Li alloying with noble metals. The Li UPD layer is formed in the interface between the noble metal and the surface films, and hence the surface films present significantly influence this process. As the surface films become more inorganic in nature (e.g., containing $Li_2O$), the Li UPD layer becomes more stable. This reflects on the more complicated case of Li electrodes, which will be discussed in the next chapter, and shows that species such as $Li_2O$ are better passivating agents for Li electrodes than are the organic Li salts formed by solvent reduction processes.

e.   Since the behavior of these systems is so strongly controlled by the surface chemistry, it is very important to characterize the various surface species formed on nonactive metal electrodes in the various solutions of interest. For this purpose, surface sensitive spectroscopic techniques should be applied.

The next section reviews some work on the identification of surface films formed on nonactive metal electrodes in different solutions of importance. Most of the work devoted so far to these nonaqueous systems relates to Li salt solutions. However, it can be speculated that an effect similar to that of the cations on the surface chemistry found for Li ions (compared with tetraalkyl ammonium cations) is expected for other cations of alkaline and alkaline earth metals (e.g., $Na^+$, $Ca^{2+}$, $Mg^{2+}$).

## IV.   IDENTIFICATION OF SURFACE FILMS FORMED ON NONACTIVE METAL ELECTRODES IN NONAQUEOUS SOLUTIONS USING SURFACE SENSITIVE SPECTROSCOPIC TECHNIQUES

### A.   Introduction

During the past 14 years, intensive work has been devoted to the study of surface film formation on noble metals in nonaqueous Li salt solutions using surface sensitive techniques. The major goal of this work had been to analyze the surface species formed at different potentials on nonactive electrode surfaces as a function of solution composition. The most successful tool has been ex situ and in situ FTIR spectroscopy [3,4,6,7,16,17]. However, other techniques such as Raman [32,33], XPS [12], and EDAX have also been applied. These systems have also been studied in parallel by impedance spectroscopy [34] and EQCM [35]. These measurements are complementary to the voltammetric studies described in the previous sections and to the spectroscopic studies described in this section. In the following sections, we review spectroscopic studies related to different solutions. These include DME, THF, 2Me-THF, PC, DN, BL and MF solutions, as well as mixtures of alkyl carbonates (EC-PC, EC-DEC, EC-DMC) and mixtures of alkyl carbonates with ethers. Special attention is given to the role of the electrolyte chosen and the presence of atmospheric contaminants. The salts include Li halides, $LiClO_4$, $LiAsF_6$, $LiSO_3CF_3$, $LiPF_6$, $LiBF_4$, $LiN(SO_2CF_3)_2$, and $LiC(SO_2CF_3)_3$. The contaminants include $H_2O$, $O_2$, $CO_2$ and some relevant alcohols.

## B. Alkyl Carbonate Solutions

Figures 11–15 present typical FTIR spectra obtained ex situ and in situ from nonactive electrodes polarized to different low potentials in alkyl carbonate solutions. These figures demonstrate the effect of potential, solvent, salt, and storage conditions of the electrodes at open circuit potential after film formation on their surface chemistry.

In Figure 11, Ag electrodes were polarized to several potentials in $LiA_6F_6$/PC solutions. When the electrodes were polarized to potentials below 1.5 V, FTIR spectra of stable surface films were obtained. The spectra are typical of $ROCO_2Li$ compounds [3,4,7,8] and are very similar to those of the PC electrolysis product precipitated as a Li salt $[CH_3CH(OCO_2Li)CH_2OCO_2Li]$ [3]. Figures 12 and 13 relate to PC solution with $LiClO_4$ and $LiBF_4$ salts, respectively [30]. The surface films were formed at low potentials, followed by storage at OCV. These figures reflect the initial formation of surface films comprised of the $ROCO_2Li$ species. In a $LiClO$ solution, a partial replacement of $ROCO_2Li$ by $Li_2CO_3$ is seen, while in a $LiBF_4$ solution, the $ROCO_2Li$ species disappear upon storage. The behavior of $LiAsF_6$ solutions is similar to that of $LiClO_4$ solutions, while the behavior of $LiPF_6$ is similar to that of $LiBF_4$ solutions. In the case of the $LiN(SO_2CF_3)_2$, $LiSO_3CF_3$ and $LiC(SO_2CF_3)_3$ solutions, the spectroscopic studies reflect pronounced involvement of these salt anions in the surface chemistry developed on the electrodes. The reduction processes of these salts are treated separately in Section IV.E.

Figure 14 shows the spectra of the surface films formed in mixtures of EC and PC solutions which are also typical of the $ROCO_2Li$ species obtained as the electrolysis products of these solvents precipitated as Li salts [7]. Figure 15 presents the spectra obtained from Ni electrodes polarized to low potentials in EC-DMC mixtures with $LiAsF_6$ and $LiPF_6$ solutions. The FTIR spectra in these two figures were measured in situ (external reflectance mode). In the case of $LiAsF_6$ solutions, at the onset potential of 1.5 V, the spectra reflect EC reduction product to $ROCO_2Li$ compounds. In the case of $LiPF_6$ solutions, the spectra indicate the

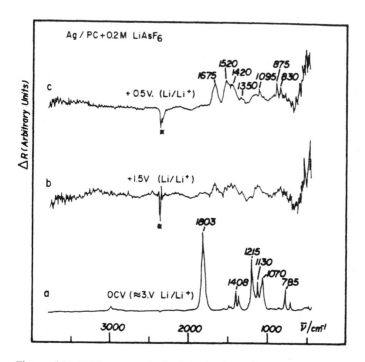

**Figure 11** FTIR spectra obtained ex situ from silver electrodes treated in PC + 0.2 $M$ LiAsF$_6$ solutions. (a) Thin layer of solvent on a silver plate. (b) Electrode removed from solution after the potential was swept from OCV (around 3 V) to 1.5 V (Li/Li$^+$). (c) Electrode removed from solution after the potential was swept from OCV to 0.5 V (Li/Li$^+$). (*) Artifact resulting from atmospheric CO$_2$ subtraction (present in the spectrometer cavity at slightly different concentrations during the sample and reference measurements) [12]. (With copyrights from Elsevier Science Ltd.)

formation of species containing P-O and P-F bonds below 2 V (Li/Li$^+$) rather than ROCO$_2$Li species, as in LiAsF$_6$ solutions [36].

The overall picture for the surface film formation in alkyl carbonate solutions obtained from the intensive surface studies (to which Figures 11–15 relate) can be summarized as follows:

1.   At the onset potential for reduction of about 1.5 V (Li/Li$^+$), alkyl carbonate–based solvents, such as PC, EC, and DMC, are reduced in the presence of Li ions on nonactive electrodes to ROCO$_2$Li compounds. The most probable compounds formed in the case of PC and EC are propylene and ethylene lithium dicarbonates, respectively. The mecha-

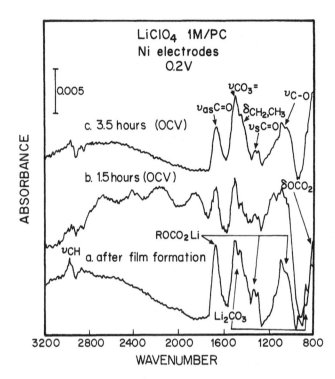

**Figure 12** Typical FTIR spectra obtained from nickel electrodes polarized to 0.2 V (Li/Li⁺) in LiClO₄ 1 *M* solution and measured ex situ, external reflectance mode (after being washed and dried). (a) The spectrum was measured after the surface films were formed. (b) The spectrum was measured after being stored for 1.5 h at OCV in solution. (c) Same as (b), storage for 3.5 h at OCV before the spectroscopic measurement [34]. (With copyright from The Electrochemical Society Inc.)

nisms for the formation of these compounds are very similar to those proposed in Scheme 4.

2. After these films are initially formed, they follow aging processes that depend on the salt and contaminants present.

3. In the presence of trace water, even at the ppm level, the $ROCO_2Li$ species undergo the following reaction [3,12,16,17]:

$$2ROCO_2Li_{(s)} + H_2O_{(sol)} \rightarrow Li_2CO_{3(s)} + 2ROH_{(sol)} + CO_{2(g)} \qquad (3)$$

Thus, the aged surface films formed in nonactive metals in alkyl carbonates contain a mixture of $ROCO_2Li$ and $Li_2CO_3$ (demonstrated in Figures 11–15).

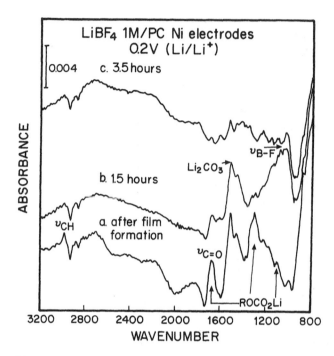

**Figure 13** Typical FTIR spectra obtained from nickel electrodes polarized to 0.2 V (Li/Li$^+$) in LiBF$_4$ 1 $M$ solution and measured ex situ, external reflectance mode (after being washed and dried). (a) The electrode was taken out of solutions soon after the surface films were formed. (b) The spectrum was measured after the electrode was stored for 1.5 h at open circuit for 1.5 h. (c) Same as (b), storage for 3.5 h at OCV before the spectroscopic measurement [34]. (With copyright from The Electrochemical Society Inc.)

4. The salt plays an important role in the aging process.
   a. In the case of LiAsF$_6$, LiClO$_4$ or Li halides (LiBr), the dominant surface species formed relate to solvent reduction. Anion reduction products such as LiCl, Li$_2$O and LiClO$_x$ (for LiClO$_4$) or LiF and Li$_x$AsF$_y$ species (for LiAsF$_6$) may also be present in small amounts on the electrode surfaces (see discussion on salt reduction processes in Section IV.E).
   b. In the case of LiSO$_3$CF$_3$, LiN(SO$_2$CF$_3$)$_2$ or LiC(SO$_2$CF$_3$)$_3$, anion reduction plays an important role in the development of the surface chemistry, as discussed in Section IV.E.
   c. In the case of LiPF$_6$ and LiBF$_4$, the salts bring about unavoidable HF contamination from hydrolysis by trace amounts of water. HF

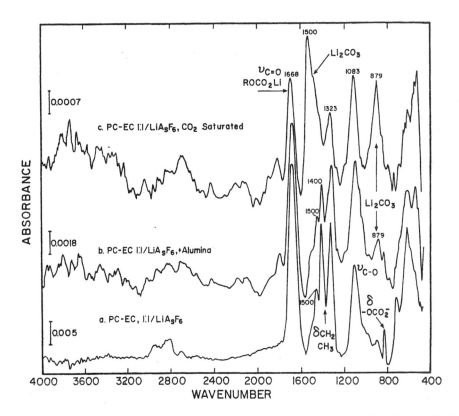

**Figure 14** FTIR spectra obtained ex situ from nickel electrodes polarized to 0 V (Li/Li⁺) in PC + EC 1:1 mixtures containing 1 $M$ LiAsF₆ (external reflectance mode). (a) Reference spectrum, uncontaminated solution. (b) The solution was stored over alumina for 24 h before the experiment. (c) CO₂ gas was bubbled through the solution for 30 min before the experiment [16]. (With copyrights from Elsevier Science Ltd.)

solubilizes the carbonates according to the following equations [34–38]:

$$ROCO_2Li_{(s)} + HF_{(sol)} \rightarrow ROCO_2H_{(sol)} + LiF_{(s)} \qquad (4)$$

$$Li_2CO_{3(s)} + 2HF_{(sol)} \rightarrow H_2CO_{3(sol)} + 2LiF_{(s)} \qquad (5)$$

Hence, aged surface films formed on nonactive electrodes at low potentials in alkyl carbonate solutions of these two salts contain LiF and other salt reduction products of the $Li_xPF_y$, $Li_xBF_y$,

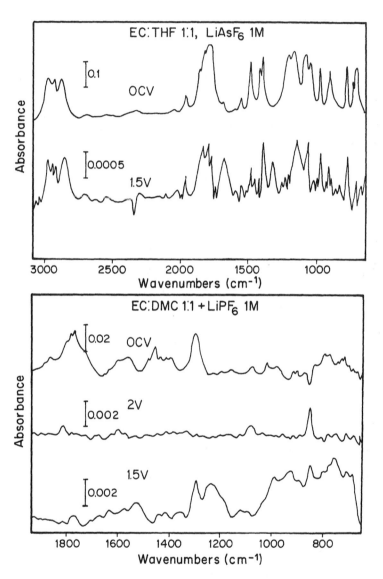

**Figure 15**  FTIR spectra measured in situ from nickel electrodes at OCV and at low potentials (versus Li/Li$^+$) in two different EC-based solutions (as indicated).

$Li_xPO_yF_z$, and $Li_xBO_yF_x$ types. The latter two species result from partial hydrolysis of the $BF_3$ or $PF_5$ species (which may also be present in these salt solutions) with trace water, followed by electrochemical reduction in the presence of $Li^+$.

d.  It should be emphasized that a critical parameter for the nature of the surface films formed on nonactive electrodes and the properties of the electrode passivation due to these surface films is the ratio between the electrode surface and the solution volume. The lower this ratio, the more pronounced is the rate of the above secondary reactions between the surface species initially formed and contaminants such as $H_2O$ and HF.

## C. Ester Solutions

We examined the representative esters, γ-butyrolactone (BL), methyl formate (MF), and methyl acetate (MA). Figures 16 and 17 show FTIR spectra measured (ex situ) from noble metal electrodes polarized to low potentials in $LiClO_4$ solutions of BL and MF, respectively [30,39]. As shown in these figures, at the onset reduction potential of around 1.3–1.2 V (Li/Li$^+$), stable surface films precipitate on the electrode surfaces. Table 1 shows the spectral analysis for the surface films formed on noble metals at low potentials in BL. The conclusion drawn from the spectroscopic study is that the major surface compound formed is the dilithiated cyclic β-keto ester, which is similar to the electrolysis product of BL in TAA salt solutions (Scheme 2).

Another possible surface constituent is $LiO(CH_2)_3COOLi$, which may be formed according to the following equations:

$$Li_2O_{(s)} + BL \xrightarrow{\text{nucleophilic attack}} LiO(CH_2)_3COOLi_{(s)} \qquad (6)$$

$$LiOH_{(s)} + BL \xrightarrow{\text{n.a.}} HO(CH_2)_3COOLi \qquad (7)$$

$$HO(CH_2)_3COOLi_{(s)} + e^- + Li^+_{(sol)} \rightarrow \tfrac{1}{2}H_{2(g)} + LiO(CH_2)_3COOLi_{(s)} \qquad (8)$$

$Li_2O$ and LiOH are formed on noble metals in these solutions by trace water reduction (in the presence of $Li^+$) below 1.5 V (Li/Li$^+$). In the case of MF, the major constituent of the surface films is lithium formate (HCOOLi) formed according to the following equations:

$$HCOOCH_3 + e^- \rightarrow HCOOCH_{3(sol)}^{\cdot-} \qquad (9)$$

$$HCOOCH_3^{\cdot-} \rightarrow HCOO^-_{(sol)} + CH_{3(sol)}^{\cdot} \qquad (10)$$

$$HCOO^-_{(sol)} + Li^+_{(sol)} \rightarrow HCOOLi_{(s)} \qquad (11)$$

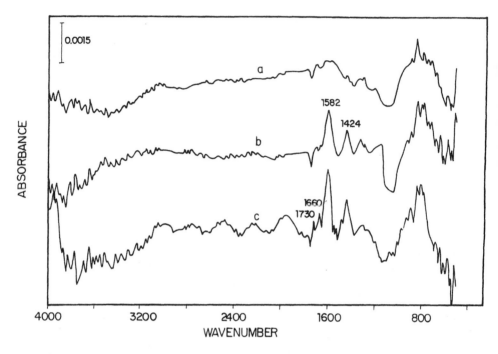

**Figure 16** Typical FTIR spectra obtained from silver electrodes treated in LiClO$_4$ 0.5 M-BL solution. The treatment included a potential sweep from OCV (3.2–3.3) to different potentials. (a) Lower potential limit was 1.35 V. (b) Lower potential limit was 1.25 V. (c) Lower potential limit was 1.0 V [30]. (With copyright from The Electrochemical Society Inc.)

In the case of methyl acetate, there is no surface film formation, since the reduction product of MA, which is lithium acetate, is soluble in the parent solvent.

## D. Ethereal and Acetal Solvents

Surface film formation on noble metal electrodes at reduction potentials was studied extensively with solutions of DME, THF, 2Me-THF, and DN. Basically, these solvents are much less reactive at low potentials than are alkyl carbonates and esters. However, in contrast to ethereal solutions of TBA$^+$ whose electrochemical window is limited cathodically by the TBA$^+$ reduction at around 0•V (Li/Li$^+$), in Li$^+$ solutions, ether reduction processes that form Li alkoxides occur at potentials below 0.5 V (Li/Li$^+$) [4]. It should be emphasized that the onset potential for surface film formation on noble metals in ethereal solutions is as high as in

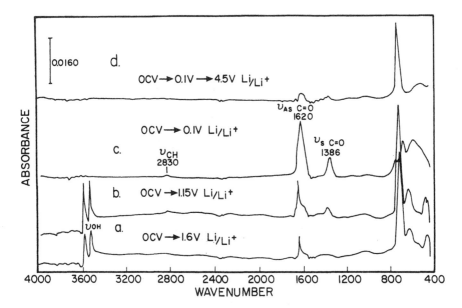

**Figure 17** FTIR spectra obtained ex situ from gold electrodes treated in MF/LiAsF$_6$ 0.5 *M* solution. The potential was swept from open circuit voltage (OCV) to predetermined potentials as indicated in spectra (a)–(d). The electrodes were held at this potential for 15 min, followed by washing (pure solvent) and drying (vacuum) [33]. (With copyrights from Langmuir.)

the more reactive ester or alkyl carbonate solvents, due to the unavoidable presence of atmospheric contaminants and possible salt anion reduction. However, a contribution of the ether reduction to the buildup of surface films is only pronounced at potentials below 0.5 V (Li/Li$^+$).

## E. Salt Anion Reduction on Noble Metal Electrodes in Nonaqueous Solutions

Most of the commonly used salts in nonaqueous systems comprise anions that are reactive and may be reduced at noble metal electrodes at low potentials. In the presence of cations such as Li$^+$, salt anion reduction may precipitate insoluble surface species on the electrodes and thus become the dominant surface film forming process. The criteria chosen here for the reactivity of the various salt anions used are the onset potential of their reduction on noble metal electrodes and to what extent their reduction on the electrodes dominates the surface film chemistry. In this respect, the commonly used salt anions can be divided into three

**Table 1** Summary of the Spectral Results Obtained from Ag and Au Electrodes Treated in Pure Li Solutions of BL, and IR Peaks of Some Reference Compounds for a Comparison (only pronounced IR bands are listed)

| Lithium butyrate CH$_3$(CH$_2$)$_2$COOLi | [β-keto ester anion structure: O Li$^+$ / C=O, C—(CH$_2$)$_3$—O$^-$ Li$^+$, lactone ring CH$_2$—CH$_2$—O] | Lithium surfaces treated in BL | Ag, Au electrodes 1.25–1 V | Ag, Au electrodes 1 V and below | Peak assignments[a] |
|---|---|---|---|---|---|
| 2693 | 2952 | 2960 | 2900–2800 Weak and broad | 2951 | $\nu_{as}$CH |
| 2874 | 2892 | 2872 | | 2871 | $\nu_s$CH |
| | | 1730 | | 1750–1730 | $\nu$COOR (ester) |
| | 1670–1650 | 1670–1650 | | 1660 | β-keto ester anion |
| 1582 | | 1585–1570 | 1590–1580 | 1590–1580 | $\nu_{as}$COOLi (carboxylate) |
| | 1520–1510 | 1510 | | 1520–1505 | β-keto ester anion |
| 1448 | | 1445–1420 | 1440–1420 | 1450–1420 | $\nu_s$COOLi (carboxylate) |
| | 1370–1330 | 1370–1330 | | | $\delta_{as}$CH3, CH$_2$ |
| | | | | 1330 | $\delta_s$CH$_3$, CH$_2$ |
| | 1259–1225 | 1249 | | 1250–1240 | C—C(C=O) β-keto ester anion |
| | 1172 | 1160 | | 1170–1150 | C—C(C=O) ester |
| | 1080–1060 | 1060 | | 1080–1060 | $\nu_{as}$COC, COLi |
| | 1040–1000 | 1032 | | | $\nu_s$COC, COLi |
| | | | | 1040–1020 | CCO (ester) |

[a] ν = stretching vibration; δ = bending vibration, as = asymmetric; s = symmetric. A peak around 1330 cm$^{-1}$ may also be related to a bending vibration of the β-keto ester anion system.
*Source:* From Ref. 30.

groups: the first group includes the halides ($Cl^-$, $Br^-$, $I^-$) which are not reducible at all (and thus are cathodically inactive). The second group includes $AsF_6^-$ and $ClO_4^-$, which are moderately reactive. The third group includes $SO_3CF_3^-$, $N(SO_2CF_3)_2^-$, and $C(SO_2CF_3)_3^-$, which are very reactive. There is a question about $BF_4^-$ and $PF_6^-$, and it is not yet clear whether the reactivity of the $PF_6^-$ anion, as reflected by the measurements described below, is intrinsic or related to contaminants such as $POF_3$ and HF, which may be unavoidably present in the $PF_6^-$ salt solutions. (The latter two compounds are the hydrolysis products of $PF_5$.)

In the case of $AsF_6^-$, there is evidence for reduction on noble metal electrodes at potentials below 1.5 V ($Li/Li^+$) [12,39,40]. Obvious products of its reduction are LiF. XPS studies of noble metal electrodes polarized to low potentials in $LiAsF_6$ solutions provide clear evidence for precipitation of LiF on the electrode surfaces at potentials below 1.5 V [12].

In the case of $LiClO_4$, there is some spectroscopic evidence of anion reduction below 1.5 V [39]. The stable surface species which may precipitate due to the $ClO_4^-$ reduction is, among others, $Li_2O$. We have no spectroscopic evidence for precipitation of stable $LiClO_x$ ($x = 1-3$) or for LiCl onto noble metals at potentials above those of Li bulk deposition. In any event, the above salt anion reduction processes *do not* dominate the overall surface film formation on nonactive electrodes at low potentials in most aprotic solvents. Thus, both anions can be considered as only moderately reactive. The onset potential for the reduction of the anions from the third group is about 2 V ($Li/Li^+$). This is clearly demonstrated in Figures 18 and 19, which show FTIR spectra measured in situ from

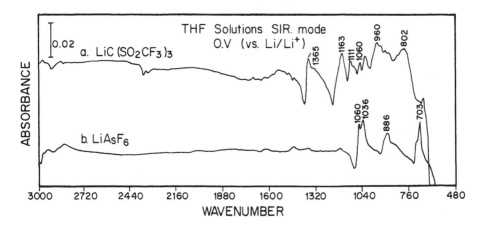

**Figure 18** FTIR spectra measured in situ from Pt deposited on NaCl (SIRFTIR mode), polarized to 0. V ($Li/Li^+$) in THF 1 $M$ solutions of the two salts, as indicated [42]. (With copyrights from Elsevier Science Ltd.)

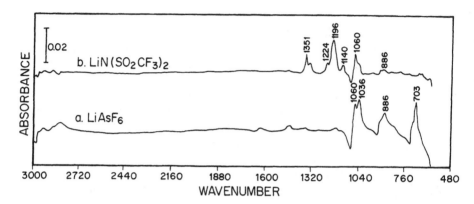

**Figure 19** FTIR spectra measured in situ from Pt deposited on NaCl (SIRFTIR mode), polarized to 0. V (Li/Li$^+$) in THF 1 $M$ solutions of the two salts, as indicated [42]. (With copyrights from Elsevier Science Ltd.)

noble metal electrodes biased to low potentials in ethereal solutions [11]. These spectroscopic studies clearly prove that below 2 V (Li/Li$^+$) the [SO$_2$-CF$_3$] group is reduced on noble metal electrodes in the presence of Li ions. Reduction products containing lithium salts of the Li$_x$SO$_y$ or Li$_x$SO$_y$CF$_z$ types precipitate on the electrode surface. Scheme 6 proposes possible surface reactions of these salts based on the spectral studies [12].

Since the reduction of these salts occurs at very high potentials, their reduction products contribute considerably to the buildup of surface films on noble

a) $LiN(SO_2CF_3)_2 + ne^- + nLi^+ \rightarrow Li_3N + Li_2S_2O_4 + LiF + C_2F_xLi_y$

b) $LiN(SO_2CF_3)_2 + 2e^- + 2Li^+ \rightarrow Li_2NSO_2\,CF_3 + CF_3SO_2Li$

c) $Li_2S_2O_4 + 6e^- + 6Li^+ \rightarrow 2Li_2S + 4Li_2O$

d) $2LiC(SO_2CF_3)_3 + 8e^- + 8Li^+ \rightarrow Li_2C_2F_4 + 2LiF + 2Li_2C\,(SO_2CF_3)_2 + Li_2S_2O_4$

e) $LiC(SO_2CF_3)_3 + 2e^- + 2Li^+ \rightarrow Li_2C(SO_2CF_3)_2 + LiSO_2CF_3$, etc.

f) $Li_2S_2O_4 + 4e^- + 4Li^+ \rightarrow Li_2SO_3 + Li_2S + Li_2O$

**Scheme 6** Possible $LiN(SO_2CF_3)_2$ and $LiC(SO_2CF_3)_3$ reduction patterns on Li.

metal electrodes polarized cathodically. This is also true for their solutions with more active solvents, such as esters and alkyl carbonates.

The case of the $PF_6^-$ and $BF_4^-$ anions is more complicated. As mentioned in Section IV.B, the presence of these anions in alkyl carbonate solvents strongly influences the surface chemistry developed on noble electrodes at low potentials. The organic- and inorganic carbonate–based surface films developed on the electrode surfaces at low potentials in alkyl carbonates dissolve in the presence of both $BF_4^-$ and $PF_6^-$ anions. However, this effect is attributed to HF contamination, as explained in Section IV.B. From electrochemical quartz microbalance studies of $LiPF_6$/PC solutions, there is evidence that the $PF_6^-$ anion is reduced at around 2 V versus $Li/Li^+$ [35]. FTIR measurements seem to indicate the formation of surface species containing P—F and/or P—O bonds. There is no similar information for $LiBF_4$ solutions. However, XPS studies of noble metals polarized cathodically in $LiBF_4$ solutions below 1 V clearly indicate the formation of surface LiF [12]. It should also be noted that ethereal solutions of $LiBF_4$ or $LiPF_6$ tend to polymerize, probably due to the unavoidable HF, $BF_3$, or $PF_5$ contamination.

In conclusion, there is evidence for anion-related surface reactions on noble metals at low potentials in solutions of both $LiBF_4$ and $LiPF_6$. However, it is difficult to decide whether these surface reactions result from intrinsic reactivity of the anions at low potentials or whether the surface chemistry is mostly influenced by reactive contaminants such as HF, $PF_5$, $BF_3$, or $POF_3$, etc.

## F. Reduction of Atmospheric Contaminants on Nonactive Metal Electrodes in Nonaqueous Solutions

In reviewing the intrinsic electrochemical behavior of nonaqueous systems, it is important to describe reactions of the most common and unavoidable contaminants. Some contaminants may be introduced by the salts (e.g., HF in solutions of the $MF_x^-$ salts; M = P, B, As, etc.). Other possible examples are alcohols, which can contaminate esters, ethers, or alkyl carbonates. We examined the possible effect of alcoholic contaminants such as $CH_3OH$ in MF and 1,2-propyleneglycol at concentrations of hundreds of ppm in PC solutions. It appears that the commonly used ester or alkyl carbonate solvents are sufficiently reactive (as described above), and so their intrinsic reactivity dominates the surface chemistry if the concentration of the alcoholic contaminant is at the ppm level. We have no similar comprehensive data for ethereal solutions. However, the most important contaminants that should be dealt with in this section, and which are common to all of these solutions, are the atmospheric ones that include $O_2$, $H_2O$, and $CO_2$. The reduction of these species depends on the electrode material, the solvent used, and their concentration, although the cation plays the most important role. When the electrolyte is a tetraalkyl ammonium salt, the reduction products of $H_2O$, $O_2$ or $CO_2$ are soluble. As expected, reduction of water produces $OH^-$ and hydrogen gas. The hydroxide ion reacts irreversibly with the tetraalkyl ammonium cations, as described above, to form the trialkyl amine, alkene and water. Thus, the voltammetric behavior of water-contaminated nonaqueous TAA salt solutions shows a more positive reduction wave at potentials between 1–1.5 V versus the $Li/Li^+$ reference potential (depending on the concentration of water) [3,7–12]. $O_2$ reduction produces $O_2^{\div}$, which can be oxidized back to $O_2$ or further reduced to $O_2^{2-}$ at lower potential, which can then react nucleophilically with electrophilic solvents (such as PC [12]). $CO_2$ is reduced in TBA solutions of aprotic solvents, such as acetonitrile, at potentials below $-1.3$ V (SCE), which correspond to potentials below 1.5 V versus $Li/Li^+$ [43]. Surface studies based on in situ FTIR indicated the formation of CO and formate anion [15]. There are reports on possible formation of oxalate $(COO^-)_2$ [44] and carbonate $(CO_3^{2-})$ [45] anions as well.

In the presence of Li ions, the picture changes completely. All the reduction processes of the above atmospheric components become irreversible and produce insoluble species that precipitate on the nonactive metal electrodes as surface films. In Section III we described typical cyclic voltammograms measured with noble metal in Li salt solutions containing trace $O_2$ and $H_2O$. We assume that the irreversible $O_2$ reduction peak that appears in these voltammograms around 2 V reflects the formation of $LiO_2$. However, we do not have solid spectral evidence for this claim. Our studies indicated that at lower potentials $Li_2O$ becomes the final product of $O_2$ reduction [12,30]. It precipitates as surface films and strongly influences the UPD of lithium on noble metal electrodes, as discussed above.

In the presence of $Li^+$, $H_2O$ is reduced to LiOH, as evidenced by spectral studies [30–46]. We have spectral evidence that LiOH is not the final stable product of $H_2O$ reduction at low potentials and that it is further reduced to $Li_2O$. It is quite possible that the LiOH surface films detected by some spectral studies on noble metal electrodes polarized in Li salt solutions result from hydration of the $Li_2O$ initially formed. Hence, in the case of $H_2O$ contamination, the water is reduced at an onset potential of 1.5 V ($Li/Li^+$) to LiOH, which is further reduced to $Li_2O$ at lower potentials. The surface film thus formed may contain $Li_2O$ in an inner layer and LiOH at different stages of hydration in outer layers. These surface films passivate the electrodes, and thus water reduction becomes kinetically limited by the rate of diffusion of water through the surface films.

It should be emphasized that all possible surface films formed on electrodes in Li salt solutions of polar aprotic solvents are permeable to water because all the above-described surface species are hygroscopic. Thus, water hydrates any surface species formed in these systems, diffuses to the metal surface, and may be reduced close to the electrode surface film interface at low potentials. Consequently, despite the apparent passivation of nonactive metal electrodes polarized to low potential in water-contaminated nonaqueous Li salt solutions, continuous water reduction is unavoidable. Its rate depends, among other things, on the concentration of water. The hydrogen produced by water reduction may remain partially adsorbed on the electrode. This is clearly seen with platinum electrodes. The voltammograms of platinum electrodes in water-contaminated nonaqueous Li salt solutions have the typical broad water reduction peak around 1.6–1.2 V (Li/Li$^+$) (which is pronounced initially and is small in steady state voltammograms, as shown in Section III.B) and a corresponding hydrogen oxidation peak around 2.2 V. This process reforms water by reaction of the H$^+$ thus formed with the Li$_2$O-LiOH species on the electrode surface. It should be noted that H$^+$ formed

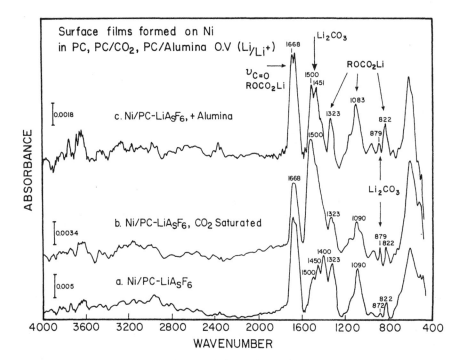

**Figure 20** FTIR spectra obtained ex situ from nickel electrodes polarized to 0 V (Li/Li$^+$) in PC + LiAsF$_6$ solutions (external reflectance mode). The electrodes were held at this potential for 15 min followed by washing (pure solvent) and drying. (a) PC + 1 $M$ LiAsF$_6$ solution, no contaminants (the reference spectrum). (b) CO$_2$ gas was bubbled for half an hour through this solution before the measurement. (c) The solution was stored over neutral Al$_2$O$_3$ for 24 h before the experiment [16]. (With copyrights from Elsevier Science Ltd.)

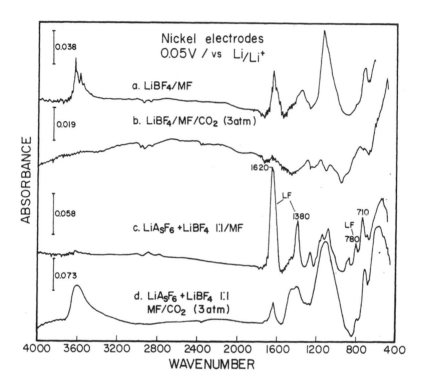

**Figure 21** FTIR spectra obtained ex situ from nickel electrodes polarized to 0.05 V (Li/Li$^+$) in various LiBF$_4$/MF solutions. (a) LiBF$_4$ 1 $M$/MF. (b) LiBF$_4$ 1 $M$/MF/CO$_2$ (3 atm). (c) LiAsF$_6$ 0.5 $M$/LiBF$_4$ 0.5 $M$/MF. (d) LiAsF$_6$ 0.5 $M$/LiBF$_4$ 0.5 $M$/MF/CO$_2$ (3 atm). The electrodes were held at this potential for 30 min followed by washing (pure solvent) and drying (vacuum) [33]. (With copyrights from Langmuir.)

by the hydrogen oxidation might attack ethereal molecules and induce polymerization. Such a phenomenon indeed occurs in 1,3-dioxolane, which partially polymerizes during cyclic voltammetric experiments in its water-contaminated solutions [47].

In the case of CO$_2$ contamination, we have strong evidence that its reduction on noble metal electrodes in nonaqueous systems in the presence of Li ions (and the absence of water) forms Li$_2$CO$_3$ and CO [17]. Figures 20 and 21 show typical FTIR spectra obtained from noble metal electrodes polarized to low potentials in CO$_2$-saturated nonaqueous Li salt solutions and provide clear evidence for Li$_2$CO$_3$ formation as the major surface species that is precipitated [15,39]. The CO$_2$ reduction mechanism for the reaction appears in the literature [43] and is described in the following equations:

$$CO_{2(sol)} + e^- + Li \rightarrow CO_2^{\cdot-} Li^+ \tag{12}$$

$$CO_{2(sol)} + CO_2^{\cdot-} Li^+ \rightarrow OCOCO_2Li^{\cdot} \tag{13}$$

$$OCOCO_2Li^{\cdot} + e^- + Li^+_{(sol)} \rightarrow Li_2CO_{3(s)} + CO_{(g)} \tag{14}$$

## V. REMARKS ON THE ELECTROCHEMICAL BEHAVIOR OF OTHER IMPORTANT NONAQUEOUS SOLVENTS

### A. Acetonitrile

It is well known that ACN reacts with active metals (Li, Ca) to form polymers [48]. These polymers are products of condensation reactions in which $ACN^{\cdot-}$ radical anions are formed by the electron transfer from the active metal and attack, nucleophilically, more solvent molecules. Species such as $CH_3C{=}N(CH_3)C{=}N^{\cdot-}$ are probably intermediates in this polymerization. ACN does not react on noble metal electrodes in the same way as with active metals. For instance, well-resolved Li UPD peaks characterize the voltammograms of noble metal electrodes in ACN/Li salt solutions. This reflects a stability of the Li ad-layers that are formed at potentials above Li deposition potentials. Hence, the cathodic limit of noble metal electrodes in ACN solutions is the cation reduction process (either TAA or active metal cations). As discussed in the previous sections, with TAA-based solutions it is possible that the electrode surfaces remain bare. When the cations are metallic (e.g., $Li^+$), it is expected that the electrode surfaces become covered with surface films originating from atmospheric contaminants reduction if the electrodes are polarized below 1.5 V ($Li/Li^+$). As Mann found [13], in the presence of Na salts the polarization of metal electrodes in ACN solutions to sodium deposition potentials leads to solvent decomposition, with evolution of $H_2$, $CH_4$ and sodium cyanide (due to reaction with metallic sodium).

### B. Dimethyl Sulfoxide

Dimethyl sulfoxide is an important solvent in nonaqueous electrochemistry due to its high polarity (dielectric constant of 47), its high donor number (29.8), and a relatively wide electrochemical window. The limiting cathodic voltages in which this solvent can be used depend on the cation used (as expected from the discussion on the cation effects on the reduction processes of the above nonaqueous solvents). Using salts of alkali metals (Li, Na, K), the cathodic limit obtained was around $-1.8 \rightarrow -2$ V versus SCE [49], whereas with tetrabutyl ammonium, the cathodic limit was as low as $-2.7 \rightarrow -3$ V versus SCE [49]. There is evidence that in the presence of Na ions, DMSO reduction produces $CH_4$ and $H_2$ on platinum electrodes [50]. It is expected that, similar to other nonaqueous solvents, in the presence of metallic cations the electrodes are passivated by precipitation of the reduction products of DMSO on the electrode, which may include sulfides and oxides of the cation. In addition, insoluble reduction products of trace $H_2O$, $O_2$, and $CO_2$ precipitate on the electrode, as discussed in previous sections. In the case of TAA salts, the low cathodic limit reported [49] might indicate that $TBA^+$ reduction becomes the cathodic limiting reaction of these solutions.

## C.  *N,N*-Dimethylformamide

Dimethylformamide is an important solvent for nonaqueous electrochemical systems due to its high polarity (dielectric constant = 37), its high donor number (26.2), and its relatively wide electrochemical window. The cathodic limiting reactions of this solvent on noble metal electrodes are not yet clear. However, the cathodic limiting voltages are reported to be around $-1.6$ V and $-2.5$ V (SCE) [13] for sodium and tetrabutyl ammonium salt solutions, respectively. With noble metal electrodes, the anodic limiting potentials are around 1.5 V (SCE) [13]. Hence, from many points of view, the behavior of DMF solutions in terms of cathodic limiting potentials, polarity, and compatibility with organic and inorganic substances is similar to that of DMSO.

## D.  Oxyhalide Solvents

Oxyhalide solvents include solvents such as thionylchloride ($SOCl_2$), sulfurile chloride ($SO_2Cl_2$), and $POCl_3$, which are also utilized as liquid cathodes in primary Li batteries. The basic reduction process of thionyl chloride may be illustrated as

$$2SOCl_2 + 4e^- \rightarrow SO_2 + S + 4Cl^- \qquad (15)$$

This occurs at potentials around 3.5 V versus $Li/Li^+$. The cathodic limit of $POCl_3$ is around 2.5 V ($Li/Li^+$), and that of sulfurile chloride is around 3.9 V ($Li/Li^+$) [51].

Table 2 lists oxyhalide solvents that may all be utilized as liquid cathodes for Li, Ca, or Mg primary batteries, along with some of their physical properties.

Although these nonaqueous solvents are highly polar and thus may be attractive as media for nonaqueous electrochemistry, their electrochemical window is very narrow, since their cathodic potential limit is high (2.5–4 V versus $Li/Li^+$). Hence, their major importance remains as cathodic active materials in primary, high energy density batteries based on active metal (Li, Mg, Ca) anodes.

**Table 2** Representative List of Oxyhalide and Thiohalide Solvents and Some of Their Physical Properties (interesting as liquid cathodes and solvents for primary Li, Ca and Mg batteries)

| | Melting point (°C) | Boiling point (°C) | Gravimetric density (g cm$^{-3}$) | Equivalent weight |
|---|---|---|---|---|
| $COCl_2$ | −127.8 | 7.56 | 1.392 | 49.46 |
| $COBr_2$ | — | 64.5 | 2.44 | 93.92 |
| $CSCl_2$ | — | 73.5 | 1.51 | 57.50 |
| $Cl(CO)_2Cl$ | −12 | 64 | 1.49 | 63.67 |
| $CH_3CClO$ | −112 | 51–2 | 1.11 | 78.50 |
| $NOBr$ | −55.5 | −2 | >1.0 | 109.92 |
| $NOCl$ | −64.5 | −5.5 | 2.99 | 65.47 |
| $NOF$ | −134 | −56 | — | — |
| $NO_2Cl$ | >−31 | 5 | 2.57 | 81.47 |
| $NO_2F$ | −139 | −63.5 | — | — |
| $POCl_3$ | 2 | 105.3 | 1.675 | 51.12 |
| $POBr_3$ | 56 | 193 | 2.822 | 95.58 |
| $PSCl_3$ | −35 | 125 | 1.635 | 56.47 |
| $PSBrCl_2$ | −30 | d.150 | 2.12 | 71.29 |
| $PSBr_3$ | 38 | d.175 | 2.85 | 100.93 |
| $SOCl_2$ | −105 | 75 | 1.655 | 59.49 |
| $SOClF$ | −139.5 | 12.2 | (2.293) | 51.26 |
| $SOF_2$ | −110 | −30 | 2.93 | 43.04 |
| $SOBr_2$ | −52 | 138 | 2.68 | 103.95 |
| $SO_2Cl_2$ | −54.2 | 69.1 | 1.667 | 67.49 |
| $S_2O_5Cl_2$ | −38 | 140 | 1.82 | 107.53 |
| $C_6H_5SO_2Cl$ | 14.5 | 246 | 1.378 | 176.62 |

*Source*: From Ref. 51.

## E. Sulfur Dioxide (SO₂)

$SO_2$ is used both as a liquid cathode for Li batteries [52] and as a nonaqueous solvent for the study of electrochemical reactions. Bard et al. made an extensive study of the electrochemical reactions in $SO_2$-based solutions at low temperature and under critical conditions [53,54]. The basic cathodic reaction of this solvent is

$$SO_2 + 2e^- \rightleftharpoons S_2O_4^{2-} \tag{16}$$

and its limiting cathodic potential is around 3 V versus Li/Li$^+$ ($\approx$ 0 versus SCE). It should be emphasized that while the cathodic limit of this solvent is relatively high (3 V versus Li/Li$^+$, which is 2–3 V above the cathodic limit of the commonly used organic aprotic solvents), its electrochemical window is wide (3 V), since its oxidation potential is very high (>6 V versus Li/Li$^+$).

## F.  Amine Solvents

In the group of amine solvents we may include ammonia ($NH_3$) and organic amines of the general form $RNH_2$, $R_2NH$, and $R_3N$, where R = alkyl or aryl group. The use of ammonia requires either pressurized systems or low temperatures, as its liquid range at atmospheric pressure is $-77.7$ to $-33.4°C$. Bard et al. [55,56] performed extensive work on electrochemistry in liquid ammonia under critical and subcritical conditions. The cathodic limiting reaction in ammonia may be the dissolution of electrons at potentials around $-2.3$ V versus mercury pool electrode, which may correspond to $\approx 1$ V versus Li/Li$^+$. However, the cathodic limit depends strongly on the salt used and the nature of the electrode. A low cathodic limit in these solutions may be achieved only when the relevant cations are reduced at low potentials (alkali metals or $R_4N^+$) and the electrode possesses high overvoltage for the liberation of hydrogen. Otherwise, hydrogen evolution may become the limiting cathodic reaction.

With TAA salts of small alkyl groups (e.g., ethyl, methyl), cation reduction is usually the limiting cathodic reaction. The anodic limiting reaction for ammonium ions is their oxidation to nitrogen and protons. It should be emphasized that atmospheric contaminants are supposed to influence the above cathodic and anodic limits of liquid ammonia, as they do for the other nonaqueous systems discussed in the previous sections.

Based on the work of Bard et al. [55,56], it appears that one can expect an accessible electrochemical window of about 2 V for solutions based on liquid ammonia. The electrochemical behavior of nonaqueous systems based on organic amines is somewhat similar to that described for liquid ammonia, except for the temperature and pressure range of operation, which, for most of the practically used amines, is an ambient one. The limiting cathode reaction of amines is also either electron dissolution or cation reduction, depending on the electrolyte used. For instance, when using a supporting electrolyte such as TBAI, the anion is not reducible and the cation's reduction potential is very low (0. V versus Li/Li$^+$, as discussed above). Indeed, the cathodic reaction in this case is electron dissolution, and thus, blue electron solutions are obtained at sufficiently low potential. For primary and secondary amines, the cathodic limit may also be limited by hydrogen evolution for electrodes in which the overvoltage for its liberation is small.

Table 3 provides typical examples, taken from Mann, for the cathodic potential limits measured in ethylene diamine and pyridine, and thus presents good examples for the expected lowest potentials at which these solutions can be utilized in nonaqueous electrochemistry.

**Table 3** Cathodic Limits of Ethylene Diamine (EDA) and Pyridine Using Dropping Mercury Electrode

| Solvent | Supporting electrolyte | Concentration ($M$) | Reference electrode | Limiting low potential (V) |
|---|---|---|---|---|
| EDA | LiCl | 0.25 | Zn-Hg/ZnCl$_2$ | −1.1 |
| EDA | TEAN | 0.1 | NCE | −2.74 |
| EDA | TEAP | 0.1 | NCE | −1.80 |
| EDA | TEAC | 0.1 | NCE | −2.12 |
| EDA | TMAB | 0.01 | NCE | −1.79 |
| EDA | TPAP | 0.01 | NCE | −1.98 |
| EDA | TPAI | 0.01 | NCE | −1.68 |
| EDA | TBAI | 0.01 | NCE | −2.35 |
| EDA | Et$_i$PrNI | 0.01 | NCE | −1.98 |
| Pyridine | TBAI | — | Ag-AgNO$_3$ | −2.3 |
| Pyridine | LiCl | 0.5 | Hg pool | −1.53 |
| Pyridine | LiClO$_4$ | 0.1 | Hg pool | −1.77 |
| Pyridine | LiClO$_4$ | 0.5 | Hg pool | −1.65 |
| Pyridine | KSCN | 0.5 | Hg pool | −1.53 |
| Pyridine | NaI | 0.1 | Hg pool | −1.18 |
| Pyridine | NaBPh$_4$ | — | Hg pool | −1.57 |

NCE = normal (1M KCl) calomel electrode, TMA = tetramethyl ammonium, TEA = tetraethyl ammonium, TPA = tetrapropyl ammonium, TBA = tetrabutylammonium, N = nitrate, B = bromide, C = chloride, I = iodide, P = perchlorate.
*Source*: From Ref. 13.

Our experience with tributylamine shows that its cathodic potential limit may be as low as 0 V Li/Li$^+$. The cathodic limiting reaction may be the reduction of the cation (for alkaline metals and TAA salts). We have evidence that tributylamine reacts with lithium to form amides (R$_x$NLi$_y$, $0 < x < 2$, $1 < y < 3$) [57], and thus it is expected that as the deposition potentials of the alkaline metals are reached, trialkylamine solvents will react with the deposits. The anodic limit of most of the trialkylamines, as well as of secondary amines, is in the 3.5–4 V range versus Li/Li$^+$. The reaction is probably the formation of tetraalkyl ammonium cations, protons, and nitrogen. Hence, the electrochemical limit of amines may range between 2–4 V (higher for the tertiary amines).

## G. Protic Nonaqueous Solvents

Extensive work is currently being carried out with protic nonaqueous systems that include acids such as acetic acid [58], formic acid [59], and methane sulfonic acid [60], as well as alcohols, such as ethanol and methanol [61,62]. Such solvents may be used for reactions in which both the reagents and the products are not reactive with protic hydrogens. The advantage of using such solvents lies, first of all, in their high polarity, which enables highly conductive solutions to be obtained. These can dissolve a large variety of organic substances that are insoluble in water. Another important advantage relates to their easy purification, handling, and relatively low cost.

Provided that supporting electrolytes of wide electrochemical windows are used (e.g., TBA salts), the cathodic limit for acetic acid reduction is around $-1.7$ V versus SCE [63], and that of formic and methane sulfonic acids is around $-0.8$ V versus SCE [64]. The cathodic limiting reaction is hydrogen evolution, thus forming the acid anion as the coproduct. The apparent electrochemical window of acetic acid is about 4 V [63], whereas that of formic acid is around 1 V [49]. For methanol and ethanol, there are reports on limiting cathodic potentials around $-2$ V versus mercury pool electrode [65], and their accessible electrochemical window is around 2 V. The cathodic limiting reactions are probably hydrogen evolution and an alkoxide formation.

## H. Miscellaneous

In this section we briefly review data on a few more solvents which are not widely used in nonaqueous electrochemistry. Nitromethane ($NO_2CH_3$) is regarded as an interesting solvent due to its high dielectric constant [37] and wide liquid range ($-29 \rightarrow 101°C$). The accessible potential window for this solvent may exceed 4.5 V with salts such as $LiClO_4$ [66]. As discussed previously, noble metal electrodes become passivated in the presence of metallic cations at low potentials, and, thus, massive solvent reduction is prevented. Hence, with salts such as $Mg(ClO_4)_2$ or $LiClO_4$, the cathodic limiting potentials may be as low as $-2.4$ V $\rightarrow -2.2$ V versus Ag/AgCl electrode [66]. In the absence of possible electrode passivation, the cathodic limiting potential for TBAP solutions is only $-1$ V (Ag/AgCl), due to the reaction which forms $NO_2$ and $CH_2$—$CH$=$NO^-$ [67].

Another solvent that should be mentioned is methylene chloride ($CH_2Cl_2$). This solvent is important for electrochemical studies of nonpolar substances and for studies at low temperatures (accessible temperature range $-97° \rightarrow 40°C$ at ambient pressure). An advantage to this solvent lies in the stability of its radical ions. The cathodic and anodic voltage limits for this solvent with TBAP as the electrolyte and noble metal electrodes (Pt, Au) are $\pm1.7$–$1.8$ V versus SCE [68].

The last solvent to be mentioned is acetone ($CH_3COCH_3$), which may be advantageous for electrochemical studies of a wide variety of nonpolar organic and inorganic substances. Its potential range is wide ($-2.4$ V $\rightarrow$ 1.6 V versus SCE for TBAP solutions with noble metal electrodes) [69].

## VI. CATHODIC BEHAVIOR OF CARBON ELECTRODES IN NONAQUEOUS SOLVENTS

### A. General

Different types of carbon are often used as electrode materials in nonaqueous systems. Among the many types of carbon available, we mention glassy carbon; graphite; graphitic carbons (e.g., carbon fibers of layered structure); disordered, high surface area, soft (graphitizable) or hard (nongraphitizable) carbons; fullerenes; ($C_{60}$, etc.), and diamonds. In reviewing these carbonaceous materials as electrodes for analytical or synthetic purposes, the following points should be taken into account.

a. The carbons are usually covered by surface functional groups which can be reduced at low potential or oxidized at high potentials and thus affect the voltammetric response of the electrochemical system studied [70].

b. Layered carbons undergo intercalation with solution species at either low or high potentials. Cations, anions and neutral molecules may undergo intercalation with graphite and graphitic carbons [28,71].

c. Disordered, high surface area carbon may also insert cations, thus forming carbon insertion compounds. Such insertion processes can be either intercalation-like, in which the inserted species enter the bulk materials, or adsorption-like, in which the inserted species are bound to the carbon's surface [72–75].

d. The nature of the intercalation/insertion processes depends very strongly on the solution's composition, especially on the ions present.

e. As for the case of nonactive metal electrodes, when carbons are polarized to low potentials, solvents, salt anions, and contaminants (e.g., $H_2O$, $O_2$, and $CO_2$) may be reduced on the electrode surface [76–81].

These reduction products may dissolve in the bulk solution or may precipitate to produce surface films. (However, this depends strongly on the type of cation in the solution.) An important issue that relates to the basic behavior of carbons in nonaqueous systems is their use as insertion anodes in high energy density, rechargeable batteries [82]. This subject is an extensive one and includes surface and material science consideration of carbonaceous material, and thus is beyond the scope of this chapter. However, some aspects of insertion carbon electrodes for batteries with respect to the above five points are discussed later in this section.

## B. Surface Chemistry of Carbon Electrodes Polarized to Low Potential in Polar Aprotic Systems and Related Potentiodynamic Behavior

Similar to the behavior of nonactive metal electrodes described above, when carbon electrodes are polarized to low potentials in nonaqueous systems, all solution components may be reduced (including solvent, cation, anion, and atmospheric contaminants). When the cations are tetraalkyl ammonium ions, these reduction processes may form products of considerable stability that dissolve in the solution. In the case of alkali cations, solution reduction processes may produce insoluble salts that precipitate on the carbon and form surface films. Surface film formation on both carbons and nonactive metal electrodes in nonaqueous solutions containing metal salts other than lithium has not been investigated yet. However, for the case of lithium salts in nonaqueous solvents, the surface chemistry developed on carbonaceous electrodes was rigorously investigated because of the implications for their use as anodes in lithium ion batteries. We speculate that similar surface chemistry may be developed on carbons (as well as on nonactive metals) in nonaqueous systems at low potentials in the presence of $Na^+$, $K^+$, or $Mg^{2+}$, as in the case of Li salt solutions. The surface chemistry developed on graphite electrodes was extensively studied in the following systems:

*Solvents*
Alkyl carbonates: PC [76,79,83], EC [83,84–86,77,78,80,81], DMC [77,78,83–86,80,81], DEC [77,80,81,83–85], EMC [87], PMC [88]
Ethers: DME [76], THF [76,83], 2Me-THF [73,83], 1,3-dioxolane [76,83]
Esters: MF[76,81,83], γ-BL[76,83,84]

*Salts*
$LiAsF_6$, $LiClO_4$, $LiBF_4$, $LiPF_6$, $LiC(SO_2CF_3)_3$, $LiN(SO_2CF_3)_2$ [76,83–89]

*Additives*
$CO_2$ [76,83–86], $H_2O$ [84], $SO_2$ [89]

Typical cyclic voltammograms of graphite electrodes in $LiAsF_6$ with ethereal or alkyl carbonate solutions are characterized by a broad irreversible peak appearing around 1.3 V versus (Li/Li$^+$) and relating to the reduction of $AsF_6^-$ [18]. Below this potential, a gradual solvent reduction takes place, appearing as an irreversible wave of increasing currents, as the potential is lowered [18]. Figure 22 shows a typical chronopotentiogram obtained from a pristine graphite electrode polarized galvanostatically to low potentials in a polar aprotic Li salt solution, as indicated [76,83–86]. Figure 23 shows, for comparison, a chronopotentio-

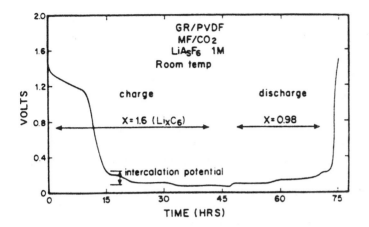

**Figure 22** Typical chronopotentiometric profile of a graphite (4% PVDF) electrode and Li counter electrode loaded with methyl formate-LiAsF$_6$ 1 $M$ solution under CO$_2$ (6 atm), galvanostatic operation C/33 h, 0.5 mA/cm$^2$ [83]. (With copyright from The Electrochemical Society Inc.)

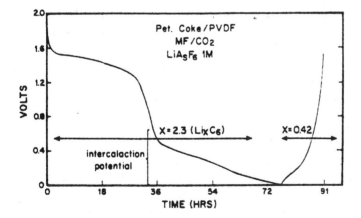

**Figure 23** Typical chronopotentiometric profile of a petroleum coke (4% PVDF) electrode. The first charge-discharge cycle of a cell with the petroleum coke electrode and Li counter electrode loaded with MF-LiAsF$_6$ 1 $M$ (6 atm, CO$_2$), galvanostatic operation, C/33 h, 0.5 mA/cm$^2$ [83]. (With copyright from The Electrochemical Society Inc.)

gram obtained from a pristine disordered carbon electrode during intercalation with lithium. The chronopotentiograms in Figures 22 and 23 are characterized by plateaus at potentials below 1.5 V (Li/Li$^+$). After these initial plateaus, the potential drop corresponds to the intercalation process with lithium (see next section). With disordered carbons, these curves reflect Li intercalation over a wide potential domain (1.5–0 V) versus Li/Li$^+$, whereas with graphitic carbons Li intercalation occurs between 0.3 and 0. V versus Li/Li$^+$. These first plateaus reflect massive reduction of solution species, both of solvent molecules and salt anions. The products precipitate as surface films that passivate the electrodes. As these surface films reach a certain thickness and the electrode becomes passivated, the potential falls toward the next intensive process available, which is Li intercalation. It should be noted that, in contrast to nonactive metal electrodes, the surface of graphite electrodes might not be stable. In many solutions the formation of surface films at potentials between 1.5 and 0.8 V versus (Li/Li$^+$) is accompanied by microexfoliation and cracking of the graphite. Gas formation and/or intercalation processes cause separation between graphene planes due to the formation of some internal pressure in the graphite crystal [18]. This leads to an increase in the surface area of the graphite as the reduction of the solution species proceeds, and thus stable passivating films cannot be formed unless a massive reduction of solution species takes place. Such situations occur in solvents such as PC [76,83,84], MF [76,83], $\gamma$-BL [84], and some ethers [76,83,84]. It should be noted, however, that this situation can be attenuated by incorporation of additives whose reduction occurs at higher potentials than those of solvent/salt anion reduction, and their reduction products block the carbon surface before massive solution reduction takes place (examples are $CO_2$ [76] and $SO_2$ [89]). Table 4 summarizes the reduction products of various solvents, salt anions, and additives on carbon electrodes and the approximate onset potentials of their reduction.

In conclusion, the above surface film formation processes are expected to occur with all types of the carbons mentioned (including doped diamonds and fullerenes) in nonaqueous solvents containing metal ion salts. Hence, when carbon electrodes are utilized for electroanalysis or electrosynthesis in such solutions at low potentials, they should be considered as modified electrodes covered with surface films that are, at least partially, electronic insulators (but may be ion conductors).

It should be noted that the reactions between the solution species and the carbons are not the only ones that determine the electrode surface chemistry. The carbon's surface usually contains functional groups, which are formed as a result of reactions of the carbon with oxygen and water during its manufacture. Thus, the surface may contain, for example, OH, COH, COOH, and NH edge groups, all of which may be reduced as the carbon electrode is polarized to low potentials [90]. These reactions can also interfere with the potentiodynamic behavior of these electrodes in electroanalytical and synthetic studies.

## C. Intrinsic Reactions of Carbonaceous Electrodes Which May Limit Their Electrochemical Window

Conducting diamonds (obtained by doping with boron [91]) are the only carbonaceous materials that are not expected to have intrinsic electrochemical activity at low potentials and can thus be considered as inert, nonactive electrode materials over a wide potential window [92]. All the other carbons have intrinsic electrochemical activity besides the surface chemistry developed by solution reduction processes, as described in the previous section. Graphite, graphitic-like, and disordered, high surface area carbons undergo intercalation/insertion reactions with cations at low potentials and with anions at high potentials. Usually, the smaller the cation, in the case of cation intercalation, the more pronounced these intercalation/insertion processes. Hence, when the electrolyte is TAA salt, these processes are either negligible or do not take place at all. However, when alkaline or alkaline earth metal ions ($Li^+$, $Mg^{2+}$, $K^+$) are present, their insertion into the carbons at low potentials may be pronounced. Special attention was focused on intercalation and insertion of lithium into carbonaceous materials due to the interest in lithiated carbons as anodes in rechargeable high energy density Li ion batteries. Lithium intercalates with graphite within the potential range 0.3–0 V ($Li/Li^+$). Figure 24 presents a typical voltammogram of graphite electrodes in a Li salt nonaqueous solution measured at very slow potential scan rate. This voltammogram has three distinctive sets of peaks that reflect phase transitions between the various intercalation stages existing for the lithiated graphite (as indicated). Hence, the electrochemical window of this system is, intrinsically, cathodically limited around 0.3 V ($Li/Li^+$), as demonstrated in Figure 24 [93].

Disordered carbons, either hard, nongraphitizable or soft, graphitizable, can insert lithium at much higher potentials than can graphite, and, in fact, the onset for Li insertion with these materials may be as high as 1.5 V ($Li/Li^+$) [72–75]. (Figure 23 provides a typical example.) In the case of glassy carbon electrodes, it appears that such insertion reactions are negligible and may be ignored.

The last types of carbons to be mentioned are fullerenes ($C_{60}$, etc.). These carbons were found to undergo about four reduction processes, thus forming $C_n^{-4}$ anions in the potential range 1–2.5 V ($Li/Li^+$) [94].

## D. Concluding Remarks

Carbon electrodes may undergo both intrinsic electrochemical reactions (insertion/intercalation) at low potentials and surface reactions with solution species, which, in some cases, cover them with passivating surface films. This, however, depends on the salt's cation. In the case of TAA, no surface film precipitation takes place, and the electrochemical window is cathodically limited by reduction of the solvents or the TAA cation (similar to the behavior of these

**Table 4** Reaction Products of Solvent, Salts, and Atmospheric Contaminants on Carbon Electrodes at Low Potentials. Some Onset Potentials versus Li/Li$^+$ for These Processes Are Indicated

| Solvent type | Specific solvent and relevant reduction potential | Dry | Contaminants | | | |
| --- | --- | --- | --- | --- | --- | --- |
| | | | $H_2O$ | $O_2$ | $CO_2$ | $SO_2$ [66] |
| Alkyl carbonates | PC < 1 V | $CH_3CH(OCO_2Li)CH_2OCO_2Li\downarrow$ propylene$\uparrow$ | $ROCO_2Li$ | $ROCO_2Li$ | $ROCO_2Li$ | $Li_2S_2O_4$ |
| | EC < 1 V | $(CH_2OCO_2Li)\downarrow$ethylene$\uparrow$ | | | + | |
| | DMC < 1 V | $ROCO_2Li\downarrow(CH_3OCO_2Li)$ | $Li_2CO_3$ | $Li_2O$ | $Li_2CO_3$ | $Li_2S$ |
| | DEC < 1 V | $CH_3CH_2OCO_2Li\downarrow + CH_3CH_2OLi\downarrow$ | (reaction of | $Li_2O_2$ | | |
| | EMC < 1 V | $CH_3OCO_2Li$; $CH_3OLi$ | $ROCO_2Li + H_2O)$ | | | |
| | PMC < 1 V | $CH_3OCO_2Li$; $CH_3OLi$, $CH_3CH_2OLi$ | $LiOH$-$Li_2O$ | | | |
| Esters | MF < 1.2 V | Mostly $HCO_2Li$, $ROLi$ ($CH_2OLi$?) | $HCO_2Li$ | Li oxides + $ROCO_2Li$ species | $HCO_2Li + Li_2CO_3$ | |
| | $\gamma$-BL < 1.2 V | $CH_3(CH_2)_2$ $COOLi$, cyclic $\beta$-keto ester anion di-Li salt | $LiO(CH_2)_3COOLi$ | | | |
| Ethers | THF < 0.5 V | $ROLi$ ($CH_3(CH_2)_3OLi$) | $LiOH$ | Li oxides + $ROLi$ species | $Li_2CO_3$ + $ROLi$ species | |
| | 2Me-THF < 0.5 V | Li pentoxides | $Li_2O$ | | | |
| | DME < 0.5 V | $ROLi$ ($CH_3OLi$) | Li alkoxides | $ROLi$ species | | |
| | 1,3-DN < 0.5 V | $CH_3CH_2OCH_2OLi$, $HCO_2Li$, Poly DN species | | | | |

| Mixtures | |
|---|---|
| EC-PC | EC reduction products dominate |
| EC-DEC | EC reduction products dominate |
| MF-EC | $HCO_2Li$, $ROCO_2Li$ species |
| MF-PC | $HCO_2Li$, $ROCO_2Li$ species |
| MF-DMC | $HCO_2Li$ dominates, $ROCO_2Li$ (minor) |
| MC-DEC | $HCO_2Li$ dominates, $ROCO_2Li$ (minor) |
| MF-ethers | $HCO_2Li$ dominates, |
| EC or PC with ethers | $ROCO_2Li$ species dominate, ROLi (minor) |
| THF-2MeTHF | THF reduction products dominate |
| Salt | |
| $LiAsF_6$ 1.3 V | LiF, $Li_xAsF_y$ |
| $LiClO_4$ < 1 V | $Li_2O$, LiCl, $LiClO_3$, $LiClO_2$, etc. |
| $LiBF_4$ | LiF, $Li_xBF_y$, $Li_xBF_yO_z$? |
| $LiPF_6$ 2 V, 0.7 V | LiF, $Li_xPF_y$, $Li_xBF_yO_z$? |
| $LiN(SO_2CF_3)_2$ < 1.5 V | LiF, $Li_2S_xO_y$, $Li_3N$, etc. |

*Source:* From Refs. 18, 76, 83–89.

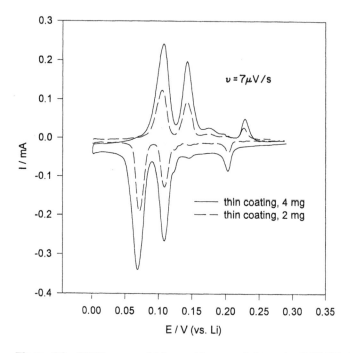

**Figure 24**   SSCV curves of thin graphite-coated electrodes (90% KS-6 powder and 10% PVF binder, 1.2 × 1.2 cm²) in 1 *M* LiAsF₆/EC:DMC (1:3).

solutions with nonactive metal electrodes) and in some cases by TAA intercalation processes. The behavior of conductive diamond and glassy carbon electrodes with both TAA and metal ion salt solutions is very similar to that of noble metals (i.e., no intrinsic electrochemical activity is observed at low potentials).

## VII.   STUDY OF SURFACE PHENOMENA ON NONACTIVE ELECTRODES IN NONAQUEOUS SYSTEMS USING IMPEDANCE SPECTROSCOPY (EIS) IN CONJUNCTION WITH ATOMIC FORCE MICROSCOPY (AFM) AND ELECTROCHEMICAL QUARTZ CRYSTAL MICROBALANCE (EQCM)

The previous section dealt with spectroscopic identification of surface films formed on nonactive metal electrodes polarized to low potentials in a variety of important polar aprotic systems and with the related voltammetric behavior. The

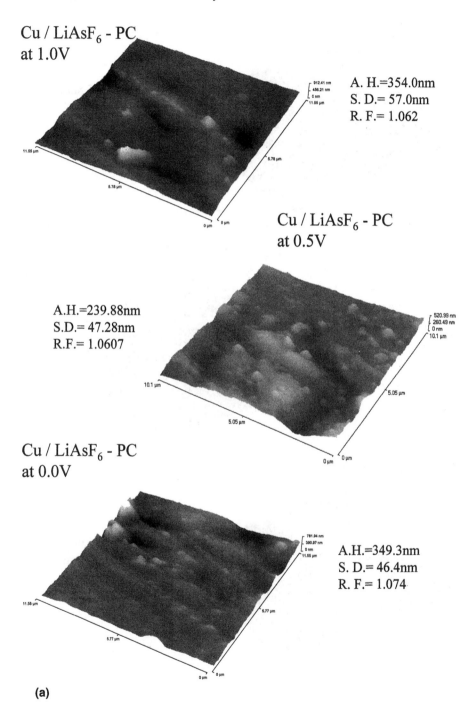

Cu / LiAsF$_6$ - PC
at 1.0V

A. H.=354.0nm
S. D.= 57.0nm
R. F.= 1.062

Cu / LiAsF$_6$ - PC
at 0.5V

A.H.=239.88nm
S.D.= 47.28nm
R.F.= 1.0607

Cu / LiAsF$_6$ - PC
at 0.0V

A.H.=349.3nm
S. D.= 46.4nm
R. F.= 1.074

**(a)**

**Figure 25** AFM images of copper electrodes polarized to the specified potentials in PC/LiAsF$_6$ solutions: (a) dry solutions, (b) wet solutions (200 ppm H$_2$O).

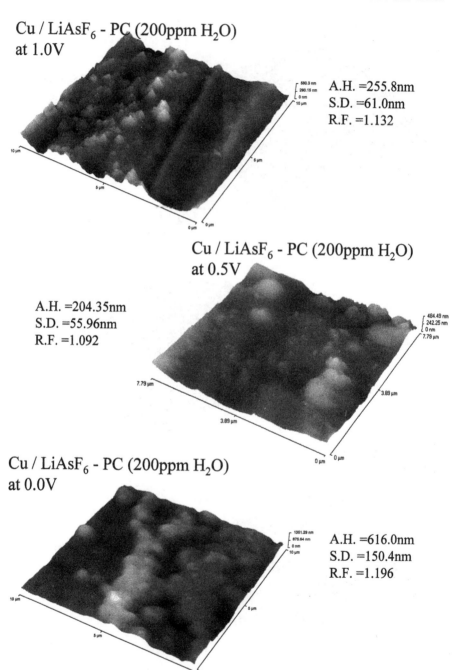

Cu / LiAsF$_6$ - PC (200ppm H$_2$O)
at 1.0V

A.H. =255.8nm
S.D. =61.0nm
R.F. =1.132

Cu / LiAsF$_6$ - PC (200ppm H$_2$O)
at 0.5V

A.H. =204.35nm
S.D. =55.96nm
R.F. =1.092

Cu / LiAsF$_6$ - PC (200ppm H$_2$O)
at 0.0V

A.H. =616.0nm
S.D. =150.4nm
R.F. =1.196

**(b)**

**Figure 25**   Continued

present section illuminates some other surface chemical aspects of these systems from other points of view: simultaneous studies of surface phenomena on nonactive electrodes in polar aprotic electrolyte solutions using EIS, EQCM, and AFM, these methods being complementary. EIS provides information on the electrical properties of the electrode-solution interface and possible surface films that precipitate on the electrode surface. From the calculation of the capacity of the surface films, knowing the dielectric constant of the surface species formed (since many surface species were identified), it is possible to estimate the thickness of the surface layers, assuming a parallel plate capacitor model for their capacitive behavior.

From the change in the impedance spectra of these systems measured at open circuit voltage after the formation of the surface films, we can establish a basis for comparison of the stability of surface films in different systems. AFM

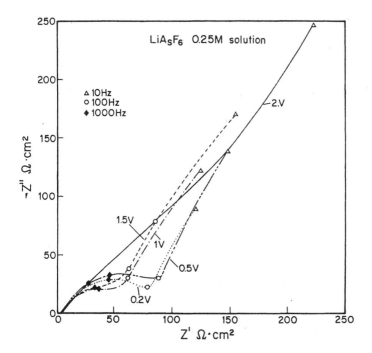

**Figure 26** Typical Nyquist plots obtained from impedance spectroscopic measurements of nickel electrodes polarized to different potentials in PC/LiAsF$_6$ 1 *M* solutions. The spectra were measured at the potentials indicated near each plot after the film formation was completed (the current reached a steady low value of *Ca.* 1 μA/cm$^2$) [34]. (With copyright from The Electrochemical Society Inc.)

provides in situ images of the surface films that show their morphology, porosity, and uniformity. EQCM provides direct information on the actual mass accumulation on the electrode surfaces due to the various surface reactions. Since both the mass accumulation and the total charge transfer involved are measured simultaneously, it is possible to characterize the various surface processes as a function of solution composition and the potential applied by mass (accumulated) per mole electrons transferred (m.p.e.). The difference between the m.p.e. values obtained in an experiment and the expected equivalent weight of the surface species

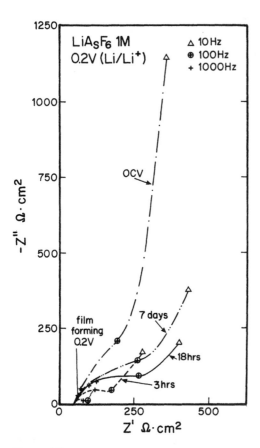

**Figure 27** Typical Nyquist plots obtained from impedance spectroscopic measurements of a nickel electrode polarized to 0.2 V (Li/Li$^+$) in PC/LiAsF$_6$ 1 $M$ solutions, which, after the surface films were formed, was allowed to stand in solution at open circuit. The spectrum before the potential was applied (OCV) is also shown. Storage time is indicated near each plot [34]. (With copyright from The Electrochemical Society Inc.)

(known from previous spectroscopic studies) is a valuable source of information on the stability of the surface films formed, and the partition of the charge transferred, between surface and solution bulk reactions.

Due to the practical implication related to the field of nonaqueous batteries, the solutions studied by this combination of methods were organic carbonates and ethereal-based solutions of Li salts (e.g., for EIS, see Refs. 34 and 95; for AFM, Refs. 96 and 97; and for EQCM, Refs. 35 and 98). Some representative results, which demonstrate the application of these methods to the study of important interfacial phenomena, are described briefly. As an example, we review a study of surface film formation on nonactive metal electrodes (Ni, Cu, Au) in PC with Li salt solutions.

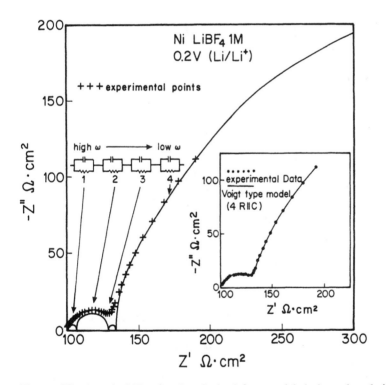

**Figure 28**  A typical Nyquist plot obtained from a nickel electrode polarized to low potentials (0.2 V versus Li/Li$^+$) in PC solutions (1 $M$ LiBF$_4$ in this case). The equivalent circuit analog of 4 $R \parallel C$ circuits in series and their separate Nyquist plots (four semicircles) are also shown. The frame in the lower right represents a typical fitting between the experimental data and this equivalent circuit analog [34]. (With copyright from The Electrochemical Society Inc.)

Figure 25 shows two sets of in situ AFM images of copper electrodes polarized to 1.5, 1, 0.5 and 0 V versus Li/Li$^+$ in dry and wet PC/LiAsF$_6$ solutions [97]. These images show very clearly the formation of surface films, which become more distinctive as the potential is lower. These images show that the outer part of these films is porous. In addition, they reflect the higher reactivity of the wet solutions (rougher surface film morphology) [97].

Figure 26 shows a family of Nyquist plots obtained by EIS from nickel

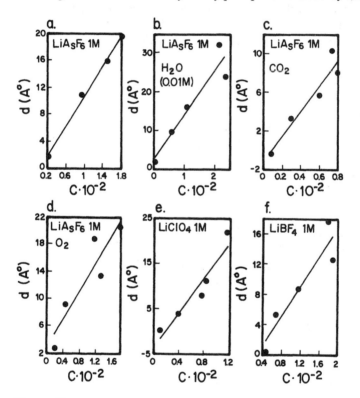

**Figure 29** Plots of the thickness of the compact part of the interphase formed on nickel electrodes in different PC solutions as a function of the charge involved in the surface films formation. Each point in these plots was obtained by polarizing a nickel electrode in the solution indicated to a certain low potential, and recording the charge accumulated due to the cathodic process (surface film formation and capacitive current). The electrode was then polarized to 2 V. The charge involved in the buildup of the surface films is nearly the difference between the charge values recorded in the two processes (Faradaic charge–capacitive charge). The thickness of the surface films (compact part layer) was calculated from the EIS measurements as explained in the text. (a) LiAsF$_6$ solution; (b) LiAsF$_6$ 1 $M$ solution contaminated with H$_2$O (0.01 $M$); (c) LiAsF$_6$ 1 $M$ solution saturated with CO$_2$; (d) LiAsF$_6$ saturated with oxygen; (e) LiClO$_4$ 1 $M$ solution; (f) LiBF$_4$ 1 $M$ solution [34]. (With copyright from The Electrochemical Society Inc.)

electrodes polarized to different potentials in a PC/LiA$_5$F$_6$ solution [34]. Figure 27 shows families of curves obtained from nickel electrodes polarized to O.2V in a PC/LiA$_5$F$_6$ solution, then stored in the same solution at open circuit conditions for different periods, followed by repeated EIS measurements at the apparent open circuit potentials [34]. The high frequency semicircle that characterizes these spectra reflects Li$^+$ migration through the surface films ($R_{\text{migration}}$ coupled with film capacitance). The curved, sloping low frequency parts of these spectra reflect the charge transfer resistance of these electrodes for reduction of the solution species, coupled with the double layer capacitance. A relevant typical model appears in Figure 28 [34] for the case of PC/LiBF$_4$ solutions.

The above spectra reflect surface film formation which becomes more pronounced as the electrode potential is lower, and its slow dissolution as the electrodes are held at open circuit (no continuous driving force for surface film forma-

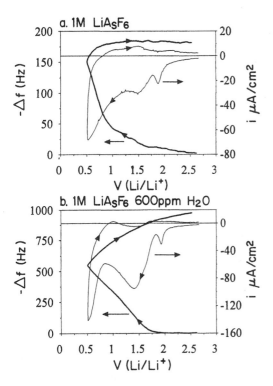

**Figure 30** Cyclic voltammograms (*I* versus *E*) and Δ*f* versus *E* measured during a first potential scan from OCV (≈2.5 V) to 0.5 V (Li/Li$^+$) and back to OCV (20 mV/s). Gold electrodes on QC, LiAsF$_6$ 1 *M* solutions in PC: (a) 20 ppm H$_2$O, (b) 600 ppm H$_2$O [35]. (With copyright from The Electrochemical Society Inc.)

tion) [34]. From the parameters of the high frequency semicircle, it was possible to estimate the relative thickness of these surface films. A good correlation between the calculated thickness and the charge involved in the formation of the surface layers was found, as demonstrated in Figure 29. The change in the calculated thickness for surface films formed at different solutions and potentials, as a function of storage time at open circuit, was found be a good measure for

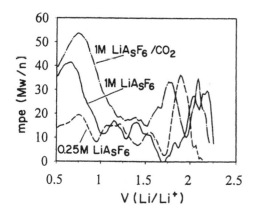

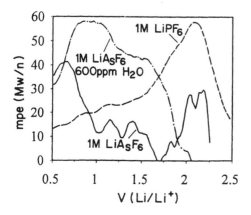

**Figure 31** M.p.e. values of the surface species formed as a function of the potential applied, calculated from the comparison between $I$ ($E$) and $d\Delta f/dE$ ($E$). Linear sweep voltammetry, 20 mV/s from OCV (3–2.5 V) to 0.5 V (Li/Li$^+$). (top) 1 $M$ LiAsF$_6$, 0.25 $M$ LiAsF$_6$ and 1 $M$ LiAsF$_6$, CO$_2$ saturated PC solutions. (bottom) 1 $M$ LiAsF$_6$ uncontaminated and water-contaminated solutions 600 ppm H$_2$O) and 1 $M$ LiPF$_6$/PC solution [D. Aurbach et al., *J. Electroanal. Chem.* 393, 43, 1995]. (With copyrights from Elsevier Science Ltd.)

comparing the stability of the various surface films formed on the electrodes in different solutions [34,95].

Finally, Figures 30 and 31 present data collected from similar systems in EQCM measurements [35]. In Figure 30, typical cyclic voltammograms obtained with gold electrodes (deposited on quartz crystals) in PC solutions are shown, together with the changes in the quartz crystal resonance vibration frequency during the surface film formation [35]. These frequency changes are translated linearly to mass accumulation. The results shown in this figure reflect the influence of the solution composition on the mass accumulation. Figure 31 presents curves of calculated mass (accumulated) per mole of electrons for the various processes versus potential. These m.p.e. values are compared with the equivalent weight of the expected surface species formed on the electrodes in these systems. The data in Figure 31 reflect reduction of trace $O_2$ to $LiO^{\pm}$ around 2 V, and reduction of PC below 1 V versus Li/Li$^+$ to $ROCO_2Li$ species (which partially react with trace water to form surface $Li_2CO_3$ ($CO_2$ and ROH are the coproducts) [12,34,35].

The brief description above demonstrates how novel techniques such as EQCM and AFM, in conjunction with a more conventional electroanalytical tool (EIS) can be applied for the study of surface film formation on electrodes, which is a fundamental phenomenon in many important nonaqueous systems.

## VIII. OXIDATION PROCESSES OF NONAQUEOUS SOLUTIONS AND THE ANODIC LIMITS OF THE ELECTROCHEMICAL WINDOWS

### A. General

Until recently, oxidation processes of nonaqueous systems were the focus of much less attention than were the reduction processes of these systems, for the following reason: one of the major driving forces for a rigorous study of intrinsic reactions of nonaqueous systems is the development of high energy density batteries based on polar aprotic electrolyte solutions. The major advantage of these solutions for battery application is the apparent stability of active electrodes (e.g., Li, Mg, Ca, Li-C, etc.) with them. However, their stability was found to be due to passivation by surface films formed by spontaneous reduction processes of solution components. Thus, understanding the cathodic reactions of many nonaqueous systems which are suitable as battery electrolyte solutions was crucial for advancing this field and providing the basic science behind the R&D of these reactive battery systems. Hence, the low potential behavior of nonaqueous systems was an important issue. The first generation of many nonaqueous batteries included cathodes whose potential range was 1.5–3.5 V versus Li/Li$^+$. Most of the nonaqueous solvents and salts of interest are not oxidized at these potentials, and thus their intrinsic anodic behavior was not sufficiently important for a rigorous investigation. Li ion battery technology emerged only in recent years. These batteries utilized $Li_xMO_y$ cathodes (M is a transition metal such as Mn, Ni, Co) whose potential range extends beyond 4 V ($Li/Li^+$). These materials were developed in order to compensate for the loss of energy density in the change from Li metal to lithiated carbon anodes. Because such high potential cathodes are used, oxidation of the relevant nonaqueous solutions becomes a crucial issue. Moreover, there are reports on the synthesis of 5-V lithiated transition metal cathodes [99], and, thus, searching for anodically stable solvents and increasing the anodic stability of existing systems is highly important.

Another important point of interest is the role of the electrode material in determining the anodic stability limit of electrochemical windows in nonaqueous systems. At potentials above 3 V versus $Li/Li^+$, many nonactive metals that are important as battery current collectors (e.g., Ni, SS and Al) may oxidize and dissolve. Thus, the anodic limit of the electrochemical system may be determined by the oxidation of the electrode metal. However, the metal cation and the salt anion (e.g., $ClO_4^-$, halides, $AsF_6^-$, $BF_4^-$, $SO_3CF_3^-$) may form insoluble salt that passivates the electrode and thus prevent massive dissolution of the electrode material at high potentials. Hence, in many cases we obtain an apparent stability of metal electrodes because of passivation. A typical case is aluminum, which is the most preferred current collector for 4 V cathodes in Li ion batteries, and which, in fact, can withstand even much higher potentials. This is because the most commonly used salts for Li batteries (which comprise fluorinated anions such as $AsF_6^-$, $PF_6^-$, and $BF_4^-$) liberate active fluoride (e.g., due to trace HF unavoidably present with these salts, or formed by hydrolysis with trace water). It is assumed that at high potentials, $AlF_3$ is formed and passivates the electrode very efficiently, preventing Al ion diffusion into the solution. Fortunately, the $AlF_3$ film is sufficiently thin to allow electron tunneling, and thus the electrical contact with the cathode active mass is maintained. Hence, when studying electrode processes in nonaqueous systems at high potentials, one has to bear in mind that the electrode reaction may occur via electron tunneling through anodic surface layers. An additional charge transfer resistance (related to the electron transfer through the passivating films) may complicate the electrochemical kinetics.

The last important point that should be mentioned in this introductory section is that, as previously mentioned, the definition of a limit of the electrochemical window in general is very vague and problematic, because in many cases the background oxidation currents are high. Hence, the definition of the anodic limit is, in most cases, arbitrary and depends on the maximal anodic background current tolerable for the specific reaction or system investigated. For instance, electrochemical mechanistic studies of reactions in which the concentration of the reactants can be high may tolerate much higher background currents than do cathode reactions in rechargeable batteries, where even very small background currents may be significant. These considerations are most important in batteries in which the electrode's area-to-solution volume ratio is high, since degradation of the solution components may be pronounced, even at low scale, low current (but continuous) oxidation processes.

## B. Review of the Anodic Limits of Nonaqueous Electrolyte Solutions

The first comprehensive set of data on the oxidation reactions of nonaqueous electrolyte solutions, as well as their positive anodic limit for use in electrochemistry, was prepared by Mann [13]. Table 5 summarizes some useful selected data from his review. This data is representative for several families of organic solvents, including ethers, organic carbonates, amides, acids, and alcohols, in terms of anodic potential limits.

From Mann's review, it is clear that the anion and the electrode material have a pronounced effect on the oxidation potentials of the nonaqueous systems. The metals to which the highest potentials can be applied in nonaqueous systems are obviously the noble metals (Pt, Au). The limiting reaction when the anions are halides ($Cl^-$, $Br^-$, $I^-$) was found to be their oxidation to the elemental form. When the anion is $ClO_4^-$, its oxidation onset at potentials above 1.5 versus Ag/$Ag^+$ may promote further intensive solvent degradation, as was found with ACN. It is important to note that using $BF_4^-$ instead of $ClO_4^-$ in ACN (which is an important and useful nonaqueous solvent in electrochemistry) extended its anodic stability limit by 2 V.

Since the time of Mann's review, more work has been conducted on the anodic stability of nonaqueous systems, promoted mostly by battery research. We review some highlights of this work. Table 6 summarizes data on the anodic limit of $LiClO_4$ solutions in a variety of solvents and solvent mixtures, including PC, ethers, and methyl formate (Ref. 100 and references therein).

The potentials in Table 5 are reported versus SCE or Ag/$Ag^+$ reference electrodes, whereas the data in Table 6 is reported versus the nonaqueous Li/$Li^+$ reference electrode. However, as shown by Auborn and Ciemieki [101], the SCE electrode potential is 3.24 V versus Li/$Li^+$, and thus the data in the two tables can be compared. The following important conclusions can be derived from these tables.

a. The early data provided by Mann relating to aqueous reference electrodes corresponds well to later data that relate to nonaqueous reference electrodes (for the same solvents and salts).

b. PC (and probably other alkyl carbonates as well) has a much higher anodic stability than do ethers (>1 V difference).

**Table 5** Selected Data on Oxidation Potentials and Reactions of Nonaqueous Systems

| Solvent | Salt anion | Working electrode | Reference electrode | Anodic limiting reaction/products | Limiting potential (V) |
|---|---|---|---|---|---|
| ACN | $ClO_4^-$ | Pt | $Ag/AgNO_3$ | $ClO_4^- \to e^- + ClO_4^{\cdot}$; $ClO_4^{\cdot} + CH_3CN \to HClO_4 + {}^{\cdot}CH_2CN$; $2CH_2CN^{\cdot} \to NCCH_2CH_2CN$ | 2 |
| ACN | $BF_4^-$ | Pt | $Ag/AgNO_3$ | Reaction unidentified. $CH_3CONH_2$ was detected | 4 |
| HAC | $ClO_4^-$, $BF_4^-$ | Pt | SCE | Unidentified[a] | 1.7–2 |
| DMF | $ClO_4^-$ | Pt | SCE | $HCOCN(CH_3)CH_2 \cdot$ Formation | 1.6 |
| DMSO | $ClO_4^-$ | Pt | SCE | Polymer formation | 0.7 |
| $CH_3OH$ | $ClO_4^-$ | Pt | $Ag/AgNO_3$ | Unidentified[a] | ≈0. |
| $CH_2Cl_2$ | $ClO_4^-$ | Pt | SCE | Unidentified[a] | 1.8 |
| PC | $ClO_4^-$ | Pt | SCE | Unidentified[a] | 1.7 |
| Pyridine | $ClO_4^-$ | Pt | $Ag/AgNO_3$ | Unidentified[a] | 1.4 |
| THF | $ClO_4^-$ | Pt | $Ag/AgClO_4$ | Unidentified[a] | 1.5–1.8 |

[a] The limiting reaction was unidentified at the time of Mann's review.
*Source:* From Ref. 13.

**Table 6**  Oxidation Potentials (V versus Li/Li$^+$) for Several Electrolyte Solutions

| Solvent | Salt | Pt | GC[a] | TAB[b] |
|---|---|---|---|---|
| PC | LiClO$_4$ | 5.3 | 5.3 | 4.7 |
| PC-DME (90/10) | LiClO$_4$ | 4.5 | 4.5 | |
| PC-DME (75/25) | LiClO$_4$ | 4.5 | 4.5 | |
| PC-DME (50/50) | LiClO$_4$ | 4.6 | 4.7 | 4.6 |
| DME | LiClO$_4$ | 4.5 | 4.5 | 4.6 |
| 2Me-THF/THF/2MeF[c,d] (49/49/2) | LiClO$_4$ | 4.1 | 4.0 | |
| 2Me-THF/THF (50/50) | LiClO$_4$ | 4.1 | 4.0 | |
| 2Me-THF/THF (90/10) | LiClO$_4$ | 3.9 | 4.0 | |
| 2Me-THF | LiClO$_4$ | 4.0 | 4.2 | |
| THF | LiClO$_4$ | 4.2 | | |
| PC LiClO$_4$ | LiAsF$_6$ | 5.6 | 5.2 | |
| PC/DME (90/10) | LiAsF$_6$ | 4.9 | 4.9 | |
| PC/DME (75/25) | LiAsF$_6$ | 4.7 | 4.8 | |
| PC/DME (50/50) | LiAsF$_6$ | 4.7 | 4.7 | |
| DME | LiAsF$_6$ | 4.6 | 4.6 | |
| 2Me-THF/THF/2-MeF (49/49/2) | LiAsF$_6$ | 4.0 | 3.9 | 4.0 |
| 2Me-THF/THF (90/10) | LiAsF$_6$ | 4.0 | 4.1 | |
| 2Me-THF/THF (75/25) | LiAsF$_6$ | 4.0 | 4.0 | |
| 2Me-THF/THF (50/50) | LiAsF$_6$ | 4.0 | 4.1 | |
| 2Me-THF | LiAsF$_6$ | 4.1 | 4.1 | |
| THF | LiAsF$_6$ | 4.2 | 4.2 | |
| 2MF (50/50) | LiAsF$_6$ | 5.0 | 4.9 | |
| MF | LiAsF$_6$ | 4.9 | | |

[a] GC = glassy carbon.
[b] TAB = Teflonized acetylene black.
[c] MF = methyl formate.
[d] 2Me-THF = 2-methyl furan.
*Source*: From Ref. 100 and references therein.

   c.    Cyclic ethers such as THF or 2Me-THF have a lower anodic stability than do open chain ethers, e.g., DME.

   d.    The anodic stability of LiAsF$_6$ solutions is slightly higher than that of LiClO$_4$ solutions (100–500 mV). This corresponds well with oxidation mechanisms proposed for several nonaqueous LiClO$_4$ solutions in which oxidation of the perchlorate anion is an initial step [102].

   e.    Similar results were obtained from Pt and glassy carbon electrodes, indicating that both electrodes may have a similar anodic stability and that no catalytic effects take place in the electrochemical oxidation [103].

Table 7 presents data taken from Herr [103] on the oxidation potential of LiClO₄ solutions in PC, γ-BL, DME, and nitromethane, with three electrode materials: Pt, carbon, and nickel. The data in Table 7 also demonstrate the higher anodic stability of organic carbonate solvents such as PC or solvents from the ester family, such as BL, over ethers such as DME. However, the values shown by Herr are generally higher than those measured by others for $ClO_4^-$ solutions. In addition, the difference between Pt and carbon electrodes in the table is surprising. The high oxidation potentials measured with Ni electrodes may be attributed to the passivation of this electrode. The data in Table 7 correlate with the data in Tables 5 and 6 and indicate further that, despite some unavoidable differences in the data obtained by different groups (due to impurities and the subjective nature of the determination of the anodic limits of these systems), the picture converges.

Table 8 compares oxidation potentials of selected solvents and their donor numbers [100]. This table demonstrates that there is a correlation between higher donor numbers and lower stability limits for oxidation. A similar correlation for a larger variety of solvents presented by Zyat'kova et al. [104] for LiAsF₆ solutions is shown in Table 9.

The oxidation potentials in Table 9, which refer to aqueous Ag/AgCl/KCl (sat.) electrodes, correspond well to the data in Tables 5–8. However, as mentioned in the introductory section, the definition of the oxidation potentials of nonaqueous systems may be very vague in many cases and may thus depend on subjective definitions (such as ''significant anodic currents''). 

Table 10 presents data taken from Ue et al. [105] for the oxidation poten-

**Table 7**   Potential Range for Various LiClO₄ Solutions[a]

| Electrolyte | Salt | Working electrode | Potential range | |
| --- | --- | --- | --- | --- |
| | | | From | To |
| Propylene carbonate | LiClO₄ | Pt | +2.3 | −2.2 |
| Propylene carbonate | LiClO₄ | Ni | +2.2 | −2.8 |
| Propylene carbonate | LiClO₄ | C | +1.5 | −1.0 |
| γ-Butyrolactone | LiClO₄ | Pt | +2.7 | −1.7 |
| γ-Butyrolactone | LiClO₄ | Ni | +1.9 | −1.3 |
| γ-Butyrolactone | LiClO₄ | C | +1.4 | −1.6 |
| 1,2-Dimethoxyethane | LiClO₄ | Pt | +1.5 | −2.0 |
| 1,2-Dimethoxyethane | LiClO₄ | Ni | +2.3 | −3.3 |
| Nitromethane | LiClO₄ | Pt | +3 | −2.4 |

[a] Reference electrode was Ag/Ag⁺.
*Source*: From Ref. 103.

**Table 8** Correlation between Donor
Numbers and Oxidation Potentials
for Selected Solvents

| Solvent | Donor number | Oxidation potentials vs. Li/Li$^+$ |
|---|---|---|
| MF | 13.8 | 4 |
| SL | 14.8 | 4.7 |
| PC | 15.1 | 4.7 |
| 2Me-THF | 19 | 4.1 |
| Diethyl ether | 19.2 | 4.2 |
| THF | 20 | 4.2 |
| DME | 20 | 4.6 |

*Source*: From Ref. 100.

tials of TEABF$_4$ solutions in a variety of polar aprotic solvents, with glassy carbon electrodes.

When the salt anion is BF$_4$$^-$, the oxidation potentials of solutions based on PC and ACN are much higher (by 1–2 V) than those of LiClO$_4$ or LiAsF$_6$ solutions, as was also shown by Mann [13]. For DMF solutions based on BF$_4$$^-$ anion or ClO$_4$$^-$, the oxidation potential (1.5 V versus SCE) was the same [105]. For

**Table 9** Decomposition Potentials of 0.5 $M$ LiAsF$_6$ Solutions in Nonaqueous Solvents versus Aqueous Ag/AgCl/KCl Sat. Reference Electrode

| Solvent | Donor number | Decomposition potentials (V) | | |
|---|---|---|---|---|
| | | $E_a$ | $-E_c$ | $\Delta E$ |
| AN | 14.1 | 1.4 | 3.17 | 4.6 |
| SL | 14.8 | 1.3 | 3.18 | 4.5 |
| PC | 15.1 | 1.3 | 3.18 | 4.5 |
| BL | 18.0 | 1.3 | 3.28 | 4.6 |
| THF | 20 | 1.0 | 3.25 | 4.2 |
| DME | 24 | 1.0 | 3.40 | 4.4 |
| DMF | 26.6 | 0.9 | 3.63 | 4.5 |
| DMA | 27.7 | 0.8 | 3.66 | 4.5 |
| DMSO | 29.8 | 0.7 | 3.71 | 4.4 |

*Source*: From Ref. 104.

**Table 10**  Limiting Reduction and Oxidation Potentials
of Organic Solvents Containing 0.65 mol dm$^{-3}$ Tetraethylammonium Tetrafluoroborate at 25°C (glassy carbon electrodes)

| Solvent | $E_{red}$ V vs. SCE | $E_{ox}$ V vs. SCE |
|---|---|---|
| Propylene carbonate | −3.0 | +3.6 |
| Butylene carbonate | −3.0 | +4.2 |
| γ-Butyrolactone | −3.0 | +5.2 |
| γ-Valerolactone | −3.0 | +5.2 |
| Acetonitrile | −2.8 | +3.3 |
| Glutaronitrile | −2.8 | +5.0 |
| Adiponitrile | −2.9 | +5.2 |
| Methoxyacetonitrile | −2.7 | +3.0 |
| 3-Methoxypropionitrile | −2.7 | +3.1 |
| N,N-Dimethylformamide | −3.0 | +1.6 |
| N-Methyloxazolidinone | −3.0 | +1.7 |
| N,N'-Dimethylimidazolidinone | −3.0 | +1.2 |
| Nitromethane | −1.2 | +2.7 |
| Nitroethane | −1.3 | +3.2 |
| Sulfolane | −3.1 | +3.3 |
| Dimethylsulfoxide | −2.9 | +1.5 |
| Trimethyl phosphate | −2.9 | +3.5 |

*Source*: From Ref. 105.

DMSO, the oxidation potential with salts based on $BF_4^-$ was higher by only 0.8 V than with solutions based on $ClO_4^-$. Hence, solvents from the alkyl carbonate, nitrile, and ester families may have a 6–8 V wide electrochemical window provided that appropriate salts of sufficiently high anodic stability are chosen as the electrolytes.

Table 11 presents data on the oxidation potentials with a variety of TAA or tetraalkyl phosphonium (TAP) salts [105]. In all cases, the anion was $BF_4^-$. The experiments were performed utilizing glassy carbon W.E.

All the oxidation potentials measured for the above solutions are around 3.5–3.85 V versus SCE. This further demonstrates the importance of the salt anion in the determination of the oxidation potentials of nonaqueous solutions of solvents with high anodic stability, such as alkyl carbonates, whereas the cations may have only a minor and secondary effect.

As further demonstrated by Gores and Barthel [106], $PF_6^-$ solutions may also possess high anodic stability, similar to $BF_4^-$ solutions, when the solvents are alkyl carbonates and nitriles.

From the data presented above, it appears that acetonitrile is the solvent

**Table 11** Limiting Reduction and Oxidation Potentials of Propylene Carbonate Electrolytes Containing 0.65 mol dm$^{-3}$ Quaternary Ammonium or Phosphonium Tetrafluoroborate at 25°C (glassy carbon W.E.)

| Electrolyte | $E_{red}$ V versus SCE | $E_{ox}$ V versus SCE |
|---|---|---|
| ME$_4$BF$_4$ | −3.10[a] | +3.50[a] |
| Me$_3$EtNBF$_4$ [b] | −3.00 | +3.60 |
| Me$_2$Et$_2$NBF$_4$ | −3.00 | +3.65 |
| MeEt$_3$BF$_4$ | −3.00 | +3.65 |
| Et$_4$NBF$_4$ | −3.00 | +3.65 |
| Pr$_4$NBF$_4$ [b] | −3.05 | +3.65 |
| MeBu$_3$NBF$_3$ | | |
| Bu$_4$NBF$_4$ [b] | −3.05 | +3.65 |
| Hex$_4$NBF$_4$ [b] | −3.10 | +3.85 |
| Me$_4$PBF$_4$ | −3.05 | +3.60 |
| Et$_4$PBF$_4$ | −3.00 | +3.60 |
| Pr$_4$PBF$_4$ | −3.05 | +3.60 |
| Bu$_4$PBF$_4$ | −3.05 | +3.80 |

[a] 0.1 mol dm$^{-3}$.
[b] Me = methyl; Et = ethyl; Pr = propyl; Bu = butyl; Hex = hexyl.
*Source*: From Ref. 105.

with the highest anodic stability. Hence, it seems appropriate to use this solvent solution for comparison of the anodic stability of salts. Such an approach was indeed initiated by Auborn and Ciemieki [101]. The anodic behavior of various salt solutions in ACN was compared using Pt electrodes. All their voltammo-grams show similar behavior: At a certain anodic potential, a steep increase in the oxidation current is observed. However, small oxidation currents are observed at much lower potentials. Hence, in order to characterize the anodic behavior of these systems, two potentials should be recorded: the onset of the rise of the anodic current and the onset of the small anodic background current as well. This is summarized in Table 12.

The onset potentials reported in Table 12 were estimated from the deflection points in the $I$ versus $E$ curves. However, if we determine the oxidation onset potential according to the potentials in which the anodic current exceeds a predetermined value, such as 0.5 mA/cm$^2$, it is then clear from Auborn's data that the anodic stability of the above salts is according to the following order: LiAsF$_6$, LiBF$_4$, LiPF$_6$ > LiClO$_4$, LiSO$_3$CF$_3$.

**Table 12**   Onset Potentials of the Oxidation Currents of Various Salt Solutions in ACN

| Salt | Onset potential (V) for anodic background current (versus SCE) | Onset potential (V) for the steep rise in the anodic currents (versus SCE) |
|---|---|---|
| $LiClO_4$ | 1.5 | 2.5 |
| $LiAsF_6$ | 1.25 | 2 |
| $LiPF_6$ | 1.2 | 2.2 |
| $LiBF_4$ | 1.3 | 2.4 |
| $TBABF_4$ | 0.6 | 2.4 |
| $LiCF_3SO_3$ | 1.2 | 2.5 |
| $Li_2B_{10}Cl_{10}$ | 0.9 | 1.35 |
| $LiAlClO_4$ | 0.8 | 1 |

*Source*: From Ref. 101.

## C.   Remarks on the Study of Oxidation Reactions of Alkyl Carbonates and Ethers

The mechanisms of the oxidation of solvents such as THF and PC were studied by several groups, utilizing FTIR and XPS spectroscopy [107–109] and on-line mass spectrometry (DEMS–differential, electrochemical mass spectroscopy [110–112]). For example, using ex situ FTIR spectroscopy, Lacaze et al. [46] showed that THF in $LiClO_4$ solutions are polymerized on electrodes biased to high potentials. The proposed mechanism involves oxidation of $ClO_4^-$ as an initial step, as shown in Scheme 7 [46,102]. ESR measurements also support such a mechanism. However, there are also suggestions for possible direct oxidation

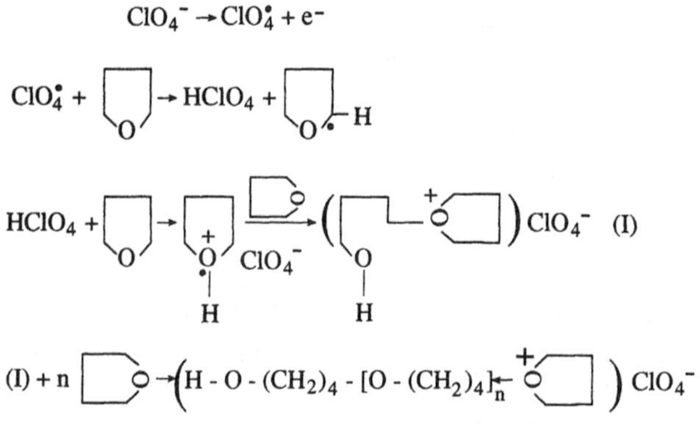

**Scheme 7**   Oxidation of THF/$LiClO_4$ solutions.

of THF via electron withdrawal and deprotonation, which forms THF radical cations as an initial step [113].

Vielstich et al. [112] studied the oxidation of PC by DEMS and FTIR spectroscopy, and Kanamura et al. studied its oxidation using in situ FTIR spectroscopy [108,109]. We also studied the oxidation of other alkyl carbonate solutions (EC-DMC and EC-DEC mixtures) using in situ FTIR spectroscopy [114].

The above-mentioned studies of the different groups lead to the following conclusions:

a. During PC oxidation (and probably other alkyl carbonates as well), evolution of $CO_2$ was clearly detected by DEMS. Ring opening and a change in the nature of the carbonyl group as an intermediate step were suggested based on in situ FTIR measurements.

b. The onset potential of $CO_2$ liberation, as well as the spectral changes observed by in situ FTIR spectroscopy, occurs at 4 to 4.5 V (Li/Li$^+$). This further demonstrates that the true onset potential for the oxidation of apparently anodically stable solvents may occur at much lower potentials than those that are usually deduced from voltammetry.

c. As shown by Eggert and Heitbaum in DEMS studies [110], when $LiClO_4$ is the electrolyte used, the anion is also involved in the oxidation process of the system at the 4–4.5 V (Li/Li$^+$) potential range, and intermediates such as $ClO_2$ and radical or molecular oxygen are formed. Traces of water were also identified (by DEMS) [110–112] among the oxidation products of PC-$LiClO_4$ solution. This product may be formed through hydrogen abstraction by an oxygen radical from the solvent molecules.

d. When the oxidation of PC solutions containing $LiClO_4$, $LiBF_4$, and $LiAsF_6$ was compared using DEMS [112], it was clear that $CO_2$ formation at potentials above 4 V (Li/Li$^+$) is a result of PC oxidation, and it was found to be much more pronounced for $LiClO_4$ solutions than for the other two salt solutions. Parallel results were obtained by the study of comparative oxidation of solutions of $LiAsF_6$, $LiClO_4$, $LiBF_4$, $LiPF_6$, and $LiSO_2CF_3$ with PC by Kanamura et al. [109], using in situ FTIR spectroscopy. It is clear from their results that, among these systems, the oxidation of PC solutions containing $LiClO_4$ as the salt is the most pronounced, and its onset potential is below 4.2 V (Li/Li$^+$).

Hence, in conclusion, studies on the oxidation of PC solutions clearly differentiate between $LiClO_4$ solutions and other salt solutions. It is clear that $ClO_4^-$ oxidation occurs in parallel to solvent oxidation, and the latter is accelerated by anion oxidation. Obvious products obtained from PC oxidation are $CO_2$ and polymeric species, probably derivatives of poly(propylene oxide). In the case of $BF_4^-$ or $PF_6^-$ salts, the oxidation products also include fluorinated derivatives of

propylene and/or propylene oxide [112]. Another solvent whose electrochemical
oxidation was investigated recently was DN [57]. We found that when this ketal
is diluted with other ethers, it exhibits high anodic stability, similar to THF (4.5
V $> E_{ox} > 4$ V versus Li/Li$^+$). However, since this solvent is an acetal, it is
highly sensitive to Lewis acids and thus can polymerize at potentials as low as
3 V versus Li/Li$^+$ in the single solvent solutions of salts such as LiClO$_4$. At these
potentials, adsorbed hydrogen on the electrode (formed by trace water reduction)
is oxidized to H$^+$, which readily initiates polymerization of dioxolane. The product of dioxolane polymerization is poly(ethylene)-poly(methylene oxide) (poly
DN). The polymerization at such low potentials ($\approx 3$ V versus Li/Li$^+$) can be
avoided by the presence of basic inhibitors such as tributylamine, which neutralizes the acids formed [47,57].

All the above data, which demonstrate the strong effect of the salt anion
on the oxidation potentials of these systems, strongly motivate rigorous studies
in which the intrinsic oxidation behavior of the solvents can be determined, with
no possible interference of anion effects. This can be done in two ways:

1. Study the electrochemical behavior of pure solvents (without electrolytes) by finding a way to overcome the huge solution resistance expected
2. Study the different solvents as reagents (small concentration) in systems in which the major component is a salt of very high anodic stability (such as LiPF$_6$ or LiBF$_4$, as shown above)

In fact, both approaches were initiated and reported in the literature.
McMillan et al. [115] conducted voltammetric studies in pure THF and PC using
microelectrodes. In this way, the effect of the solution resistance (even if very
high) on the voltammetric behavior may be negligible. In these studies, THF was
found to be anodically stable up to 4.2 V (Li/Li$^+$), and PC was found to be
stable up to 5.5 V (Li/Li$^+$). However, some small background current could be
measured at potentials higher than 4.5 V (Li/Li$^+$). This further demonstrates the
problem of a clear determination of an electrochemically stable window of these
solvents, and proves that it may be somewhat subjective. There was also a preliminary report on studies of the voltammetric behavior of solvents in solid matrices
composed of Li salts [116].

## D.  Concluding Remarks

While the above sections provide some useful and converging data on the anodic
stability of a variety of nonaqueous systems, there is still a lot of work to be
done in this area. In most cases, the anodic reactions of polar aprotic systems and
their mechanisms are not clear. In addition, the onset potentials for the oxidation
reactions of many systems depend on the salt, the electrode materials and impuri-

ties. Hence, the definition of the anodic limits of many nonaqueous systems becomes dependent on the specific system studied and its tolerance to anodic background currents, which may or may not interfere with the major electrochemical reaction under focus. Consequently, when the determination of the anodic stability limit of any specific system is imported, each system has to be examined thoroughly by itself as a prerequisite study, despite the data that can be collected from the literature.

## ABBREVIATIONS

| | |
|---|---|
| $\gamma$BL | $\gamma$-Butyrolactone |
| 2MeF | 2Methyl furane |
| 2Me-THF | 2Methyl tetrahydrofurane |
| ACN | Acetonitrile |
| AES | Auger electon spectroscopy |
| AFM | Atomic force microscope |
| Bu$_4$P | Tetrabutyl phosphonium cation |
| CH$_2$Cl$_2$ | Methylene chloride |
| DEC | Diethyl carbonate |
| DEE | Diethoxyethane |
| DEE | Diethyl ether |
| DEMS | Differential electrochemical mass spectroscopy |
| DHgE | Dropping mercury electrode |
| DMC | Dimethyl carbonate |
| DMF | *N,N*-Dimethylformamide |
| DMSO | Dimethyl sulfoxide |
| DN | 1,3-dioxolane |
| EC | Ethylene carbonate |
| EDA | Ethylenediamine |
| EDAX | Energy dispersive analysis by X-rays |
| EIS | Electrochemical impedance spectroscopy |
| EMC | Ethylmethyl carbonate |
| E$_{ox}$ | Oxidation potential |
| EQCM | Electrochemical quartz crystal microbalance |
| $E_{red}$ | Reduction potential |
| ESR | Electron spin resonance |
| Et$_4$N | Tetraethyl ammonium cation |
| FTIR | Fourier transform infrared spectroscopy |
| GC | Glassy carbon |
| GCMS | Gas-chromatograph-mass spectroscopy |
| HAc | Acetic acid |

| Hx$_4$N | Tetrahexyl ammonium cation |
| MA | Methyl acetate |
| Me$_4$N | Tetramethyl ammonium cation |
| Me$_4$P | Tetramethyl phosphonium cation |
| MeBu$_3$N | Methyl tributyl ammonium cation |
| MF | Methyl formate |
| NaBPh$_4$ | Sodium tetraphenylborate |
| NCE | Normal calomel electrode |
| NM | Nitromethane |
| ·NMR | Nuclear magnetic resonance |
| PC | Propylene carbonate |
| PMC | Propyl methyl carbonate |
| ppm | Parts per million |
| Pr$_4$P | Tetrapropyl phosphonium cation |
| PVdF | Poly(vinylidene difluoride) |
| SCE | Saturated calomel electrode |
| SL | Sulfolane |
| TAA | Tetraalkyl ammonium cation |
| TAB | Teflonized acetylene black |
| TBA$^+$ | Tetrabutyl ammonium cation |
| TBABF$_4$ | Tetrabutyl ammonium tetrafluoroborate |
| TBAI | Tetrabutyl ammonium iodide |
| TBAP | Tetrabutyl ammonium perchlorate |
| TBAPF$_6$ | Tetrabutyl ammonium hexafluorophosphate |
| THF | Tetrahydrofurane |
| TPA | Tetrapropyl ammonium cation |
| UPD | Under potential deposition |
| UPDS | Under potential deposit stripping |
| XPS | X-ray photoelectron spectroscopy |

## REFERENCES

1. G. Froyer, H. Ollivier, C. Cherrot and A. Siore, *J. Electroanal. Chem.* **327**, 159 (1992).
2. N. Vinokur, B. Miller, Y. Avygigal and R. Kalish, *J. Electrochem. Soc.* **143**, L238 (1996).
3. D. Aurbach and H. E. Gottlieb, *Electrochimica Acta* **34**, 141 (1989).
4. Y. Gofer, R. Barbour, Y. Lou, D. Tryk, J. Jayne, G. Chottiner and D. A. Scherson, *J. Phys. Chem.* **99**, 11741 (1995).
5. H. Gerischer and D. Wagner, *Ber. Bunsenges. Chem.* **92**, 1325 (1998).
6. D. M. Kolb, M. Prasnyski and H. Gersicher, *J. Electroanal. Chem.* **54**, 25 (1974).

7. L. G. Irr, *Electrochim. Acta 29*, 1, (1984).

8. M. B. Ardalayan and Y. M. Povarov, *Elektrokhimia 21*, 96 (1985).

9. A. Gaillet and G. Demange-Guerin, *J. Electroanal. Chem. 40*, 69 (1972).

10. B. Burrows and S. Kirkland, *J. Electrochem. Soc. 115*, 1164 (1968).

11. P. Zelenay, M. Winnicka-Maurin and J. Sobkowski, *J. Electroanal. Chem. 278*, 361 (1990).

12. D. Aurbach, M. L. Daroux, P. Faguy and E. Yeager, *J. Electroanal. Chem. 297*, 225 (1991).

13. C. K. Mann, in *Electroanalytical Chemistry*, Ed. A. J. Bard, Marcel Dekker, New York (1970), Vol. 3.

14. R. T. Atansoski, H. H. Law, R. C. Macintosh and C. W. Tobias, *Electrochim. Acta 32*, 877 (1997).

15. D. Aurbach, Y. Ein-Eli and A. Zaban, *J. Electrochem. Soc. 141*, L155 (1993).

16. D. Aurbach, Y. Gofer, M. Ben-Zion and P. Aped. *J. Electroanal. Chem. 339*, 451 (1992).

17. W. Behrendt, G. Gattow and M. Drager, *Z. Anorg. Allg. Chem. 397*, 237 (1973).

18. D. Aurbach, M. D. Levi, E. Levi and A. Schechter, *J. Phys. Chem. B. 101*, 2195 (1997).

19. V. Svetlicic and E. Kariv-Miller, *J. Electroanal. Chem. 209*, 91–100 (1986).

20. E. Kariv-Miller, P. B. Lawin and Z. Vajtner, *J. Electroanal. Chem. 195*, 435 (1985).

21. E. Kariv-Miller, C. Nanjundiah, J. Eaton and K. E. Swenson, *J. Electroanal. Chem. 167*, 141 (1984).

22. D. T. Sawyer, G. Chlericato, C. T. Angelis, E. J. Nanni and T. Tsuchiya, *Anal. Chem. 54*, 1720 (1982).

23. T. A. Lorenzola, B. A. Lopez and M. C. Giordano, *J. Electrochem. Soc. 136*, 1359 (1983).

24. A. A. Frimer, "Organic Reactions Involving Super Oxide Anion," in *The Chemistry of Functional Groups, Peroxides*, Ed. S. Patai, Wiley, New York (1983), Ch. 14.

25. A. J. Bard and L. A. Faulkner, *Electrochemical Methods, Fundamentals and Applications*, Wiley, New York (1980), Ch. 6.

26. M. V. Chankashvili, O. O. Denisova and T. Agladze, *Elektrokhimia 18*, 318 (1982).

27. E. Peled, in J. P. Gabano (Ed.), *Li Batteries*, Academic Press, London (1983), Ch. 3.

28. G. Pistoia, Ed., "Lithium Batteries: New Materials, Developments and Perspectives," in *Industrial Chemistry Library*, Vol. 5, Elsevier, Amsterdam (1994), Ch. 1.

29. D. Aurbach, Y. Malik, A. Meitav and P. Dan, *J. Electroanal. Chem. 282*, 73 (1990).

30. D. Aurbach, *J. Electrochem. Soc. 136*, 906 (1989).

31. X. K. Xing, P. Abel, R. McIntyre and D. Scherson, *J. Electroanal. Chem. 216*, 261 (1987).

32. D. Aurbach, M. L. Daroux, P. Faguy, A. Wilkinson and E. Yeager, Proceedings of the Meeting of the Electrochemical Society, Toronto, May 1985, The Electrochemical Society, Inc., Pennington, N.J., Extended Abstracts, p. 760 (1985).

33. M. Odziemkowski, M. Krell and D. E. Irish, *J. Electrochem. Soc. 139*, 3052 (1992).

34. D. Aurbach and A. Zaban, *J. Electrochem. Soc. 141*, 1808 (1994).

35. D. Aurbach and A. Zaban, *J. Electrochem. Soc., 142*, L108 (1995).

36. D. Aurbach, A. Schechter, B. Markovsky, Y. Cohen, I. Weissman and M. Mosh-

kovich, Proceedings of the Symposium on Materials for Electrochemical Energy Storage and Conversion, Li Batteries, Capacitors and Fuel Cells, MRS Meeting, 1–5 December, 1997, Boston, MA, The Materials Research Society, Warrendale, PA (1998).

37. S. Shirashi, K. Kanamura and Z. Takehara, *J. Appl. Electrochem.* 25, 584 (1995).
38. K. Kanamura, H. Tamura, S. Shiraishi and Z. Takehara, *J. Electroanal. Chem.* 894, 49 (1995).
39. D. Aurbach and Y. Ein-Eli. *Langmuir* 8, 1845 (1992).
40. V. Koch, *J. Electrochem. Soc.* 126, 181 (1979).
41. D. Aurbach, A. Zaban, O. Chusid and I. Weissmann, *Electrochim. Acta* 39, 51 (1994).
42. D. Aurbach, O. Chusid and I. Weissman, *Electrochim. Acta* 41, 747 (1996).
43. I. Taniguchi, in "Modern Aspects of Electrochemistry," Eds. B. E. Conway, J. O'M. Bockris and R. E. White, Plenum Press, New York (1990), Vol. 20.
44. U. von Kaiser and E. Heitz, *Ber. Bunsenges. Phys. Chem.* 77, 818 (1973).
45. P. A. Christensen, A. Hammett, A. V. G. Muir and N. A. Freeman, *J. Electroanal. Chem.* 288, 197 (1996).
46. J. E. Dubois, G. Tourillon and P. C. Lacaze, *J. Electrochem. Soc.* 125, 1257 (1978).
47. O. Youngman, P. Dan and D. Aurbach, *Electrochim. Acta* 35, 639 (1990).
48. D. Aurbach and R. Skaletsky, Proceedings of the Symposium on Primary and Secondary Lithium Batteries, 1990. The Electrochemical Society softbound series PV 91-3, The Electrochemical Society, Inc., Pennington, N.J. (1991), pp. 429–442.
49. M. Kolthoff and T. B. Reddy, *J. Electrochem. Soc.* 108, 980 (1961).
50. M. C. Giordano, J. C. Bazan and A. L. Arvia, *Electrochim. Acta* 11, 741 (1966).
51. C. R. Schlaikjer, in *Li Batteries*, Ed. J. P. Gabano, Academic Press, London (1983), Ch. 6.
52. David Linden, *Handbook of Batteries*, 2nd ed., McGraw Hill (1995), Ch. 14.1.
53. C. R. Cubrena, E. Garcia and A. J. Bard, *J. Electroanal. Chem.* 260, 457 (1989).
54. E. Garcia, J. Kwak and A. J. Bard, *J. Inorg. Chem.* 27, 4377 (1988).
55. C. Combellas, H. Marzouck and A. Theibault, *J. Appl. Elec.* 21, 267 (1991).
56. Z. Feimeng, C. Christophe and A. J. Bard, *J. Am. Chem. Soc.* 114, 11004 (1992).
57. D. Aurbach, Y. Gofer and M. Ben-Zion. *J. Power Sources* 39, 163 (1992).
58. I. Bergman and J. C. James, *Trans. Faraday Soc.* 48, 956 (1952).
59. S. D. Ross, M. Finkelstein and R. C. Peterson, *J. Org. Chem.* 31, 128 (1966).
60. S. Wawzonek, R. Berkey and D. Thomson, *J. Electrochem. Soc.* 103, 513 (1956).
61. T. Tsuru, S. K. Gogia and R. Kammel, *J. Appl. Electrochem.* 27, 209 (1997).
62. S. H. Cohen, R. T. Iwamoto and G. M. Kleinberg, *J. Chem. Phys.* 67, 1275 (1963).
63. G. B. Bachman and M. J. Astle, *J. Am. Chem. Soc.* 64, 1303 (1942).
64. T. P. Pinfold and F. Sebba, *J. Am. Chem. Soc.* 78, 5193 (1956).
65. W. Rogers, Jr., and S. M. Kipnes, *Anal. Chem.* 27, 1916 (1955).
66. J. D. Voorhies and E. J. Schurdak, *Anal. Chem.* 34, 939 (1962).
67. G. Cauquis and D. Serve, *Bull. Soc. Chim. France*, 302 (1966).
68. J. Phelps, K. S. V. Santhanam and A. Bard, *Private Communication*, 1967.
69. J. F. Coetzee and W.-S. Siao, *Inorg. Chem.* 2, 14 (1963).

70. E. Peled, C. Menachem, D. Bar-Tov and A. Melman, *J. Electrochem. Soc. 143*, L4 (1996).
71. X. Z. Shu, R. S. McMillan and J. J. Murray, *J. Electrochem. Soc. 140*, 922 (1993).
72. Y. Matsumura, S. Wang and J. Mondori, *Carbon 33*, 1457 (1995).
73. W. Xing and J. R. Dahn, *J. Electrochem. Soc. 144*, 1195 (1997).
74. N. Tamaki, A. Satoh, T. Takahisha and M. Kanda, *Electrochim. Acta 42*, 2537 (1997).
75. Y. Liu, S. Xue, T. Zheng and J. R. Dahn, *Carbon 34*, 193 (1996).
76. D. Aurbach, O. Youngman Chusid, Y. Carmeli, M. Babai and Y. Ein-Eli. *J. Power Sources 43*, 47 (1993).
77. C. Menachem, E. Peled, L. Burstein and Y. Rosenberg, *J. Power Sources 68*, 277 (1997).
78. D. Bar Tov, E. Peled and L. Burstein, Proceedings of the 1997 Joint International Meeting, Paris (1997) PV 97-2, The Electrochemical Society and the International Society of Electrochemistry, pp. 159–160.
79. Y. Matsumura, S. Wang and J. Mondori, *J. Electrochem. Soc. 142*, 2914 (1995).
80. S. Mori, H. Asahina, H. Suzuki, A. Yoney and K. Yokota, *J. Power Sources 68*. 59–64 (1997).
81. H. Momose, H. Honbo, S. Takenchi, K. Nishimura, T. Toriba, Y. Muraraka, Y. Kozono and H. Midera, *J. Power Sources 68*, 208 (1997).
82. B. Scrosati, *J. Electrochem. Soc. 139*, 2776 (1992).
83. D. Aurbach, Y. Ein-Eli, O. Chusid, M. Babai, Y. Carmeli and H. Yamin. *J. Electrochem. Soc. 141*, 603 (1994).
84. D. Aurbach, Y. Ein-Eli, B. Markovsky, Y. Carmeli, H. Yamin and S. Luski, *Electrochim. Acta 39*, 2559 (1994).
85. D. Aurbach, B. Markovsky, A. Schechter, Y. Ein-Eli and H. Cohen. *J. Electrochem. Soc. 143*, 3809 (1996).
86. D. Aurbach, A. Zaban, A. Schechter, Y. Ein-Eli, E. Zinigrad and B. Markovsky. *J. Electrochem. Soc. 142*, 2873 (1995).
87. D. Aurbach, A. Schechter, B. Markovsky, Y. Ein-Eli and V. Koch, *J. Electrochem. Soc. 143*, L273 (1996).
88. Y. Ein-Eli, S. F. McDevitt, B. Markovsky, A. Schechter and D. Aurbach, *J. Electrochem. Soc. 144*, L180 (1997).
89. Y. Ein-Eli, S. R. Thomas and V. R. Koch, *J. Electrochem. Soc. 144*, 1159 (1997).
90. Y. Ein-Eli and V. R. Koch, *J. Electrochem. Soc. 144*, 2968 (1997).
91. R. Tenne, K. Patel, H. Hashimoto and A. Fugisima, *J. Electroanal. Chem. 347*, 409 (1993).
92. L. F. Li, D. Totir, B. Miller, G. Chottiner, A. Argoitia, J. C. Angus and D. Scherson, *J. Electrochem. Soc. 145*, L85 (1998).
93. M. D. Levi and D. Aurbach, *J. Electroanal. Chem. 421*, 79 (1997).
94. B. S. Guillous, W. Kutner, M. T. Jones and K. M. Kadish, *J. Electrochem. Soc. 143*, 550 (1996).
95. D. Aurbach and A. Zaban, *J. Power Sources 54*, 289 (1995).
96. D. Aurbach and Y. Cohen, *J. Electrochem. Soc. 143*, 3525 (1996).
97. D. Aurbach and Y. Cohen, *J. Electrochem. Soc. 144*, 3355 (1997).
98. D. Aurbach and M. Moshkovich, *J. Electrochem. Soc. 145*, 2629 (1998).

99.   Y. Ein-Eli and W. F. Howard, Jr., *J. Electrochem. Soc. 144*, L205 (1997).
100.  F. Ossala, G. Pistoia, R. Seeber and P. Ugo, *Electrochim. Acta 33*, 47 (1988).
101.  K. T. Ciemieki and J. J. Auborn, in *Proceedings of the Symposium on Lithium Batteries*, Ed. A. N. Dey, ECS Meeting, Washington, The Electrochemical Society, Inc., softbound series PV 84-1, The Electrochemical Society, Inc., Pennington, N.J. (1984), p. 363.
102.  D. H. Jang and S. M. Oh, *J. Electrochem. Soc. 144*, 3344 (1997).
103.  R. Herr, *Electrochim. Acta 35*, 1257 (1990) and
104.  L. A. Zyat'kova, V. N. Atanasev, G. A. Krestov and T. V. Ivanova, *Elektrokhimiya 29*, 946 (1993).
105.  M. Ue, K. Ida and S. Mori, *J. Electrochem. Soc. 141*, 2989 (1994).
106.  H. J. Gores and G. Barthel, *Pure Appl. Chem. 67*, 919 (1995).
107.  K. Kanamura, S. Toriyama, S. Shiraishi and Z. Takehara, *J. Electrochem. Soc. 142*, 1383 (1995).
108.  K. Kanamura, S. Toriyama, S. Shiraishi, M. Ohashi and Z. Takehara, *J. Electroanal. Chem. 419*, 77 (1996).
109.  K. Kanamura, S. Toriyama, S. Shiraishi and Z. Takehara, *J. Electrochem. Soc. 143*, 2548 (1996).
110.  G. Eggert and J. Heitbaum, *Electrochim. Acta 31*, 1443 (1986).
111.  O. Olter, J. Willsou, G. Eggert and J. Heitbaum, ECS Spring Meeting, Toronto. The Electrochemical Society, Inc., softbound series PV 85-1, The Electrochemical Society, Inc., Pennington, N.J. Extended Abstracts, p. 884 (1985).
112.  B. Rasch, E. Cattaneo, P. Novak and W. Vielstich, *Electrochim. Acta 36*, 1397 (1991).
113.  A. N. Dey and E. J. Rudd, *J. Electrochem. Soc. 121*, 1294 (1974).
114.  D. Aurbach and M. Moshkovich, ''On the oxidation of electrolyte solutions for Li batteries,'' in preparation.
115.  S. A. Campbell, C. Bowes and R. S. McMillan, *J. Electroanal. Chem. 284*, 195 (1990).
116.  V. Koch, Covalent Assoc. Inc. USA, personal communication.

# 5

# The Chemical and Electrochemical Reactivity of Nonaqueous Solvents Toward Metallic Lithium

**Daniel A. Scherson**
*Case Western Reserve University, Cleveland, Ohio*

## I. INTRODUCTION

A better understanding of the electrochemical reactivity of nonaqueous solvents may be expected to have a pronounced impact in the further optimization of rechargeable Li-based energy storage devices [1]. Progress in this area has been largely hampered by problems associated with the presence of adventitious impurities, mostly water and oxygen, which are very difficult to remove without decomposing the solvent. Although unimportant to preparative organic synthesis, the presence of such contaminants at levels on the order of a fraction of a ppm often lead upon reduction to the formation of species capable of reacting with solution constituents, which can in turn combine with intrinsic reaction products derived from the solvent and the electrolyte generating sparingly soluble films of complex stoichiometry adhered to the electrode surface. Rather serendipitously, however, these layers are key to the operation of lithium electrodes in battery applications by providing a barrier that protects the metal beneath from undergoing further corrosion and a medium for the facile transport of $Li^+$ [2]. It is, thus, of utmost interest to gain further insight into the chemistry and electrochemistry of lithium in nonaqueous solvents, as it might lead to the rational design of Li passive films in terms of structure and composition, which may improve the capacity, rates of charge and discharge and cycle life characteristics of secondary lithium batteries.

This chapter reviews two strategies developed in our laboratories for the study of the chemical and electrochemical reactivity of lithium under conditions which greatly minimize effects due to impurities. The first is based on the assembly and characterization of Li-solvent interfaces in ultrahigh vacuum (UHV), employing a nominally inert metal as a support [3,4], whereas the second takes advantage of the exceedingly low vapor pressure of certain Li-salt polyethylene oxide (PEO) solutions to perform conventional electrochemical experiments in UHV environments [5,6].

## II. ASSEMBLY AND CHARACTERIZATION OF Li-SOLVENT INTERFACES IN ULTRAHIGH VACUUM

The procedure implemented at CWRU for exploring the chemical reactivity of organic solvents toward metallic lithium relies on the vapor deposition of a Li film onto a nominally inert clean support followed by condensation of the solvent from the gas phase to form the substrate/lithium/solvent interface. Information regarding various aspects of such interfaces may be obtained via a combination of electron- and photon-based spectroscopic techniques, including Auger electron and X-ray photoelectron spectroscopies (AES and XPS, respectively), Fourier transform infrared reflection absorption spectroscopy (FT-IRRAS), temperature programmed desorption (TPD) and work function ($\Delta\Phi$) measurements. The assignment of spectral features can be assisted by measuring the properties of plausible reaction products formed directly on the modified substrate surface by adapting well-known synthetic organic chemistry routes to UHV environments. This overall strategy offers a number of advantages:

1. UHV affords conditions of utmost cleanliness for the assembly and characterization of such highly reactive interfaces, making it possible to examine the effects of gas phase impurities on the reaction pathways.
2. The combined use of electron- and photon-based spectroscopic techniques, including TPD, $\Delta\Phi$ and other methodologies, enables various aspects of the interface to be probed with a high degree of sensitivity and specificity.
3. The direct correspondence between $\Delta\Phi$ and the electrochemical potential [7] suggests that the potential drop across the Li–frozen solvent interface resembles closely that of an unpolarized Li electrode in contact with a liquid solution involving the same solvent; hence, the reactivity of Li toward the frozen solvent, except for kinetic hindrances derived from differences in temperature, should approach that of a Li electrode immersed in such a solution at open circuit (i.e., in situ) conditions.

This section describes in detail key aspects of each step involved in these experiments, focusing some attention on the nature of lithium-substrate and solvent-substrate interactions that may be of significance to the interpretation of results obtained with the more complex interfacial system of relevance to this work.

## A. Experimental Aspects

### 1. Instrumentation

The assembly and characterization of lithium-solvent interfaces can be performed in conventional UHV chambers equipped with surface analytical techniques, including those involved in the preparation of clean substrate surfaces. Measurements at CWRU were conducted in three custom-designed UHV chambers operating at pressures of ca. $2 \cdot 10^{-8}$ Pa (1 Pa $= 7.5 \cdot 10^{-3}$ torr). One of these chambers (A) is equipped with a hemispherical electron energy analyzer (Vacuum Science Workshop, VSW HA100), an electron gun (VSW EG5) for AES, an ion gun (PHI 04-161) for $Ar^+$ sputtering, low energy electron diffraction (LEED) optics (PHI 120) and a computer-controlled quadrupole mass spectrometer (Dycor M200M) for the analysis of residual gases and TPD [8]. The second chamber (B) was constructed by Varian and incorporates a single pass cylindrical mirror analyzer (CMA, Varian 981-2607) with a coaxial electron gun (Varian 981-2613) for the acquisition of AES spectra, and a Dycor mass spectrometer [5]. X-ray photoelectron spectroscopy (XPS) experiments were carried out in a separate custom-designed, multitechnique Perkin-Elmer system (C) equipped with an Al/Mg dual X-ray source and a Perkin-Elmer 10-360 hemispherical energy analyzer [9].

All XPS multiplex spectra were collected at a pass energy of 35.75 eV using the Mg anode operating at 400 W. XPS data analyses were performed using a Perkin-Elmer 5000 series software package. Peak positions and areas were determined by curve-fitting the spectral features with asymmetric Gaussian-Lorenztian functions and the energies calibrated with a Au—Ag—Cu standard built into the sample holder. AES data were collected in a derivative mode with beam voltages of ca. 3 kV, beam currents of ca. 5 μA and a beam diameter of 1–1.5 mm.

Specimens were mounted on transferable copper or stainless steel holders mounted on precision XYZ rotary manipulators with provisions for the crystal temperature to be varied between ca. 100 and 1500 K. Samples were heated either resistively or by electron bombardment from the rear, and their temperature monitored with an alumel-chromel thermocouple either spot-welded to the edge of the sample or affixed by other means.

Fourier transform infrared spectroscopy (FTIR) measurements in the reflection absorption mode were performed in a small UHV chamber designed to fit in the sample compartment of an IBM-98 Brucker FTIR spectrometer (see Figure 1A) [10]. Species were adsorbed and/or condensed on metal films vapor-deposited in situ on the surface of a quartz crystal microbalance (QCM) for micro-gravimetric determinations. The QCM was mounted on a variable-temperature, Au-plated Cu block, which houses a second (reference) QCM, attached in turn to a liquid He cold finger. The unique capabilities of this apparatus make it possi-ble to correlate quantitatively various spectral characteristics, including peak widths and intensities, with the absolute mass of adsorbed and/or condensed ma-terial over a wide range of coverages and temperature. As shown in Figure 1B, the IR beam enters the UHV chamber through one of the differentially pumped Polytran NaCl windows (5), impinging at an angle of about 74° with respect to the surface normal onto the surface of the sample/mirror/QCM assembly (S/M/QCM, see below) placed at the focal plane of the spectrometer. Two additional flat, gold-plated stainless steel mirrors (23 & 24) installed on fully adjustable stainless steel mounts (see below) are used to direct the beam out of the chamber across the second differentially pumped Polytran window (13) and into the detec-tor module (DM) through the second KBr window (12). A wire grid polarizer placed in the DM close to the focal plane allows only *p*-polarized light to reach the detector. This nonfocusing, three-mirror arrangement increases slightly the optical path, decreasing to about half the net (unpolarized) throughput due primar-ily to restrictions in the size of the third mirror imposed by the dimensions of the UHV chamber.

Specimens were prepared by a combination of metal vapor deposition and adsorption and/or condensation of molecular gases directly on the front face of the polished gold-coated copper block covered partially by a flush-mounted gold-coated QCM or S/M/QCM, kept in place by a Cu-Be spring (27), which faces the evaporator/doser collimator (25) arrangement (see Figure 1C). Although the S/M/QCM assembly is attached to a cold finger designed to work at temperatures of ca. 4 K, measurements have only been attempted using liquid nitrogen as the refrigerant. The S/M/QCM assembly is heated with a nichrome film vapor deposited onto a sapphire substrate (28) mounted on the rear of the gold-plated Cu block, providing a working temperature range from about 80 to 350 K.

For the measurements presented in this work, the sample/mirror surface was first coated with a rather thick layer of gold (ca. 50 nm) vapor-deposited from a tungsten basket prior to collecting spectral data. This approach makes it possible to bury material present on the surface from previous experiments and provides a highly reflecting substrate for conducting subsequent measurements. A collimator (25) was attached to the flange that houses the metal sources to restrict the line of sight of the metal vapor (and molecular gas emerging from the doser) to an area only slightly larger than the sample/mirror assembly.

Individual single beam FTIR spectra were obtained by co-adding 1600 con-secutive interferometric scans with 8 cm$^{-1}$ resolution. All spectra reported in

UPPER SECTION                         A

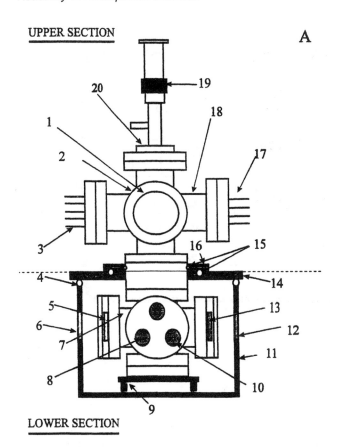

LOWER SECTION

**Figure 1** Schematic of a UHV chamber specially designed to fit in the sample compartment of a Brucker IR-98 FTIR spectrometer. (A)(1) roughing manifold (not shown); (2) ion pump (not shown); (3) ion gauge; (4) & 15 Viton O-rings; (5, 13) differentially pumped Polytran windows; (6, 12) KBr windows; (7) five-way cross; (8) doser feedthrough; (9) height/tilt adjustable table; (10) evaporator feedthrough; (11) FTIR sample compartment (SC); (14) sealing plate; (16) sealing annulus; (17) electrical feedthrough; (18) six-way cross; (19) cryostat; (20) rotatable flange. (B)(21) IR beam path; (22) sample/mirror/QCM assembly; (23) second mirror; (24) third mirror; (25) collimator. (C)(26) Au-coated QCM (front); (27) Au-coated Be-Cu clip; (28) sample heater (on rear surface); (29) Pt thermometer; (30) Au-coated QCM (back). (From Ref. 10.)

B

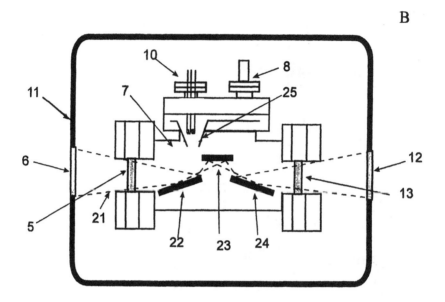

C

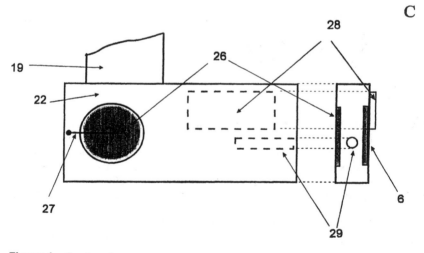

**Figure 1** Continued

this work are displayed as raw (unsmoothed) normalized difference spectra, i.e., $\Delta R/R = (R_{ref} - R)/R_{ref}$, using the spectrum of clean, freshly evaporated gold recorded at the same temperature ($R_{ref}$) as a reference. In this fashion, contributions due to impurities present on windows and mirrors are subtracted out. The integrated areas under the spectral features of interest were determined by fitting the data either with one or two Gaussians or Gaussian/Lorentzian combinations.

The frequency drift of the QCM at about 90 K (ca. 0.83 Hz/min) was larger than that observed at ca. 300 K (0.14 Hz/min). This is due to the better stability of the system at room temperature and to the fact that the crystals were cut to minimize their thermal coefficient at ca. 300 K. Based on the characteristics of the QCM used in these studies, a 1 Hz frequency change corresponds to 24 ng/cm$^2$. The uncertainty in the mass determination was estimated from the magnitude of the fluctuations of the QCM frequency (or period) readings during spectral acquisition, i.e., ca. 0.5 Hz or 12 ng/cm$^2$, which is well below the range of masses examined in this work.

The measured mass can, in principle, be converted into equivalent monolayers based on the molecular weight ($M_w$) of the species in question and the projected molecular area, $M_a$; e.g., for THF, $M_w = 72$ amu and $M_a = 20\text{--}40 \times 10^{-16}$ cm$^2$/molecule, which yields a value of ca. 0.06--0.03 $\mu g_{THF}$ per monolayer. Since neither the molecular orientation nor the actual surface area of the specimen are known, due to the rough character of gold films deposited on cold substrates, masses will be given in terms of $\mu g/cm^2$.

## 2. Substrate Preparation

Clean substrate surfaces for nonoptical experiments were prepared by a series of Ar$^+$-sputtering/thermal annealing cycles to remove surface impurities and restore atomic smoothness, respectively, using AES and/or XPS to determine cleanliness and LEED to assess surface ordering for single crystal specimens. Most of the work so far reported has involved nominally unreactive substrates, namely Ag and Au.

## 3. Lithium Deposition

Lithium was evaporated from thoroughly degassed SAES Getters sources onto the desired clean substrate [3], usually kept at low temperatures, ca. 100--120 K, although other methods involving Li/Al alloys have, more recently, been introduced [11]. For very clean Li films, the AES and XPS spectra, shown in Figures 2 and 3, respectively, display features at 52 eV, attributed to the KVV AES transition [12], and at 54.9 eV (XPS), ascribed to the photoemission of electrons from the Li(1$s$) orbital, characteristic of metallic Li [13]. For most of the experiments, AES spectra of Li deposits were only acquired in the energy regions of interest, i.e., 20--90 eV for Li, 200--390 eV for C and 450--520 eV for O, as well

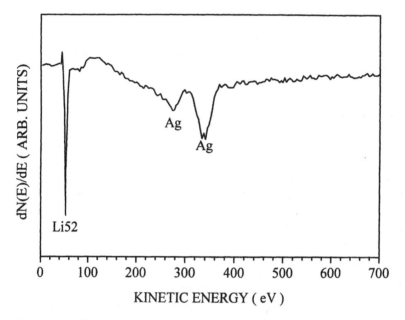

**Figure 2**   AES spectrum of a vapor deposited Li layer (94%) on clean Ag(poly) at 130 K. (From Ref. 3.)

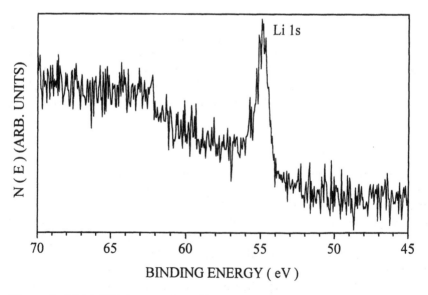

**Figure 3**   Li(1s) XPS spectrum for a lithium film deposited at 130 K. (From Ref. 3.)

as those of the substrate, immediately prior to dosing the solvent to decrease the time the highly reactive surface is exposed to background gases and thereby decrease the risk of contamination.

Lithium coverages are reported throughout in terms of percent of AES signal (%Li) defined as $(I_{Li}/S_{Li})/(I_{Li}/S_{Li} + I_{sub}/S_{sub})$, where $I_i$ and $S_i$ are the peak-to-peak amplitude and sensitivity factor of (one of) the most prominent AES features of element $i$, respectively [14], and the subscript "sub" represents the metal substrate. For example, in the case of Ag, the peak at 351 eV was used as a reference for Li coverage determinations.

Although somewhat approximate, the thickness of the Li film, assuming a perfectly uniform deposit, can be estimated based on the homogeneous attenuation of the substrate AES signal, via the equation $I = I_0 e^{-x/k}$ [15], where $k$ is the energy dependent mean free path of AES electrons; e.g., for Au, $k$ is ca. 1 nm at 239 eV. The density of Li is 0.53 g/cm$^3$ [16], which corresponds to an atomic density of $4.57 \times 10^{22}$ atoms/cm$^3$; hence, a 78% Li layer would have a thickness of about 1.6 nm or ca. 8 monolayers (ML).

## 4. Solvent Condensation

Solvent layers were condensed directly from the gas phase using a glass multicapillary array placed directly in front of the substrate kept at cryogenic temperatures. Precise monitoring of the magnitude of the dose was provided by a capacitance manometer (MKS, Model 390HA) attached to a small calibrated volume allowing specimens to be exposed to high molecular fluxes of pure or mixed gases without raising appreciably the background pressure.

Two solvents with widely different physical properties were examined in our initial studies: (1) propylene carbonate (PC), a viscous, high dielectric constant solvent [17], which has found wide application in Li-based battery technology, either mixed with a second nonaqueous cosolvent [18] or as a plasticizer in lithium-ion conducting solid polymer electrolytes [19], and (2) tetrahydrofuran (THF), a low viscosity, low dielectric constant solvent used to decrease the solvent viscosity and thereby increase the diffusion coefficient of ions in the media. Perdeuterated THF ($C_4D_8O$ or TDF, $M_w = 80$ amu) and PC ($C_4O_3D_6$ or $d_6$-PC, $M_w = 108$ amu) obtained from Icon Corp. were used in our experiments to distinguish hydrogen derived from these species from that present as the major background gas in UHV, a factor that simplifies considerably the analysis of TPD data.

Both gases were purified by freeze-thaw cycles followed by pumping to remove high vapor pressure residual gases. In the case of $d_6$-PC, the temperature of the container during evacuation was kept at 200 K using an external alcohol/dry-ice bath. This procedure was also effective in removing water, as evidenced by the gradual reduction in intensity of the $m/e = 17$ and 18 fragments in the mass spectra of the $d_6$-PC gas after each cycle. Gaseous purified $d_6$-PC kept in the stainless steel doser lines decomposed over a few hours, generating CO and $CO_2$. Furthermore, the TPD of $d_6$-PC (1 L) left in the doser for a few days condensed on a Au substrate yielded a peak with $m/e = 80$ TPD centered at ca. 190 K, which was not observed for freshly purified $d_6$-PC.

During the dosings all filaments were turned off to avoid decomposition of the material. Such degradation effects are illustrated in Figure 4, which compares the $m/e = 80$ TPD spectra obtained after dosing a clean Au(poly) surface never exposed to Li at 150 K with the mass spectrometer, ion gauge and electron gun filaments off to 0.23 L $d_6$-PC (curve a) and to 0.1 L $d_6$-PC (curve b). As clearly indicated, the hot filaments induce the partial fragmentation of $d_6$-PC, yielding TPD features not found for the neat species.

Fragmentation patterns of intact deuterated solvents were determined either

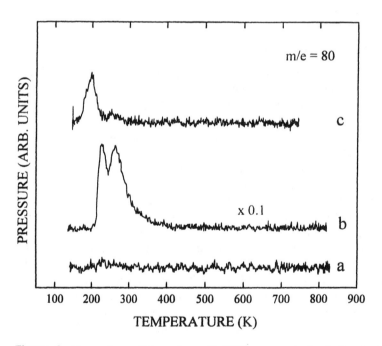

**Figure 4**  Comparison of the $m/e = 80$ TPD spectra obtained after dosing a clean Au(poly) surface with $d_6$-PC at 150 K with the filaments of the mass spectrometer, ion gauge and electron gun turned off (0.23 L $d_6$-PC, curve a) and on (0.1 L $d_6$-PC, curve b). The $m/e = 80$ TPD spectra shown in curve c were obtained by condensing ultrapurified $d_6$-PC (1 L), which had been left in the doser for several days, onto the gold surface. Heating rate: 3 K/s. (From Ref. 4.)

from the TPD spectra of thick layers of the materials, e.g., 10 to 30 L exposure, condensed on a clean substrate never exposed to Li at ca. 120 K (see below), or after flushing both the doser and the UHV chamber to desorb CO and other gases from the chamber walls until a stable mass spectrum was obtained. The relative abundance of the fragments observed after either of these procedures was assumed to be characteristic of the pure solvent and regarded as a reference standard for the analysis of TPD data.

## 5. Assembly of Solvent/Li/Substrate Interfaces

Solvent/lithium interfaces are formed by sequentially depositing lithium onto the clean substrate and, following rapid AES characterization, condensing the solvent onto it, as shown schematically in Figure 5. An array of techniques, including XPS, AES, TPD and FTIR, are then used to examine their electronic, structural and vibrational properties. Additional insight into the effects induced by Li was also obtained from studies of the behavior of THF and PC condensed on bare, clean, nominally unreactive substrates such as Ag and Au.

## 6. Other Considerations

*Electron-Beam Damage[3a]*

Exposure of thick layers of TDF (on the order of several monolayers) condensed on either Ag(poly) or Li/Ag(poly) at 135 K to the (AES) electron beam for as little as 1 min led to development of a dark spot at the beam position clearly visible with the naked eye. No significant amounts of carbon could be found with AES anywhere on the specimen upon raising the temperature above 300 K, except in the area originally probed by the beam at 135 K. This indicates that the layer of TDF is irreversibly damaged by the electron beam (at least for the current

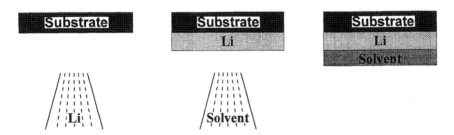

**Figure 5** Schematic representation of the assembly of supported lithium/solvent interfaces in ultrahigh vacuum (UVH).

settings used), limiting considerably the usefulness of AES for the characterization of molecular species.

The high lability of TDF, and numerous other organic compounds as well, to electron beam damage is also illustrated by the results of TPD experiments in which TDF/Ag(poly) specimens were heated using electron bombardment from the rear. Curve A, Figure 6, shows a typical TPD spectrum of *m/e* = 48 (the largest of all peaks) for a TDF layer condensed at ca. 130 K on bare Ag(poly). As clearly indicated, very poorly defined features could be found above 300 K, not only for *m/e* = 48 but also for other *m/e* values, including 80, 78, 48, 46 and 30 (not shown in this figure). Very different results were obtained for essentially identical experiments in which the temperature was raised using resistive heating or by flowing hot $N_2$ into the cryostat, for which the corresponding TPD (for *m/e* = 48 and all other fragments) showed no discernible features at $T >$ 300 K (see curve B in Figure 6).

*X-Ray Beam Damage[3a]*

Exposure of thick layers of TDF on Ag(poly) (see curves 1 in Figure 7) to Mg-$K_\alpha$ X-rays at 400 W for 2.5 h, while maintaining the sample at ca. 130 K, yielded

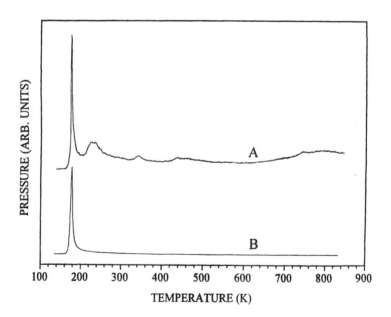

**Figure 6**  Comparison between TPD (*m/e* = 48) spectra for TDF condensed at 130 K on a Li-covered Ag(poly) surface using electron bombardment from the rear (curve A) and resistive heating (curve B). Heating rate: 5 K/s. (From Ref. 3.)

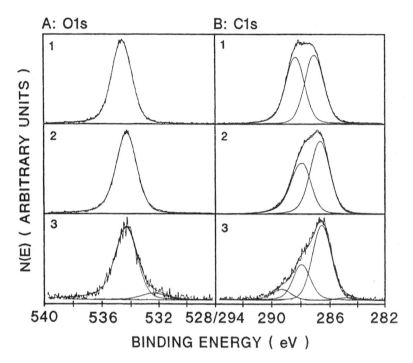

**Figure 7** C(1$s$) (right panel) and O(1$s$) (left panel) XPS spectra for a freshly condensed layer of TDF (ca. 5 nm thick) on Ag(poly) at 130 K before (curves 1) and after (curves 2) exposure to X-ray radiation from a Mg source for 2.5 h. Curves 3 show the corresponding XPS spectra obtained for the X-ray exposed specimen after it had been warmed up to room temperature using hot nitrogen. (From Ref. 3.)

a $C_{COC}/C_{CC}$ peak ratio, where COC and CC refer to carbon in C—O—C bonds and carbon in C—C bonds, respectively, of about 0.75, which is much lower than that observed for freshly condensed TDF layers (see curves 2 in Figure 7). More noticeable was the shift in the C(1$s$) XPS peak positions toward lower binding energies and the emergence of an additional spectral feature of much smaller intensity at higher binding energies, which are indicative of major structural modifications.

Perhaps the most compelling evidence for the X-ray induced decomposition of bulk layers of TDF is the fact that a substantial amount of TDF-derived features remained on the surface after the same X-ray irradiated specimen was warmed to room temperature (see curves 3, Figure 7). This behavior is unlike that observed for TDF films condensed on clean Ag and not exposed to the X-ray beam until after the specimen had been heated to ca. 300 K by flowing hot $N_2$ through the cryostat (vide infra), in which case clean Ag was obtained.

*Lithium Alloy Formation*

Considerable care must be exercised when using metal substrates capable of forming alloys with Li. For example, the TPD of layers of THF condensed on Ag and Au which had been exposed to Li in previous experiments, but which displayed no detectable Li in the AES, yielded TPD patterns quite different than those obtained on the same substrate metals never placed in contact with Li.

## B.  Lithium-Substrate Interactions

Figure 8 shows a series of AES spectra obtained at increasing temperatures (see values in each curve) for a 92% atomic percent Li layer by AES deposited on Ag(poly) wafer cooled to 113 K over a range of ca. 500 K. As discussed above, the peak at 52 eV is ascribed to the CVV AES transition of metallic Li, whereas the spectral features at around 270, 303, 375 eV are attributed to Ag [20]. Noticeably absent in the spectra recorded in the low temperature range is a peak at 505 eV characteristic of oxygen. The emergence of an oxygen signal as the temperature was raised is due to the gradual contamination of the surface (probably adventitious CO adsorption) during the relatively long times required for data acquisition.

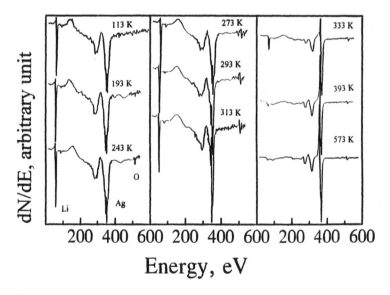

**Figure 8**  Series of Auger electron spectra recorded at increasing temperatures for a 92% Li/Ag specimen prepared at 113 K in UHV.

A plot of the Li AES signal as a function of temperature $(T)$ constructed from these data (see open circles in Figure 9) revealed a sudden drop in the amount of Li at ca. 350 K. Similar observations have been reported by Parker [21] for Li adsorption on Ag(111) using AES and $\Delta\Phi$ measurements. Also shown in Figure 9 (see solid circles) are AES data obtained by Wang and Ross [22] for Li vapor deposited on Ni(111) using the same methodology. The sharp decrease in the Li AES signal in this case occurs ca. 200 K higher than that observed for Ag and, therefore, closer to the bulk Li evaporation temperature. Multilayer Li desorption temperatures in the range 450–480 K have also been observed for Li films condensed on Ru(001) using TPD [23,24]. The strikingly different behavior displayed by Ag is due to its high miscibility with Li, with which it forms alloys of different compositions [25]. It may thus be concluded from the data shown in Figure 9 that Li diffusion into Ag, rather than evaporation, is responsible for the disappearance of Li from the surface region.

The tailing of the Li AES signal versus $T$ for the Li/Ag system at $T > 350$ K may be due in part to specific Li/Ag surface interactions of the type observed for the Li/Ni(111) and Li/Ru(001) interfaces, which account for the presence of submonolayer coverages on these substrates at temperatures much higher than its boiling point.

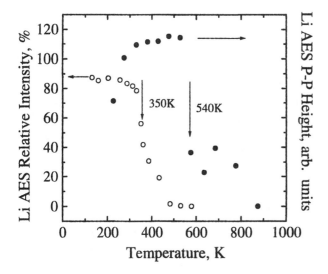

**Figure 9** Plots of the Li AES signal as a function of temperature for the Li/Ag system constructed from the data in Figure 8 (right ordinate, open circles). The solid circles (right ordinate) represent similar data reported by Wang and Ross for Li vapor deposited on Ni(111) [22]. (From Ref. 22.)

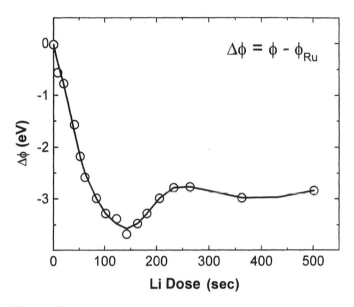

**Figure 10** Dependence of the work function ($\Delta\Phi$) on Li adsorption for Ru(001) surfaces. (From Ref. 27.)

Surface phase diagrams for alkali metals on Ni(111) derived from the analysis of LEED patterns display at least two ordered phases as a function of alkali metal coverage, $\theta_{alk}$, for $\theta_{alk} < 1/3$ [26]. In fact, the critical coverage for the formation of a second metal overlayer was found to be as low as 1/3 for K/Ni(111) and even lower for Cs/Ni(111). Furthermore, $\Delta\Phi$ as a function of $\theta_{alk}$ adsorbed on higher work function metals, such as Ni and Ag, drops to values below $\Phi_{alk}$, the work function of the alkali metal in bulk form, for $\theta_{alk}$ as small as 0.3, and increases thereafter to reach $\Phi_{alk}$ at $\theta_{alk}$ ca. 1.3. This behavior is shown in Figure 10 for Li on Ru(001) [27].

## C. Solvent-Substrate Interactions

Bare and, rather surprisingly, Li covered substrates (see below) displayed no affinity for TDF at room temperature [3]. This was clearly evidenced by the results obtained with AES, XPS, and TPD, for which no species could be detected either on or desorbing from the surface after dosings as high as 100 L. For this reason, all experiments, including those involving $d_6$-PC, were performed by condensing solvents at temperatures in the range 120–140 K onto the selected substrate.

## 1.  TDF/Ag(poly) [3]

*XPS*

The C(1*s*) and O(1*s*) XPS spectra for a thick (ca. 5 nm), freshly condensed layer of TDF on Ag(poly) at ca. 130 K, shown in curves 1 in the right and left panels in Figure 7, respectively, displayed characteristic C(1*s*) features at 287.0 (C in C—C bonds) and 288.2 eV (C in C—O—C bonds), and one oxygen peak at 534.6 eV. The relative integrated areas of these features were in good agreement with the TDF stoichiometry.

*TPD*

A series of *m/e* = 48 TPD spectra obtained for TDF exposures of 5 (curve A), 10 (curve B), 15 (curve C), 25 (curve D), 50 (curve E) and 100 L (curve F) to Ag(poly), kept at ca. 130 K, is shown in Figure 11. For 5 L, the TPD spectrum exhibited a single peak, $\alpha$, for which the maximum temperature ($T_\alpha$) was found to shift to lower values as the coverage was increased. The area under this peak ($A_\beta$) reached a limiting value for 15 L. For higher exposures, the TPD spectra displayed a second peak, $\beta$, with $T_\beta = 172 \pm 3$ K independent of coverage. In contrast to the behavior observed for peak $\alpha$, $A_\beta$ increased monotonically with exposure and became dominant above exposures of 50 L.

   As shown in Table 1, the fragmentation pattern of peak $\beta$ was identical to that observed for gas phase TDF (see above). However, a similar analysis of the fragments observed for peak $\alpha$, also given in that table, revealed systematic deviations of the peak ratios with respect to those obtained for neat TDF. These observations clearly indicate that peak $\beta$ is due to the simple sublimation of bulk-like condensed TDF, whereas peak $\alpha$ can be ascribed to TDF bound directly to bare Ag(poly). No evidence for carbon or oxygen impurities could be detected by AES after raising the temperature to 850 K, which corresponds to the end of the TPD scan.

## 2.  TDF/Au(poly) [10]

*FTIR*

A series of normalized reflectance spectra recorded for films of THF adsorbed and/or condensed on a clean Au substrate at 93 K, while monitoring simultaneously the mass with QCM, $m_{THF}$, are shown in panels A and B of Figure 12 [9]. Except for their relative intensities (see below), all of the spectral features observed for the heavier deposits (top two curves), including the peaks at 838, 908, 921, 1058, 1179, 1441, 2849, 2924 and 2947 cm$^{-1}$, are characteristic of frozen THF (at 93 K) [27]. A plot of $m_{THF}$ recorded during this entire experiment

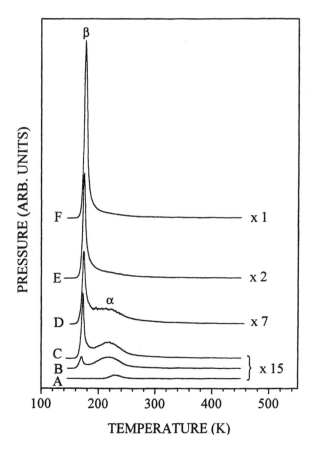

**Figure 11**   Series of TPD $m/e = 48$ spectra obtained for TDF at exposures of 5 L (curve A), 10 L (curve B), 15 L (curve C), 25 L (curve D), 50 L (curve E), 100 L (curve F). Heating rate: 5 K/s. (From Ref. 3.)

is shown in Figure 13, where the asterisks represent the short intervals involved in the opening and closing of the leak valve for THF dosing.

A plot of the integrated intensity of the cyclic C—O—C asymmetric stretch ($I_{C-O-C(asy)}$) at 1064 cm$^{-1}$ normalized by $I_{C-O-C(asy)}$ for the largest coverage versus $m_{THF}$, based on the data in panel B, Figure 12, respectively, is shown in Figure 14. The best fit to the peak for $m_{THF} = 0.096$ and 2.76 µg/cm$^2$ (which correspond to the spectra in curves a and f in Figure 12B, respectively) are given in panels A and B, Figure 15, respectively. The uncertainties in the normalized

**Table 1** Temperature Programmed Desorption Peak Areas Normalized by the Area of $m/e = 48$ for High and Low[a] Exposure of TDF to Ag(poly) and Li-Covered Ag(poly)

| Specimen | $m/e$ | | | | | |
|---|---|---|---|---|---|---|
| | 48 | 46 | 30 | 80 | 78 | 28 |
| vapor TDF | 100 | 53 | 44 | 16 | 16 | 18 |
| Ag high | 100 | 55 | 49 | 14 | 14 | 18–38 |
| Ag low | 100 | 64 | 57 | 12 | 12 | |
| Li/Ag high | 100 | 55 | 48 | 16 | 16 | |
| Li/Ag low | 100 | 64 | 55 | 11 | 12 | 100–140 |

[a] High and low in this context refer to TDF exposures for which the TPD spectra were dominated by the low ($\beta$) and high ($\alpha$) temperature features, respectively. The uncertainties in the magnitude of the fragment ratios derived from a statistical analysis of the results of many independent measurements does not exceed 7% for all species; for example, for Ag at high exposure the area under the $m/e = 30$ peak was $49 \pm 3$, except for $m/e = 28$, for which the error was much greater (ca. 50%).

$I_{C\text{-}O\text{-}C(asy)}$ yielded values ranging from 1% for the lowest coverage down to 0.3% for the highest coverage. It may be noted that for the lowest value of $m_{THF}$ the uncertainty was assumed to be as large as the unusual (and yet unexplained) peak observed in the plot. The results shown in Figure 14 indicate that there is a marked decrease in the slope of the normalized $I_{C\text{-}O\text{-}C(asy)}$ versus $m_{THF}$ curve as the amount of THF increases. Although no satisfactory explanation for this phenomenon has yet been found, saturation effects do not seem to play a role [10].

A more detailed examination of the overall spectral behavior revealed that as $m_{THF}$ increased the ratio of the peak heights of the 1064 and 921 cm$^{-1}$ peaks ascribed to C—O—C cyclic asymmetric and symmetric stretches, respectively, and the relative intensities of the CH stretching modes at 2924 cm$^{-1}$ (asymmetric) and 2854 cm$^{-1}$ (symmetric) modes also increased.

These observations provide clear evidence that THF molecules in contact with the gold adopt a net orientation with respect to the surface normal, which is different than molecules in the upper layers. This phenomenon is due most probably to specific adsorbate/substrate interactions, including distortions derived from surface electric fields. The same arguments have been invoked to explain drastic changes in the relative intensities of spectral features for alkanes condensed on Pt(111) as a function of the thickness of the adsorbed layer [29]. The relative intensities of the most prominent peaks were still found to differ from those obtained for very thick THF layers (not shown here). This effect suggests that THF microcrystals appear to grow along a preferential axis with respect to the underlying substrate.

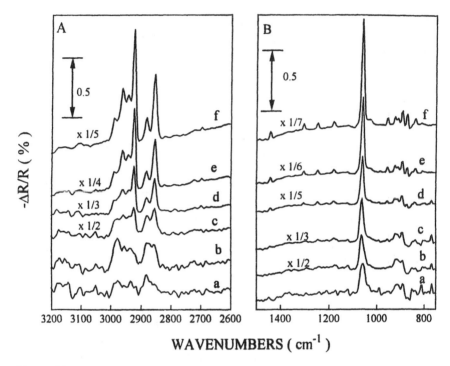

**Figure 12** Series of normalized reflectance spectra in the regions 2600–3200 cm$^{-1}$ (A) and 750–1500 cm$^{-1}$ (B) for films of THF condensed on a clean Au substrate at 93 K as a function of $m_{THF}$ determined from simultaneous dual QCM measurements: (a) 0.1, (b) 0.17, (c) 0.34, (d) 0.67, (e) 1.46, (f) 2.76 $\mu g_{THF}/cm^2$. The reflectance spectra recorded for the bare Au substrate at the same temperature was used as a reference. (From Ref. 10.)

Also noteworthy is the much larger width of the 1064 cm$^{-1}$ peak for the lowest coverage (ca. 20 cm$^{-1}$) compared to the thicker film (ca. 10 cm$^{-1}$). A representative example of peak broadening in vibrational spectroscopy of molecular species adsorbed on metal surfaces has been provided by Harris et al. [30], who monitored with sum frequency generation the C—H stretching mode of methyl thiolate (CH$_3$S) adsorbed at 300 K on Ag(111) in UHV. In particular, a freshly prepared CH$_3$S/Ag(111) interface displayed a peak at 2918 cm$^{-1}$ with about twice the width as that obtained after annealing for 7 h at room temperature. Therefore, it seems possible that a transient effect may also be involved in the THF/Au system.

Further insight into the nature of the Au/THF interactions was gained based on a series of temperature programmed desorption (TPD) experiments involving a Au(poly) foil described in the next subsection.

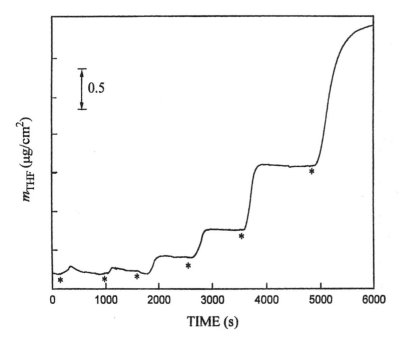

**Figure 13**  Plot of $m_{THF}$ as a function of time recorded during the entire experiment in Figure 12. The asterisks represent the short time interval associated with the THF dosing. (From Ref. 10.)

*TPD[3]*

In analogy with the results obtained for THF on a Ag(poly) foil, the $m/e = 80$, 48, 30 TPD at moderate coverages (4.5 L), the TPD spectra obtained at a heating rate of 3 K/s displayed two peaks centered at 181 ($\alpha$) and 131 K ($\beta$; see Figure 16). For low exposures, only peak $\alpha$ could be detected, whereas as the coverage increased, peak $\beta$ was largely dominant. These observations are consistent with peaks $\alpha$ and $\beta$ being attributed to adsorbed and bulklike material, respectively. The latter is consistent with observations made in the FTIR spectra, for which the features attributed to bulklike THF disappeared upon raising the temperature to 143 K.

### 3. $d_6$-PC/Au(poly) [4]

*TPD*

All major fragment ions of $d_6$-PC condensed on Au(poly) never exposed to Li for dosages in the range 0.35 to 3 L displayed two clearly defined sets of peaks

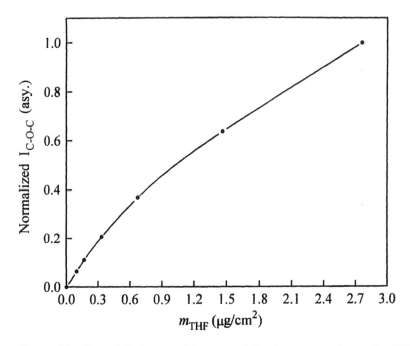

**Figure 14**  Plots of the integrated intensity of the ring asymmetric stretch of THF at 1064 cm$^{-1}$ vs. $m_{THF}$ obtained from the data in Figure 12. The errors are too small to be shown. (From Ref. 10.)

in the TPD, centered at about 230 and 260 K, denoted as $\alpha$ and $\beta$, respectively. As shown for $m/e = 90$ in Figure 17, all such fragments showed a large monotonic increase in the intensity of peak $\alpha$ with exposure; hence, this feature can be attributed to the desorption of intact bulklike $d_6$-PC from the surface. For small $d_6$-PC exposures, e.g., 0.2 L (curve a in Figure 17), the TPD displayed only peak $\beta$, for which the rate of growth with exposure was much smaller than that observed for peak $\alpha$. On this basis, peak $\beta$ can be ascribed to the desorption of $d_6$-PC molecules adsorbed directly on the Au(poly) surface. Some indication that the species associated with peak $\beta$ is activated or partially fragmented $d_6$-PC is given by the differences in the relative intensities of the most prominent fragments for peaks $\alpha$ and $\beta$ (see Table 2). A graphical illustration of this phenomenon is shown for the TPD of $m/e = 90$ and 34 for a $d_6$-PC exposure of 0.35 L in curves a and b, Figure 18, respectively, where the intensity of peak $\alpha$ compared to that of peak $\beta$ is much greater for $m/e = 34$ than for $m/e = 90$.

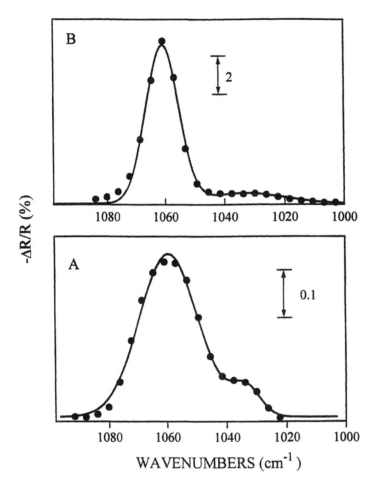

**Figure 15** Spectral deconvolution of the 1064 cm$^{-1}$ feature for samples with $m_{THF}$ = 0.096 (A) and 2.76 $\mu g_{THF}/cm^2$ (B), shown in curves a and f in Figure 12B, respectively. (From Ref. 10.)

## 4. $d_6$-PC/Ag(poly) [4]

Peaks of essentially the same shape and, therefore, derived from the same species, were observed for $m/e$ = 48, 44, 46, 30, 18, and 4 between 500 and 600 K for $d_6$-PC condensed on Ag(poly) surfaces prepared under the same conditions described in the previous subsection for Au(poly).

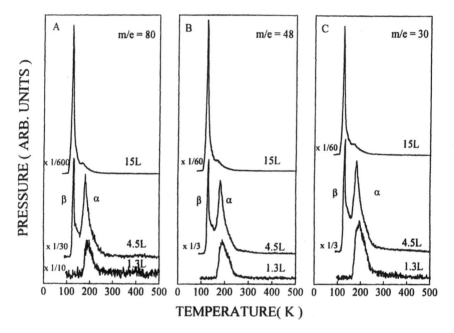

**Figure 16**  Series of $m/e$ = 80 (A), 48 (B), and 30 (C) TPD spectra for three different dosages of TDF (1.3 L, 4.5 L and 15 L) on a polycrystalline Au foil at ca. 100 K obtained in chamber A. Heating rate: 3 K/s. (From Ref. 10.)

## 5.  THF/Al(111) [31]

*XPS/UPS*

Interactions between thick (ca. 10–20 nm) films of THF condensed on Al(111) as a function of temperature have been studied by Ross et al. at the Lawrence Berkeley National Laboratory (LBNL) by using both XPS and UPS [31]. The C(1s) and O(1s) XPS spectra, as seen in the left and right panels of Figure 19, showed at 135 K features resembling those observed at CWRU for TDF/Ag(poly), as described in section II.C.1.

As the sample was warmed past the melting point of the film (165 K), clear Al(2p) XPS features could be identified in the spectrum consistent with the sublimation of the THF film. Furthermore, the binding energies of the C(1s) and O(1s) peaks shifted uniformly by ca. 0.6 eV toward lower values, an effect attributed by the investigators to the charging of the rather insulating character of the thick THF film, which disappears as the material evaporates. Also noted was a decrease in the C/O integrated peak areas from 4.2, as predicted from the chemical formula of THF, to 3.3, which remained constant up to 470 K. Despite the

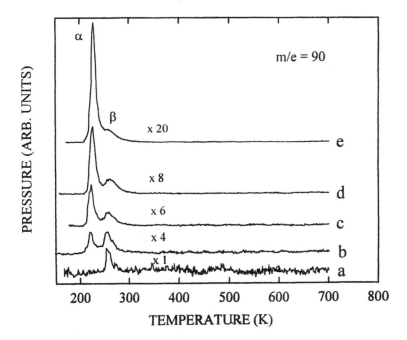

**Figure 17** $m/e = 90$ TPD spectra of $d_6$-PC adsorbed on Au(poly) at ca. 160 K for exposures of (a) 0.2, (b) 0.35, (c) 0.15, (d) 3, (e) 6 L. Heating rate: 3 K/s. (From Ref. 4.)

**Table 2** Relative[a] Peak Areas[b] of the Fragments Observed for the Low ($\alpha$) and High ($\beta$) Temperature Features in the Temperature Programmed Desorption of PC-$d_6$ Adsorbed on Pristine Au(poly) at ca. 120 K

| Relative peak areas | Peak $\alpha$ | Peak $\beta$ |
|---|---|---|
| Fragment | | |
| $m/e = 180$ | $0.90 \pm 0.50$ | $0.25 \pm 0.10$ |
| $m/e = 90$ | $3.70 \pm 0.70$ | $1.60 \pm 0.60$ |
| $m/e = 64$ | $50.5 \pm 8.00$ | $53.60 \pm 7.70$ |

[a] The values represent the average and standard deviations for the areas under the specified $m/e$ TPD feature normalized by the area under the $m/e = 34$ peak obtained in six independent measurements.

[b] The peak areas were determined by fitting the TPD curves using a combination of a Gaussian and a Lorentzian. Fits to the formulas for first and second order desorption kinetics were also attempted but did not provide significantly better statistical results.

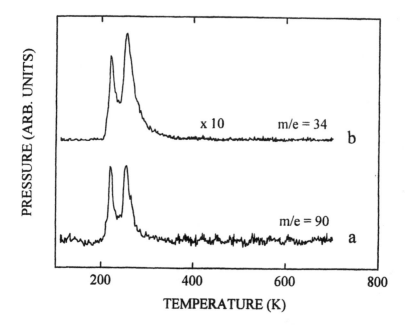

**Figure 18** $m/e = 90$ (curve a) and $m/e = 32$ (curve b) TPD spectra for 0.35 L $d_6$-PC adsorbed on Au(poly) at ca. 150 K. Heating rate: 3 K/s. (From Ref. 4.)

drop in the C/O ratio, the He(II) UPS yielded THF-type features over much of this temperature range (200–400 K), leading the investigators to suggest the formation of a polymeric form of THF.

At temperatures higher than 470 K, there was a significant loss of carbon from the surface, as evidenced by the decrease in the mostly carbidic XPS C(1s) peak. In addition, the O(1s) signal was characteristic of an oxide, rather than of an ether-type functionality. Support for the presence of $Al_2O_3$ was obtained from the emergence of a second higher binding energy feature in the Al(2p) XPS spectrum.

## 6. PC/Al(111) [31]

*XPS/UPS*

The group at LBNL also examined the interaction of PC with Al(111). The results of this study are summarized in Figure 20, which displays C(1s) (left panel) and O(1s) (right panel) XPS spectra as a function of temperature. Thick films of PC

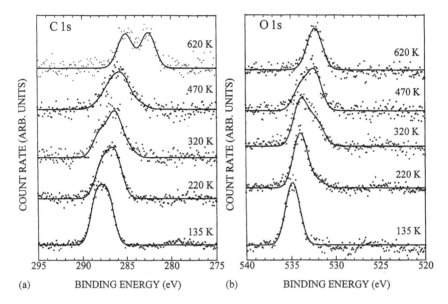

**Figure 19** (a) C(1s) XPS spectra acquired as a function of temperature for THF/Al(111). (b) O(1s) XPS spectra acquired as a function of temperature for THF/Al(111) sample dosed with THF at 135 K. (From Ref. 31.)

on Al(111) showed the same binding energy shifts in the XPS spectra as those found for THF. These shifts disappeared following desorption of the thick, bulk-like PC. Unlike the behavior found for THF/Al(111), however, a significant amount of unreacted PC was detected on the surface past its melting point. This phenomenon cannot be explained based on the model put forward for THF, as PC does not undergo polymerization. Ross et al. theorized that a portion of the PC reacts with Al(111) through extraction of the carbonyl oxygen, thereby forming a new compound that prevents some of the PC from desorbing.

Subsequent heating of the specimen led to a continuous extraction of carbonyl oxygen atoms from the interface until, at about room temperature, no unreacted PC is left on the Al(111) surface. Based on the analysis of their XPS spectra, the investigators proposed formation of an alkoxide on the Al(111) surface, represented as $AlOCHCH_3CH_2OAl$. Upon heating to 470 K, the film became completely deoxygenated, leaving elemental carbon and $CH_x$-type fragments on the surface. Further heating to 620 K yielded similar products to those formed for THF/Al(111) under the same thermal protocol, i.e., a mixture of $Al_2O_3$ and $Al_4C_3$.

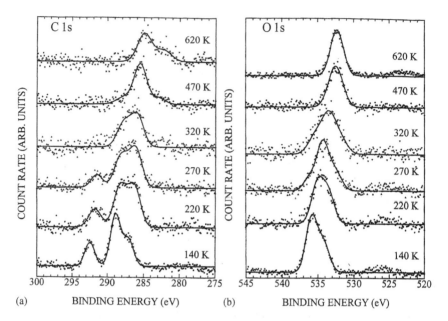

**Figure 20** (a) C(1s) XPS spectra obtained for PC/Al(111) versus temperature. (b) O(1s) XPS spectra obtained for PC/Al(111) versus temperature. (From Ref. 31.)

## D. Characterization of Solvent/Lithium/Substrate Interfaces

### 1. TDF/Li/Ag [3]

*XPS*

The Li(1s), O(1s) and C(1s) regions obtained at ca. 300 K for a TDF layer which had been first condensed at ca. 130 K on Li/Ag(poly) and later warmed to room temperature (using hot nitrogen) to desorb bulklike material are shown in Figure 21. Particularly noteworthy is the fact that the C(1s) and O(1s) binding energies associated with this specimen are lower than those found for bulk TDF, and, in the case of C, the two large features display very different integrated areas. This indicates that the interactions between TDF and Li are strong enough to induce substantial modifications to the cyclic ether structure. This is in contrast with the behavior observed for a single monolayer of TDF adsorbed on Ag(poly) at 130 K, in which case the XPS (recorded at that temperature) was essentially identical to bulklike TDF and the AES, TPD and XPS spectra obtained after raising the temperature to 300 K (using hot nitrogen), which yielded no evidence for the presence of carbon or oxygen on the surface. This lack of residual impurities

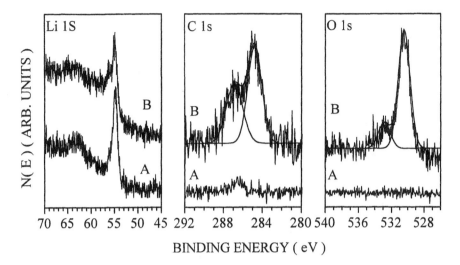

**Figure 21** Li(1*s*), C(1*s*) and O(1*s*) XPS regions obtained on clean Li (curve A) and after adsorption and subsequent desorption (using hot nitrogen) of condensed layers of TDF (ca. 50 L) at 130 K (curve B). (From Ref. 3.)

offers strong proof that the damage induced by short X-ray exposures is negligible, and therefore that the spectral changes observed in the case of the TDF/Li/ Ag(poly) specimen are promoted by the Li overlayer. It may also be noted that the C(1*s*) binding energies associated with the X-ray damaged TDF layer are higher than those found for the TDF/Li/Ag(poly) interface.

The fact that the sticking coefficient of TDF on Li (vide supra) at ca. 300 K was found to be negligible points to a high activation energy for gas phase adsorption; however, once this barrier is overcome by, for example, forcing molecules on the surface via condensation (as was the case in these experiments), more complex pathways may be favored. Far more revealing information was obtained from the results of TPD experiments described in detail below.

*TPD*

The $m/e = 48$ TPD spectra for TDF condensed at 130 K on Ag(poly) and Li/ Ag(poly) at coverages low enough for peak $\alpha$ to be the dominant feature are shown in curves A and B, Figure 22, respectively. As indicated, the width of the peak for the Li/Ag(poly) specimen was larger than that for Ag(poly) with a tail extending beyond room temperature.

Both Li covered Ag(poly) and bare Ag(poly) yielded essentially identical ratios for $m/e = 80, 78, 48, 46, 30$. However, for Li/Ag(poly) the $m/e = 28$

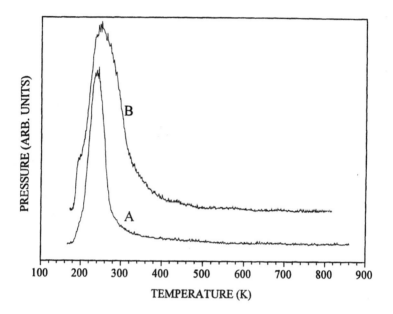

**Figure 22** TPD ($m/e = 48$) spectra for TDF condensed at 130 K on bare Ag(poly) (curve A) and Li-covered Ag(poly) (curve B). Heating rate: 5 K/s. (From Ref. 3.)

peak for low exposures was five times larger than that observed for neat TDF. As was the case with bare Ag after TPD, the amount of carbon on the surface, as determined by the appropriate AES amplitude ratio (vide supra), $0.45 \pm 0.03$, was essentially negligible. In contrast, the corresponding ratio for the oxygen AES feature was slightly higher than that for clean Ag. This difference, however, does not appear to originate from TDF, as similar values were observed for Li/Ag surfaces after heating to 800 K without prior exposure to TDF, and therefore may be attributed to a very small oxygen contamination within the lithium layer.

The most striking effect derived from the presence of Li on the silver surface is the appearance of a sharp $m/e = 4$ TPD peak at 600 K associated with $D_2$ (see curve A in Figure 23) not observed on bare Ag(poly) (see curve B in Figure 23). Experiments in which Li/Ag surfaces at 300 K were exposed to $H_2$ (generated purposely by the titanium sublimation pump) yielded an $m/e = 2$ TPD feature in the same temperature range. No such peak was observed for identical experiments involving bare Ag(poly) surfaces. Based on these observations and the fact that the peak temperature is very similar to that at which LiH is expected to undergo thermal decomposition, it can be concluded that this TPD feature corresponds to the thermal decomposition of LiD [32,33]. Further arguments in support of this view may be found in the original literature [3].

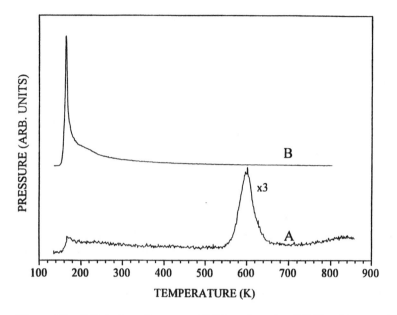

**Figure 23** TPD ($m/e$ = 4) spectra of TDF condensed at 130 K on Li-covered Ag(poly) (curve A) and on Ag(poly) (curve B). Heating rate: 5 K/s. (From Ref. 3.)

Since the amount of carbon and oxygen left on the surface after TPD was essentially negligible, and given that the possibility of C and O diffusion into Ag seems unlikely, it can be concluded that TDF (THF) reacts with Li at temperatures below ca. 350 K (following the trailing edge of the TPD peak associated with larger masses) to form LiD as the only reaction product left on the surface after heating to 350 K.

Additional insight into the possible nature of the compounds generated either prior to or during LiD formation was gained by examining the nature of the desorbed products. Based on a thorough survey of the cracking pattern literature, one of the most likely candidates is 1-butene-3,4-epoxy, $C_4D_6O$. This compound lacks a parent peak ($m/e$ = 76), but displays the same major fragments found for TDF [34]. Another possibility involves the formation of $C_3D_6$ (mass = 48) and $CD_2O$ (mass = 32), for which all the fragments observed are in agreement with literature data [32]. The main virtue of this latter model is that $CD_2O$ can lose $D_2$ and thus account for the rather large amounts of CO in the TPD spectrum. It must be stressed, however, that the presence of an $m/e$ = 80 peak in the TPD spectra of TDF at low coverages indicates that about two-thirds of the TDF overlayer desorb intact from the surface.

In view of the similarities between the TPD behavior of TDF on Ag and

Li/Ag (see Table 1), it seems conceivable that the thermal desorption pathways on both surfaces are the same except that in the latter case deuterium reacts with Li to form LiD.

## 2.  $d_6$-PC/Li/Au(poly) [4]

In contrast to the results obtained with clean Ag(poly) surfaces, experiments involving very low exposures of $d_6$-PC (so as to avoid contributions due to bulklike species) on Li/Au(poly) surfaces at ca. 120 K yielded no detectable TPD peaks with $m/e = 90$ and 64, the largest $d_6$-PC fragments, in the whole temperature range examined. This indicates that PC in contact with Li gives rise to the activation and, at sufficiently high temperatures, to the complete decomposition of the species on the surface. Marked changes in the shape and positions were also found for other $m/e$ values compared to their intact $d_6$-PC analogs.

The $m/e = 44$ TPD spectrum for $d_6$-PC(4L)/Li/Au(poly) (see curve b in Figure 24), for example, was characterized by a broad peak extending from about

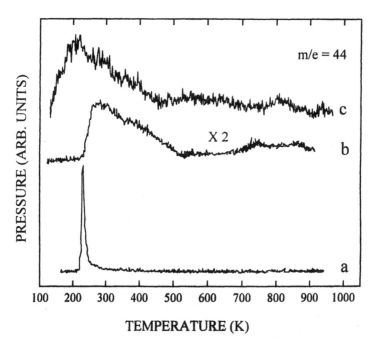

**Figure 24**   TPD ($m/e = 44$) spectra for 4 L of $d_6$-PC condensed on a bare (curve a) and a Li-covered Au(poly) (curve b). Heating rate: 3 K/s. Curve c was obtained from a layer of lithium butyl carbonate synthesized in UHV on the same substrate under otherwise identical conditions (see text for details). (From Ref. 4.)

230 to 520 K and two smaller peaks centered about 750 and 860 K. Peaks with precisely the same shape and temperature, although with about one-eighth of the intensity, were found for $m/e = 28$, which corresponds almost exactly to the ratios observed with our mass spectrometer for the cracking of gas phase $CO_2$. This indicates that all the clearly detectable $m/e = 44$ features in the TPD spectra are indeed derived from carbon dioxide.

One possible explanation for the broad $m/e = 44$ feature may be found in the thermal decomposition of a lithium alkyl carbonate produced by the reaction between PC and Li, for which $CO_2$ would be released at a much lower temperature than the corresponding inorganic carbonate. The formation of such a species has been suggested by Aurbach and co-workers on the basis of in situ and ex situ external reflection FTIR measurements performed in PC-based electrolytes [18]. Support for this assignment was obtained from experiments in which a genuine alkyl carbonate was prepared in UHV by exposing to $CO_2$ a layer of lithium alkoxide formed by the adsorption of an alcohol onto the Li surface, as described in Section I.E.

The smaller high temperature $m/e = 44$ peaks observed in curve b (Figure 24) can be attributed to the thermal decomposition of $Li_2CO_3$. Support for this view was obtained from experiments involving small exposures of Li/Au(poly) to $CO_2$, for which the TPD showed two $m/e = 44$ peaks (with corresponding peaks at $m/e = 28$) at about the same temperatures.

Also found in this series of experiments were features for $m/e = 4$ and 18 centered at about 600 K. A more detailed discussion of the possible species responsible for these peaks will be given later. AES spectra recorded immediately after each TPD run was completed yielded large amounts of oxygen and carbon. The presence of lithium could not be clearly discerned, however, due to the interference of Au AES features in the low energy region.

## 3.   $d_6$-PC/Li/Ag(poly) [4]

### *TPD*

A series of additional experiments was performed with a Ag(poly) foil [rather than a Au(poly)] foil to examine in more detail the nature of the reaction products of $d_6$-PC with metallic Li, and possible substrate effects. According to data compiled in the literature (Table 3), all of these fragments, except $m/e = 4$, are consistent with, albeit not unique to, ethylene oxide, e.g., acetaldehyde. No features could be identified for $m/e = 32$ and 64, indicating that propylene oxide, if produced, yields signals too small to be detected. Furthermore, no differences were found between the peak shapes and temperatures obtained for these experiments and those observed using Au(poly); hence, the reaction pathway does not seem to be affected by the nature of the substrate. Based on the behavior found for BuOLi, for which the series of high temperature peaks are found in the range

**Table 3**  Fragmentation Patterns of Selected Organic Molecules[a,b,c]

| Compound | Fragmentation pattern | | | | | |
|---|---|---|---|---|---|---|
| 1-n-Butanol | 31(34)[d] | 56(64) | 41(46) | 43(50) | 27(30) | 42(48) |
| (CH$_3$(CH$_2$)$_3$OH) | CH$_2$OH | C$_4$H$_8$ | C$_3$H$_5$ | C$_3$H$_7$ | C$_2$H$_3$ | C$_3$H$_6$ |
| | (100)[e] | (81) | (62) | (60) | (50) | (31) |
| Propylene oxide | 28(32) | 58(64) | 27(30) | 31(34) | 29(30,34) | 43(46) |
| (CH$_3$CHOCH$_2$) | C$_2$H$_4$ | C$_2$OH$_6$ | C$_2$H$_3$ | CH$_3$O | CHO,C$_2$H$_5$ | C$_2$OH$_3$ |
| | (96.9) | (64.0) | (46.2) | (34.5) | (28.3) | (23.3) |
| Ethylene oxide | 44(48) | 29(30) | 15(18) | 43(46) | 42(44) | 14(16) |
| (CH$_2$OCH$_2$) | C$_2$OH$_4$ | CHO | CH$_3$ | C$_2$OH$_3$ | C$_2$OH$_2$ | CH$_2$ |
| | (100) | (82.6) | (52.5) | (21.6) | (12.6) | (12.1) |
| Acetaldehyde | 29(30) | 44(48) | 43(46) | 42(44) | 28(28) | 41(42) |
| (CH$_3$CHO) | CHO | C$_2$H$_4$O | C$_2$H$_3$O | C$_2$H$_2$O | CO | C$_2$HO |
| | (100) | (88) | (50) | (15) | (9) | (6) |

[a] *Index of Mass Spectral Data*, ADM11, American Society for Testing and Materials, 1969.
[b] *Advances in Mass Spectrometry*, vol. 1, J. D. Waldron, Ed., Pergamon, 1959, p. 349.
[c] *Interpretation of Mass Spectra*, F. W. McLafferty, University Science Books, 1980.
[d] The numbers in parentheses are the masses of the fragments in perdeuterated form.
[e] The numbers in parentheses below the chemical identity of the fragment are relative abundances
of the nondeuterated materials.

$550 < T < 600$ K, it seems quite plausible that ethylene oxide is derived from the thermal decomposition of a lithium alkoxide (most probable propoxide) produced by the (lower temperature) release of $CO_2$ from the corresponding Li alkyl carbonate. In fact, the fragments observed for the thermal decomposition of BuOLi were consistent with at least two different epoxides.

The presence of a prominent $m/e = 4$ peak not associated with any of the species proposed above could be due to a reaction between metallic Li and PC to yield lithium hydride. Although the same type of process was proposed for the reaction between THF and Li (see above), the temperature observed in this case seems too high for LiH decomposition. A more likely source of dihydrogen (or $D_2$) in the TPD is the thermally induced dehydrogenation of one (or more) adsorbed reaction products.

## 4.  THF/Li/Al(111) [35]

*XPS*

Ross et al. implemented the same techniques introduced much earlier by the group at CWRU to investigate the interaction of condensed THF (as well as PC, see Section E) films on vapor-deposited Li layers on Al(111). Plots of the amount of Li as determined from AES versus temperature reported in Ref. [35] are similar in appearance to those shown in Figure 9 in this work for Li vapor deposited on Ag, except that the onset temperature for formation of the Li alloy is lower for Al than for Ag. On this basis, it is not possible in the case of Li/Al(111) to have metallic Li in contact with the solvent layer at room temperature and, therefore, the relevance of data obtained using Al as a substrate to study passive films on Li electrodes may not be immediate.

XPS spectra for the C(1s) and O(1s) for THF layers condensed on Li/Al(111) are shown in the left and right panels of Figure 25, respectively. It is

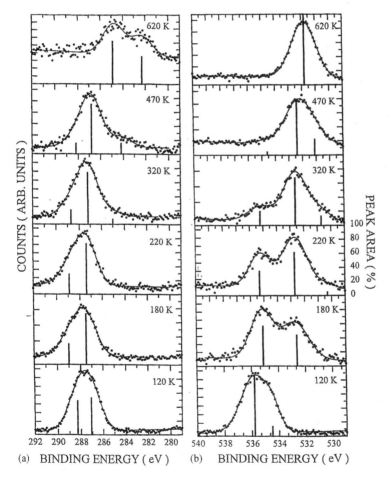

**Figure 25** (a) C(1s) XPS spectra for THF/Li-Al(111) versus temperature. (b) O(1s) XPS spectra for THF/Li-Al(111) versus temperature. (From Ref. 35.)

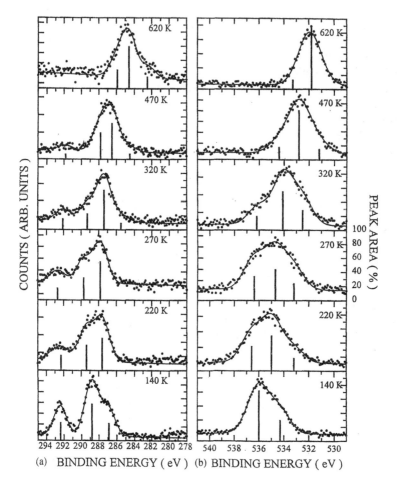

**Figure 26** (a) C(1s) XPS spectra for PC/Li-Al(111) versus temperature. (b) O(1s) XPS spectra for PC/Li-Al(111) versus temperature. (From Ref. 35.)

important to stress that the data collected at 320 K are different than those obtained earlier for TDF/Li/Ag(poly) by the group at CWRU at room temperature. In particular, the O(1s) signal in curve b, Figure 25 (right panel) shows two prominent peaks at 535.2 and 532.5 eV. Two similar features were also reported by Wang et al. [3] as depicted in curve b (right panel), Figure 21, except that the energies shown are about 2 eV lower than those obtained by Zhuang et al. [35]. The C(1s) spectra reported by the group at LBNL consisted of two closely spaced peaks centered at 288.3 and 287.1 eV. These features are not only closer than the two peaks observed by Wang et al. [3] but their binding energies (at 287 and 285 eV) are significantly higher than those in curve b, Figure 21 (middle panel). These discrepancies are at present difficult to reconcile. Nevertheless, Ross et al. concluded on the basis of their data that Li, unlike Al(111), activates THF at about 120 K, leading at higher temperatures (ca. 180 K) to ring opening and subsequent polymerization. Moreover, as the temperature was further increased, chain-terminating reactions ensued, accompanied by the evolution of hydrocarbon gas or gases, such as ethylene and propylene, yielding at 320 K a surface composed exclusively of alkoxide.

## 5.  PC/Li/Al(111) [35]

*XPS*

Condensation of PC on vapor-deposited Li thin films on Al(111) at 140 K, i.e., well below the melting point of PC, gave rise to XPS C(1s) and O(1s) features virtually identical to those of PC condensed on Al(111) as described in Section C. As shown by the C(1s) and O(1s) XPS spectral data in Figure 26, rather large changes in chemical environment could be identified as the temperature of the specimen was raised. The investigators attributed these changes to the formation of a Li alkyl carbonate species, probably derived from propylene carbonate on the surface, before the melting point of the PC is reached. Upon further increase in temperature, the alkyl carbonate was found to decompose to form a Li alkoxide between 270–320 K. This proposed pathway is consistent with that suggested earlier by the group at CWRU based on the TPD behavior of the corresponding alkyl carbonate synthesized in situ by sequential deposition of reactants to be described in detail in Section II.E. Also, in accordance with the results reported by this latter group is the formation of small, i.e., <5%, amounts of Li carbonate on the surface.

## E.  Organic Synthesis in Ultrahigh Vacuum [4]

The primary aim of these experiments was to examine the TPD spectra of a genuine lithium alkyl carbonate formed on the surface of Au(poly) to determine whether the thermal decomposition of this UHV-synthesized material exhibits an $m/e = 44$ TPD spectra which resembles that observed in the TPD of $d_6$-PC adsorbed on Li/Au(poly). The synthetic pathway employed is based on the sequential condensation of an ultrapurified alcohol (*n*-butanol, BuOH) onto a Li/Au(poly) surface followed by exposure to $CO_2$ to form the desired alkyl carbonate.

High performance liquid chromatography (HPLC) grade BuOH (Aldrich) was purified by adding a few drops of liquid Na-K alloy to remove residual water, followed by a series of freeze-thaw cycles. The fragmentation pattern of BuOH was determined by condensing 20 ML of the material onto a clean Au foil in UHV and monitoring via TPD the peaks associated with the bulk desorption (ca. 200 K at a heating rate of 3 K/s). Nine masses (74, 56, 44, 43, 41, 31, 28, 18, and 2) were investigated in two different runs.

### 1.  *n*-Butanol Adsorption on Li/Au(poly)

The TPD of a BuOH(20 L)/Li(17 ML)/Au(poly) surface, prepared by condensing ultrapurified BuOH on a Li/Au(poly) surface kept at 120 K, revealed, in addition to the features characteristic of bulk BuOH at 200 K, peaks for $m/e$ = 57 (curve a), 44 (curve b), 43 (curve c) and 2 (curve d), centered at 580 K (see Figure 27). Also noteworthy is the presence of a second $m/e$ = 2 peak at 435 K. A reaction pathway consistent with these data involves the dehydrogenation of BuOH to form the corresponding lithium butoxide (BuOLi) and lithium hydride, which at a sufficiently high temperature undergoes thermal decomposition to yield elemental Li and dihydrogen:

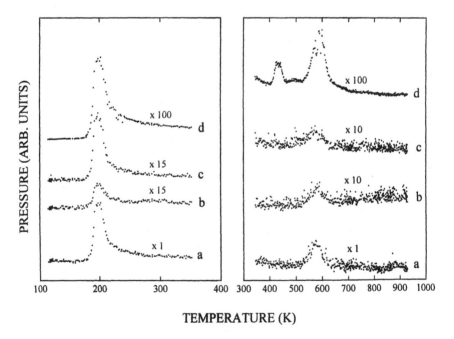

**Figure 27**  TPD spectra of a BuOH(20 L)/Li(17 ML)/Au(poly) surface prepared by condensing ultrapurified BuOH (20 L) on a Li(17 ML)/Au(poly) substrate at 120 K for $m/e$ = 57 (curve a), 44 (curve b), 43 (curve c) and 2 (curve d). The high temperature features have been expanded for clarity. (From Ref. 4.)

$$CH_3-CH_2-CH_2-CH_2OH + 2Li$$
$$\rightarrow CH_3-CH_2-CH_2-CH_2OLi + LiH \quad (1)$$

$$2LiH \xrightarrow{\text{heat}} H_2 + Li \quad (2)$$

Support for this model was provided by the results of experiments in which a freshly prepared Li/Au(poly) surface was exposed to $H_2$ at 140 K. As shown in Figure 28, the $m/e = 2$ TPD in this case showed a clearly defined peak at about 440 K, attributed to the thermal decomposition of LiH [see Eq. (2)], which is very close to the peak temperature of the small feature observed in curve d in Figure 27.

BuOH/Li/Au(poly) surfaces assembled at 120 K, heated to ca. 200 K to desorb bulklike BuOH, and cooled back to 120 K prior to the initiation of the TPD, showed the same high temperature features as those of nonheated surfaces prepared in the same fashion. A more detailed analysis of the high temperature TPD features will be given later in this section.

The AES spectra of the surface after the TPD experiments were completed

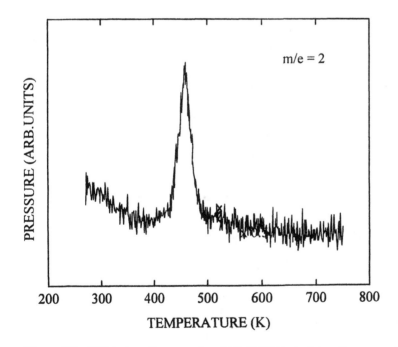

**Figure 28** TPD ($m/e = 2$) spectra for a Li(2.4 ML)/Au(poly) surface exposed to 162 L of $H_2$ at ca. 300 K. Heating rate: 3 K/s. (From Ref. 4.)

showed large amounts of oxygen (60%) and carbon (15%) with no detectable Li
AES features.

## 2.  Carbon Dioxide Adsorption on a BuOH/Li/Au(poly) Surface

Figure 29 shows TPD spectra for $m/e$ = 74 (a), 43 (b), 44 (c), 28 (d) and 2 (e)
for a surface prepared by condensing 21 L of BuOH at 150 K on Li(15 ML)/
Au(poly), heating to 200 K for 3 min to remove (most of the) excess bulk BuOH,
and further exposing to carbon dioxide first (74 L) at 200 K and later at 130 K
(34 L). A few minutes were allowed to elapse prior to the acquisition of the TPD
data to pump excess gas phase $CO_2$. As indicated in this figure, and in curve c
of Figure 24, the $m/e$ = 44 spectrum is characterized by a broad peak in the
temperature range 200–400 K and contains two distinct contributions:

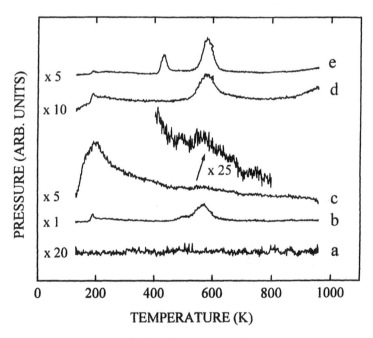

**Figure 29**  TPD spectra for $m/e$ = 74 (curve a), 43 (curve b), 44 (curve c), 28 (curve
d) and 2 (curves e) for a surface prepared by condensing 21 L of BuOH at 150 K on
Li(15 ML)/Au(poly), heating to 200 K for 3 min to remove (most of the) excess bulk
BuOH, and further exposing to carbon dioxide first (74 L) at 200 K and later at 130 K
(34 L). A few minutes were allowed to elapse prior to the acquisition of the TPD data
to pump excess gas phase $CO_2$. (From Ref. 4.)

1. A feature in the $m/e = 44$ and $m/e = 28$ traces at a temperature much lower than that required for BuOH bulk sublimation, which can be ascribed to the desorption of $CO_2$ adsorbed (and not reacted with BuOLi) on bare or Li-modified Au sites on the surface. In fact, experiments involving $CO_2$ adsorption on bare Au(poly) at ca. 100 K yielded a sharp $m/e = 44$ TPD peak centered at ca. 135 K.

2. A broad, highly asymmetric $m/e = 44$ peak in the region $200 < T < 450$ K. It may be noted that the ratio of $m/e = 44$ to $m/e = 43$ for bulk BuOH is very small; therefore, the expected contribution of this material to the $m/e = 44$ spectra at about 200 K is essentially negligible.

The most probable reaction pathway consistent with the latter spectral contribution is given by

$$CH_3-CH_2-CH_2-CH_2OLi + CO_2$$
$$\rightarrow CH_3-CH_2-CH_2-CH_2OCO_2Li \quad (3)$$

which decomposes thermally over the range $200 < T < 400$ K to release $CO_2$ and regenerate BuOLi. In fact, the presence of peaks with $m/e = 43, 44$ and 2 at ca. 580 K is in accordance with the behavior found for the UHV-synthesized BuOLi on the surface. It is possible that part of the original BuOLi may have not reacted with $CO_2$; however, the $m/e = 44$ peak is sufficiently large to be attributed to conversion of a sizable fraction of BuOLi to the corresponding organic carbonate.

Attempts were made to identify possible gas phase products of the thermal decomposition of supported BuOLi by monitoring via TPD a number of other $m/e$ values. All peaks detected in these experiments (in order of relative abundance, $m/e = 43, 44, 57, 41, 42$, and 2) had the same shape and temperature as those discussed in previous sections.

A more precise assignment could, in principle, be based on the relative intensities of the different peaks. Although much of that information has been compiled in the literature, those earlier values are not quantitative due to the fact that experimental parameters, such as the electron impact energy, relative $m/e$ sensitivities and the specific design of the instrument often cause marked changes in the observed peak abundances. It should be possible to measure each species with the same instrument and under the same conditions as those involved in the TPD experiments; such measurements, however, have not as yet been pursued.

## F. Effect of Atmospheric Impurities on the Reaction between PC and Li [4]

A series of experiments was conducted to explore the possible influence of carbon dioxide, dioxygen and water on the reactivity of PC toward Li covered Au(poly).

Mixtures of $d_6$-PC and $CO_2$ yielded much larger $m/e = 44$ peaks in the high, compared to the low, temperature range, indicating an increase in the amount of $Li_2CO_3$ compared to that of lithium alkyl carbonate. This may not be surprising as $CO_2$ would be expected to react with Li, leaving less available metal to react with PC. No net increase in the amount of $Li_2CO_3$ could be observed in the corresponding TPD of Li/Au(poly) surfaces exposed to $d_6$-PC/$O_2$ mixtures; however, some, as yet unexplained changes were noticed in the overall shape of the broad $m/e = 44$ feature.

## G. Future Prospects

The ultrahigh vacuum (UHV) assembly and characterization of Li/solvent layers supported on nominally inert substrates provides optimum conditions for examining the reactivity of nonaqueous solvents toward metallic Li. Although studies have been restricted to tetrahydrofuran (THF) and propylene carbonate (PC), this overall experimental methodology appears of sufficient versatility to be readily applied to a wide range of solvents and other species displaying measurable vapor pressure. A better identification of the reaction mechanism may be expected to be obtained by making systematic use of highly specific vibrational probes, such as FTIR, high resolution electron energy loss spectroscopy (HREELS) and sum frequency generation (SFG), which are gaining in popularity for the study of interfaces in UHV. Such a multitechnique approach is currently being pursued in the author's laboratory with a custom-designed UHV chamber with capabilities for both FTIR and SFG in addition to AES, LEED, XPS and TPD. Continuing efforts in this area, including those at Lawrence Berkeley National Laboratory (LBNL) [36] using spectroscopic ellipsometry in UHV, is expected to provide quantitative information regarding structural and chemical aspects of lithium/solvent interfaces of fundamental and technological interest.

## III. ELECTROCHEMISTRY IN ULTRAHIGH VACUUM

The advent of a growing variety of polymeric materials displaying high ionic mobilities at relatively low temperatures has opened new prospects for the development of improved electrical energy storage and energy generation devices [19]. Polymers exhibiting both chemical stability toward metallic Li and high $Li^+$ conductivities have received much attention in recent years because of their potential use in high power, high energy density secondary (rechargeable) batteries [37]. Polyethylene oxide may be regarded as the most widely studied materials of this type, owing to its relatively high inertness toward reduction and rather facile $Li^+$ transport at temperatures as low as 55°C in the absence of cosolvents or other additives [38].

In the course of an investigation aimed at characterizing in situ the short-term chemical stability of ultrapurified PEO-based electrolytes toward metallic Li, a series of attenuated total reflection Fourier transform infrared (ATR-FTIR) spectroelectrochemical measurements were attempted under reduced pressure, on the order of a fraction of a Pa, to minimize effects due to gas phase contaminants [39]. Such an approach did not only accomplish the desired objective, but also raised the possibility of conducting such electrochemical experiments under even lower pressures. To this end, the degassing properties of ultrapurified PEO, $LiClO_4$ and, later, $LiClO_4$/PEO films were carefully examined, using first a turbo molecular pump and ultimately an ion pump in a UHV quality chamber equipped with a mass spectrometer. No observable increase in the base pressure could be detected with any of these specimens, even for temperatures as high as 70°C, indicating that their degassing rates were lower than ca. $10^{-7}$ Pa · L/s · cm². This finding may be regarded as highly significant in that it identifies $LiClO_4$/PEO as an ideal solution for carrying out conventional electrochemical experiments in UHV and avoids the transfer of well-characterized specimens to a high pressure compatible auxiliary chamber, as required for electrochemical experiments in aqueous media. This section describes in some detail the methods and procedures involved in these measurements and summarizes the most important results obtained to date.

## A. Experimental Aspects

### 1. Ultrahigh Vacuum Instrumentation

Electrochemical experiments under UHV were carried out in either chamber A or B, described in detail in Section I.A.1, at pressures of ca. $2 \cdot 10^{-8}$ Pa (1 Pa = $7.5 \cdot 10^{-3}$ torr), with hydrogen, water, carbon monoxide and carbon dioxide as the main background gases. The electrochemical cell assembly (vide infra) was introduced into and retrieved from the main chamber (MC) through an independently pumped antechamber (AC) using a long magnetically coupled manipulator (MCM1) isolated electrically from the main chamber by a ceramic nipple (CN) (see Figure 30).

### 2. Ultrapurification of PEO

Commercially available PEO (Aldrich, 600,000 molecular weight) was purified by placing a 2% w/w aqueous solution of the material into thoroughly rinsed dialysis bags (Sigma Chemical Co.), which were then sealed and immersed in a large beaker filled with ultrapure water (18 MΩ). After replacing the water several times over a period of days, the PEO solutions were poured into a scrupulously clean flask and evaporated slowly to dryness at 50°C using a Rotavap. This procedure yielded a film that was stored in capped brown bottles in a high

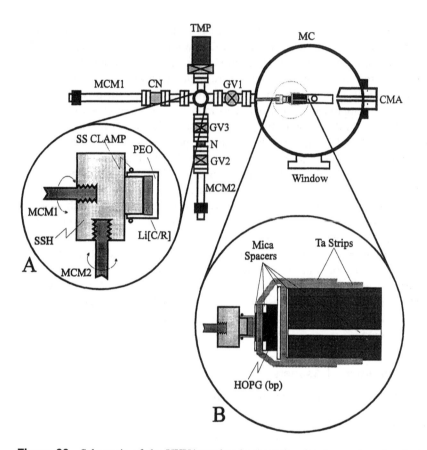

**Figure 30** Schematic of the UHV/antechamber/transfer chamber system for electrochemical measurements involving solid polymer electrolytes. Insert A provides an exploded view of the HOPG(bp) sample holder and Li[C/R]/PEO(LiClO$_4$) stainless steel holder (SSH) arrangement attached to both magnetically coupled manipulators. Insert B shows in detail the assembled HOPG(bp)/PEO(LiClO$_4$) cell in the UHV chamber. MCM = magnetically coupled manipulator; GV = gate valve; N = nipple; CN = ceramic nipple; SSH = stainless steel holder; TMP = turbomolecular pump. (From Ref. 6.)

quality glove box (vacuum atmospheres) equipped with water and oxygen sensors. All subsequent operations, including the preparation, casting and drying of PEO/Li salt films and the assembly of the electrochemical cell (see Section II.A.4) were performed in the same glove box to avoid exposure of the materials to the laboratory atmosphere.

## 3.  Preparation of PEO/Li Salt Films

PEO/Li salt films were prepared by dissolving ultrapurified PEO (12 g) and the purified lithium salt, e.g., $LiClO_4$ (0.81 g), in 120 mL of acetonitrile (AN), to obtain a PEO/Li$^+$ molar ratio of ca. 36:1. A small aliquot of the PEO/Li salt solution was placed on a glass plate and then spread with a film applicator (Gardner AR-5312) adjusted to ca. 0.76 mm gap. After evaporation, the films thus formed were allowed to dry in a temperature-controlled vacuum desiccator installed inside the glove box at ca. 25°C for 6 h. The temperature was then gradually raised while under vacuum to 100°C over a period of about a day, held at that value for 48 h and then allowed to cool to room temperature. This overall procedure yielded smooth PEO/Li salt films 50 to 65 μm thick, which were stored in sealed polyethylene bags.

## 4.  Electrochemical Cell for UHV Measurements

The UHV compatible electrochemical cell incorporates a small circular foil of freshly scraped metallic Li (0.3 cm$^2$ in diameter), used as both counter- and reference electrode, Li[C/R], placed onto the flat end of a specially designed stainless steel holder (SSH) (see Figure 30). A round section of a $LiClO_4$(PEO) film (see below) of larger diameter is placed on top of the Li foil and around the SSH (see insert A in Figure 30) and then held tightly in position by a small stainless steel ring. The SSH is screwed onto the end of a long magnetically coupled manipulator (MCM2) and then isolated from the glove box atmosphere with a UHV-quality gate valve (GV2). This transfer arm (TA) is then removed from the glove box and attached via a nipple (N) and a gate valve (GV3) to an antechamber (AC) connected in turn via another gate valve to the main UHV chamber. After evacuation, the AC, including TA, was baked at ca. 150°C for about 10 h to remove water and gas phase impurities trapped in the PEO film and/or adsorbed on the chamber walls. The antechamber was then allowed to cool down and the Li[C/R](PEO)/SSH assembly was then transferred from MCM1 to MCM2 using a second tapped hole in SSH with the PEO/Li foil facing the UHV chamber port normal to the MCM1 axis. A schematic of SSH with the two attached MCMs is shown in insert A of Figure 30. In this fashion, the electrochemical cell could be easily formed by translating MCM1 into the main UHV chamber and placing the PEO/Li assembly directly parallel and against the working electrode surface. A schematic of the assembled cell is shown in insert B of Figure 30 (shown for a HOPG sample). More recently, a nickel foil was mounted on the holder around the electrode to allow the PEO film to be preelectrolyzed just prior to the actual experiments. This procedure involves polarization of the nickel electrode at potentials slightly positive to Li bulk deposition for a period of time long enough to achieve exhaustive reduction of residual impurities in the film.

After assembly, the sandwich-type cell is heated to 50–70°C and maintained at a fixed selected value during acquisition of electrochemical data. The potential is controlled using an RDE 3 Pine potentiostat equipped with a built-in signal generator. Cyclic voltammograms were recorded at scan rates of 5 to 100 mV/s. All potentials are reported with respect to Li[C/R].

## B. Electrochemical Properties of Well-Characterized Electrode Materials in PEO-Based Solutions

This section summarizes some of the most significant electrochemical results obtained to date for selected electrodes cleaned and characterized under UHV in PEO-lithium-based solutions, and include nonalloy (Ni)- and alloy-forming metals (Ag and Al), a noninteracting substrate (boron-doped diamond, BDD) and a material capable of intercalating $Li^+$ (graphite). It is expected that the information herein contained will serve to illustrate the power of this methodology for the study of highly reactive interfaces.

### 1. Single Crystal Nickel

Nickel is among the very few metals with which Li does not form alloys even at moderately high temperatures, ca. 450 K [25,40], and, hence, provides unique opportunities for exploring fundamental aspects of the electrochemistry of Li, including its electrodeposition, without complications derived from the formation of alloys or other phases.

The Ni(111) and Ni(110) crystals (Cornell Laboratory) used in this study were circular disks of ca. 0.8 and 1 cm² cross-sectional area, respectively, oriented to ~1° by Laue back-diffraction, mounted on transferable holders [5,41]. Specimens were cleaned by $Ar^+$ sputtering (0.5–1.5 keV)/thermal annealing (850–1175 K) cycles, and characterized by AES and LEED. A K-type thermocouple spot-welded to the edge of the crystals was used to monitor their temperature. All measurements were carried out under UHV in chamber A.

The first and second voltammetric cycles obtained for nominally clean Ni(111) in pre-electrolyzed LiClO₄/PEO at 333 K at a scan rate of 5 mV/s are shown in dashed and solid lines in panel A of Figure 31, respectively. The peak labeled as a, at ca. 1.3 V versus Li[C/R], observed during the first scan in the negative direction, may be attributed to water reduction [42] and/or to a reaction of Li UPD with traces of adsorbed sulfur on the surface (see Section II.B.3). Its relatively small size, however, suggests that the bulk or surface concentration of these impurities is indeed very small. Also found during this first scan was a sharp feature centered at about 0.5 V induced by adventitious, adsorbed CO (peak b) on the surface, a phenomenon to be further discussed later in this section. The voltammogram obtained upon reversing the scan at a potential $E_{rev}$ very close to

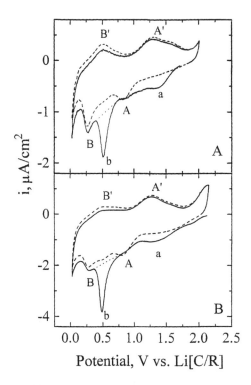

**Figure 31** First (solid line) and second (dashed line) voltammetric cycles for nominally clean Ni(111) (A) and Ni(111)-$c(4 \times 2)$CO surfaces (B) in LiClO$_4$/PEO at 333 K. Scan rate: 5 mV/s. The somewhat arbitrary baseline used for the coulometric analysis of peak b is shown by dotted lines.

Li bulk deposition showed two clearly defined peaks, B′ and A′, with maxima at 0.5 and 1.3 V, respectively. During the second (and subsequent) voltammetric scan(s), peaks a and b virtually disappeared, yielding clearly defined A,A′ and B,B′ features. The complementarity between the two latter sets of peaks was verified by scanning the potential over narrower ranges, as shown by the thin solid and dotted lines in Figure 32. This behavior is characteristic of a general phenomenon known as metal underpotential deposition (UPD), which involves adsorption/charge transfer of metal ions on metal electrodes of higher work function at potentials more positive than those expected based on bulk Nernstian thermodynamics and, therefore, can be attributed in this case to Li UPD on Ni.

Coulometric analysis of these voltammetric features, using the somewhat arbitrary background shown in dotted lines in Figure 31, yielded a total charge $Q_{A+B}$ of ca. 40 µC/cm$^2$ and $Q_{A'+B'}$ of ca. 50 µC/cm$^2$. Assuming that the discharge

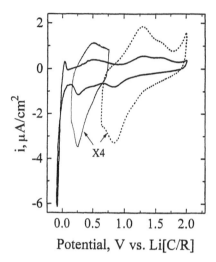

**Figure 32** Steady state voltammogram for the surface in Figure 31A showing the complementarity of peaks A-A′ and B-B′. Other conditions are the same as those specified in Figure 31. (From Ref. 54.)

of $Li^+$ upon UPD is complete (vide infra), these charges correspond to Li coverages, $\theta_{Li}$, of 0.13 and 0.17, respectively. These values are significantly smaller than those expected for a full, hexagonally closed packed monolayer of Li on Ni(111), assuming the same Li-Li Interatomic distance found in bulk Li, and suggest that Li bulk deposition occurs before the Ni(111) surface is fully covered with Li. Two independent lines of evidence seem to support this view:

1. The critical coverage for the formation of a second alkali metal layer on Ni(111), based on LEED analysis, has been found to be as low as 1/3 for K/Ni(111) and even lower for Cs/Ni(111) [26].
2. The work function of alkali (and other electropositive) metals adsorbed on higher work function metals, such as Ni and Ag, as a function of $\theta_{alk}$, drops to values below $\Phi_{alk}$, the work function of the alkali metal in bulk form, for $\theta_{alk}$ as small as 0.3 and increases thereafter to reach $\Phi_{alk}$ at $\theta_{alk}$ of about 1.3.

Since $\Phi$ *tracks* quantitatively the applied potential in electrochemical environments [7], it seems reasonable to conclude that bulk Li electrodeposition can commence when the Ni(111) surface is covered by only a fraction of a monolayer, as the experimental data presented above appears to indicate. At least two important issues may have to be carefully considered before this suggested model can be validated: the state of charge of adsorbed Li is not only unknown, but it may also vary as a function of coverage [43], and the coadsorption of other species, such as water and other common solvents, has been found to increase the work function of, for example, K/Pt(111), especially [44] for small $\theta_{alk}$.

Although somewhat speculative, the presence of two (or perhaps three) Li UPD/stripping features is consistent with information derived from LEED for K, Cs, and on a more preliminary basis Li, adsorbed on Ni(111) in UHV [26], from which the surface phase diagram shows at least two ordered phases as a function of $\theta_{alk}$ for $\theta_{alk} < 1/3$.

A loop was observed in the voltammetric curve obtained by extending $E_{rev}$ to values negative to 0.0 V versus Li[C/R] (not shown in this figure) followed by a sharp stripping peak in the scan in the positive direction, without additional contributions to other processes, except those derived from UPD. This behavior is characteristic of phase nucleation and growth and may thus be attributed to the kinetically facile deposition of Li without evidence for alloy formation, in agreement with the UHV results of Wang and Ross [22] and the phase diagram for Li-Ni [25].

In the case of *nominally* clean Ni(110), the first scan in the negative direction (dotted line) initiated at the open circuit potential (ca. 1.8 V), shown in panel A of Figure 33, yielded a very large feature consisting of at least three clearly identifiable peaks or shoulders. At least two of these peaks disappeared during the second (solid line in this figure) and subsequent cycles, giving rise to two sets of complementary peaks (dotted and dash-dot curves in Figure 34) somewhat reminiscent of those found for Ni(111) after the first scan in the negative direction (Figure 31) and attributed to Li UPD. Evidence that the shoulder at 1.25 V in the first scan in the negative direction may be related to a process associated with the presence of CO was obtained from experiments in which Ni(110) was exposed to 30 L CO, to be described in detail below.

Prompted in part by the results obtained in the previous section, the effects of adsorbed carbon monoxide and sulfur on the voltammetric behavior of single crystal Ni surfaces were examined in more detail.

## 2. CO-Modified Single Crystal Nickel Surfaces

Sharp LEED patterns consistent with a $c(4 \times 2)$CO superstructure could be readily obtained by exposing clean Ni(111) to 30 L CO in UHV at room temperature as reported in the literature [45–51]. Also in accordance with information published elsewhere [52] was the increase in the background intensity of the rectangular Ni(110) LEED pattern following dosing, attributed to the presence of an amorphous CO overlayer. No AES spectra or LEED patterns were recorded after CO adsorption prior to the electrochemical experiments, to avoid electron-induced decomposition [53].

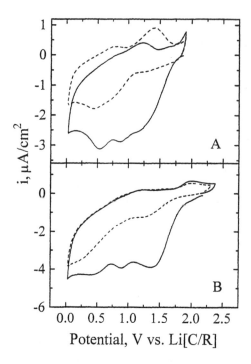

**Figure 33** First (solid line) and second (dashed line) voltammetric cycles of nominally clean Ni(110) (A) and Ni(110) exposed to 30 L CO (B) in $LiClO_4$/PEO at 333 K. Scan rate: 5 mV/s. The peak at the more positive potential, attributed to reactions involving Li and CO, appears to contain contributions due to sulfur (see text).

*Ni(111)*

Exposure of a clean Ni(111) to CO to form a $c(4 \times 2)$ CO superstructure yielded in the first linear scan in the negative direction a more pronounced peak b (solid line, panel B, Figure 31) compared to that in panel A in the figure, providing rather conclusive evidence that the feature in question is derived from the presence of CO on the surface. The potential at which this peak occurs is sufficiently negative for Li to UPD on Ni, as evidenced by the two sets of complementary peaks A,A′ and B,B′ shown more clearly in the second scan (dotted line) [54]; hence, it seems reasonable to assume that the effect observed originates from interactions between coadsorbed UPD Li and CO on the surface.

Support for this view is provided by the work of Jaensch et al. [27,55], who examined Li-CO coadsorption on Ru(001) and Ag/Ru(001). Based on TPD, $\Delta\Phi$, and metastable quenching (MSQ) electron emission spectroscopic tech-

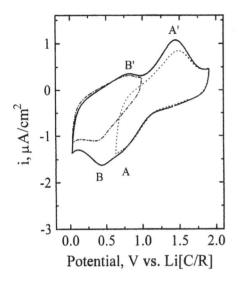

**Figure 34** Steady state voltammogram for the surface in Figure 33A showing the complementarity of peaks A-A' and B-B'. Other conditions are the same as those specified in Figure 33.

niques, these authors concluded that coadsorbed CO and Li interact to generate, depending on the surface stoichiometry and temperature ($T$), at least two different species. In particular, for small $\theta_{alk}$ and $T = 80$ K, a Li-CO complex with stoichiometry $Li_n(CO)_{2n}$ appears to form, which undergoes decomposition at $T > 750$ K. The TPD of CO exposed to Li-modified Ru(001) at 400 K was consistent with $Li_2O$ and the polymeric Li-CO complex as being the most likely products. Evidence for such polymeric species has also been obtained for CO adsorbed on Li [56]. It seems, therefore, plausible that the same species are produced by the reaction between UPD Li and preadsorbed CO in the electrochemical experiments. Unfortunately, both the amount of reactive Li on the surface, as well as the stoichiometry of the species produced, are not known with certainty; hence, no reliable information can be obtained from a simple coulometric analysis. Nevertheless, it is significant that the charge under peak b, using a rather arbitrary baseline (see dotted lines in panels A and B of Figure 31), increases by about a factor of 2 for Ni(111)-$c$(4 × 2)CO surfaces (41 $\mu$C/cm$^2$) compared to CO-contaminated Ni(111) (20 $\mu$C/cm$^2$), for which the amount of CO would be expected to be smaller. Furthermore, the atomic density for Ni(111) is $1.86 \times 10^{15}$ atoms/cm$^2$, which is equivalent to a charge density of ca. 300 $\mu$C/cm$^2$, assuming one electron per surface atom. Within the uncertainties in the electrode area, and

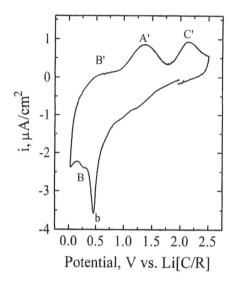

**Figure 35** Same as Figure 31B under extremely clean conditions. Note the presence of a well-defined peak C′ (see text for details).

considering that the coverage of CO for a $c(4 \times 2)$ superstructure is $1/2$, the charge associated with peak b in panel B of Figure 31 would correspond to about 0.3 electron per CO molecule.

The disappearance of peak b in the second and subsequent cycles appears consistent with the irreversible formation of a superficial layer of $Li_2O$, an electrochemically inert material which would consume oxygen from CO, leaving the remaining carbon in a yet to be identified state. Another aspect of these results that requires further investigation is the presence of a peak with an onset potential of ca. 1.8 V observed during the first and subsequent scans in the positive direction upon reversal of the scan at 0.05 V. As shown by the results obtained for a very clean Ni(111)-$c(4 \times 2)$CO surface (no noticeable peak a) in Figure 35, this feature, denoted C′, is centered at about 2.1 V, and, thus, very close to that reported earlier for oxygen-modified Ni(poly) in the same media [54]. Whether oxygen in this case is derived from monomeric or polymeric CO of the type proposed by Jaensch et al. [55] (vide supra) or from the reduction of PEO or perchlorate ion still remains to be explored.

*Ni(110)*

As shown in panel B of Figure 33 (dotted line), Ni(110) surfaces exposed to 30 L CO yielded a clearly defined peak centered at about 1.4 V. Particularly striking is the very distorted character of the voltammogram obtained during the second (see solid line in this panel) and subsequent scans, which may be indicative of irreversible changes on the surface induced at least in part by reactions between Li and CO. Also to be considered are the effects of sulfur to be discussed in the following section, which introduce yet additional complications to the interpretation of these data.

## 3.  Sulfur-Modified Ni Single Crystal Surfaces

Sulfur-modified nickel surfaces were obtained by annealing the single crystal specimens to a high enough temperature to induce surface segregation [57,58]. Clean Ni(110) surfaces heated at ca. 1070 K for several hours yielded AES spectra displaying, in addition to Ni features, a prominent S peak (19% by AES), without additional contributions due to other elements, and LEED patterns showing additional center spots on the basic rectangular array, resembling a distorted hexagonal lattice. These latter results are in harmony with earlier LEED reports by Demuth and Rhodin [57], and with more recent UHV-STM findings by Backer and Horz [58], who describe this superstructure as a $c(2 \times 2)S$, suggesting that the sulfur atoms are chemisorbed in twofold hollow sites on Ni(110). Extensive annealing of clean Ni(111) in UHV was also found to yield significant amounts of sulfur (14% by AES; see Figure 36) and an increased background in the LEED

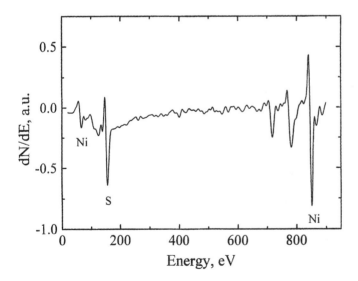

**Figure 36**  Auger electron spectra of an extensively annealed Ni(111) surface showing a rather large amount of superficial sulfur (14%).

pattern; however, none of the various ordered superstructures observed by LEED [57] (or UHV-STM) [58–61] upon dosing Ni(111) with H₂S in UHV could be readily discerned.

*Ni(111)*

A very distinct voltammetric peak at 1.25 V versus Li[C/R] could be found for sulfur-modified Ni(111) specimens during the first cycle (solid line, Figure 37), which, as in the case of CO, disappeared during the second cycle to yield a curve virtually identical to that obtained for *nominally* clean Ni(111) after the same treatment (dotted lines, Figure 37). It seems reasonable to suggest, from a chemical viewpoint, that sulfur reacts with Li UPD to form an electrochemically inert Li-S surface layer, which, based on the recent results obtained for S adsorbed on Li using soft X-ray photoemission, could be Li₂S [62]. In fact, the charge under the sulfur-induced voltammetric feature curve is about 247 μC/cm², which would correspond to close to one electron per surface or, equivalently, two electrons per S atom on the surface, based on the size of AES S signal obtained for Ni(110)-c(2 × 2)S (vide infra) for which the coverage is ½. It may be noted that the onset potential for this feature occurs at about the same value as that of peak a in Figure 37, suggesting that the latter may not be due to water, but to small

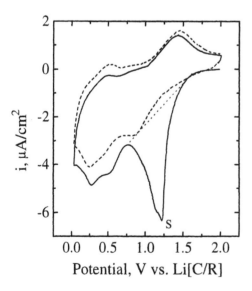

**Figure 37** First (solid line) and second (dashed line) voltammetric cycles of a sulfur contaminated Ni(111) surface described in the caption to Figure 36.

amounts of sulfur remaining on the surface after the cleaning procedure not detectable by AES under the measurement conditions.

*Ni(110)*

In analogy with the behavior found for S-modified Ni(111), an additional feature of rather similar nature could be found for Ni(110)-$c(2 \times 2)$S during the initial scan (solid line, panel A, Figure 38). Also shown in that figure (dotted line) is the subsequent voltammetric cycle. Further evidence for sulfur as being responsible for this effect was obtained by placing the LiClO$_4$(PEO) film on an area adjacent to, although slightly overlapping, the first spot, in which case the intensity of the voltammetric peak decreased, yielding only a shoulder on the feature attributed to Li UPD (panel B, Figure 38). As before, the solid and dashed lines

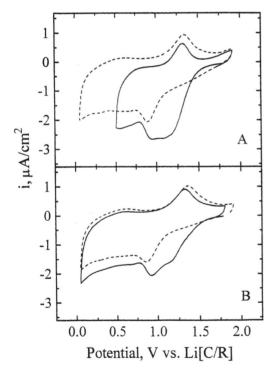

**Figure 38** (A) First (solid line) and second (dash line) voltammetric cycles of a Ni(110)-$c(2 \times 2)$S in LiClO$_4$/PEO at 333 K (A). Scan rate: 5 mV/s. The results in (B) were obtained under the same conditions on another slightly overlapping spot on the same specimen (see text for details).

represent the first and second cycles. Based on a comparison between the first voltammograms in Figures 37 and 38 (solid lines), it is possible that the first scan obtained for a CO-saturated Ni(110) may also contain contributions due to residual sulfur on the surface.

## 4. Polycrystalline Silver

Studies involving silver electrodes were aimed in part at gaining insight into the physicochemical phenomena responsible for improvements in the performance of carbon-based lithium-intercalation electrodes for energy storage applications induced upon incorporation of this metal as an additive [63], which include enhanced rechargeability and shifts in the discharge potential toward more negative values compared to ordinary graphites. Also of interest was to compare various aspects of Li UPD on a Li alloy-forming substrate and thus contrast the results with those obtained with Ni.

For these experiments, a polycrystalline silver foil (Alfa Products, m3N, 0.127 mm thick) cleaned by multiple $Ar^+$ sputtering/thermal annealing cycles [54] was used as a substrate.

As shown in the upper panel of Figure 39 (see dotted line), the first linear scan in the negative direction of a Ag foil surface free of detectable impurities

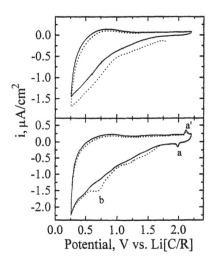

**Figure 39** First two cyclic voltammetry curves for clean Ag in $LiClO_4(PEO)$ recorded at a temperature of 323 K (upper panel) and 333 K (lower panel) initiated at the measured open circuit potential; $v = 5$ mV/s.

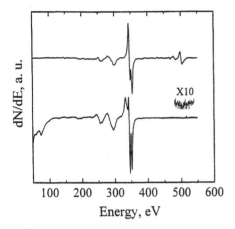

**Figure 40** Auger electron spectra of a clean Ag(poly) foil (lower curve) and of an oxygen-contaminated silver foil (upper curve).

(see AES in Figure 40, upper trace) recorded at $T = 323$ K was characterized by a featureless sloping background, which became particularly pronounced at potentials of $E < 1.0$ V. Upon reversing the potential at 0.25 V the resulting scan displayed a very broad feature centered at 0.8 V without any additional peaks at higher potentials. A virtually identical voltammetric behavior was observed in the second (solid line in this panel) and subsequent cycles within the same voltage limits.

Similar cyclic voltammetric curves were also obtained in experiments performed at a slightly higher temperature, i.e., 333 K, under otherwise the same conditions (see lower panel, Figure 39), except for the presence of a small peak b at 0.75 V during the first scan in the negative direction (dotted line), which disappeared after the first cycle, and for what seems to be a surface-bound redox couple at about 2.0 V $(d,d')$ also observed by other workers in liquid nonaqueous solutions [64,65].

As clearly evidenced by these results, the charge obtained by integration of the current over the entire scan in the negative direction, $Q_-$, is considerably larger than that found in the corresponding subsequent scan in the positive direction, $Q_+$, regardless of the potential at which the scan was reversed ($E_{rev}$), and especially for $E_{rev} > 0.0$ versus Li[C/R]. For example, for the solid curves in panels A and B in Figure 39, $Q_-/Q_+ = 8.9$ and 4.4, respectively. At least two effects could account for this charge imbalance: irreversible reduction reactions involving the solvent, the electrolyte and/or residual impurities present therein, and alloy formation. Calculations based on reasonable values for the diffusion coefficients of relatively small species in PEO ($10^{-7}$ cm$^2$/s) and the actual thickness of the polymeric solution (ca. 10 μm, when molten) indicate that the pre-electrolysis time (20 min) should be sufficient to remove impurities capable of undergoing reduction at potentials of 25 mV versus Li[C/R], such as water and oxygen. More likely are reactions involving the solvent and/or the electrolyte; for example, the far less pronounced charge imbalance observed in LiI/PEO solutions using Ni electrodes suggests that perchlorate ion undergoes reduction at these negative potentials. However, the single most important contribution to the charge imbalance appears to be the formation of one or more Li/Ag alloys known to exist at room temperature (see Section II.B) [12a]. This phenomenon would account in part for the negative currents observed at potentials positive to Li bulk deposition as a fraction of Li in the alloy would migrate deeper into the Ag lattice and unavailable for stripping. Direct evidence in support of this view may be found in the temperature dependent AES study of Li vapor deposited on Ag in UHV described in Section II.B, for which the onset of Li diffusion into Ag was found to occur at temperatures slightly higher than 300 K.

After several voltammetric scans in the range 0.25–2.10 V versus Li[C/R] at 333 K, the negative limit ($E_{rev}$) was extended beyond the potential associated with Li bulk deposition, first to −0.1 V (dotted line, Figure 41) and then to −0.2 V (solid line in Figure 37). A well-defined peak centered at 0.5 V could be identified in the scan in the negative direction without a clear counterpart for $E_{rev} = -0.1$ V (dotted line). In the case of $E_{rev} = -0.2$ V, however, the subsequent linear scans in the positive direction displayed three peaks (A′, B′ and C′) not

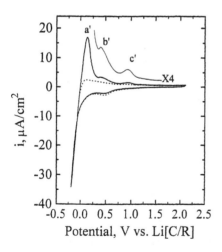

**Figure 41**  Cyclic voltammograms obtained after reversing the potential first at −0.10 (dotted line) and then at −0.20 V (solid line, see text for details) at 333 K; ν = 5 mV/s (see text for details).

observed for voltammetric curves for $E_{rev} > 0$. Several interesting observations could be made based on a more detailed analysis of these data: the $Q_-/Q_+$ ratio was 1.65–2.0, i.e., much smaller than that observed for more positive $E_{rev}$ values; the charge under peaks C and C′ is on the order of ca. 200 $\mu C/cm^2$, i.e., close to one electron per surface site. In fact, a peak at ca. 1.0 V versus $Li/Li^+$ in the scan in the positive direction has been reported by Aurbach et al. [42] and Kolb et al. [66] and attributed by these authors to Li UPD on Ag. The potential of the peak is in very good agreement with that predicted by the Kolb-Gerischer UPD/ work function correlation, lending support to this assignment. Two factors may account for the emergence of peaks C and C′ only after polarizing the Ag electrode to potentials negative to Li bulk deposition: the near surface region becomes enriched in Li, and/or reactions take place between Li and the solution yielding irreversibly adsorbed species on the surface. Both these effects may stabilize the UPD layer and thereby reduce the rates of Li diffusion into the bulk. Peak B′ has a charge similar to that of C′ and could, in principle, correspond to a second UPD stripping, for which the corresponding deposition would occur at potentials too close to Li bulk deposition for a feature to be clearly identified. Lastly, peak A′ appears consistent with the stripping of bulk Li, although contributions from a Li/Ag alloy cannot be ruled out (see Section II).

*Oxygen-Contaminated Ag(poly) Surfaces*

Sputtering (1 keV $Ar^+$ at 10 $\mu A/cm^2$ for 4 h) of a fresh Ag(poly) foil failed to remove oxygen from the specimen, as evidenced by the presence of a large O peak at 515 eV in the AES spectra (see upper curve, Figure 40), which corresponds to ca. 26% oxygen on the surface. This treatment, however, eliminated superficial carbon, as judged from the magnitude of the height ratios of the AES peaks at 266 and 303 eV, i.e., 0.44 [54].

The cyclic voltammetry of this oxygen-contaminated Ag specimen was examined in $LiClO_4$(PEO) under similar conditions ($T = 333$ K; $v = 5$ mV/s) to identify the effects of this common impurity on the electrochemical response of clean silver and the results are shown in Figure 42. Two small peaks at 1.3 V (a) and 0.7 V versus Li[C/R] (b) on top of a sloping featureless background were found during the first scan in the negative direction (see dotted line) initiated at the open circuit potential (ca. 1.7 V). Reversal of the scan at 0.2 V gave rise to a prominent feature (D′) centered at ca. 1.35 V. Although peaks a and b disappeared in the second scan in the negative direction (see solid line in Figure 38), peak D′ clearly increased in magnitude. The charge under the oxygen-derived feature, using the dotted line in this figure as a background, amounts to about 27 $\mu C/cm^2$; hence, if one assumes a charge equivalent of a single electron per surface atom of 250 $\mu C/cm^2$ for silver the amount of oxygen would be of about 11%.

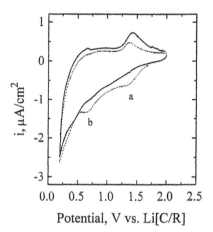

**Figure 42** First (dotted line) and second (solid line) cyclic voltammetric curves for the oxygen-contaminated Ag specimen described in the caption to Figure 40 (upper curve) in $LiClO_4(PEO)$ recorded at 333 K. Other conditions are specified in the caption to Figure 39.

It may thus be concluded from these results that voltammetric measurements of the type described in this work are sufficiently sensitive to detect oxygen on silver at submonolayer coverages and that Ag specimens prepared by the methods specified in the previous section render surfaces free of oxygen impurities.

## 5.  Boron-Doped Diamond

Hydrogen-terminated boron-doped diamond (BDD) has emerged as a unique electrode material displaying, among other characteristics, extraordinary low chemical and electrochemical reactivity over a wide potential range allowing studies to be performed in PEO/Li salt solutions without interference from underpotential deposition, alloy formation and/or intercalation processes [67].

Boron-doped diamond films (ca. $10^{20}$ B atom/$cm^3$) were grown on clean, unpolished diamond macles, i.e., twinned, flat triangular diamond platelets, with the twin plane parallel to the two triangular (111) faces, by the hot filament CVD method, using a source gas mixture of overall composition 0.5% $CH_4$ and $2 \cdot 10^{-3}\%$ trimethyl boron in $H_2$. Additional details of the growth method are given elsewhere [68,69]. During deposition, macles were kept at a temperature of 800°C, placed 0.5 cm away from four hot (2000°C) tungsten filaments, and rotated at 1/3 rpm to improve homogeneity. Films obtained after 7–24 h deposition displayed a sharp Raman scattering peak at 1332 $cm^{-1}$ [70], characteristic of diamond, with no evidence of graphitic, $sp^2$ carbon (ca. 1580 $cm^{-1}$). Atomic

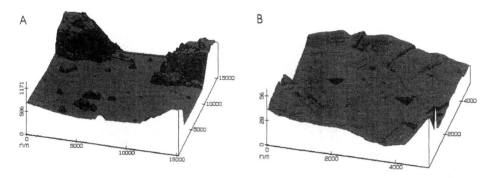

**Figure 43** Atomic force microscopy (AFM) images of boron-doped diamond on diamond (BDD/D) for two magnifications: (A) 15 μm × 15 μm; (B) 5 μm × 5 μm. (From Ref. 67.)

force microscopy (AFM) images of BDD/D films recorded in air (Nanoscope II, Digital Instruments) using $Si_3N_4$ microfabricated cantilevers with a force constant of 0.58 N/m, revealed rather large protrusions evenly distributed on a fairly smooth background (see panels A (15 μm × 15 μm) and B (5 μm × 5 μm) in Figure 43. These prominent features are believed to arise from penetration twins, which are a common occurrence when growing diamond on (111) surfaces [71].

A large Ni foil (1 cm × 1.5 cm, Alfa, 99.995% purity) with a triangular opening slightly smaller than the dimensions of the triangle-shaped macle (2.5 mm per side; see Figure 44) was used to mount the BDD on diamond (BDD/D) specimens onto a specially designed stainless steel holder [72]. This holder is used to transfer samples in and out of a UHV chamber equipped with both preparation (thermal annealing) and characterization (Auger electron spectroscopy, AES) capabilities [5,6]. The AES spectra of BDD/D ($i_{beam}$ = 10 μA, electron beam diameter = 1 mm) showed, after annealing in UHV at 700°C for 1 h, a pre-edge carbon AES peak at 268 eV, characteristic of diamond [70], with only a very small trace of oxygen (see Figure 45) and as expected no traces of B, for which the bulk concentration would be less than 0.1%.

The first cyclic voltammogram of BDD/D in PEO/LiClO4 recorded at 60°C and 5 mV/s showed a virtually featureless trace in the region 0.05–2.1 V versus Li[C/R], down to the most negative end (see dotted line in Figure 46). Although the current during the negative-going scan appears larger than that observed during the positive counterpart, especially at the more negative limits, the overall shapes of these curves are consistent with an increase in the interfacial capacity as the potential becomes more negative, to the extent that faradaic contributions can be excluded. For BDD films grown on Si in aqueous media at much more positive potentials, the interfacial capacity decreases monotonically, as the

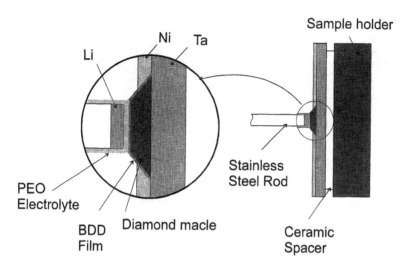

**Figure 44** Schematic of the experimental arrangement for electrochemical measurements of hydrogen-terminated boron-doped diamond (BDD) films in a Li$^+$-based solid polymer electrolyte in ultrahigh vacuum (UHV). (From Ref. 67.)

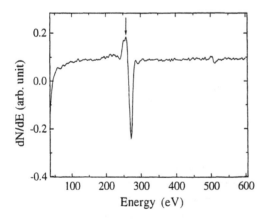

**Figure 45** Auger electron spectra (AES) of a hydrogen-terminated boron-doped diamond film supported on a diamond macle (see text for details). (From Ref. 67.)

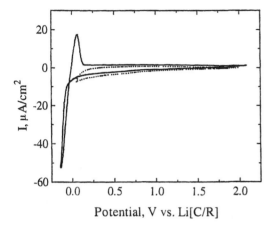

**Figure 46** First (dotted line) and subsequent (solid line) cyclic voltammogram of hydrogen-terminated boron-doped diamond film supported on a diamond macle in PEO/LiClO$_4$ acquired in ultrahigh vacuum (UHV) at ca. 60°C. Scan rate: 2 mV/s (see text for details). (From Ref. 67.)

potential becomes more negative [74]. Taken together, these results suggest that the interfacial capacity of BDD may exhibit a minimum at a potential between 2.5 and 3.5 V versus Li/Li$^+$. The same qualitative behavior has been found for the basal plane of graphite both in aqueous [75] and nonaqueous electrolytes [76], where the minimum has been ascribed to the minimum in the density of states near the Fermi level [76]. It must be stressed, however, that a quantitative determination of actual interfacial capacities can be made most reliably by employing impedance spectroscopy techniques in a suitable cell configuration.

During a subsequent cycle, the negative potential limit was extended to $-0.3$ V versus Li[C/R] (see solid line in Figure 46), yielding features characteristic of those expected for the deposition and stripping of bulk Li with only very minor contributions due to the kinetically hindered reduction of PEO and perchlorate ion.

## 6. Polycrystalline Aluminum

The ability to perform electrochemical experiments under UHV conditions makes it possible to explore the adsorptive characteristics of highly reactive metals on highly reactive substrates. The UPD of Li on Al appears particularly interesting as Al has been considered as a storage matrix for Li in rechargeable high and low temperature battery applications [77–86]. In addition, recent theoretical calculations of Schmickler and Lehnert [87,88] have yielded estimates of the magnitudes of Li and Na UPD shifts on low index Al single crystal surfaces, which have not as yet been verified experimentally.

Measurements were conducted in UHV chamber B using an Al foil (Alfa, 0.13 mm thick, m5N in purity), cleaned by a series of Ar$^+$ (600 eV) sputtering and annealing cycles to 600°C.

## *Al(poly) Foil*

The AES spectra of an aluminum foil prepared by the methods described in the previous section (see Figure 47) displayed prominent peaks in the region 50 to 100 eV characteristic of the clean metal, including its major AES feature at 68 eV and the bulk/surface plasmon losses at 52 eV [89]. Except for the peaks at 200 and 215 eV, attributed to Ar embedded into the lattice during sputtering, no other features could be discerned up to the highest energies examined. Particularly noticeable is the absence of the oxygen peak at ca. 505 eV, which could not be detected by AES even at a much higher sensitivity (see Figure 47), setting an upper limit for this impurity of less than ca. 0.01 monolayer (ML) equivalent.

The first cyclic voltammetric scan in the negative direction obtained with a clean Al foil (dotted line in Figure 48), was initiated at the open circuit potential (OCV), i.e., 1.75 V, which is about 0.7 V lower than that reported by Dey [90] for slightly oxidized aluminum. Unlike similar data collected for far less reactive metal surfaces, such as Au and Ni, the fact that the current down to ca. 1.0 V

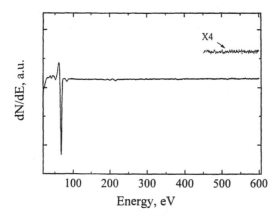

**Figure 47**  Auger electron spectra of a clean Al foil. The peaks at 200 and 215 eV are attributed to Ar embedded into the lattice during sputtering. Spectra obtained at higher sensitivity in the range 450 to 600 eV (see insert) show no trace of oxygen at 505 eV.

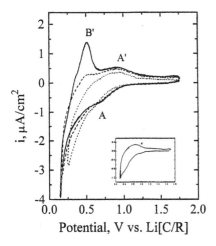

**Figure 48** First (dotted line), second (dashed line) and third (solid line) cyclic voltammograms recorded in sequence for a clean Al foil reversed at increasingly negative potentials, i.e., 0.25, 0.15 and 0.1 V vs. Li[C/R]. This series was initiated at the open circuit potential, i.e., 1.75 V.

is negligible may not be attributed to the absence of water or other atmospheric impurities in the PEO solution, as such impurities would simply react with Al to yield Al oxides or hydroxides without contributing to the external current. Nevertheless, based on the pre-electrolysis time, the thickness of the film and assuming reasonable values for the diffusion coefficients of species in PEO invoked in Section II.C.4, the amount of reducible impurities in the film following this procedure would indeed be negligible and certainly far below that required to form a monolayer on the Al surface.

The increase in the current observed at potentials more negative than 1 V may be ascribed to UPD of Li into Al (see below) and to the onset of Li diffusion into Al to form the corresponding alloy. Reversal of the scan at a potential $E_{rev}$ of ca. 0.25 V yielded a small and rather broad stripping peak (A′) centered at ca. 0.8–0.9 V, followed by a region in which the current was positive and constant. During the second scan in the negative direction, a clearly identifiable shoulder was observed centered at ca. 0.7 V, whereas the corresponding stripping curve for $E_{rev} = 0.2$ was characterized by the emergence of a new feature, B′, and a much better resolved A′ peak. In the subsequent scan for which $E_{rev}$ was slightly more negative, the intensity of B′ markedly increased, without affecting that of A′. Also noticeable as $E_{rev}$ was decreased was the increase in the (constant) current during the scan in the positive direction for potentials higher than 1.25 V. The complementarity of peaks A and A′ was demonstrated in an independent experiment in which the potential scan was restricted between 0.5–1.3 V (see insert in Figure 48), lending credence to the view that these features are derived from the UPD of Li on Al.

Experiments in which $E_{rev}$ was only slightly larger than 0.0 V yielded very large B′ peaks (see, for example, Figure 48), as would be expected for the formation of one of several bulk Li/Al alloys. It is based on this behavior that Al has been considered as an alternative Li storage electrode for battery applications without the problems associated with dendrite formation found for pure Li electrodes [83]. Further insight into this phenomenon was gained from measurements involving Li deposition on Al films vapor deposited on BDD to be described in the following section.

The position of peak A′ in Figure 48, i.e., 0.9 V, is consistent with Kolb-Gerischer's UPD correlation [91], which predicts a UPD shift corresponding to half the difference between the work functions $\Phi$ of the substrate (sub) and adsorbed (ads) in their bulk forms; i.e., $\Delta U_p = \frac{1}{2}(\Phi_{sub} - \Phi_{ad})$. According to values reported in the literature [92,93] (i.e., $\Phi_{Al} = 4.2$ and $\Phi_{Li} = 2.25$ eV), $\Delta U_p$ would be ca. 0.98 eV, which is also in good agreement with calculations based on density functional theory and the jellium model reported by Lehnert and Schmickler [87,88] for the adsorption of $sp$ group elements at submonolayer coverages on the low index faces of Al. For the case of Li, the theoretical UPD shifts on Al(111) (with honeycomb structure), Al(100) and Al(110) were found to be 0.15, 0.64 and 1.78, respectively. On this basis, the substrate used in our experiments can be better represented as an Al(100) and/or Al(110) surface rather than Al(111) [85,86]. Possible contributions due to the latter surface cannot be ignored, however, as the predicted shift would be so small and overlap with currents originating from Li/Al alloy formation.

## 7.  Basal Plane of Highly Oriented Pyrolytic Graphite

Lithium-intercalation carbon electrodes are being considered as the most promising anodes for rechargeable lithium battery applications [94,95]. In addition to their improved safety characteristics and ease of manufacturing, the overall morphology of many of these materials is only slightly affected by their state of charge, i.e., amount of lithium in the lattice. This behavior is unlike that observed with metallic lithium, which during recharge (lithium deposition) develops dendrites and/or other fibrillar deposits which can puncture the separator, leading to internal shorts [96].

The basal plane of highly oriented graphite [HOPG(bp)] was selected for these initial experiments because of its structural simplicity and ease of preparation. Specimens in the form of square wafers, ca. 1 cm$^2$ area (Advanced Ceramics Corp., Parma, OH), cut with a razor knife to form a circular protruding section of about 0.7 cm$^2$ (see insert B in Figure 30) were used as substrates. Prior to their introduction into the UHV chamber (B), the wafer was attached via a special holder to the main manipulator and a new basal plane exposed using adhesive tape. Two electrically isolated Ta strips were used to heat the HOPG resistively during the cleaning procedure (up to 1000 K) and the electrochemical experiments (325 K). The temperature was monitored with a K-type alumel/chromel thermocouple junction inserted on the side of the HOPG(bp) (not shown in the figure) and controlled manually by adjusting the output of the power supply. Electrical connection to the HOPG was achieved with a wire (not shown in the figure) spot-welded to a 0.10 cm² piece of Ta trapped between the circular mica spacer and the back of the specimen (not shown in the figure). Just before transferring the HOPG specimen to the main chamber, a new HOPG basal plane was exposed by cleaving the surface with adhesive tape. A series of heating cycles was performed up to 1000 K to remove water and oxygen from the HOPG(bp), as evidenced by AES.

The AES spectra of HOPG(bp) obtained after the thermal treatment, immediately before the Li[C/R]/LiClO$_4$/PEO assembly was introduced into the main chamber (see curve A in Figure 49) displayed features characteristic of clean graphitic carbon. After the cell had been formed and the temperature had reached 55°C, a series of cyclic voltammograms were acquired at 80 mV/s between the open circuit potential (ca. 2.9 V) and 30 mV versus Li[C/R]. The first and subsequent scans in the negative direction displayed in each case monotonically increasing currents attributed to lithium intercalation at potentials more negative than ca. 1.25 V versus Li[C/R] with no evidence for staging [97]. The lack of any additional features at potentials positive to the onset of lithium intercalation in the first voltammetric cycle is consistent with the absence of trace water and/or oxygen impurities in this ultrapurified film and is in agreement with the results obtained recently for the underpotential deposition of lithium on polycrystalline gold from the same electrolyte in UHV [5]. All scans in the positive direction, following reversal of the scan at 30 mV, showed a clearly defined peak ascribed to lithium deintercalation at potentials between 1 and 1.5 V versus Li[C/R]. A characteristic voltammogram of the HOPG(bp) ultraclean PEO(LiClO$_4$) interface obtained in this type of experiment, which corresponds in this case to a third cycle, is given in curve A of Figure 50. Essentially identical features were found in parallel experiments conducted with a sandwich-type cell involving the same constituents in the glove box, as shown in curve B of Figure 50. This provides unambiguous evidence that the electrochemical behavior observed in UHV is characteristic of the PEO(LiClO$_4$)/HOPG(bp) interface and, hence, not affected in any discernible way by the ultralow pressures.

The charge associated with the first scan in the negative direction was always larger than that observed upon reversing the scan at the negative limit, an effect that became less pronounced as the cycling was continued or as the scan rate was increased. A charge imbalance, particularly during the first charging

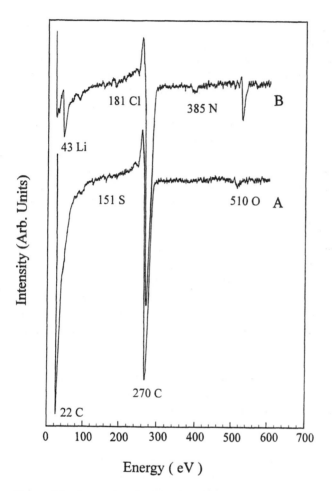

**Figure 49** Comparison between the AES spectra of HOPG(bp) obtained after the thermal treatment, immediately before the Li[C/R]/PEO(LiClO$_4$) assembly was introduced into the main chamber (curve A) and after electrochemical intercalation of Li$^+$ (curve B, see text for details). (From Ref. 6.)

cycle, is almost universally found in the case of lithium intercalation into a variety of high area carbons in liquid electrolytes. Such a phenomenon, referred to as irreversible capacity loss, has often been attributed to the formation of a film at the interface derived from irreversible reactions between intercalated lithium and the electrolyte, including both the salt and the solvent, and affects adversely battery performance. In situ spectroscopic techniques may be required to determine

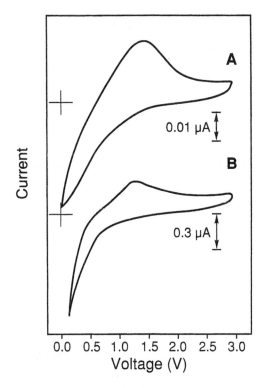

**Figure 50** Comparison between the cyclic voltammogram for the intercalation and deintercalation of $Li^+$ from $PEO(LiClO_4)$ obtained in UHV (curve A) and in a high quality glove box (curve B) at 55°C. Scan rate: 80 mV/s. The difference in the current scales is due to the higher area of contact achieved for the cell in the glove box, i.e., 0.75 $cm^2$ vs. 0.04–0.06 $cm^2$ in UHV. (From Ref. 6.)

whether diffusion of $Li^+$ into the bulk lattice or chemical reactions or both are responsible for the charge imbalance under this very controlled conditions.

Immediately after the series of voltammograms in UHV had been recorded, the HOPG(bp) was polarized at ca. 30 mV versus Li[C/R]. After a few minutes, during which the current was on the order of only a few $\mu A/cm^2$, the cell was disassembled under potential control by simply retracting MCM1.10. This process may be regarded as equivalent to a procedure known as immersion in liquid electrolytes. No signs of carbon or PEO could be found by visual inspection on the PEO or HOPG(bp) surface, respectively. An AES peak at 43 eV, ascribed to intercalated lithium in HOPG(bp), which might include some contribution due to lithium-based surface impurities (see below), could be clearly identified in the spectra recorded at 48°C while the specimen was cooling to room temperature (see curve B in Figure 49).

A quantitative analysis of the AES spectra based on the homogeneous attenuation model indicates that the total AES signal associated with O, Cl, S, and N nitrogen is below 8%. According to these data, the relative amount of O to Cl, corrected by their respective sensitivity factors, is close to 4 to 1. Since Cl is not found in pristine HOPG(bp), both O and Cl could be attributed to a trace of perchlorate ion from the electrolyte adhered to the electrode surface. As is well known, perchlorate ions are labile to reduction by electron and X-ray beams; hence, the possibility of chloride being on the surface cannot be excluded. Sulfur at very low levels was found in the majority of HOPG(bp) specimens examined in this laboratory before lithium intercalation and, therefore, is not derived from impurities in the solution. The fact that the total amount of impurities is so low is not consistent with the presence of a film of any significant thickness on the $Li^+$-intercalated/HOPG(bp) surface following immersion; however, the possibility of a reaction taking place between $Li^+$-intercalated/HOPG(bp) and the PEO/$LiClO_4$ film to yield a chemically modified polymeric material that would become an integral part of the solution phase must also be considered.

## C.  Future Prospects

The processes described illustrate the extraordinary potential this methodology offers for the study of a much wider variety of physical electrochemical phenomena involving well-defined, highly reactive interfaces under carefully controlled conditions. Areas being actively pursued in our laboratories exploit the high versatility of conventional vapor and chemical vapor deposition techniques for the modification of electrode surfaces, as exemplified by the studies of CO and S adsorbed on nickel described in Sections II.B.2 and II.B.3. For example, advantage has been taken of the ability of clean nickel surfaces to catalyze the formation of graphitic layers via the thermal activation of ethylene or other gas mixtures, which can then be used as electrodes to examine lithium intercalation and deintercalation processes. In addition, such graphitic overlayers can be further functionalized by using gas phase exposure to activated oxygen, nitrogen and other species imparting electrodes yet to be explored electrochemical characteristics. The ability to prepare such complex, albeit rather well-defined, graphitic structures in terms of elemental composition may be crucial to the further understanding of the so-called irreversible capacity loss of Li-intercalation carbon electrodes. As described in the technical literature, this phenomenon refers to the much larger charge, i.e., storage capacity, observed during the first, compared to subsequent, charging steps during cycling, and is believed to be associated in part to irreversible reactions involving lithium and highly reactive sites on carbon or impurities in the form of chemically bound surface functional groups. Further efforts toward unveiling the nature of these processes rely on the electron-induced graphitization of boron-doped diamond surfaces, which is expected to produce arrays of small graphite crystals decorating the exposed area. This strategy appears especially intriguing as the interpretation of cyclic voltammetry data may not be complicated by substrate-induced effects as when Li undergoes underpotential deposition on nickel. The rather featureless voltammetric characteristics of BDD in PEO/Li salt solutions identify this material as an ideal support for studies of alloy formation in which thin layers of the base metal can be vapor-deposited on BDD by conventional techniques and then used as electrodes for electrodeposition of a second alloy forming metal. This approach has been recently implemented for studies involving Li-Al alloys in which Al layers of different thicknesses were directly evaporated on BDD and their electrochemical properties subsequently examined in $PEO(LiClO_4)$ solutions. A particularly challenging problem this overall strategy faces is a better identification of the nature of the processes responsible for the charge imbalance observed even for nonalloying substrates such as nickel or BDD. Although it appears certain that perchlorate reduction plays a significant role, the slow decomposition of the polymer itself cannot at this time be fully excluded. Spectroelectrochemical experiments involving attenuated total reflection in UHV might provide much of this needed information; however, a number of technical difficulties must be overcome before these measurements could be fully realized. Despite these challenging hurdles, electrochemistry in UHV will continue to provide valuable insights into certain aspects of interfacial behavior which would be difficult, if not impossible, to achieve with more conventional techniques.

## ACKNOWLEDGMENTS

Much of the work performed at CWRU and described in this article was generously supported initially by the Department of Energy through a subcontract from the Lawrence Berkeley Laboratory, and more recently by the Department of Energy, Office of Basic Energy Science. Additional funding was provided by Eveready Battery Company, Westlake, Ohio. I also wish to express my appreciation to Mrs. Dana Totir, Mr. Yibo Mo, and Mr. Louie Rendek, all graduate students in my group, for their invaluable help during preparation of this manuscript, and to Prof. Gary Chottiner of the Physics Department at CWRU for the years of fruitful and, indeed, enjoyable collaboration. His close involvement with each and every aspect of this research made it possible to turn the methods and results summarized in this chapter from an idea into a continuous series of successful accomplishments.

## REFERENCES

1. Megahed, S.; Scrosati, B. *J. Power Sources*, **1994**, *51*, 79–104.
2. Croze, F.; Scrosati, B. *J. Power Sources*, **1993**, *43/44*, 9–19.
3. Wang, K.; Chottiner, G. S.; Scherson, D. A. *J. Phys. Chem.*, **1993**, *97*, 11075.
4. Zhuang, G.; Chottiner, G. S.; Scherson, D. A. *J. Phys. Chem.*, **1995**, *99*, 7009.
5. Gofer, Y.; Barbour, R.; Luo, Y.; Tryk, D. A.; Jayne, J.; Chottiner, G. A.; Scherson, D. A. *J. Phys. Chem.*, **1995**, *99*, 11739.
6. Gofer, Y.; Barbour, R.; Luo, Y.; Tryk, D. A.; Jayne, J.; Chottiner, G. S.; Scherson, D. A. *J. Phys. Chem.*, **1995**, *99*, 11797.
7. Rath, D. L.; Kolb, D. M. *Surf. Sci.*, **1981**, *109*, 641.
8. Herrera-Fierro, P.; Wang, K.; Wagner, F. T.; Moylan, T. E.; Chottiner, G. S.; Scherson, D. A. *J. Phys. Chem.*, **1992**, *96*, 3788.
9. Howells, A.; Harris, T.; Sashikata, K.; Chottiner, G. S.; Scherson, D. A. *Sol. State Ionics*, **1997**, *94*, 115.
10. Zhuang, G.; Chottiner, G. A.; Bae, I. T.; Hwang, E.; Scherson, D. A. *Rev. Sci. Instrum.*, **1994**, *65*, 2494.
11. (a) Schorn, R. P.; Hintz, E.; Musso, S.; Schweer, B. *Rev. Sci. Instrum.*, **1989**, *60*, 3275; (b) Esposto, F. J.; Griffiths, K.; Norton, P. R.; Timsit, R. S. *J. Vac. Sci. Technol.*, **1994**, *A12*, 3245.
12. (a) Zehner, D. M.; Clausing, R. E.; McGuire, G. E.; Jenkins, L. H. *Sol. State Comm.*, **1973**, *13*, 681; (b) Jackson, A. J.; Tate, C.; Gallon, T. E.; Bassett, P. J.; Matthew, J. A. D. *J. Phys. F: Metal Physics*, **1975**, *5*, 363; (c) Schowengerdt, F. D.; Forrest, J. S. *Scan. Electron Micro.*, **1983**, *11*, 543.
13. (a) Hoenigman, J. R.; Kell, R. G. *Surf. Sci.*, **1984**, *18*, 207; (b) Zavadil, K. R.; Armstrong, N. R. *Surf. Sci.*, **1990**, *230*, 61.
14. Davis, L. E.; MacDonald, N. C.; Palmberg, P. W.; Riach, G. E.; Weber, R. E. *Handbook of Auger Electron Spectroscopy*, Physical Electronics Division, Perkin-Elmer Corp: Eden Prairie, MN, 1978.
15. Feldman, L. C.; Wagner, J. W. *Fundamentals of Surface and Thin Film Analysis*, North-Holland: Amsterdam, 1986.
16. Weast, R. C. *CRC Handbook of Chemistry and Physics*, 57th ed., CRC Press: Cleveland, 1976–1977.
17. Izutsu, K.; Kolthoff, I. M.; Fujinaga, I.; Hattori, M.; Chantooni, M. K. Jr. *Anal. Chem.*, **1977**, *49*, 503.
18. Aurbach, D.; Gofer, Y. *J. Electrochem. Soc.*, **1992**, *139*, 3529.
19. Scrosati, B., Ed. *Applications of Electroactive Polymers*, Chapman & Hall: New York, 1993.
20. (a) Cardillo, M. J.; Becker, G. E.; Hamann, D. R.; Serri, J. A.; Whitman, L.; Mattheiss, L. E. *Phys. Rev.*, **1983**, *B28-2*, 494; (b) Campbell, C. T.; Paffett, M. T. *Surf. Sci.*, **1984**, *143*, 517; (c) Musket, R. G.; McLean, W.; Colmenares, C. A.; Makowiecki, D. M.; Siekhaus, W. J. *Appl. Surf. Sci.*, **1982**, *10*, 143.
21. Parker, S. D. *Surf. Sci.*, **1985**, *157*, 261.
22. Wang, K.; Ross, P., Jr. *J. Electrochem. Soc.*, **1995**, *142*, L95.
23. Semancik, S.; Doering, D. L.; Madey, T. E. *Surf. Sci.*, **1986**, *176*, 165.

24. Rodriguez, J. A.; Hrbek, J.; Yang, Y. W.; Kuhn, M.; Sham, T. K. *Surf. Sci.*, **1993**, *293*, 260.
25. Massalski, T. B. *Binary Alloy Phase Diagram*, 2nd ed., Vol. 3, 1990.
26. Chandavarkar, S.; Diehl, R. D.; Fake, A.; Jupille, J. *Surf. Sci.*, **1989**, *211/212*, 432.
27. Jänsch, H. J.; Huang, C.; Ludviksson, A.; Martin, R. M. *Surf. Sci.*, **1994**, *315*, 9.
28. Palm, A.; Bissell, E. R. *Spectrochim. Acta*, **1960**, *16*, 459.
29. Chesters, M. A.; Gardner, P.; McCash, E. M. *Surf. Sci.*, **1989**, *209*, 89.
30. Harris, A. L.; Rothberg, L.; Dahr, L.; Levinos, N. J.; Dubois, L. H. *J. Chem. Phys.*, **1991**, *94*, 2438.
31. Wang, K.; Ross, Jr., P. N. *Surf. Sci.*, **1996**, *365*, 753.
32. Powell, G. L.; McGuire, G. E.; Easton, D. S.; Clausing, R. E. *Surf. Sci.*, **1974**, *46*, 345.
33. Holcombe, C. E., Jr.; Powell, G. L.; Clausing, R. E. *Surf. Sci.*, **1972**, *30*, 561.
34. *Index of Mass Spectral Data*, AMD11, 1969, American Society for Testing and Materials.
35. Zhuang, G.; Wang, K.; Ross, P. N., Jr. *Surf. Sci.*, **1997**, *387*, 199.
36. Zhuang, G.; Ross, P. N., Jr.; Kong, F.; McLarnon, F. *J. Electrochem. Soc.*, **1998**, *145*, 159.
37. A collection of articles in the area of polymer electrolytes may be found in *Electrochim. Acta*, **1992**, *37*.
38. Abraham, K. M. *Electrochim. Acta*, **1993**, *38*, 1233.
39. Zhuang, G.; Wang, K.; Chottiner, G.; Barbour, R.; Luo, Y.; Bae, I. T.; Tryk, D.; Scherson, D. *J. Power Sources*, **1995**, *54*, 20.
40. Nash, P., Ed. *ASM Handbook, Binary Alloy Phase Diagrams*, Vol. 2, p. 188, ASM International, 1991.
41. Chottiner, G. S.; Jennings, W. D.; Pandya, K. I. *J. Vac. Sci. Technol. A*, **1987**, *5*, 2970.
42. Aurbach, D.; Daroux, M.; Faguy, P.; Yeager, E. *J. Electroanal. Chem.*, **1991**, *297*, 225.
43. Hoelzl, J.; Fritsche, L. *Surf. Sci.*, **1991**, *247*, 226.
44. Villegas, I.; Weaver, M. J. *Electrochim. Acta*, **41**, 661(1996).
45. Trenary, M.; Uram, K. J.; Yates, J. T., Jr. *Surf. Sci.*, **1985**, *157*, 512.
46. Netzer, F. P.; Madey, T. E. *J. Chem. Phys.*, **1982**, *76*, 710.
47. Trenary, M.; Uram, K. J.; Bozso, F.; Yates, J. T., Jr. *Surf. Sci.*, **1984**, *146*, 269.
48. Froitzheim, H.; Koehler, U. *Surf. Sci.*, **1987**, *188*, 70.
49. Edmonds, T.; Pitkethly, R. C. *Surf. Sci.*, **1969**, *15*, 13.
50. List, F. A.; Blakely, J. M. *Surf. Sci.*, **1985**, *152/153*, 463.
51. Wang, K.; Chottiner, G. S.; Scherson, D. A. *J. Phys. Chem.*, **1992**, *96*, 6742; **1993**, *97*, 10108.
52. Madden, H. H.; Kuppers, J.; Ertl, G. *J. Chem. Phys.*, **1973**, *58*, 3401.
53. (a) Madden, H. H.; Ertl, G. *Surf. Sci.*, **1973**, *35*, 211; (b) Madden, H. H. *J. Vac. Sci. Technol.*, **1976**, *13*, 228.
54. Li, L. F.; Gofer, Y.; Totir, D.; Chottiner, G.; Scherson, D. A. *Electrochim. Acta*, **1998**, in press.
55. Jänsch, H. J.; Huang, C.; Ludviksson, A.; Redding, J.; Metiu, H.; Martin, R. M. *Surf. Sci.*, **1989**, *222*, 199.

56. Rodriguez, J. A.; Hrbek, *J. Phys. Chem.*, **1994**, *98*, 4061 and references therein.
57. Demuth, J. E.; Rhodin, T. N. *Surf. Sci.*, **1974**, *45*, 249.
58. Backer, R.; Horz, G. *Vacuum*, **1995**, *46*, 1101.
59. Harte, S. P.; Vinton, S.; Lindsay, R.; Hakansson, L.; Muryn, C. A.; Thornton, G.; Dhanak, V. R.; Robinson, A. W.; Binsted, N.; Norman, D.; Fischer, D. A. *Surf. Sci.*, **1997**, *380*, L463.
60. Ruan, L.; Stensgaard, I.; Lagsgaard, E.; Besenbacher, F. *Surf. Sci.*, **1993**, *296*, 275 and references therein.
61. Maurice, V.; Kitakatsu, N.; Siegers, M.; Marcus, P. *Surf. Sci.*, **1997**, *373*, 307.
62. Shek, M. L.; Sham, T. K.; Hrbek, J.; Xu, G. Q. *Appl. Surf. Sci.*, **1991**, *48/49*, 332.
63. Takamura, T.; Nishijima, Y.; Sekine, K. *Extended Abstract of ECS Meeting*, Vol. 97-1, abst. # 74, Montreal, 1997.
64. Aurbach, D. *J. Electrochem. Soc.*, **1989**, *136*, 906.
65. Gofer, Y.; Li, L. F.; Scherson, D. A. unpublished results.
66. Kolb, D. M.; Przasnyski, M.; Gerischer, H. *J. Electroanal. Chem.*, **1974**, *54*, 25.
67. Li, L. F.; Totir, D.; Miller, B.; Angus, J.; Chottiner, G.; Scherson, D. A. *J. Am. Chem. Soc.*, **1997**, *119*, 7875, and references therein.
68. Argoitia, A.; Angus, J. C.; Wang, L.; Ning, X. I.; Pirouz, P. *J. Appl. Phys.* **1993**, *73*, 4305.
69. Argoitia, A.; Martin, H.; Rozak, E. J.; Landau, U.; Angus, J. C. in *Diamond for Electronic Applications*, Dreifus, D. L.; Collins, A.; Humphrey, T.; Das, S. K.; Pehrsson, P., Eds. Mat. Res. Soc. Symp. Proc. 416, 349, MRS, Pittsburgh, PA, 1996.
70. Knight, D. S.; White, W. B. *J. Mater. Res.*, **1988**, *4*, 385.
71. Angus, J. C.; Cassidy, W. D.; Wang, L.; Wang, Y.; Evans, E.; Kovach, C.; Tamor, M. A. Mat. Res. Soc. Symp. Proc. 383, 45, MRS, Pittsburgh, PA, 1995.
72. Chottiner, G. S.; Jennings, W. D.; Pandya, K. I. *J. Vac. Sci. Technol.*, **1987**, *5A*, 2970.
73. Pepper, S. V. *Appl. Phys. Lett.*, **1981**, *38*, 344.
74. Swain, G. M. Private communication.
75. (a) Randin, J. P.; Yeager, E. *J. Electrochem. Soc.*, **1971**, *118*, 712; (b) Randin, J. P.; Yeager, E. *J. Electroanal. Chem.*, **1972**, *36*, 257; *ibid.*, **1975**, *58*, 313.
76. Gerischer, H.; Mcintyre, R.; Scherson, D.; Storck, W., *J. Phys. Chem.*, **1987**, *91*, 1930, and references therein.
77. Selman, J. R.; DeNuccio, D. K.; Sy, C. J.; Steunenberg, R. K. *J. Electrochem. Soc.*, **1977**, *124*, 1160.
78. Rao, B. M. L.; Francis, R. W.; Christopher, H. A. *J. Electrochem. Soc.*, **1977**, *124*, 1490.
79. Wen, C. J.; Boukamp, B. A.; Huggins, R. A. *J. Electrochem. Soc.*, **1979**, *126*, 2258.
80. Epelboin, I.; Froment, M.; Garreau, M.; Thevenin, J.; Warin, D. *J. Electrochem. Soc.*, **1980**, *127*, 2100.
81. Baranski, A. S.; Fawcett, W. R. *J. Electrochem. Soc.*, **1982**, *129*, 901.
82. Jow, T. R.; Liang, C. C. *J. Electrochem. Soc.*, **1982**, *129*, 1429.
83. Maskell, W. C.; Owen, J. R. *J. Electrochem. Soc.*, **1985**, *132*, 1602.
84. Lantelme, F.; Iwadate, Y.; Shi, Y.; Chemla, M. *J. Electroanal. Chem.*, **1985**, *187*, 229.

85. Kumagai, N.; Kikuchi, Y.; Tanno, K. *J. Appl. Electrochem.*, **1992**, *22*, 620.
86. Kumagai, N.; Kikuchi, Y.; Tanno, K.; Lantelme, F.; Chemla, M. *J. Appl. Electrochem.*, **1992**, *22*, 728.
87. Schmickler, W. *Chem. Phys.*, **1990**, *141*, 95.
88. Lehnert, W.; Schmickler, W. *J. Electroanal. Chem.*, **1991**, *310*, 27.
89. Bottomley, D. J.; Lupke, G.; Bloch, J.; van Driel, H. M.; Timsit, R. S. *Appl. Surf. Sci.*, **1992**, *62*, 97.
90. Dey, A. N. *J. Electrochem. Soc.*, **1971**, *118*, 1547.
91. Gerischer, H.; Kolb, D. M.; Przasnyski, M. *Surf. Sci.*, **1974**, *43*, 662.
92. Trasatti, S. *J. Electroanal. Chem.*, **1971**, *33*, 351.
93. Weast, R. C., Ed. *CRC Handbook of Chemistry and Physics*, 54th ed., CRC: Cleveland, 1974.
94. Abraham, K. M. *Electrochim. Acta*, **1993**, *38*, 1233.
95. Tarascon, J. M.; Guyomard, D. *Electrochim. Acta*, **1993**, *38*, 1221.
96. Scrosati, B. *J. Electrochem. Soc.*, **1992**, *139*, 2776.
97. Dahn, J. R.; Sleigh, A. K.; Shi, H.; Reimers, J. N.; Zhong, Q.; Way, B. M. *Electrochim. Acta*, **1993**, *38*, 1179.

# 6

# The Electrochemical Behavior of Active Metal Electrodes in Nonaqueous Solutions

**Doron Aurbach**
*Bar-Ilan University, Ramat-Gan, Israel*

## I. INTRODUCTION

A major part of the work with nonaqueous electrolyte solutions in modern electro-chemistry relates to the field of batteries. Many important kinds of novel, high energy density batteries are based on highly reactive anodes, especially lithium, Li alloys, and lithiated carbons, in polar aprotic electrolyte systems. In fact, a great part of the literature related to nonaqueous electrolyte solutions which has appeared during the past two decades is connected to lithium batteries. These facts justify the dedication of a separate chapter in this book to the electrochemi-cal behavior of active metal electrodes.

Active metal electrodes are defined as electrodes which react spontaneously with the solution components at open circuit potential. With the exception of miscellaneous nonaqueous systems based on aromatic nonpolar solvents, de-scribed in Chapter 1 (e.g., Gileadi et al. [1,2]), most electrochemical nonaqueous systems are based on polar solvents. These contain C-halogen (Cl, Br), C—O, C—N, C—S, or C—P bonds that can readily be reduced by alkali metal, and probably by alkaline earth metals as well. In addition, the commonly used salts comprise anions such as $ClO_4^-$, $BF_4^-$, $AsF_6^-$, $PF_6^-$, $SO_3CF_3^-$, $N(SO_2CF_3)_2^-$, $C(SO_2CF_3)_3^-$, which can also be attacked by alkali and alkaline earth metals. Thus, both classes of metals fit into the category of active electrode materials.

As discussed in Chapter 4, solvents such as alkyl carbonates and esters and salt anions such as $AsF_6^-$, $ClO_4^-$, and $PF_6^-$ may be reduced on noble metal electrodes at potentials $>1.5$ V versus $Li/Li^+$ [3–5]. Hence, looking through the standard electrochemical thermodynamic tables [6], metals such as aluminum, zinc, and iron, as well as many other possible transition metals, may react spontaneously with commonly used polar aprotic solvents and salt anions. There are indeed reports on the reactive behavior of aluminum [7] and iron [8,9] in nonaqueous, polar aprotic systems. In addition, one must take into account the unavoidable presence of atmospheric contaminants such as water and oxygen in polar aprotic systems. This opens up another frontier of possible spontaneous interactions between metallic electrodes and solution components. In this chapter we concentrate on lithium, magnesium, calcium, and aluminum electrodes, with special emphasis on the first three metals. This is due to their importance as anodes for nonaqueous, high energy density batteries, which has placed them at the center of so many R & D efforts and publications over the past two decades. Aluminum has also been mentioned as an anode material for nonaqueous batteries. However, the major interest of nonaqueous aluminum electrochemistry relates to Al electroplating.

Basic issues such as surface reactions, surface film formation, passivation, ionic and electronic transport phenomena through surface films, problems in uniformity of deposition and dissolution processes, correlation between surface chemistry, morphology, and electrochemical properties are common to all active metal electrodes in nonaqueous solutions and are dealt with thoroughly in this chapter. It is believed that many conclusions related to Li, Mg, Ca, and Al electrodes can be extended to other active metal electrodes as well.


## II. REMARKS ON THE COMMON ELECTROCHEMICAL BEHAVIOR OF ACTIVE METAL ELECTRODES

Active metals are always covered by surface films, which are inevitably formed by their spontaneous reactions with atmospheric components. It should be emphasized that exposure of the fresh surface of alkali, alkaline earth metals, metals of the third group (e.g., aluminum), as well as many transition metals, to atmospheres of dry rooms, glove boxes, and even high vacuum, will lead to their fast coverage by new surface layers [10]. The surface films which naturally cover active metals are a combination of the metal oxide [11], hydroxide [12], and carbonate [13]. Metal oxides such as $Li_2O$ react readily with $CO_2$ to form $Li_2CO_3$ on the oxide's surface [14]. Hydration of the metal oxides forms an outer layer of metal hydroxide. The former species may also react with $CO_2$ to form $M_x(CO_3)_y$ and/or $M_xH_y(CO_3)_z$. Alkali metals such as Li react spontaneously with atmospheric nitrogen to form the metal nitride [15]. When active metal electrodes are introduced into the electrolyte solutions, the first interaction taking place is between the pristine films and the solution components. There are several possibilities:

1. The surface species dissolve in the solution due to penetration of solvent molecules through grain boundaries, followed by physical desorption of surface species as a result of their interaction with the solvent [16].

2. The surface films react chemically with solution species, thus leading to their dissolution as reaction products [17]. Surface species such as oxides, hydroxides, and nitrides may be highly nucleophilic, while many polar aprotic solvents are highly electrophilic. Hence, chemical dissolution of pristine surface films on active metals in solutions is a very probable route [18].

3. Once dissolution of the pristine surface films is possible, it opens the door for spontaneous reactions between the active metal and solution species [16–18]. Hence, a partial, or even completed, replacement of the pristine surface films by new films originating from solution species takes place. This process, however, stops, as the new films are thick and sufficiently compact to prevent the tunneling of electrons through them.

4. Hence, the equilibrium situation of many active metal electrodes in solution involves the dynamic deposition-dissolution of surface species related to solution components in a steady state, which leaves the surface films at a constant average thickness.

5. It should be emphasized that when water is present in solution, it can hydrate surface species and thus diffuse through the films closer to the active metal-film interphase, where they can be reduced [19]. There are only a few surface species that can be regarded as impermeable to water. Thus, whenever the solutions contain water, the water should react with the active metal during prolonged storage via a mechanism which involves $H_2O$ diffusion through hydrated surface films. These processes should gradually increase the thickness of the surface films, enriching them with metal hydroxide species. However, it is expected that part of the metal hydroxide may be further reduced to metal oxide. Thus, in the presence of water, the surface films should contain $MO_x$ in the inner layer (metal side) and $M(OH)_y$ in the outer layer (solution side) [20].

6. The surface films covering the active metals are formed by a gradually decreasing driving force. When the fresh, active metal is exposed either to reactive gases or to solution species, its reactivity is maximal, and, thus, reduction of atmospheric or solutions components takes place under highly nonselective conditions. Any bare metal surface is readily

covered. As the film thickens, electron tunneling through it is more difficult, and thus, further reduction takes place under a weaker driving force. Finally, reduction of atmospheric and solution species can only continue sporadically at scattered and occasional sites on the outer part of the surface films, where local defects in the films allow for some electron tunneling.

This situation, together with the dissolution-deposition cycles of surface species occurring at steady state, as described above, leads to a porous structure of the outer part (solution side) of the surface films. Hence, we expect that surface films in active metals in solutions should comprise a compact part, close to the metal side of the surface films, and a porous layer at the solution side [21,22]. The various situations described above for surface film formation on active metal electrodes in solutions are illustrated in Figures 1–4.

Hence, in conclusion, the electrochemical behavior of active metal electrodes is always controlled by the above-described surface films and the transport properties of electrons and ions through them. We can see the following possibilities.

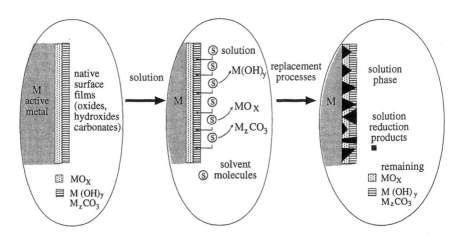

**Figure 1** A schematic view of replacement of native surface films on active metals by new ones in solutions. It should be noted that in reality the borders between the layers which comprise the surface films are not distinctive, and their microstructure is mosaic-like. This, and the subsequent figures, are attempts to emphasize the fact that the composition and structure of the surface films change as a function of their distance from the active metal.

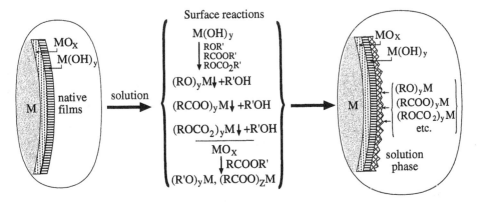

**Figure 2** A schematic view of formation of multilayer surface films by secondary reactions of native films with solution species.

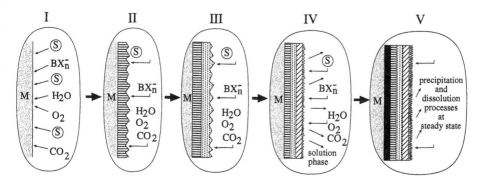

**Figure 3** A schematic view of formation of multilayer surface films on active metals exposed fresh to solution phase. Stage I: Fresh surface-nonselective reactions; Stage II: Initial layer is formed, more selective surface film formation continues; Stage III: Formation of multilayer surface films; Stage IV: Highly selective surface reactions at specific points; partial dissolution of surface species; Stage V: Further reduction of the surface species close to the active metal, deposition-dissolution of surface species at steady state; the surface film is comprised of a multilayer inner compact part and an outer porous part.

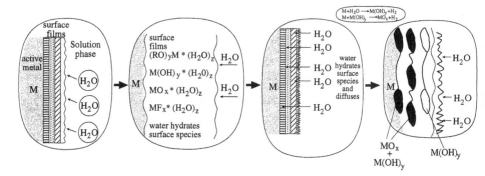

**Figure 4**  A schematic description of the influence of water in solutions on the structure of surface films covering active metal electrodes.

## A.  The Films Are Electronically Insulating but Conduct the Ions of the Active Metal

This allows metal dissolution-deposition processes to occur at a relatively low overpotential. A typical example for such behavior is lithium in many polar aprotic systems. The nature of the surface films in these cases determines the following possibilities:

1.  Dissolution and deposition processes occur uniformly via the surface films and beneath them. Hence, the surface films are sufficiently flexible to follow the changes in the metal-film interface and thus continuously protect the active surface from reactions with the solution species. Such behavior is ideal for rechargeable battery applications (see Figure 5).

2.  Dissolution forms an unstable metal interface, which leads to a breakdown of surface films: Thus, the cracked film allows the solution to reach the active surface, where it is repaired by further reaction of solution species in the cracks. This behavior is termed ''breakdown and repair mechanism'' [23] (Figure 6).

3.  Deposition is nonuniform. Thus, dendrites are formed as active metal deposits emerge from the surface films. These deposits induce nonuniform current distribution that enhances the phenomenon, as illustrated in Figure 7. These dendrites react readily with solution species since they are bare active metal deposits and thus become covered with surface films. In narrow parts and bottlenecks, these surface reactions can disconnect the dendrite electronically from the bulk. This phenomenon

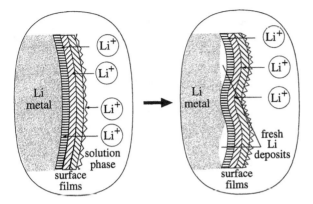

**Figure 5** A schematic description of homogeneous electrodeposition of active metal beneath passivating surface films.

is a major reason for the loss of active mass in rechargeable batteries based on active metal anodes [24–26].

4. Deposition of the active metal may occur by electron tunneling through partially insulating or semiconducting species. Thus, metal deposits which are not directly connected electrically with the bulk are formed. Such a type of deposition forms an isolated island of deposited metal embedded in the surface films [27]. This possibility is also illustrated in Figure 7.

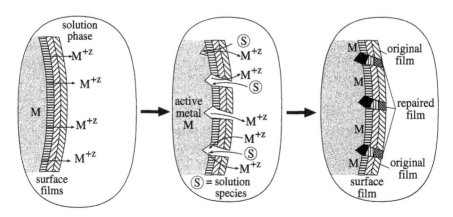

**Figure 6** A schematic view of active metal dissolution via a breakdown and repair mechanism of the surface films.

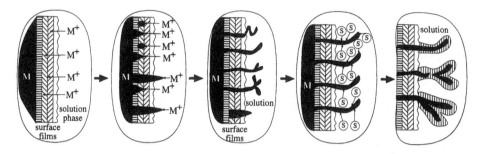

**Figure 7**   A schematic view of dendritic deposition of active metals.

## B.   The Active Metal Is Covered with Surface Films That Are Electronically and Ionically Insulating

This is the case for magnesium and calcium electrodes whose cations are bivalent. The surface films formed on such metals in a wide variety of polar aprotic systems cannot transport the bivalent cations. Such electrodes are blocked for the metal deposition [28–30]. However, anodic processes may occur via the "breakdown and repair mechanism." Due to the positive electric field, which is the driving force for the anodic processes, the film may be broken and cracked, allowing metal dissolution. Continuous metal dissolution creates an unstable situation in the metal-film and metal-solution interfaces and prevents the formation of stable passivating films. Thus, once the surface films are broken and a continuous electrical field is applied, continuous metal dissolution may take place at a relatively low overpotential (compared with the high overpotential required for the initial breakdown of the surface films). Typical examples are calcium dissolution processes in several polar aprotic systems [31].

In conclusion, a key feature of active metal electrodes in nonaqueous solutions is their coverage by surface films which control their electrochemical behavior. Hence, transport phenomena in these films deserve special attention.

## III.   TRANSPORT PHENOMENA AND KINETIC CONSIDERATION IN SURFACE FILMS COVERING ACTIVE METAL ELECTRODES IN SOLUTIONS

## A.   Kinetics of the Growth of Surface Films

When a fresh active metal is exposed to a polar aprotic solution whose components are reduced on the active surface to form insoluble metal salts, a surface film grows via a corrosion process. The driving force for this process is the difference between the redox potentials of the active metal and the solution species $\Delta V_{\text{M-S}}$. As a first approximation, we can assume a homogeneous surface film and Ohm's law, connecting the corrosion current density $i_{\text{corr}}$ and $\Delta V_{\text{M-S}}$.

Hence,

$$i_{\text{corr}} = \frac{\Delta V_{\text{M-S}}}{\rho(\text{film}) \cdot l(t)} \tag{1}$$

where $\rho(\text{film})$ is the surface film's resistivity for electron tunneling (assuming homogeneous condition), and $l(t)$ is its thickness (which grows in time). Assuming that all the reduction products precipitate on the active metal surface, then

$$\frac{dl}{dt} = K i_{\text{corr}} \tag{2}$$

$K$ is the proportionality constant that depends on the molecular size of the surface species and their density of packing on the surface. Combining Eqs. (1) and (2) and integrating with the boundary condition $t = 0 \rightarrow l = 0$ yields

$$l = \left( \frac{2K\Delta V_{\text{M-S}}t}{\rho(\text{film})} \right)^{1/2} \tag{3}$$

which is the well-known parabolic growth of the surface films [32]. When the active metal exposed to solution is already covered by initial surface films, and hence at $t = 0$, $l = l_0$, then

$$l = \left( l_0^2 + \frac{2K\Delta V_{\text{M-S}}t}{\rho(\text{film})} \right)^{1/2} \tag{4}$$

Indeed, a nearly parabolic dependence of surface film growth in exposure time was found for lithium and calcium in thionyl chloride solutions [33,34]. In any event, the above description is only a first approximation. In reality, the films are not necessarily homogeneous, and, as described in the previous section, water contamination may play an important role, and possible dissolution-deposition cycles of the surface species, as well as their possible secondary reactions with solution species, may considerably distort the above parabolic dependence of $l$ (film) versus $t$.

## B. Conductance Mechanisms in Surface Films on Active Metal Electrodes

We can assume that as the surface films formed on active metals in solutions reach a certain thickness, they become electronic insulators. Hence, any possible electrical conductance can be due to ionic migration through the films under the electrical field. The active electrodes are thus covered with a solid electrolyte interphase (the SEI model [35]), which can be anionic or cationic conducting or both.

For a classical SEI electrode such as lithium, the surface films formed on it in most of the commonly used polar aprotic systems conduct Li ions, with a transference number ($t_+$) close to unity. As stated earlier the surface films on active metals are reduction products of atmospheric and solution species by the active metal. Hence, these layers comprise ionic species that are inorganic and/or organic salts of the active metal. Conducting mechanisms in solid state ionics have been dealt with thoroughly in the past [36–44]. Conductance in solid ionics is based on defects in the medium's lattice. Figure 8 illustrates the two common defects in ionic lattices: interstitial (Frenkel-type) defects [37] and hole (Schottky-type) defects [38].

In the former case, the ions migrate among the interstitial defects, which may be relevant only to small ions such as $Li^+$. This leads to a transference number close to 1 for the cation migration. In the other case, the lattice contains both anionic and cationic holes, and the ions migrate from hole to hole [39]. The dominant type of defects in a lattice depends, of course, on its chemical structure as well as its formation pattern [40–43]. In any event, it is possible that both types of holes exist simultaneously and contribute to conductance. It should be emphasized that this description is relevant to single crystals. Surface films formed on active metals are much more complicated and may be of a mosaic and multilayer structure. Hence, ion transport along the grain boundaries between different phases in the surface films may also contribute to conductance in these systems.

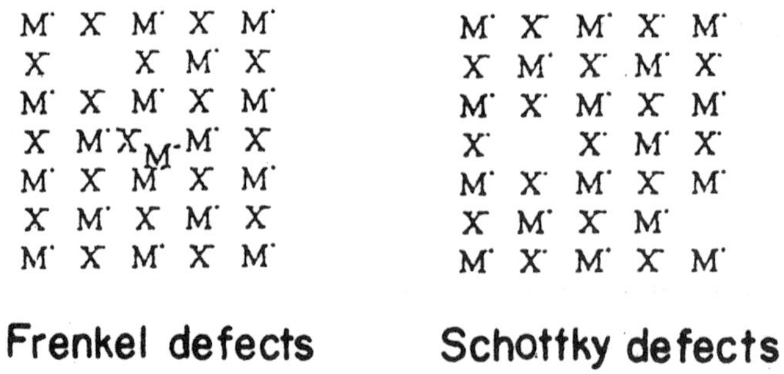

**Frenkel defects**          **Schottky defects**

**Figure 8**  Schematic presentation of Frenkel- and Schottky-type lattice defects.

## C. Kinetics of Active Metal Electrodes Covered by Surface Films: Current-Potential Relationship

The kinetics of the simplest solid electrolyte interphase (SEI) electrode should include three stages: charge transfer in the solution-film interface, ion migration through the surface films, and then charge transfer in the film-metal interface. It is reasonable to assume that the ion migration is the rate-determining step. Thus, it may be possible to use the basic equation (5) for ionic conductance in solids as the starting point [32,36,45]:

$$i = 4zFan\nu \exp\left(\frac{-W}{RT}\right)\sinh\left(\frac{azFE}{RT}\right) \tag{5}$$

where $a$ is the jump's half-distance, $\nu$ is the vibrational frequency in the lattice, $z$ is the ion's charge, $W$ is the energy barrier for the ion jump, $n$ is the ion's concentration, $E$ is the magnitude of the electric field, and $F$ is the Faraday number.

When all of the potential falls on the surface films, then

$$\eta = \eta_{SEI} = El \tag{6}$$

where $l$ is the film's thickness. At equilibrium $\eta = 0$, so the net current is zero; the exchange current is

$$i_0 = 2zFa\nu n \exp\left(\frac{-W}{RT}\right) \tag{7}$$

At high electrical field, $azF\eta > RTl$, and thus a Tafel-like behavior is obtained:

$$i = i_0 \exp\left(\frac{azF\eta}{RTl}\right) \tag{8}$$

At low electrical field, Eq. (8) can be linearized, and thus an ohmic behavior is obtained:

$$i = \frac{4.6i_0\eta}{b} \tag{9}$$

where $b$ is the analog of the Tafel slope extracted from Eq. (8):

$$b = \frac{2.3RTl}{azF} \tag{10}$$

Hence, the average resistivity of the surface films can be extracted as

$$\frac{\rho}{A} = \frac{R_{film}}{l} = \frac{b}{4.6i_0l} = \frac{RT}{2azFi_0}, \qquad A = \text{electrode surface area} \qquad (11)$$

where $R_{film} = \eta/I$ is the surface film resistance for ionic conductance, extracted from Eq. (9), and $I = iA$.

Table 1 presents some typical average resistivity values of surface films on active metal electrodes in polar aprotic systems.

Hence, it appears that active metal electrodes in solutions (which are cov-

**Table 1** Average Composition and Resistivities of Surface Films on Active Metal Electrodes

| Metal | Solvent system | Electrolyte | Surface film composition | Resistivity $\Omega \cdot cm$ | Ref. |
|---|---|---|---|---|---|
| Li | $SOCl_2$ | $LiAlC_4$ | LiCl | $10^7-10^8$ | 32 |
| Li alloys | $SOCl_2$ | $LiAlC_4$ | LiCl | $10^8-10^9$ | 32 |
| Li | $SOCl_2$ | $Li_2B_{12}Cl_{12}$ | LiCl | $10^7-10^8$ | 32 |
| Li | $SO_2 + AN$ | $Li_2S_{8.4}$ | $Li_2S$ | $10^8$ | 32 |
| Li | THF | LiBr | $Li_2S_2O_4$ | $1-4 \times 10^7$ | 32 |
| Mg | $SOCl_2$ | $Mg(AlCl_4)_2$ $Mg(FeCl_4)_2$ | $MgCl_2$ | $10^9$ | 32 |
| Ca | $SOCl_2$ | $Ca(AlCl_4)_2$ $Ca(FeCl_4)_2$ | $CaCl_2$ | $10^8-10^{10}$ | 32 |
| Ca | $SOCl_2$ | $Ca(AlCl_4)_2$ | $CaCl_2$ | $10^{10}$ | 32 |
| Li | PC | $LiAsF_6$ | $ROCO_2Li$ LiF, $Li_2CO_3$ | $1-2 \times 10^8$ | 47 |
| Li | PC | $LiClO_4$ | $ROCO_2Li$ $Li_2CO_3$ | $1-2 \times 10^8$ | 47 |
| Li | PC | $LiAsF_6$ $H_2O$ 200 ppm | $ROCO_2Li$ $Li_2CO_3$ $LiOH-Li_2O$ | $1-2 \times 10^8$ | 47 |
| Li | PC | $LiPF_6$ 1 $M$ | $ROCO_2Li$ LiF | $1-5 \times 10^{10}$ | 48 |
| Li | 1,3-dioxolane | $LiAsF_6$ | ROLi PolyDN $HCO_2Li$ | $1-2 \times 10^8$ | 49 |
| Li | THF | $LiAsF_6$ | ROLi | $1-1.5 \times 10^8$ | 50 |
| Li | EC-DMC | $LiAsF_6$ | $ROCO_2Li$ | $1-4 \times 10^8$ | 51 |
| Li | EC-DMC | $LiPF_6$ 1 $M$ | LiF $ROCO_2Li$ | $1-5 \times 10^{10}$ | 51 |

ered by surface films) behave electrochemically, similar to the usual classical electrochemical systems (Butler Volmer type behavior [46]).

## IV. SURFACE FILM FREE ACTIVE METAL ELECTRODES

The most common active metal electrodes in nonaqueous systems are indeed the surface film controlled ones described above. However, there are some exceptional cases in which surface film free active electrodes in nonaqueous solutions can be obtained. Three types of such systems are blue solutions, aluminum nonaqueous-aprotic plating systems, and magnesium electrodes in Grignard reagent solutions.

## A. Blue Solutions

Alkali metals can dissolve in ammonia and, to a lesser extent, in ultrapure amines and ethers [52], according to the following equation:

$$M^0 \xrightarrow{\text{RNH2, ROR}'} M^+(\text{solv.}) + e^-(\text{solv.}) \tag{12}$$

This process enables a surface film free active electrode to be obtained in a blue solution of solvated electrons. As explained by Peled [32], an equilibrium is reached at a certain concentration of the metal ions and electrons according to the following equation:

$$E_M^\circ + \frac{RT}{F} \ln a_M{}^+ = E_{e^-}^\circ - \frac{RT}{F} \ln a_{e^-} \tag{13}$$

A similar situation may be obtained when alkali metals are immersed in ultrapure ethers containing benzophenone [53]. The metal thus dissolves via formation of stable ketal radical anions in solution (and metal ions as well). It should be emphasized that the above processes occur even when the active metal is initially introduced into the solution covered by surface films (due to reactions with atmospheric contaminants). We assume that electron tunneling through the films enables the initiation of the dissolution process. This process breaks the film on the metal (as metal is depleted beneath the rigid surface film), thus enabling solvent molecules to reach the active surface and solvate more electrons. This increases the metal solubilization and the further breakdown of the initial surface films. Hence, an equilibrium between a bare metal and the blue solution can finally be reached, as explained above (Eq. 13).

## B.  Aluminum Nonaqueous-Aprotic Plating Baths (see also Section VIII)

Gileadi et al. showed that it was possible to obtain aluminum ion-conducting nonpolar aprotic solutions [54–56]. Aromatic solvents such as toluene and ethylbenzene can dissolve $AlBr_3$ or $Al_2Br_6$ and LiBr or KBr, thus forming clusters of $M^+ AlBr_4^-$ or $M^+Al_2Br_7^-$ ($M^+$ is the alkali metal ion) which conduct ions by a hopping-type mechanism (in contrast to the ion transport via frictional motion in polar systems). In such solutions it is possible to plate aluminum and thus form a surface film's free active metal electrode in aprotic media (provided that the aprotic media are sufficiently pure and do not contain atmospheric and/or protic contaminants) [57].

## C.  Magnesium Electrodes in Grignard Reagent Solutions (see also Section VI)

It is well known that it is possible to deposit and dissolve magnesium at relatively low overpotential in highly pure ethereal solutions containing Grignard reagents: RMgCl, RMgBr, etc. (R is the alkyl or aryl group) [58]. The interfacial behavior of magnesium electrodes in these solutions has not yet been investigated thoroughly. However, it is quite probable that magnesium does not react with ethers as lithium does [50]. The Grignard reagents react readily with protic contaminants and oxygen, and thus the magnesium deposited in these solutions may be regarded as a bare metal. This situation may be the reason why Mg dissolution and deposition occur at all in these systems.

## V.  LITHIUM ELECTRODES

### A.  Overview of Li Electrode Behavior

Lithium is obviously one of the most reactive electrode materials. Its low molecular weight and low redox potential make it highly attractive for use as an anode in high energy density batteries. Intensive studies of Li electrodes in polar aprotic electrolyte systems for more than three decades have led to successful R & D, producing commercial lithium batteries. During the seventies, commercial primary Li batteries appeared on the market. During the eighties, efforts were focused on the R & D of rechargeable Li batteries. These efforts led to only a few commercially important rechargeable batteries based on Li metal anodes and liquid electrolyte solutions, the MoLi cell ($Li-MoS_2$) [59] and the Re-Li cell ($Li-TiS_2$) [60]. These efforts continued intensively in the nineties and produced the Li ion rechargeable battery technology (the "rocking-chair" type cells) [61].

In the field of Li metal anode-based cells, the nineties also saw the development of the Tadiran Li-Li$_x$MnO$_2$ [62] and the Moltech Li-S technologies [63] and the Li-SO$_2$ rechargeable cells [64]. This intensive research over a period of about 30 years leads to the conclusion that lithium metal is the electrode material which is the major focus of attention in modern nonaqueous electrochemistry. Consequently, a great deal of knowledge has been accumulated on the behavior of Li electrodes, which sheds light on the behavior of many other active metal electrodes in aprotic media. Li surfaces produced in dry atmosphere are covered by initial, native surface films composed of Li$_2$O in the inner layer and Li$_2$CO$_3$ and LiOH (depending on the degree of humidity) on the outer layer. Immersion of such Li surfaces in solution may lead to considerable changes in the composition of the surface films, depending on the solution composition [18,48–51]. The following situations can be considered:

1.  In the presence of acidic species such as HF in LiPF$_6$ solutions, the Li oxides are replaced by LiF.
2.  In the presence of highly electrophilic solvents such as $\gamma$-butyrolactone ($\gamma$-BL), Li$_2$O and LiOH attack solvent molecules nucleophilically to form organic Li salts (e.g., $\gamma$-oxybutyrate in the case of $\gamma$-BL).
3.  In the presence of water, the Li oxide is readily hydrated, so water diffuses through the Li oxide and then is reduced by the active metal at sites close to the Li-film interface. This process is continuous and leads to thickening of the Li$_2$O films as H$_2$O is reduced to Li$_2$O and H$_2$ as the final and most stable products.
4.  Electrochemical dissolution of lithium leads to a breakdown of the initial surface films and their replacement by surface species resulting from the reduction of solution components.
5.  Electrochemical deposition of lithium usually forms a fresh Li surface which is exposed to the solution phase. The newly formed surface reacts immediately with the solution species and thus becomes covered by surface films composed of reduction products of solution species. In any event, the surface films that cover these electrodes have a multilayer structure [49], resulting from a delicate balance among several types of possible reduction processes of solution species, dissolution-deposition cycles of surface species, and secondary reactions between surface species and solution components, as explained above. Consequently, the microscopic surface film structure may be mosaic-like, containing different regions of surface species. The structure and composition of these surface films determine the morphology of Li dissolution-deposition processes and, thus, the performance of Li electrodes as battery anodes. Due to the mosaic structure of the surface

films on Li which may induce highly nonuniform current distribution, Li deposition in most of the commonly used polar aprotic systems is dendritic. Li dendrites react readily with solution species since they possess a highly reactive fresh Li surface. Dendrite formation is a major reason for the failure of Li electrodes as anodes in rechargeable battery systems.

Hence, a key point in the study of the behavior of Li electrodes is the correlation among the surface chemistry (determined by the solution composition), morphology, and reversibility during repeated deposition-dissolution cycles. Some highlights in these studies are outlined below:

1.  The interest in Li batteries began in 1958 [65] and intensified during the seventies. The work of Bro and Selim [66], Jasinski [67] and the EIC group (Abraham and Brummer et al.) [68] should be acknowledged. These three groups contributed a great deal to the understanding of the correlation between solution composition, Li electrode morphology, and cycling efficiency. These groups also pointed out the problems and directions in the development of rechargeable batteries with Li metal anodes.
2.  Intensive work was carried out in the seventies by Dey et al. [69–71], Koch et al. [72–74], Jansta et al. [75,76], and Thevenin et al. [77–80] on the chemical analysis of the surface films formed on Li electrodes in solutions. These efforts, however, were based on XPS, AES, and some indirect methods.
3.  The SEI model for the surface films on Li and other active metal electrodes proposed by Peled [35] in 1979 contributed a great deal to the understanding of the electrochemical behavior of these systems.
4.  Judicious application of micropolarization techniques [81], as well as ellipsometry [82] and XRD [83], enabled analysis of the multilayer structure of the surface films formed on lithium in solutions.
5.  Introduction of FTIR spectroscopy for the analysis of the surface films formed on Li in electrolyte solutions by Yeager et al. [84,85], and further intensive use of this technique by Aurbach et al. [86,87], enabled more precise identification of the chemical composition of the surface films formed on lithium electrodes in a large variety of electrolyte solutions.
6.  Microelectrode techniques were applied and enabled the direct study of kinetics of the $Li/Li^+$ couple [88,89].
7.  Irish and Odziemkowski [90,91] developed a method to directly study the time constants of surface film formation on fresh Li surfaces.

8.  Impedance spectroscopy was intensively used and enabled the elucidation of the electrical properties of the Li-solution interface in a large variety of electrolyte solutions [49,92–100].
9.  Rigorous use of XPS by Kanamura et al. enabled a further understanding of the multilayer structure of the surface films formed on Li in solutions, as well as the aging processes of the surface films [101–105].
10. The effect of the stack pressure on the morphology of Li electrodes was studied by Wilkinson et al. [106]. This issue is highly important, as applying stack pressure on Li electrodes considerably suppresses dendrite formation during Li deposition.
11. Spectroscopy in UHV systems applied for the study of the interaction between polar aprotic solvents and Li surfaces prepared by deposition in ultrahigh vacuum by Scherson et al. [107] also contributed a great deal toward understanding the fundamental behavior of these systems.
12. The application of novel in situ spectroscopic techniques for the study of Li electrodes in solutions should also be acknowledged. These include FTIR spectroscopy [108], atomic force microscopy (AFM) [109], electrochemical quartz crystal microbalance (EQCM) [110], Raman spectroscopy [111], and XRD [83].
13. A great deal of effort was dedicated to the study of Li electrodes in polymeric electrolyte systems [112–115]. These can serve as alternatives for the liquid electrolyte solutions in which dendrite formation is such a severe problem.

Based on the above studies, the chemical and physical structure, as well as the electrical properties of surface films formed on lithium electrodes in most of the commonly used polar aprotic electrolyte systems, are well established. The correlation between the surface chemistry, morphology and reversibility in charge-discharge cycling in a large variety of systems is also well understood. It should be noted that during the past three decades, Li electrodes have been tested in hundreds of combinations of solvents, salts, and additives in the continuous effort to find suitable electrolyte solutions for rechargeable Li batteries with Li metal anodes. The solvents have included $SO_2$, ethers, esters, acetals, and alkyl carbonates and their combinations. The salts have included $LiAsF_6$, $LiPF_6$, $LiBF_4$, $LiClO_4$, $LiSO_3CF_3$, $LiN(SO_2CF_3)_2$, $LiC(SO_2CF_3)_3$, and $LiBR_4$ (R = alkyl or aryl group). $LiAsF_6$ was found to be the best electrolyte for Li metal anode/liquid electrolyte solution rechargeable Li batteries. Several systems in which Li cycling efficiency is very high were found. These include 1,3-dioxolane solutions [68,116], EC-based solutions [117], and a combination of THF, 2Me-THF, and methyl furan [118].

The next sections deal with particular aspects of the above studies, as well as their practical consequences, related to the use of Li anodes in rechargeable batteries.

## B.  Remarks on the Literature of Li Electrochemistry

The electrochemistry of lithium has had excellent coverage in the literature of modern electrochemistry. In this section, we provide examples of literature sources for Li electrochemistry. An important source of information on Li electrochemistry and Li batteries in general are the proceedings volumes of annual or biannual international symposia and conferences on Li batteries. A partial list of these seminars follows.

1.  Each of the annual fall meetings of the Electrochemical Society over the past 18 years has included a special seminar on Li batteries and related technology. Most of these symposia have produced books that include full presentation papers. These books are published annually, within the framework of the PV series, by the Electrochemical Society, Inc. (e.g., PV 80-4, 81-4, 84-1, 87-1, 88-6, 90-5, 91-3, 92-15, 93-24, 94-4, 94-28, 97-18).

2.  Every second year there is an international meeting on Li batteries, presentations from which are published as peer-reviewed articles in special volumes of the *Journal of Power Sources* [e.g., volumes 43–44 (1993), 54 (1995), 68 (1997) related to the 6 IMLB (Münster, Germany, May 1992), 7 IMLB (Boston, USA, May 1994), and 8 IMLB (Nagoya, Japan, June 1996), respectively].

3.  The Annual Battery Seminar in Florida, which takes place at the beginning of March and which is organized by S. P. Wolsky Florida Education Seminars, Inc., and N. Marincic Redox, Inc., also produces a valuable proceedings volume.

4.  The Annual International Power Sources Symposium that takes place in Cherry Hill, NJ in June also produces a valuable proceedings volume which contains important, updated material on Li batteries.

5.  The Annual Battery Conference in Japan produces an extended abstract volume which presents a good overview of the most recent news in research related to Li batteries in Japan.

6.  The annual fall Materials Research Society (MRS) meetings have, in recent years, included symposia on battery technology and materials. These seminars include highly updated information on R & D related to Li batteries and produce proceedings volumes on the presentations, appearing as peer-reviewed, full papers.

7.  The special journal *Progress in Batteries and Solar Cells* (which does

not relate to international meetings or symposia), is published annually by JEC Press, Inc., Brunswick, OH. A large portion of its volumes contains updated material on Li battery technology.

Other important sources of information are two comprehensive books on Li batteries:

1. *Li Batteries*, J.P. Gabano, Editor, Academic Press, New York (1983).
2. *Li Batteries: New Materials, Developments and Perspectives*, G. Pistoia, Editor, Elsevier, Amsterdam (1994).

Over the years, review articles on Li electrochemistry and Li batteries have appeared in the literature. Table 2 lists important review articles that have been published over the past two decades.

Thousands of articles have been published on Li electrochemistry and Li battery technology in leading electrochemistry journals, most of which cannot be covered in a single chapter. However, we list representative papers on Li electrochemistry, divided into different subjects:

*Solvent Effects.* PC [84,128–131]; EC [51,132–136]; THF [50,85,137–139]; 2Me-THF [50,140–143]; 1,3-DN [116,144–148]; glymes [149]; $\gamma$-BL [150,151]; MF [4,152]; alkyl carbonate-ethers mixtures [153–157].

*Salt Effects.* Comparing $LiAsF_6$, $LiPF_6$, $LiSO_3CF_3$, and $LiN(SO_2CF_3)_2$ in PC [48]; comparing $LiClO_4$, $LiBF_4$, $LiAsF_6$ and $LiPF_6$ in alkyl carbonates [16,17,159]; comparing $LiAsF_6$ and $LiClO_4$ in PC [47]; $LiAsF_6$, $LiN(SO_2CF_3)_2$,

**Table 2** Some Review Articles on Li Electrodes and Li Batteries

| Authors | Sources | Year | Ref. |
|---|---|---|---|
| Besenhard, J.O. and Eichinger, G.J. | *J. Electroanal. Chem.* | 1976 | 119 |
| Brummer, S.B. | *Mater. Adv. Batteries* | 1980 | 121 |
| Armand, M. | *Mater. Adv. Batteries* | 1980 | 122 |
| Abraham, K.M. | *J. Power Sources 7* (1–43) | 1981–2 | 120 |
| Hughes, M. et al. | *J. Power Sources* | 1984 | 124 |
| Abraham, K.M. | *J. Power Sources* | 1985 | 125 |
| Barthel, J. & Gores, H.J. | *Chemistry of Non-aqueous solutions*, G. Mamantov and A.I. Popov (eds.), VCH | 1994 | 123 |
| Bagotsky, V.S. & Skundin, A.M. | *Russian J. Electrochem.* | 1995 | 126 |
| Aurbach, D. | *J. Power Sources* | 1995 | 86 |
| Bruce, P.G. | *Phil. Trans. R. Soc. Lond. A* | 1996 | 127 |
| Aurbach, D. | *J. Power Sources* | 1997 | 87 |

and $LiC(SO_2CF_3)_3$ in 1,3-DN and THF [158]; $LiAsF_6$, $LiBF_4$, $LiClO_4$ and $LiN(SO_2CF_3)_2$ in glymes [149]; Li boranes [160–163].

*Additive Effects.* $CO_2$ [4,131,164]; $H_2O$ [4,20,47,50]; HF [165]; metallic additives [166]; fluorine compounds [167]; organic additives [168–170]; arsenic compounds [50].

*Morphological Studies.* Li electrodes in PC, THF, $\gamma$-BL solutions of $LiClO_4$ and $LiAsF_6$ by SEM [19]; examples for dendrite formation (SEM) [24–26,171]; application of AFM for morphological studies of Li electrodes [109, 172]; application of STM for morphological studies of Li or Li-C electrodes [173].

*Special Spectroscopic Studies of Li Electrodes.* XPS [16,17,101,102,159, 174]; FTIR (ex situ) [4,18,84,85,150,153,175]; FTIR (in situ) [108,158,176,177]; in situ Raman [178,179]; in situ XRD [83]; in situ ellipsometry [82]; AES [180, 181]; TPD, and mass spectrometry in UHV systems [182].

*Special Electrochemical Measurements of Li Electrodes.* Micropolarization [21,81]; EQCM [183,184]; application of microelectrodes for studies of the $Li/Li^+$ couple [88,89]; fast OCV measurements (passivation kinetics) [90,91]; Li cycling efficiency [168–170]; impedance spectroscopy (EIS) [32,35,47,49, 77–80,92–100,185,186].

## C. Identification of the Surface Films Formed on Lithium in Different Electrolyte Systems and the Related Surface Chemistry

### 1. Introduction

From the early studies of Li electrodes in polar aprotic electrolyte solutions during the seventies, it became clear that the electrochemistry of Li electrodes is controlled by the surface films covering them. Thus, identification of the composition and structure of these films and their correlation to the solution composition and to the performance of Li electrodes as anodes in battery systems became a central issue in the study of these systems. Day and Rudd [69] and Dousek et al. [76–77] identified $Li_2CO_3$ as the major reduction product of PC by lithiated graphite and Li/Hg amalgam, respectively, using bulk spectroscopic techniques. Hence, they concluded that $Li_2CO_3$ is the major component in surface films formed on lithium in this solvent. Koch et al. [72–74] analyzed the surface films formed on lithium surfaces in ethereal solvents containing $LiAsF_6$ salt (the so-called brown film [72]) as some kind of arsenic oxide polymer which has alkoxy and fluoroalkoxy groups. In parallel, XPS and AES were used by Muller et al. [82,83], and Thevenin et al. [77–80] for surface analysis of Li electrodes after their immersion in solutions. These measurements suggested that the surface films formed on Li in solvents, such as PC, contain both organic and inorganic species. Kerr proved by GC analysis of ethereal solutions in which Li electrodes were cycled (charge-discharge) that alkoxy species (ROLi) are formed by reaction between the active metal and ethers [195]. Nazri and Muller [83] introduced in situ XRD measurements of Li electrodes by which they analyzed both organic and inorganic surface species on Li in solutions.

All the above studies indicated clearly that reduction of solvent, salt, and additives (e.g., $H_2O$) by Li contribute together to the buildup of the surface films on lithium in solutions. It should be emphasized that XRD, XPS, and AES studies of Li electrodes, as well as the indirect identification of surface species from studies of reactions of lithiated graphite or Li/Hg amalgam with electrolyte solutions, could not provide specific enough information on the chemical composition of the surface films. Moreover, application of XPS for Li electrodes may induce secondary surface reactions. Visible changes appear on Li surfaces during XPS measurements. More specific information on the composition of the surface layers formed on Li could be obtained by surface-sensitive FTIR spectroscopy that was introduced into this field in the middle of 1985 by Yeager et al. [84,85,178], and which is a nondestructive technique.

The experimental consideration for the performance of ex situ and in situ FTIR spectroscopic studies of Li electrodes was reported in detail in Refs. 48, 85, 108, 131, 157, 175 and 176. A schematic description of the FTIR measurement modes is shown in Figure 9. Surface Raman could also be considered as a promising technique for the analysis of the surface layers in lithium electrodes (and can be applied in situ). It also provides information on the vibrational states of materials and, thus, the identity of functional groups. However, we found that this technique is destructive for Li surfaces since the laser beam causes visible decoloration of Li surfaces during Raman measurements [187].

The identification of the various surface compounds formed on lithium in different solutions, reported in the next section, was based on surface-sensitive FTIR spectroscopy as a major tool. A library of reference FTIR spectra of model compounds was prepared, to which actual spectra measured from Li surface treated in solution could be compared [86,87]. Important complementary information, especially on inorganic species (e.g., salt reduction products), was obtained from EDAX and XPS.

## 2. Identification of Surface Films Formed on Lithium Electrodes in Alkyl Carbonate Solutions

As a starting point, the usefulness of FTIR spectroscopy for studying the complicated Li surface chemistry is demonstrated. Figures 10 and 11 show FTIR spectra obtained from Li surfaces prepared fresh in propylene carbonate solutions of $LiClO_4$, $LiBF_4$, and $LiPF_6$. One spectrum in each figure was measured from an electrode after 2 h of storage, while the second one was measured from an identical electrode after storage for two days. (The experimental considerations are described in Ref. 175.)

# Surface Studies (FTIR)

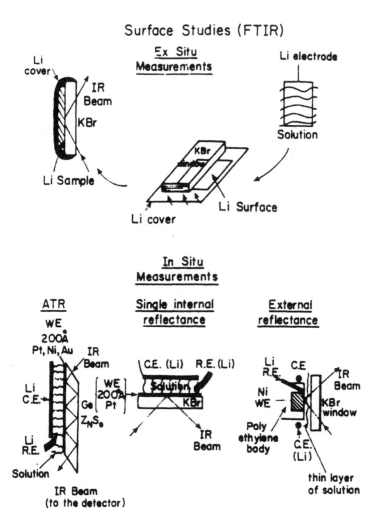

**Figure 9** Schematic presentation of the ex situ and in situ FTIR measurements of Li electrodes [176]. (With copyrights from Elsevier Science Ltd., 1998.)

Figures 12 and 13 present a pair of similar spectra from similar measurements and an experimental setup obtained from Li electrodes aged in EC and DMC solutions, respectively. Figures 14 and 15 show reference FTIR spectra of the electrolysis products of PC and EC, respectively, in tetrabutyl ammonium salt solutions, isolated as Li salts. The spectra in Figures 14 and 15 are typical of $ROCO_2Li$ species [188]. The relevant compounds were identified as

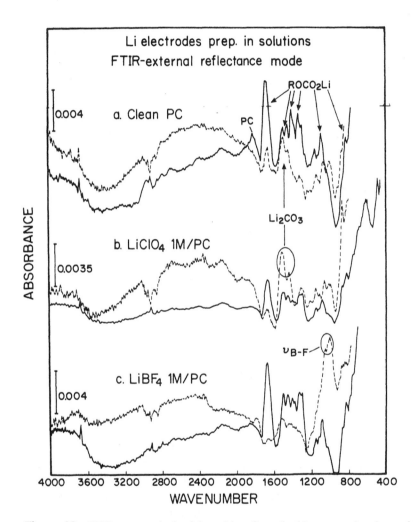

**Figure 10** FTIR spectra obtained from Li surfaces freshly prepared and stored in PC solutions. The surfaces were protected with KBr windows and measured using external reflectance mode at a grazing angle. (a) Pure solvent; (b) LiClO$_4$ 1 $M$ solutions; (c) LiBF$_4$ 1 $M$ solution. Solid lines, 2 h of storage; dashed lines, 2 days of storage [175]. (With copyrights from The Electrochemical Society, Inc., 1998.)

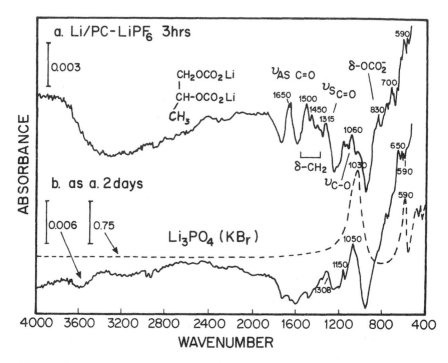

**Figure 11**  FTIR spectra obtained from Li electrodes (ex situ, external reflectance mode, protected with KBr plates) prepared fresh in PC-LiPF₆ solution and stored for different periods: (a) 3 h; (b) 2 days, and (---) a spectrum of Li₃PO₄ pelletized with KBr (transmittance) for comparison [190]. (With copyrights from Elsevier Science Ltd., 1998.)

$CH_3CH(OCO_2Li)CH_2OCO_2Li$ and $(CH_2OCO_2Li)_2$, respectively [131,189]. From comparing Figures 14 and 15 with the spectra related to short storage periods in Figures 10–13, it is clear that quite similar products are the major surface species that comprise the surface films formed initially on lithium in these solutions. The second spectra in Figures 10, 12, and 13, obtained from Li after prolonged storage, all have pronounced peaks of $Li_2CO_3$, as indicated [189]. XPS measurements of Li electrodes prepared and treated in similar experiments are in line with these conclusions [159]. It was concluded from these studies that $Li_2CO_3$ is formed on Li surfaces in these solutions due to a reaction with the $ROCO_2Li$ species, with trace water unavoidably present in solutions, rather than by direct, two-electron reduction of the alkyl carbonate by lithium [84,131,189]. However, we cannot exclude the possibility that during storage the $ROCO_2Li$ surface species initially formed are further reduced by the active metal by a slow charge transfer ($e^-$ and $Li^+$) that forms $Li_2CO_3$ and the relevant alkylene. Hence, the enrichment of the

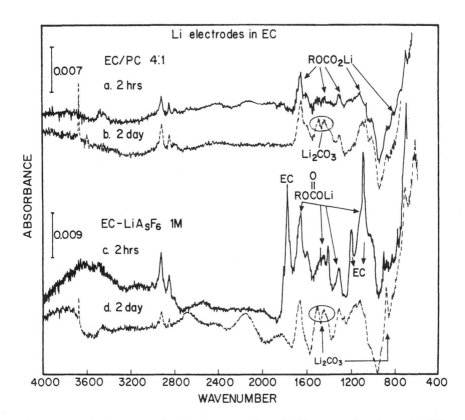

**Figure 12** FTIR spectra obtained from Li surfaces freshly prepared and stored in EC solutions (external reflectance, at a grazing angle, protected with KBr windows). (a,b) EC-PC 4:1 mixture; (c,d) EC-LiAsF₆ solution. Solid lines (a,c) 2 h of storage; dashed lines (b,d) 2 days of storage [175]. (With copyrights from The Electrochemical Society Inc., 1998.)

surface films formed on lithium in alkyl carbonates with $Li_2CO_3$ upon storage may be due to both continuous reactions of the $ROCO_2Li$ with trace water and slow charge transfer. The formation of $ROCO_2Li$ species as the major, initial reduction product of alkyl carbonate molecules by Li was verified by the careful experiments carried out in UHV by Scherson et al. [182], using TPD and mass spectroscopy. It should be noted that the conclusions of early work in this field were that $Li_2CO_3$ is the major reaction product between Li and PC. This was based on studies of reactions of Li/Hg amalgam with PC [75–76], reactions of PC with lithiated graphite [69–71], and the use of XPS [77–80] or XRD [83] techniques. This early work ignored the possibility that the $Li_2CO_3$ detected was

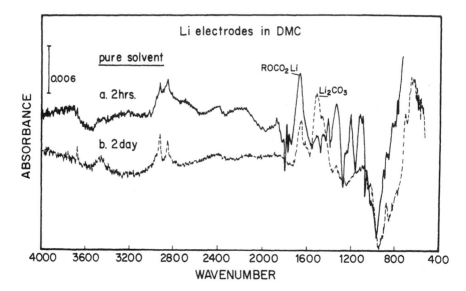

**Figure 13** FTIR spectra obtained from Li surfaces freshly prepared and stored in DMC (external reflectance, grazing angle, protected with KBr windows). (a) Solid line, 2 h of storage; (b) dashed line, 2 days of storage [175]. (With copyrights from The Electrochemical Society Inc., 1998.)

not a product of direct reduction of PC by lithium, but rather a product of a secondary reaction between ROCO$_2$Li and trace H$_2$O, which is unavoidably present in solutions in any practical inert atmosphere.

As discussed in Ref. 84, Li/Hg amalgam cannot be a model system for solid Li surfaces, because reduction of solution species on the liquid Li/Hg interface does not produce stable surface films. Thus, a massive solvent reduction may occur on Li/Hg in which each solvent molecule reacts directly with the bare active surface. In such a situation, PC and EC are indeed reduced directly to Li$_2$CO$_3$ [84,131]. However, ROCO$_2$Li species are major reduction products of PC and EC on Li/Hg as well. It should be noted that when the Li is initially covered by native surface films (Li$_2$O, Li$_2$CO$_3$), the situation is more complicated. Only part of the native surface films may be replaced upon storage in the solutions; thus, in such a case the nature of the surface films remains more inorganic than in the case of fresh Li surfaces [101–105]. In any event, upon Li deposition or dissolution, the replacement of the native surface films by solution reduction products is fast and pronounced, and the above-described surface chemistry is very relevant to practical Li anodes in batteries.

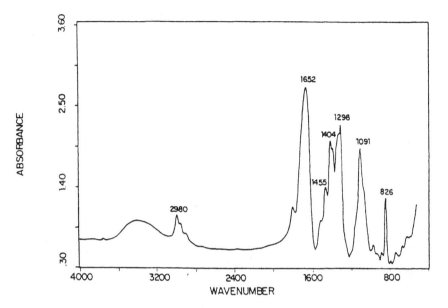

**Figure 14** FTIR spectrum obtained from the major product of PC electrolysis precipitated as lithium salt pelletized with KBr (transmittance mode) [189]. (With copyrights from Elsevier Science Ltd., 1998.)

As described in Section V.C.6, the salt used also considerably influenced the surface chemistry developed on Li in these solutions. Except for the halides, all of the commonly used salt anions that contain at least one element in a high oxidation state are reduced on the Li surface to form insoluble surface species. In the case of $LiAsF_6$ and $LiClO_4$, the surface chemistry of Li in their solutions in alkyl carbonates is dominated by solvent reduction, as described above. Other salts, e.g., $LiPF_6$, bring with them contaminants such as HF, which, when present, reacts with the carbonates to form LiF and their mother acids (i.e., $H_2CO_3$ or $ROCO_2H$). Hence, upon storage in HF containing alkyl carbonate solutions, the surface films covering lithium become rich in LiF because of the carbonates that are solubilized [16,101–105]. This is accurately reflected by the spectra of Figure 11, which show that after prolonged storage in $PC/LiPF_6$ solution the surface films lack carbonates [190].

Schemes 1 and 2 provide the most probable reduction patterns of alkyl carbonate solvents on Li surfaces (with references included). Table 3 provides a summary of the various surface species formed on lithium in the commonly used Li battery electrolyte solutions (with references included). Scaling the rela-

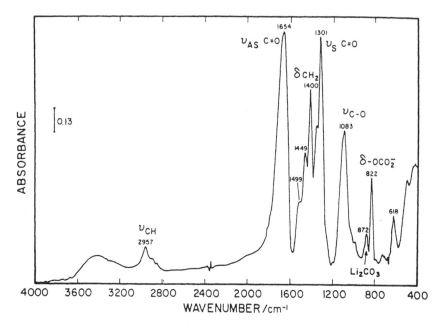

**Figure 15** FTIR spectrum of the electrolysis product of EC in THF/0.5 $M$ TBAP solution on gold, isolated as Li salt (pelletized with KBr) [131]. (With copyrights from Elsevier Science Ltd., 1998.)

tive reactivities of the various alkyl carbonates toward lithium is difficult because of two determining factors that cannot be separated easily: (1) the inherent reactivity of the solvents as electrophiles and (2) the Li passivity obtained by the solvent reduction products which precipitate as passivating surface films on the Li surface. In general, EC is the most reactive solvent in the alkyl carbonates' list (Table 3). Consequently, in mixtures containing EC as a solvent at concentrations above 25%, its reduction products are dominant in the surface films. The inherent reactivity of the cyclic alkyl carbonate (EC, PC) is higher than that of the open chain molecules, due to internal strain. However, the reduction products of EC and PC are Li alkylene dicarbonates, whose solubility is much lower than that of the $ROCO_2Li$ species formed by reduction of the open chain alkyl carbonates. Thus, Li electrodes are apparently more stable and better passivated in PC- or EC-based solutions than in the open chain alkyl carbonate-based solutions. Among the open chain alkyl carbonates, we should expect that the degree of electrophilicity decreases as the R group becomes larger, and thus the expected order of their inherent reactivity toward lithium should be DMC > EMC > DEC PMC. However, since $CH_3OLi$ and $CH_3OCO_2Li$ (which are reduction products

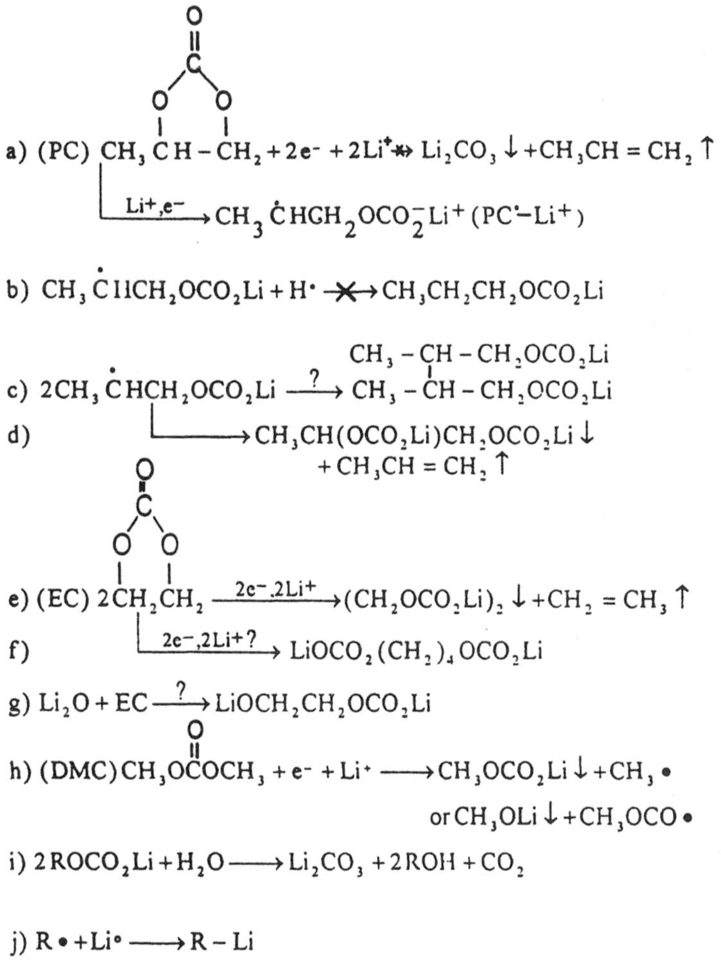

$$\underset{\displaystyle \text{O}}{\overset{\displaystyle \text{O}}{\underset{\displaystyle |}{\overset{\displaystyle \|}{\underset{\displaystyle \text{O}}{\overset{\displaystyle \text{C}}{\phantom{.}}}}}}$$

a) (PC) $CH_3\,\overset{|}{C}H-\overset{|}{C}H_2 + 2e^- + 2Li^+ \nrightarrow Li_2CO_3 \downarrow + CH_3CH=CH_2 \uparrow$

$\quad\quad\xrightarrow{\;Li^+,e^-\;} CH_3\,\dot{C}HGH_2OCO_2^-Li^+ \;(PC^-\text{-}Li^+)$

b) $CH_3\,\dot{C}IICH_2OCO_2Li + H^\cdot \nrightarrow CH_3CH_2CH_2OCO_2Li$

c) $2CH_3\,\dot{C}HCH_2OCO_2Li \xrightarrow{\;?\;} \begin{array}{l} CH_3 - CH - CH_2OCO_2Li \\ \phantom{CH_3 -} | \\ CH_3 - CH - CH_2OCO_2Li \end{array}$

d) $\phantom{2CH_3\,\dot{C}HCH_2OCO_2Li} \xrightarrow{\phantom{aaa}} CH_3CH(OCO_2Li)CH_2OCO_2Li \downarrow$
$\quad\quad\quad\quad\quad\quad\quad\quad + CH_3CH=CH_2 \uparrow$

$$\underset{\displaystyle \text{O}}{\overset{\displaystyle \text{O}}{\underset{\displaystyle |}{\overset{\displaystyle \|}{\underset{\displaystyle \text{O}}{\overset{\displaystyle \text{C}}{\phantom{.}}}}}}$$

e) (EC) $2\overset{|}{C}H_2\overset{|}{C}H_2 \xrightarrow{\;2e^-,2Li^+\;} (CH_2OCO_2Li)_2 \downarrow + CH_2=CH_2 \uparrow$

f) $\phantom{2CH_2CH_2} \xrightarrow{\;2e^-,2Li^+?\;} LiOCO_2(CH_2)_4OCO_2Li$

g) $Li_2O + EC \xrightarrow{\;?\;} LiOCH_2CH_2OCO_2Li$

h) (DMC) $CH_3O\overset{\displaystyle \text{O}}{\overset{\displaystyle \|}{C}}OCH_3 + e^- + Li^+ \longrightarrow CH_3OCO_2Li \downarrow + CH_3 \bullet$
$\quad\quad\quad\quad\quad\quad\quad\quad\quad \text{or } CH_3OLi \downarrow + CH_3OCO \bullet$

i) $2ROCO_2Li + H_2O \longrightarrow Li_2CO_3 + 2ROII + CO_2$

j) $R\bullet + Li^\circ \longrightarrow R-Li$

**Scheme 1** Possible reduction patterns of alkyl carbonates on Li [159].

of DMC, EMC, and PMC) are less soluble in the alkyl carbonates than are the same species with ethyl or propyl groups, Li is much better passivated and is apparently more stable in DMC than in the other three solvents. In DEC, Li dissolves (no passivation at all), and in EMC and PMC the Li stability is similar. (Li corrodes visibly during prolonged storage in EMC and PMC, while in DMC or any EC- or PC-based solution Li is apparently stable during prolonged storage.) Due to the dianionic nature and the compact structure of the EC reduction product $(CH_2OCO_2Li)_2$, Li passivation is highly efficient in EC-based solutions.

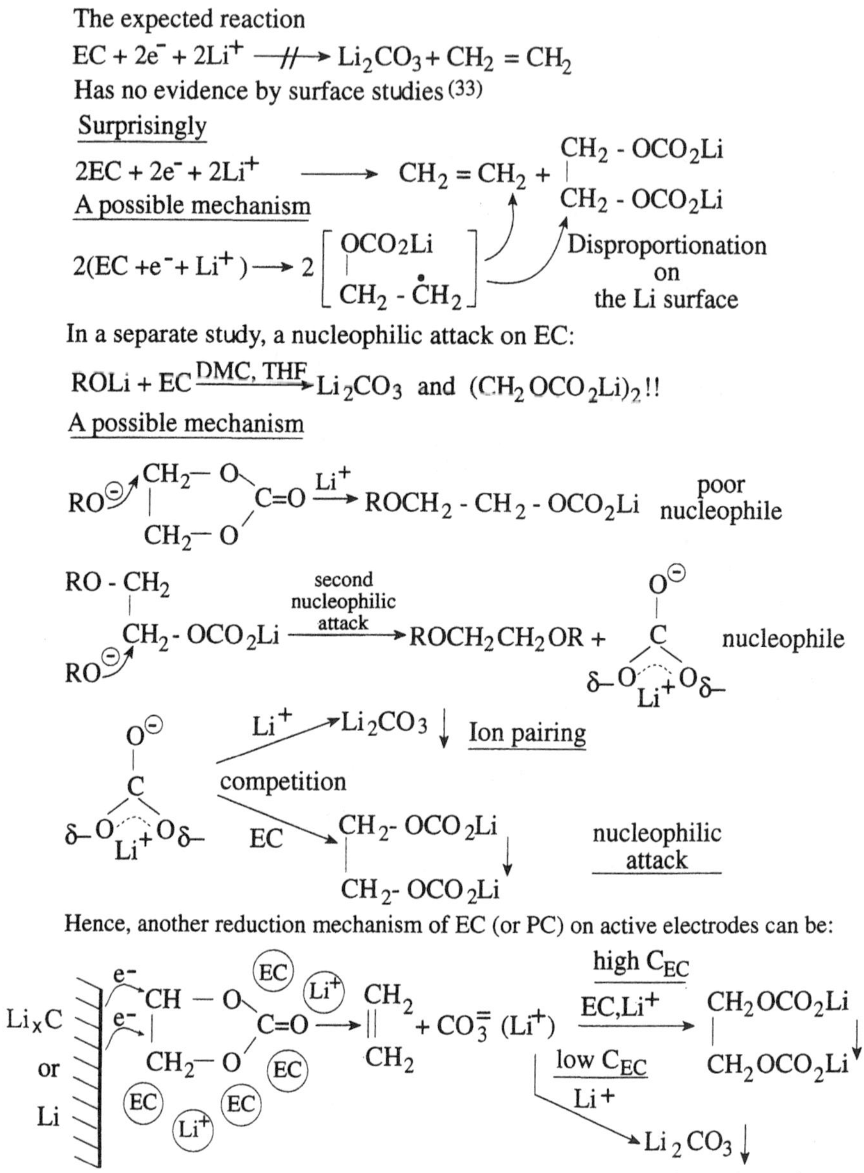

The expected reaction

$$EC + 2e^- + 2Li^+ \xrightarrow{\quad//\quad} Li_2CO_3 + CH_2 = CH_2$$

Has no evidence by surface studies [33]

Surprisingly

$$2EC + 2e^- + 2Li^+ \longrightarrow CH_2 = CH_2 + \begin{array}{l} CH_2 - OCO_2Li \\ | \\ CH_2 - OCO_2Li \end{array}$$

A possible mechanism

$$2(EC + e^- + Li^+) \longrightarrow 2 \begin{bmatrix} OCO_2Li \\ | \\ CH_2 - \overset{\bullet}{C}H_2 \end{bmatrix} \xrightarrow{\text{Disproportionation}} \begin{array}{c} \text{on} \\ \text{the Li surface} \end{array}$$

In a separate study, a nucleophilic attack on EC:

$$ROLi + EC \xrightarrow{\text{DMC, THF}} Li_2CO_3 \text{ and } (CH_2OCO_2Li)_2 \text{!!}$$

A possible mechanism

$$RO^{\ominus} \overset{\nearrow CH_2 - O}{\underset{CH_2 - O}{|}} C=O \xrightarrow{Li^+} ROCH_2 - CH_2 - OCO_2Li \quad \begin{array}{c} \text{poor} \\ \text{nucleophile} \end{array}$$

$$\begin{array}{l} RO - CH_2 \\ | \\ CH_2 - OCO_2Li \\ RO^{\ominus} \end{array} \xrightarrow[\text{attack}]{\substack{\text{second} \\ \text{nucleophilic}}} ROCH_2CH_2OR + \begin{array}{c} O^{\ominus} \\ | \\ C \\ \delta^- O \overset{\cdots}{\underset{Li^+}{}} O\delta^- \end{array} \quad \text{nucleophile}$$

$$\begin{array}{c} O^{\ominus} \\ | \\ C \\ \delta^- O \overset{\cdots}{\underset{Li^+}{}} O\delta^- \end{array} \xrightarrow[\text{competition}]{Li^+ \to Li_2CO_3 \downarrow \;\; \text{Ion pairing}}$$

$$EC \searrow \begin{array}{l} CH_2 - OCO_2Li \\ | \\ CH_2 - OCO_2Li \end{array} \downarrow \quad \begin{array}{c} \text{nucleophilic} \\ \text{attack} \end{array}$$

Hence, another reduction mechanism of EC (or PC) on active electrodes can be:

$$Li_xC \text{ or } Li \quad \begin{array}{l} CH - O \\ | \\ CH_2 - O \end{array} C=O \longrightarrow \begin{array}{c} CH_2 \\ || \\ CH_2 \end{array} + CO_3^= (Li^+) \xrightarrow[\substack{\text{low } C_{EC} \\ Li^+}]{\substack{\text{high } C_{EC} \\ EC, Li^+}} \begin{array}{l} CH_2OCO_2Li \\ | \\ CH_2OCO_2Li \end{array} \downarrow$$

$$\searrow Li_2CO_3 \downarrow$$

**Scheme 2**  EC (PC) reduction mechanisms [191].

As presented in Scheme 2, solutions containing a low concentration of EC as a cosolvent (in ethers or open chain alkyl carbonates) channel EC reduction on Li to produce $Li_2CO_3$ as an important surface species [191]. The inorganic carbonate is also an excellent passivating agent due to the dianionic and compact structure. Hence, the combination of $Li_2CO_3$ and $(CH_2OCO_2Li)_2$ formed in Li surfaces in EC solutions stabilizes the active metal in solutions very efficiently. Indeed, EC-based solutions are the best for Li ion battery systems [192,193] and have been recognized as highly promising for rechargeable Li batteries with Li metal anodes [194].

## 3. Identification of Surface Films Formed on Lithium in Esters

The most important esters in connection with Li batteries are γ-butyrolactone (BL) and methyl formate (MF). Li is apparently stable in both solvents due to passivation. Electrolysis of BL on noble metal electrodes produces a cyclic β-keto ester anion which is a product of a nucleophilic reaction between a γ-butyro-lactone anion (produced by deprotonation in position α to the carbonyl) and another γ-BL molecule. FTIR spectra measured from Li electrodes stored in γ-BL indicate the formation of two major surface species: the Li butyrate and the dili-thium cyclic β-keto ester dianion. The identification of these products and related experimental work is described in detail in Refs. 150 and 189. Scheme 3 shows the reduction patterns of γ-BL on lithium surfaces (also see product distribution in Table 3). In the presence of water, the LiOH formed on the Li surfaces due to $H_2O$ reduction attacks the γ-BL nucleophilically to form derivatives of γ-hydroxy butyrate as the major surface species [18] [e.g., $LiO(CH_2COOLi)$]. We have evidence that γ-BL may be nucleophilically attacked by surface $Li_2O$, thus forming $LiO(CH_2)_3COOLi$, which substitutes for part of the surface Li oxide [18]. MF is reduced on Li surfaces to form Li formate as the major surface species [4]. $LiOCH_3$, which is also an expected reduction product of MF on Li, was not detected as a major component in the surface films formed on Li surfaces in MF solutions [4]. The reduction paths of MF on Li and their product analysis are presented in Scheme 3 and Table 3.

## 4. Identification of Surface Films Formed on Li in Ethereal Solvents

The surface chemistry of Li in ethereal solutions has been studied rigorously by several groups. Koch et al. [72–74] studied the surface layers formed on Li in THF and 2Me-THF solutions of $LiAsF_6$, which appear as brown films covering the electrodes. They concluded that these films are polymers of lithiated arsenic oxide that contain additional functional groups such as $F^-$, $RO^-$, etc. However, these conclusions did not result from direct spectroscopic studies of Li surfaces. Important, indirect information on the composition of the surface films formed

**Table 3** Reaction Products of Solvents, Salts and Atmospheric Contaminants with Lithium

| Solvent type | Specific solvent | Ref. | Dry | Contaminants Additives | | | |
|---|---|---|---|---|---|---|---|
| | | | | H₂O | O₂ | CO₂ | HF |
| Alkyl carbonates | PC | 84,175 | CH₃CH(OCO₂Li)CH₂OCO₂Li & propylene | ROCO₂LI | ROCO₂Li | ROCO₂Li | LiF |
| | EC | 131,175 | (CH₂OCO₂Li)₂ & ethylene | Li₂CO₃ | | + | +H₂CO₃ |
| | DMC | 51,175 | ROCO₂Li (CH₃OCO₂Li) | (reaction of ROCO₂Li + H₂O) | Li₂O | Li₂CO₃ | + ROCO₂H |
| | DEC | 84,132 | CH₃CH₂OCO₂Li + CH₃CH₂OLi | | Li₂O₂ | | |
| | EMC | 360 | CH₃OLi, CH₃OCO₂Li | LiOH-Li₂O | | | |
| | PMC | 361 | CH₃OLi, CH₃OCO₂LI, CH₃CH₂CH₂OL, CH₃CH₂CH₂OCO₂Li | | | | |
| Esters | MF | 4 | Mostly HCO₂Li, ROLi(CH₃OLi?) | HCO₂Li | Li oxides + HCO₂Li | HCO₂Li + Li₂CO₃ | LiF |
| | γ-BL | 150,189 | CH₃(CH₂)₂COOLi, cyclic β-keto ester anion, di-Li salt | LiO(CH₂)₃COOLi | Li oxides ROCO₂Li species | RCOOLi + Li₂CO₃ | LiF |
| Ethers | THF | 85 | ROLi CH₃(CH₂)₃OLi | LiOH | Li oxides | Li₂CO₃ | |
| | 2Me-THF | 140 | Li pentoxides | Li₂O | | | |
| | DME | 85 | ROLi(CH₃OLi) | Li alkoxides | ROLi species | ROLi species | |
| | 1,3-DN | 144 | CH₃CH₂OLi(CH₂OLi)₂ Poly DN species | | | | |
| | DEE | 149 | CH₃CH₂OLi(CH₂OLi)₂ | | | ROCO₂Li | |
| | Diglyme (DG) | 149 | CH₃OLi, CH₃OCH₂CH₂OLi, (CH₂OLi)₂ | | | | |

| Mixtures | | | | | | |
|---|---|---|---|---|---|---|
| EC-PC | 131 | EC reduction products dominate | ROCO$_2$Li Li$_2$CO$_3$ Li$_2$O-LiOH | Li$_2$O ROCO$_2$Li | Li$_2$CO$_3$ ROCO$_2$Li | LiF |
| EC-DEC | 132 | | | | | |
| EC-DMC | 51 | | | | | |
| EC-EMC | 360 | | | | | |
| MF-EC | 261 | HCO$_2$Li, ROCO$_2$Li species | ROCO$_2$Li Li$_2$CO$_3$, HCO$_2$Li LiOH-Li$_2$O HCO$_2$Li, Li$_2$O-LiOH | Li$_2$O ROCO$_2$Li HCO$_2$Li | ROCO$_2$Li HCO$_2$Li Li$_2$CO$_3$ | LiF |
| MF-PC | | | | | Li$_2$CO$_3$ + solvent reduction products | |
| MF-DMC | 261 | HCO$_2$Li dominates, ROCO$_2$Li (minor) | LiOH | | Li$_2$CO$_3$ + solvent reduction products | LiF |
| MF-DEC | | | LiO(CH$_2$)$_3$COOLi | | Li$_2$CO$_3$ + solvent reduction products | |
| MF-ethers | 261 | HCO$_2$ dominates | | | Li$_2$CO$_3$ + solvent reduction products | |
| EC or PC with ethers | 153 | ROCO$_2$Li species dominate ROLi (minor) | Li$_2$CO$_3$, ROCO$_2$Li LiOH-Li$_2$O | | | |
| THF-2Me-THF | 50 | THF reduction products dominate | | | | |
| Salt | | | | | | |
| LiAsF$_6$ | 84,48 | LiF, Li,ASF$_y$ | | | | |
| LiClO$_4$ | 84,4,144 | Li$_2$O, LiCl, LiClO$_3$, LiClO$_2$, etc. | | | | |
| LiBF$_4$ | 159,48 | LiF, Li,BF$_y$, Li,BF$_x$O$_z$? | | | | |
| LiPF$_6$ | 48,159 | LiF, Li,PF$_y$, Li,PF$_y$O$_z$? | | | | |
| LiSO$_3$CF$_3$ | 48,159 | Li,S,O$_z$, LiF, RCF$_3$Li, etc. | | | | |
| LiN(SO$_2$CF$_3$)$_2$ | 158,48 | LiF, Li,S,O$_z$, Li$_3$N, RCF$_3$Li, Li$_2$NSO$_2$CF$_3$, etc. | | | | |

PC = propylene carbonate; EC = ethylene carbonate; DMC = dimethyl carbonate; DEC = dimethyl carbonate; MF = methyl formate; $\gamma$-BL = $\gamma$-butyrolactone; THF = tetrahydrofuran; 2Me-THF = 2-methyltetrahydrofuran; DME = dimethoxyethane; 1,3-DN = 1,3-dioxolane; EMC = ethyl methyl carbonate; PMC = propyl methyl carbonate; DEE = diethoxyethane; DG = CH$_3$OCH$_2$CH$_2$OCH$_2$CH$_2$OCH$_3$.

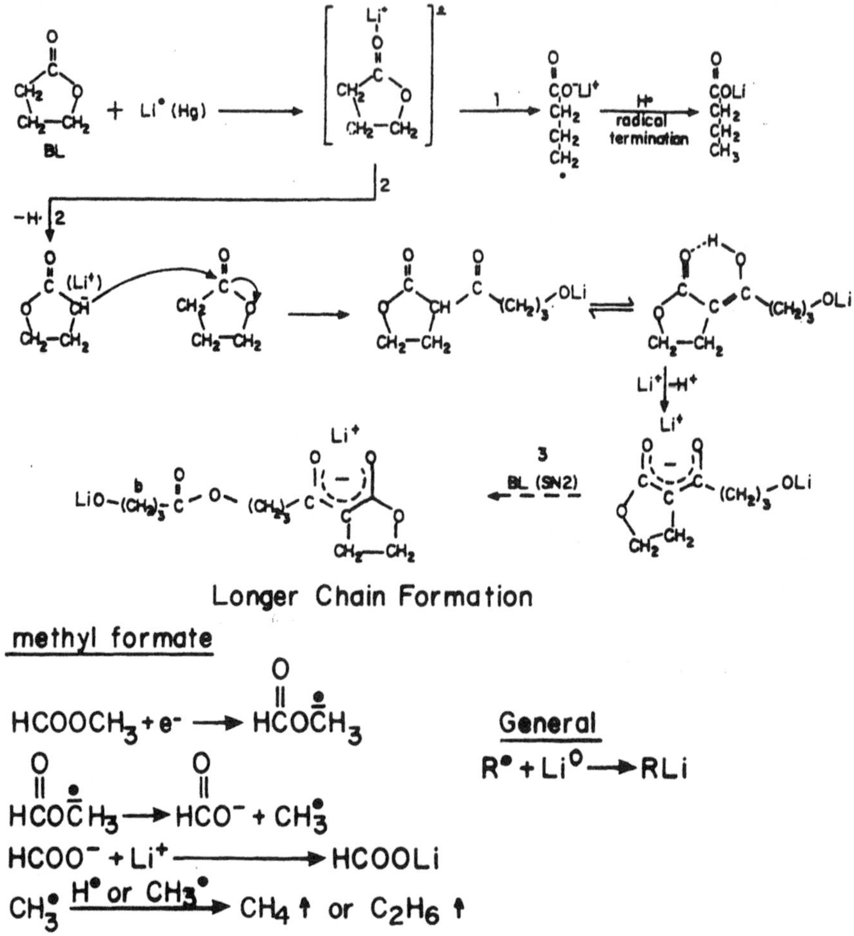

**Longer Chain Formation**

**methyl formate**

$$HCOOCH_3 + e^- \longrightarrow H\dot{C}OCH_3$$

**General**

$$R^\bullet + Li^0 \longrightarrow RLi$$

$$H\dot{C}OCH_3 \longrightarrow HCO^- + \dot{C}H_3$$

$$HCOO^- + Li^+ \longrightarrow HCOOLi$$

$$\dot{C}H_3 \xrightarrow{H^\bullet \text{ or } CH_3^\bullet} CH_4 \uparrow \text{ or } C_2H_6 \uparrow$$

**Scheme 3** Ester reduction mechanisms on lithium γ butyrolactone and methyl formate [189,4].

on Li in ethers was obtained by Kerr [195], who analyzed solutions of 2Me-THF in which Li electrodes were cycled by alcohols such as pentanol and isopentanol. Ethers are thus reduced by Li to alkoxy species. Further studies by surface-sensitive FTIR spectroscopy provided direct evidence that ethers are reduced by lithium to form surface films comprising Li alkoxy species [50,85,140], as demonstrated in Figure 16.

Figure 16 compares ETIR spectra measured from Li surfaces freshly prepared and then stored in pure ethyl glyme (EG) and diglyme (DG) with spectra measured from Li electrodes on which thin films of model Li alkoxide layers were deposited (by reacting Li surfaces with the parent alcohols). It is clear from the spectra of Figure 16 that the ether linkage is attacked by Li and that alkoxy species are thus formed. In EG, a major surface species formed on Li is $LiOCH_2CH_3$, while in DG a major surface species formed on Li is $LiOCH_2CH_2OCH_3$ [149]. Table 3 summarizes major components in the surface films formed on Li in ethers, and Scheme 4 provides the most probable reduction paths of commonly used ethers on Li surfaces. We should mention important studies by Scherson et al. [196] and Ross et al. [197] on the reactions between THF and Li in UHV systems. The former group found that fully deuterated THF reacts on Li to form LiD as a major product. The other group found evidence for more complicated surface reactions in which THF may polymerize and/or break down into smaller fractions. In any event, Li alkoxy species were found to be the major components that comprise the surface layer formed on lithium in THF[197]. An ethereal system that received special attention was a mixture of THF, 2Me-THF, 1–1.5 $M$ $LiAsF_6$, and 2Me-furan, due to the high Li cycling efficiency that could be obtained in these solutions [198]. We found that the surface chemistry of Li in these systems was dominated by THF reduction to form Li alkoxides, and by $LiAsF_6$ reduction to form LiF and $Li_xAsF_y$ species (see Section V.C.6) [50]. It was concluded that the positive effect of 2Me-furan on the performance of Li electrodes in these systems should not be attributed to an impact on the surface chemistry but rather to bulk effects. It was assumed that 2Me-furan reacts with Lewis acid species formed in the cathode side, neutralizes them, and thus prevents a detrimental polymerization of the cyclic ethers (induced by Lewis acids) [199].

Another system of special importance is that of solutions of 1,3-dioxolane (DN). It was found that Li morphology in charge-discharge cycling is very smooth and that Li cycling efficiency is very high in $DN/LiClO_4$ solutions [68,200,201]. This was attributed to a unique surface chemistry of Li in these solutions [68]. It was suggested that the solvent, the salts, and trace water react on the Li surface in these solutions to form species such as LiCl, LiH, LiOH, $Li_2O$ and, possibly, some solvent reduction products [202,203]. Further studies of Li electrodes in DN solutions by FTIR spectroscopy (both ex situ [116,144,145] and in situ [158] modes) clarified the picture. It appears that a quite complicated surface chemistry is developed on lithium in DN solutions. Table 3 summarizes the major components of the surface films formed, and Scheme 5 suggests the possible reaction paths for these products. Species such as $CH_3CH_2OCH_2OLi$ and Li formate were identified [144,204]. Based on the above studies, we suggested that the surface films formed on lithium in DN solu-

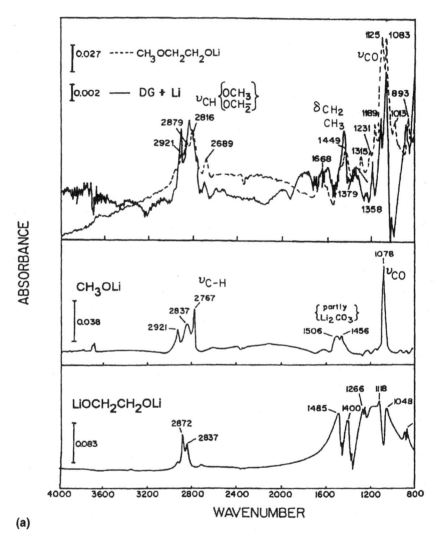

**Figure 16** FTIR spectra obtained ex situ (external reflectance mode) from Li surfaces prepared and stored in ethers of the glyme family. Reference spectra of Li alkoxides prepared as thin layers on Li surfaces are also presented [149]. (With copyrights from Elsevier Science Ltd., 1998.) (a) Diglyme (DG, $CH_3OCH_2CH_2OCH_2CH_2OCH_3$): (upper solid line) DG + Li; (dashed line) $CH_3OCH_2CH_2OLi$ on Li; spectra of $CH_3OLi$ and $(LiOCH_2)_2$ on Li are also presented (as indicated). (b) Ethyl glyme (EG, $CH_3$, $CH_2OCH_2CH_2OCH_2CH_3$): (upper solid line) EG + Li; (dashed line) $CH_3CH_2OLi$ on Li; (lower solid line) $(LiOCH_2)_2$ on Li.

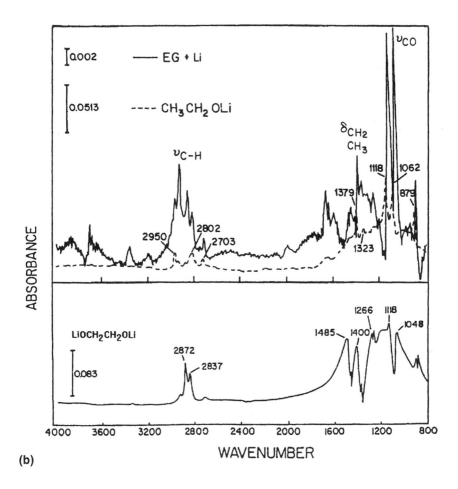

(b)

tions also contain oligomers of DN with —OLi edge groups. XPS studies of Li electrodes prepared and stored in 1,3-dioxolane solutions support these conclusions [159]. It should be noted that 1,3-dioxolane as a strained (five-membered-ring) cyclic acetal is highly sensitive to trace protons or any other Lewis acid contaminants. It polymerizes readily in the presence of trace acidic species via a cationic mechanism to form poly-DN [145]. When such a polymerization occurs, it has a detrimental effect on the reversibility of Li electrodes in these solutions. It was possible to obtain highly stable $LiAsF_6$ and $LiClO_4$ solutions of DN in the presence of 2Me-furan and KOH [205], or in the presence of a base such as tributylamine (TBA). $LiAsF_6$ solutions in DN, stabilized by TBA, were found a).    $R\text{-}O\text{-}R` + e^- + Li^+ \rightarrow R`OR^{\doteq}Li^+$

b).    $R`OR^{\doteq}Li^+ \rightarrow ROLi + R`\bullet$ <u>or</u> $R`OLi + R\bullet$

c).    $R\bullet \xrightarrow{H\bullet} RH$ <u>or</u> $2R\bullet\bullet \rightarrow R_2$ <u>or</u> $R\bullet \xrightarrow{Li^0} RLi$

d).    For instance, (EG) $CH_3CH_2OCH_2CH_2OCH_2CH_3 + Li^+ + e^- \rightarrow$

<u>$CH_3CH_2OLi \downarrow + \bullet CH_2CH_2\text{-}OCH_2CH_3$</u> <u>and</u>

$CH_3CH_2\bullet + CH_3CH_2OCH_2CH_2OLi \downarrow$

e)     DME $CH_3OCH_2CH_2OCH_3 + 2Li^+ + 2e^- \rightarrow 2CH_3OLi\downarrow + CH_2{=}CH_2\uparrow$

f)     THF $\xrightarrow{Li^0} CH_3CH_2CH_2CH_2OLi$ (and/or $Li\text{-}(CH_2)_4OLi$)

g)     2Me-THF $\xrightarrow{Li^0} CH_3(CH_2)_4OLi + CH_3\overset{\overset{\displaystyle CH_3}{|}}{C}HCH_2CH_2OLi$ (and/or $Li\text{-}ROLi$)

**Scheme 4**   Possible ether reduction patterns on Li [50,85,140,149].

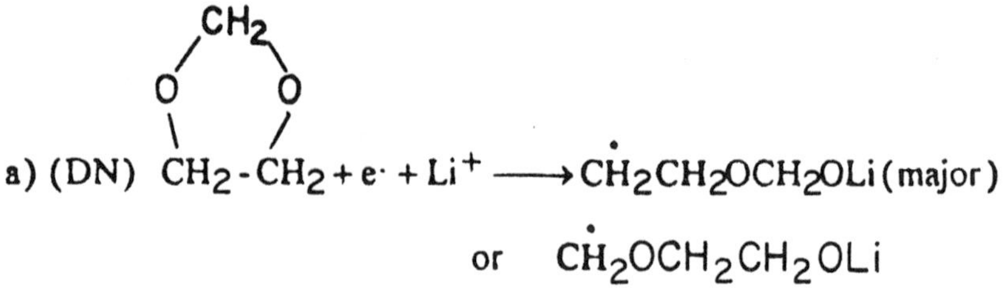

a) (DN) $\overset{\displaystyle \underset{O \quad\quad O}{\diagup CH_2 \diagdown}}{CH_2\text{-}CH_2} + e^- + Li^+ \longrightarrow \dot{C}H_2CH_2OCH_2OLi$ (major)

$\qquad\qquad\qquad$ or $\quad \dot{C}H_2OCH_2CH_2OLi$

b) $\bullet CH_2CH_2OCH_2OLi \xrightarrow{H\bullet} CH_3CH_2OCH_2OLi \downarrow$

$\qquad\qquad\qquad$ or $CH_3CH_3\uparrow + HCO_2Li \downarrow$

c) $\bullet CH_2CH_2OCH_2OLi \xrightarrow{Li^0} LiCH_2CH_2OCH_2OLi$

d) $ROLi + nDN \xrightarrow{\text{polymerization}} R-(OCH_2CH_2-OCH_2)_nOLi \downarrow$

**Scheme 5**   1,3-Dioxolane reduction patterns on Li.

to be suitable for rechargeable Li batteries and are already used in practical rechargeable Li-Li$_x$MnO$_2$ batteries [206].

In conclusion to this section, we should emphasize that since ethers are less reactive with lithium than the alkyl carbonates or the esters (reviewed in previous sections), salt and contaminant reduction may be highly important in affecting the surface chemistry developed on lithium in ethers (compared with alkyl carbonate and esters).

## 5. Identification of Surface Films Formed on Lithium in Sulfur-Containing Solvents

A few sulfur-containing solvents are important for Li battery systems, and their Li surface chemistry has been investigated. Of special importance are thionyl chloride (SOCl$_2$) and SO$_2$. Both solvents are used in practical primary Li batteries [207]. SO$_2$ has also been found to be a suitable solvent for rechargeable Li battery systems [208]. Sulfuryl chloride (SO$_2$Cl$_2$) has also been mentioned as a solvent for primary Li batteries [207]. These three solvents are also the major cathodic active materials in these batteries. The electrolyte used in these three inorganic solvents is usually LiAlCl$_4$ [207]. Scheme 6, which is based on Refs. 207, 209–214, shows possible reduction patterns of these solvents. It appears that in SOCl$_2$ and SO$_2$Cl$_2$ the major surface species formed on Li is LiCl, whereas in SO$_2$ the major surface species are Li$_2$S$_2$O$_4$ and possibly Li$_2$S, Li$_2$SO$_4$, and Li$_2$O [212–213]. Hill et al.[211] suggested that in SO$_2$/LiAlCl$_4$ solutions, LiCl and an amorphous complex LiAlCl(SO$_2$)$_3$ are also formed [211].

Another solvent that has received some attention in connection with rechargeable Li batteries is sulfolane [214]. The surface chemistry of lithium electrodes in 3-methyl sulfolane/LiAsF$_6$ solutions was studied by Yen et al. [215] using XPS and FTIR. Their analysis of the surface chemistry of Li electrodes in these electrolyte systems is summarized in Scheme 7.

## 6. Contribution of Salts to the Surface Chemistry Developed on Li Electrodes in Polar Aprotic Systems

The most relevant salts that may be used in connection with Li electrochemistry, in general, and Li batteries, in particular, include LiCl, LiBr, LiI, LiClO$_4$, LiAsF$_6$, LiBF$_4$, LiBR$_4$ (R = organic group, aryl or alkyl), LiPF$_6$, LiSO$_3$CF$_3$, LiN(SO$_2$CF$_3$)$_2$, LiC(SO$_2$CF$_3$)$_3$, and LiAlCl$_4$. The last one is used in inorganic solvents such as SO$_2$, SOCl$_2$, and SO$_2$Cl$_2$, while the remaining ones are mostly relevant to the organic polar aprotic solvents. This list may be divided into several groups:

1. The Li halides are the most inert toward lithium, since their anions cannot be further reduced. In any event, one should take into account the possibility that even unreactive anions can be entangled in the Li surface films, thus

a) <u>Li-SO$_2$</u> [211-213]

$$6Li + SO_2 \rightarrow 2Li_2O + Li_2S$$

$$Li_2O + SO_2 \rightarrow Li_2SO_3$$

$$2Li + 2SO_2 \rightarrow Li_2S_2O_4$$

$$4Li + Li_2S_2O_4 \rightarrow 2Li_2O + LiOS\text{-}SOLi$$

$$4Li + 2SO_2 \rightarrow Li_2SO_4 + Li_2S$$

$$Li_2SO_4 + Li_2S \rightarrow Li_2SSO_3 + Li_2O$$

$$3Li + 3SO_2 + AlCl_4^- \rightarrow [AlCl(SO_2)_3]^- + 3LiCl$$

b. <u>Li-SOCl$_2$</u> [207]

$$4Li + 2SOCl_2 \rightarrow S + SO_2 + 4LiCl$$

$$8Li + 3SOCl_2 \rightarrow 2S + Li_2SO_3 + 6LiCl$$

$$8Li + 4SOCl_2 \rightarrow S_2Cl_2 + Li_2S_2O_4 + 6LiCl$$

$$Li_2O + SOCl_2 \rightarrow 2LiCl + SO_2$$

c. <u>Li - SO$_2$Cl$_2$</u> [207]

$$2Li + SO_2Cl_2 \rightarrow SO_2 + 2LiCl$$

**Scheme 6**   Reactions of lithium with inorganic sulfur compounds.

affecting their properties, despite their lack of intrinsic reactivity [186,190]. While LiBr or LiI may be regarded as practical electrolytes in highly polar solvents, LiCl is not a practical electrolyte in most of the solvents of interest, including alkyl carbonates, esters, ethers, and oxyhalides. In several cases, as described below, it is, rather, a passivating agent when formed on Li surfaces. It should be noted that I$^-$ is also a problematic anion for alkyl carbonate solvents. It has been found that I$^-$ reduces PC. It seems that the mechanism involves both redox transfer and nucleophilic reactions. I$_2$, $CH_3CH\!=\!CH_2$, ROCO$_2$Li, and iodized open chain alkyl carbonate species have been analyzed as the reaction products of PC and LiI [48].

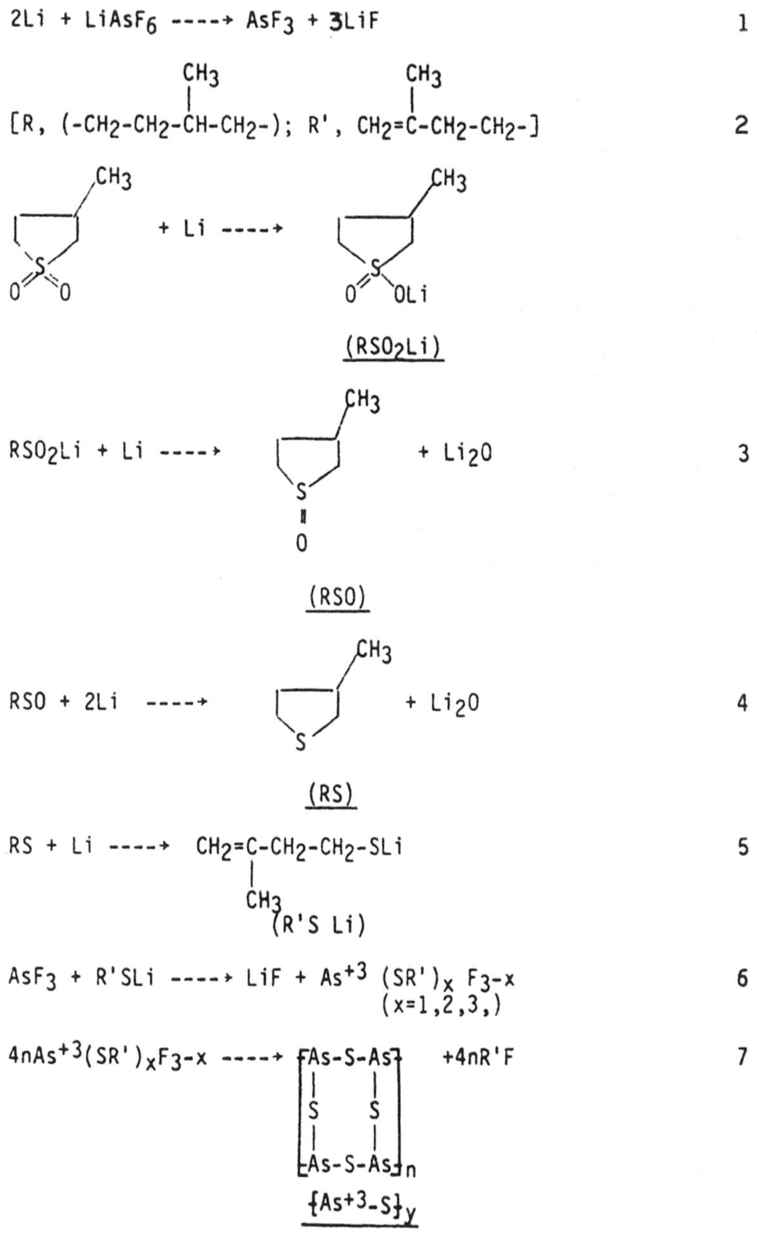

$$2Li + LiAsF_6 \longrightarrow AsF_3 + 3LiF \qquad\qquad 1$$

$$[R, (-CH_2-CH_2-\overset{\overset{\displaystyle CH_3}{|}}{CH}-CH_2-); \quad R', \ \overset{\overset{\displaystyle CH_3}{|}}{CH_2=C}-CH_2-CH_2-] \qquad 2$$

(RSO$_2$Li)

$$RSO_2Li + Li \longrightarrow \qquad + Li_2O \qquad\qquad 3$$

(RSO)

$$RSO + 2Li \longrightarrow \qquad + Li_2O \qquad\qquad 4$$

(RS)

$$RS + Li \longrightarrow CH_2=C-CH_2-CH_2-SLi \qquad\qquad 5$$
$$\underset{(R'S\ Li)}{\overset{|}{CH_3}}$$

$$AsF_3 + R'SLi \longrightarrow LiF + As^{+3}(SR')_x F_{3-x} \qquad\qquad 6$$
$$(x=1,2,3,)$$

$$4nAs^{+3}(SR')_xF_{3-x} \longrightarrow \qquad +4nR'F \qquad\qquad 7$$

$$\{As^{+3}-S\}_y$$

**Scheme 7** Reaction mechanisms for LiAsF$_6$/3-methyl sulfolane with Li [215].

2. $LiAlCl_4$ should also be mentioned separately. Li electrodes in oxyhalide solvents containing this salt develop surface films that contain insoluble LiCl.

3. The third group involves $LiClO_4$ and $LiAsF_6$. These two salts are usually not acidic and are only moderately reactive on the Li surfaces, depending on the solvents. In ethers, their reduction processes are pronounced and may dominate the Li surface chemistry, while in the more reactive solvents such as alkyl carbonates and esters, solvent reduction processes usually dominate the Li surface chemistry. $LiClO_4$ reduction on the Li surfaces contributes LiCl and $Li_2O$ as obvious surface species [77–80]. However, there is some evidence for the formation of species of the $LiClO_x$ type $(4 > x)$ [216]. $LiAsF_6$ reduction contributes LiF and species of the $Li_xAsF_y$ type as obvious surface species [101–105,159,217] (see Scheme 8).

4. The fourth group comprises the acidic salts, $LiBF_4$ and $LiPF_6$. These salts always contain unavoidable HF contamination. $LiPF_6$ may slowly decompose to LiF and $PF_5$, which readily hydrolyze with trace water to $POF_3$ and HF. Both species are highly reactive on the Li surfaces, and surface LiF and $Li_xPOF_y$ species are thus formed. HF, when present in alkyl carbonates or ethereal solutions, solubilizes carbonate and alkoxy species formed on the lithium surfaces

a) $LiAsF_6 + 2Li^+ + 2e^- \rightarrow 3LiF\downarrow + AsF_3$ (sol)

b) $AsF_3 + 2 \times Li^+ + 2 \times e^- \rightarrow Li_xAsF_{3-x}\downarrow + XLiF\downarrow$

c) $PF_6^- + 3Li^+ + 2e^- \rightarrow 3LiF\downarrow + PF_3$

d) $LiPF_6 \rightleftharpoons LiF + PF_5$

e) $PF_5 + H_2O \rightarrow PF_3O + 2HF$

f) $PF_5 + 2 \times Li^+ + 2 \times e^- \rightarrow Li_xPF_{5-x}\downarrow + XLiF\downarrow$

g) $PF_3O + 2 \times Li^+ + 2 \times e^- \rightarrow Li_xPF_{3-x}O\downarrow + XLiF\downarrow$

h) $BF_4^- \xrightarrow{Li^+,e^-} LiF\downarrow, Li_xBF_y\downarrow$ (in general)

i) $LiClO_4 + 8Li^+ + 8e^- \rightarrow 4Li_2O + LiCl$

j) $LiClO_4 + XLi^+ + Xe^- \rightarrow LiClO_{(4-\frac{1}{2}x)} + \frac{1}{2}XLi_2O$. $(X = 2, 4, 6)$

**Scheme 8** Possible $LiClO_4$, $LiBF_4$, $LiPF_6$, and $LiAsF_6$ reduction patterns (Li and Li-carbon) [48,84,85,132,140,144,149,159].

in these solutions, as described in Section V.C.1, via an acid-base reaction. This enriches the surface films with LiF (which replaces the organic Li salts) [101–105]. Hence, in $LiBF_4$ and $LiPF_6$ solutions, LiF becomes a major surface species upon storage. In addition, species of $Li_xPF_y$, $Li_xPO_yF_z$, $Li_xBF_y$, and $Li_xBO_yF_z$ may be formed [159]. XPS measurements of Li electrodes stored in $LiPF_6$ and $LiBF_4$ solutions indicate the formation of such compounds on the Li surfaces [101–105,159].

   5.   The fifth group includes the salts containing the $SO_2CF_3$ group [$LiSO_3CF_3$, $LiN(SO_2CF_3)_2$ and $LiC(SO_2CF_3)_3$]. Ex situ and in situ FTIR, XPS, and EDAX studies of Li electrodes stored in solutions of these salts reveal that they are highly reactive with lithium. Figure 17 presents a good example for that. It shows FTIR spectra obtained in situ from a platinum electrode polarized to $Li/Li^+$ potential in solutions of $LiAsF_6$, $LiN(SO_2CF_3)_2$, and $LiC(SO_2CF_3)_3$ in THF. While the spectrum related to the $LiAsF_6$ solution is typical of THF reduction products (ROLi), the other two spectra are dominated by strong absorption of the salts' anion reduction products [158]. The salts of this group have two centers of reactivity: the $—CF_3$ and the $—SO_2^-$ groups. The former group reacts with lithium to form LiF and $—CF_xLi_y$, while the other group may be reduced to a variety of Li—S—O species, including $Li_2S_2O_4$, $Li_2S$, $Li_2O$, etc. [212,213,

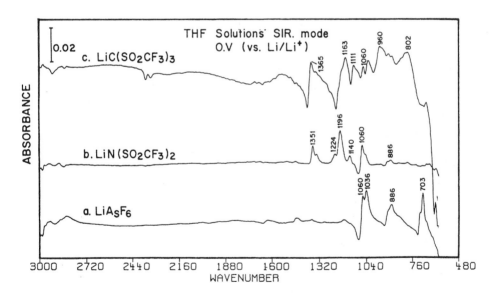

**Figure 17**   FTIR spectra measured in situ from Pt deposited on NaCl (single internal reflectance mode), polarized to 0.V ($Li/Li^+$) in THF 1 $M$ solutions of three Li salts (as indicated) [158]. (With copyrights from Elsevier Science Ltd., 1998.)

a) $LiN(SO_2CF_3)_2 + ne^- + nLi^+ \rightarrow Li_3N + Li_2S_2O_4 + LiF + C_2F_xLi_y$

b) $LiN(SO_2CF_3)_2 + 2e^- + 2Li^+ \rightarrow Li_2NSO_2\,CF_3 + \; CF_3SO_2Li$

c) $Li_2S_2O_4 + 10e^- + 10Li^+ \rightarrow 2Li_2S + 4Li_2O$

d) $LiC(SO_2CF_3)_3 + 2e^- + 2Li^+ \rightarrow Li_2C(SO_2CF_3)_2 + LiSO_2CF_3$, etc.

e) $Li_2S_2O_4 + \; 4e^- + 4Li^+ \rightarrow Li_2SO_3 + Li_2S + Li_2O$

f) $2LiSO_3CF_3 + 2Li^+ + 2e^- \rightarrow 2Li_2SO_3 + C_2F_6$

g) $R\text{-}CF_3 + 2Li^+ + 2e^- \rightarrow RCF_2Li + LiF$

h) $Li_2SO_3 + 6Li^+ + 6e^- \rightarrow Li_2S + 3Li_2O$.

**Scheme 9**  Possible $LiN(SO_2CF_3)_2$, $LiSO_3CF_3$, and $LiC(SO_2CF_3)_3$ reduction patterns (Li and Li-carbon) [48,158,159].

215]. Hence, the Li surfaces in these salt solutions are rich in a variety of salt anion reduction products. Scheme 9 provides possible reduction patterns for these salts on Li surfaces, as well as their product distribution.

In conclusion we emphasize that the final chemical composition of the surface films covering Li electrodes in solutions is the result of a delicate balance between several competing reduction processes of solvent, salt anion, and contaminants. In this section we emphasize the role of salt reduction processes. However, each solution has to be studied separately since the above balance is determined by factors such as salt concentration, the nature of the solvent, contamination level, etc., all of which differ from system to system. Table 4 presents element analysis by XPS of surface films formed on freshly prepared Li electrodes in several solutions. It demonstrates how the impact of salt reduction on the Li surface chemistry depends on the solution composition and the storage time. (The atomic percentage of fluorine on the surface is a good measure for the impact of the salt.)

## 7.  Role of Contaminants

We should emphasize that the Li surface chemistry may be strongly influenced by reactive contaminants in the ppm level. In this section, we mention the impact of a few important and commonly abundant impurities. These include the atmospheric components $O_2$, $N_2$, $CO_2$, and $H_2O$ (unavoidably present in all nonaqueous solutions and in any inert atmosphere, even of the highest quality), HF (which

**Table 4** Element Analysis (by XPS) of Li Electrodes Prepared and Stored in Alkyl Carbonate Solutions

| Solvent | Salt | Storage | %F | | %C | | %O | | %Li | | Other elements (%) | |
|---|---|---|---|---|---|---|---|---|---|---|---|---|
| | | | I | S | I | S | I | S | I | S | I | S |
| PC | LiAsF$_6$ | 2 h | 1.0 | 1.9 | 23.6 | 20.2 | 32.0 | 36.1 | 43.4 | 41.9 | | |
| PC | LiAsF$_6$ | 3 days | 2.1 | 1.3 | 40.4 | 17.5 | 37.5 | 30 | 20 | 51.3 | As (0.02) | As (0.0s) |
| PC | LiPF$_6$ | 2 h | 6.8 | 9.8 | 30.2 | 20.8 | 32.6 | 36.2 | 30.4 | 33.2 | P (0.2) | P (0.3) |
| PC | LiPF$_6$ | 3 days | 9.7 | 14.3 | 25.7 | 17 | 28.50 | 31.3 | 36 | 37.1 | B (0.2) | B (0.2) |
| PC | LiPF$_6$ | 2 h | 2.2 | 1.8 | 25.1 | 13.6 | 39.2 | 42.7 | 33.1 | 41.6 | B (0.2) | B (0.2) |
| PC | LiBF$_4$ | 3 days | 14.6 | 21.7 | 29 | 14.8 | 25.4 | 21.6 | 30.9 | 41.7 | S (0.6), N (0.3) | S (0.5), N (0.2) |
| PC | LiN(SO$_2$CF$_3$)$_2$ | 3 days | 1.9 | 0.9 | 28.7 | 25.5 | 41.7 | 50.4 | 26.6 | 25.6 | P (1.4) | P (0.7) |
| 1:1 EC:DMC | LiPF$_6$ | 3 h | 25.7 | 31.1 | 16.3 | 5.6 | 12.4 | 13.1 | 44.2 | 49.6 | P (3.5) | P (2) |
| 1:1 EC:DMC | LiPF$_6$ | 3 days | 39.3 | 42.1 | 11.8 | 6.1 | 8.6 | 5.6 | 36.8 | 44.3 | P (1.6), N (0.1) | P (11) |
| 1:1 EC:DMC + TBA | LiPF$_6$ | 3 days | 25.7 | 37.8 | 18.3 | 5.8 | 17.2 | 10 | 37 | 45.1 | | |
| 1:1 EC:DMC | LiAsF$_6$ | 3 days | 23.8 | 33.8 | 27.2 | 11.2 | 21.1 | 11.3 | 27.5 | 43 | As (0.5) | As (0.7) |
| 1:1 EC:DMC | LiAsF$_6$ | 3 days | 10.3 | 9.8 | 21 | 12.6 | 25.9 | 25.7 | 42.7 | 51.7 | As (0.2) | As (0.2) |
| 1:1 EC:DMC | LiAsF$_6$ | 3 h | 5.0 | 4.3 | 20.8 | 9.7 | 32.3 | 29.7 | 42 | 56.3 | As (0.2) | As (0.1) |

Mole percentage of the various elements appearing in XPS spectra measured before sputtering (I) and after removal of about 30–50 Å of the surface layer (S). Salt concentration was 1 M.
*Source:* From Ref. 159.

is unavoidably present in $LiPF_6$ and $LiBF_4$ solutions), and some miscellaneous contaminants such as protic species, $AsF_3$, $PF_5$, and $SO_2$. It should be noted that some of the above contaminants may be used as desirable additives in solutions, in order to improve the performance of Li anodes in batteries (due to their impact on the Li surface chemistry).

Li reacts readily with $N_2$ to form $Li_3N$. We have had an indication that other Li-N compounds of different stoichiometry can also be formed [15,18]. However, these reactions only take place in an $O_2$-free atmosphere. Hence, in a usual glove box atmosphere from which $O_2$ is continuously removed, while $N_2$ may exist in a concentration of hundreds of ppm, freshly prepared Li surfaces may always be covered by $Li_3N$ film. This film has a typical absorption in the IR, appearing as a broad peak around $680$ cm$^{-1}$ [18]. $Li_3N$ is a strong base and nucleophile which can further react with alkyl carbonate and ester solvents [18]. In solutions, the reaction between Li and $N_2$ is much less important, due to competition with other reduction processes (of solution species).

Li reacts readily with $O_2$ to form a $Li_2O$ film as the most stable product. However, $Li_2O_2$ and $LiO_2$ can also be formed at the outer part of such a film. The saturation concentration of $O_2$ at room temperature in organic polar aprotic solutions is estimated in the order of $10^{-3}$–$10^{-4}$ $M$ [218]. Nevertheless, FTIR spectroscopic studies of pairs of Li electrodes treated identically in the same solutions, with the only difference being that one solution was saturated with oxygen, revealed that the presence of $O_2$ (even in such low concentrations) influences the Li surface chemistry in alkyl carbonates and ethereal solutions [19,176].

Lithium reacts with $H_2O$ to form LiOH, which can be further reduced to $Li_2O$. Hence, at a high concentration of $H_2O$ in solutions, the surface layers covering Li electrodes may comprise an inner layer of $Li_2O$ and an outer layer of LiOH at different degrees of hydration [17,20]. $H_2O$ hydrates most of the surface species that may be formed on lithium. Hence, there is no passivation against water reduction. $H_2O$ in solutions diffuses through the surface films toward the lithium film interface and is reduced within the film. Consequently, surface films formed on Li in any water-contaminated solutions are rich in $Li_2O$ and LiOH. Both are nucleophiles which can react further with electrophilic solvents such as esters and alkyl carbonates. For instance, they both react with $\gamma$-butyrolactone to form $LiO(CH_2)_3COOLi$ and $HO(CH_2)_3COOLi$, respectively [18]. LiOH reacts with PC to form $Li_2CO_3$ and $ROCO_2Li$ [84] [probably $CH_3CH(OCO_2Li)CH_2OH$]. In addition, $H_2O$ reacts with $ROCO_2Li$ species formed by alkyl carbonate reduction to form $Li_2CO_3$, $CO_2$, and ROH [84,188] (Scheme 1).

$CO_2$ dissolved in organic polar aprotic solutions reacts with lithium to form surface $Li_2CO_3$. The coproduct is probably CO [219]. The mechanism is shown in Scheme 10 [220]. One of the sources of $CO_2$ in organic polar aprotic systems is the solvent's decomposition. For instance, drying alkyl carbonates such as EC or PC on activated alumina leads to their decomposition to $CO_2$ and derivatives

$$CO_2 + e^- + Li^+ \rightarrow \overset{\bullet}{C}O_2Li$$

$$\overset{\bullet}{C}O_2Li + CO_2 \rightarrow O=\overset{\bullet}{C}-O-CO_2Li$$

$$O=\overset{\bullet}{C}-O-CO_2Li + e^- + Li^+ \rightarrow CO\uparrow + Li_2CO_3\downarrow$$

$$2Li\,OH\downarrow + CO_2 \rightarrow Li_2CO_3\downarrow + H_2O$$

$$Li_2O\downarrow + CO_2 \rightarrow Li_2CO_3\downarrow$$

$$ROLi\downarrow + CO_2 \rightarrow LiOCO_2Li\downarrow$$

**Scheme 10** Possible $CO_2$ reaction patterns on Li surfaces [131].

of ethylene or propylene oxide [131]. It should be emphasized that $Li_2CO_3$ is one of the most efficient passivating agents for Li surfaces. As described further in the next sections, Li electrodes covered with $Li_2CO_3$ may possess an improved reversible behavior and cyclability.

Another important contaminant is HF (which is unavoidably present in $LiPF_6$ and $LiBF_4$ solutions). It reacts readily with lithium to form LiF and hydrogen gas. It also reacts with surface $ROCO_2Li$, $Li_2CO_3$, and ROLi to form surface LiF and solution phase $ROCO_2H$, $H_2CO_3$, and ROH, respectively [86].

$LiPF_6$ and $LiAsF_6$ solutions may also contain contaminants such as $PF_5$, $AsF_3$, and $AsF_5$. These contaminants also react with lithium to form surface species of the $Li_xMF_y$ type [86,87,159] (M = P, As). In addition, these contaminants are strong Lewis acids that react with ethereal solvents and polymerize them [199].

The last contaminant that we mention in this section is $SO_2$. As already proven, this species may be a desirable additive in organic electrolyte solutions for Li and Li ion batteries [221]. It was found that the addition of a small amount of $SO_2$, even at a subpercent level, influences the surface chemistry of lithiated carbon electrodes. It was possible to detect $Li_2S_2O_4$ surface species on lithiated carbon electrodes treated in alkyl carbonate solutions (e.g., EC-DMC) containing $SO_2$ [221].

## 8. Miscellaneous

In this section we mention several solvents that were not covered in the previous sections and which are of some importance to this field. Three important organic polar aprotic solvents—acetonitrile (AN), dimethyl sulfoxide (DMSO), and

$N,N$-dimethyl formamide (DMF)—are too reactive with lithium. The first solvent reacts with lithium to produce a species of the $CH_3C(=N^-Li^+)CH_2CN$ type, $CH_4$ and LiCN [222]. Lithium is not passivated in this solvent and becomes covered with a thick polymeric-like layer (probably a condensation product of AN). However, this solvent is a desirable cosolvent for Li-SO$_2$ batteries [223]. It may also be used as a cosolvent in Li batteries containing organic electrolyte solutions as well, provided that the other solvent is highly reactive toward Li, and its reduction products are highly passivating, as is the case for AN/EC [224] and AN/PC [225] solutions. There are also reports on DMSO as a possible cosolvent for Li battery electrolyte solutions [226]. Based on the analysis of the surface chemistry of lithium in SO$_2$ and sulfolane mentioned above [211,212,213,215], it is expected that DMSO may be reduced on Li to a number of products, including Li$_2$S, Li$_2$O and condensation products of DMSO.

We studied the surface chemistry of lithium in tributylamine (TBA) and found that the C—N bond is cleaved on the active surface, thus forming surface Li-amide species $(CH_3CH_2CH_2CH_2)_{3-x}NLi_x$ ($x = 1,2,3$) [116].

The last electrolyte system to be mentioned in connection with lithium electrodes is the room temperature chloroaluminate molten salt. (AlCl$_3$:LiCl 1-$R'$-3$R''$-imidazolium chloride. $R'$ and $R''$ are alkyl groups, usually methyl and ethyl, respectively.) These ionic liquids were examined by Carlin et al. [227–229] as electrolyte systems for Li batteries. They studied the reversibility of Li deposition-dissolution processes. It appears that lithium electrodes may be stable in these systems, depending on their acidity [227]. It is suggested that Li stability in these systems relates to passivation phenomena. However, the surface chemistry of lithium in these systems has not yet been studied.

## D. Electrochemical Behavior and Interfacial Properties of Li Electrodes

A number of electrochemical techniques were applied for the electrochemical analysis of Li electrodes in a large variety of electrolyte solutions. These include chronopotentiometry [230–233], potentiodynamic measurements (cyclic voltammetry) [88,89], transient methods (micropolarization) [81], fast OCV measurements [90,91] and impedance spectroscopy (EIS) [92–100]. It should be noted that electrochemical analysis of Li electrodes is very complicated for the following reasons:

1. The electrochemical behavior of lithium electrodes usually relates to Li deposition-dissolution processes that may considerably change the electrode area and surface properties during the measurements.
2. The surface film formation processes are very fast, and their properties change rapidly while the films are built up. Hence, it is almost impossi-

ble to follow the surface film formation processes by standard electro-chemical techniques.

3. Li is always covered by native surface films that are replaced in solu-tions by layers originating from reduction of solution species. These replacement processes may have prolonged time constants and are ac-companied by secondary reactions between surface species and solu-tion components. Hence, it may be difficult to measure these systems at a true steady state condition.

4. It is very difficult to study the intrinsic kinetics of the Li/Li$^+$ couple due to the above-mentioned fast reactions of Li with the solution species.

However, part of the above problems may be overcome by careful experi-mental considerations. These may include

1. The use of fast transient methods (micropolarization) that do not change the electrodes during measurements by very much

2. The use of microelectrodes for the study of fast, film free Li deposition-dissolution processes that take place faster than the film formation pro-cesses

3. Studying Li surfaces freshly prepared in solutions

The first approach was used by Geronov et al. [21,81], who applied galva-nostatic transients to Li electrodes in PC solutions. Thus, analyzing parameters such as surface film resistivity and capacity, they could conclude from their mea-surements that the surface films comprise inner, compact, and porous outer (solu-tion side) parts. Similar conclusions were obtained by others as well [234].

The use of galvanostatic transients enabled the measurement of the poten-tiodynamic behavior of Li electrodes in a nearly steady state condition of the Li/film/solution system [21,81]. It appeared that Li electrodes behave potentio-dynamically, as predicted by Eqs. (5)–(12), Section III.C: a linear, Tafel-like, log $i$ versus $\eta$ dependence was observed [Eq. (8)], and the Tafel slope [Eq. (10)] could be correlated to the thickness of the surface films (calculated from the overall surface film capacitance [21,81]). From measurements at low overpoten-tials, $I_0$, and thus the average surface film resistivity, could be measured according to Eq. (11), Section III.C [21,81]. Another useful approach is the fast measure-ment of open circuit potentials of Li electrodes prepared fresh in solution versus a normal Li/Li$^+$ reference electrode [90,91,235]. While lithium reference elec-trodes are usually denoted as Li/Li$^+$, the potential of these electrodes at steady state depends on the metal/film and film/solution interfaces, as well as on the Li$^+$ concentration in both film and solution phases [236]. However, since Li elec-trodes in many solutions reach a steady state stability, their potential may be regarded as quite stable within reasonable time tables (hours → days, depending on the system's surface chemistry and related aging processes).

Hence, when a Li surface is prepared fresh in solution, its OCV versus an aged Li electrode in the same solution changes rapidly in time. This transient behavior is unique to each solution, and, thus, comparison of the OCV versus $t$ curves of freshly prepared Li electrodes in different solutions provides a good basis for the evaluation of the kinetics of Li passivation and the role of different solution compositions. An important consequence of these studies is that the passivation of fresh lithium in polar aprotic electrolyte systems may be completed within less than 1 s [90]. The intrinsic behavior of the Li/Li$^+$ couple could only be investigated by the use of microelectrodes. Fast potentiodynamic measurements could then be applied to Li electrodes in solutions in which Li deposition processes could be faster than the corrosion of the Li deposits resulting from fast reactions with solution components. This approach was applied by Pletcher et al. for ethereal solutions [88] and PC solutions [237–238], by Farrington and Xu for ethers and polyethers [239,240] and by Verbrugge et al. [89] for PC solutions.

A Tafel-like behavior for the intrinsic redox behavior of the Li/Li$^+$ could be measured, providing $i_0$ values of 31.6 mA/cm$^2$ (at 25°C) and $\beta = 0.67$ [89]. These high values of the $i_0$ of the Li/Li$^+$ couple mean that the charge transfer resistance ($R_{CT}$) of Li electrodes is negligible compared with the surface film resistance (for Li$^+$ migration).

The most commonly used method for the electrochemical studies of Li electrodes was impedance spectroscopy (EIS). Table 5 provides a partial listing of papers published during the past two decades dealing with the EIS of Li electrodes. However, the following precautions must be taken into account in the application of EIS to Li electrochemistry and the data analysis:

1. The measurements have to be taken at a stable open circuit potential while the electrode is at a steady state.
2. The current distribution has to be completely uniform. A parallel plate cell configuration is important. Edge effects should be eliminated.
3. The Li surface preparation is very important. Immersion of Li electrodes covered by native films leads to complicated surface film replacement processes that may form a highly nonhomogeneous metal-solution interphase. In situ electrochemical surface preparation by dissolution or deposition may form very rough surfaces whose impedance spectra may be difficult to interpret properly. Hence, it seems that the most preferred way of studying the electrochemical behavior of a Li electrode in a specific solution is by using Li surfaces freshly and smoothly prepared in solutions.
4. The last comment relates to the data analysis and the choice of appropriate models for impedance spectra. As shown by Orazem et al. [241], each single impedance spectrum can be fitted by a number of equivalent circuit analogs. Hence, the choice of a model has to be based on

**Table 5** Representative Bibliography on Impedance Spectroscopy of Li Electrodes

| Author | Year | Electrolyte system | Remarks | Ref. |
|---|---|---|---|---|
| Peled | 1984 | Cyclic ethers, 2Me-THF, THF, 1,3-DN/LiClO$_4$, LiI. | Passivating film comprising two layers | 22 |
| Aurbach | 1994 | PC/LiBr, LiPF$_6$, LiBF$_4$, LiSO$_3$CF$_3$ | Pronounced salt effect. | 48 |
| | | LiN(SO$_2$CF$_3$)$_2$ | High impedance of LiF films. | |
| Zaban | 1996 | PC, 2Me-THF, EC-DMC, DN, THF/LiAsF$_6$ | Generalized model for impedance spectra | 49 |
| Zaban | 1996 | THF, 2Me-THF mixtures/LiAsF$_6$ | General behavior | 51 |
| Thevenine | 1987 | PC/LiClO$_4$ | Models of the surface films | 77 |
| Aurbach | 1994 | PC/LiAsF$_6$, LiClO$_4$, CO$_2$, H$_2$O | Application of Voigt-type analog | 186 |
| Pejovnik | 1991 | SOCl$_2$/LiAlCl$_4$ | Space charge concept | 98 |
| Pejovnik | 1992 | SOCl$_2$/LiAlCl$_4$ | Time and frequency domains | 99 |
| Pejovnik | 1993 | SOCl$_2$/LiAlCl$_4$ | General behavior | 100 |
| Matsuda | 1992 | PC/LiClO$_4$ | Effect of additives: benzene, methyl furan | 92 |
| Takami | 1992 | EC-2Me-THF/LiPF$_6$ | General behavior and models | 93 |
| Churikov | 1997 | PC, DME/LiClO$_4$ | Li-Sn, Li-Cd, Li-Sn-Cd alloys | 94 |
| Ishikawa | 1996 | PC/LiClO$_4$ | Li-Al, Li-Sn alloys | 95 |
| Lerner | 1996 | PEM/LiClO$_4$ | Interfacial behavior with conducting polymers | 96 |
| Liebenow | 1996 | PC/LiClO$_4$ | Li coated with polyvinyl pyridine | 97 |
| Xue | 1994 | PEO-PC/LiClO$_4$ | Interfacial behavior with conducting polymers | 242 |
| Osaka | 1995 | PC-based (LiClO$_4$) | Effect of CO$_2$ | 243 |
| Messina | 1995 | PC, EC, DMC, DME/LiPF$_6$, LiN(SO$_2$CF$_3$)$_2$, LiSO$_3$CF$_3$ | Effect of solution and cycling | 244 |
| Pistoia | 1990 | PC/LiClO$_4$, PC-DMC/LiClO$_4$, EC-2Me-THF/LiAsF$_6$, THF-2Me-THF/LiAsF$_6$ | Effect of solution and storage time | 245 |
| Shembel | 1987 | PC, DMSO, SO$_2$/LiClO$_4$ | Effect of cosolvents | 246 |
| Maclean | 1989 | 2Me-THF/LiAsF$_6$ | Passivation studies | 247 |
| Povarov | 1983 | PC, BL, THF/LiClO$_4$ | General behavior | 248 |
| Shembel | 1985 | PC-based | Semiconducting properties of the surface films | 249 |
| Moshtev | 1984 | PC-based | Effect of surface preparation | 250 |
| Zaban | 1994 | PC-LiClO$_4$/LiAsF$_6$ | Effect of surface preparation | 251 |

a solid physicochemical understanding of the system studied that can eliminate this ambiguity. As discussed above and in previous sections, the EIS of Li electrodes at steady state condition is supposed to probe mostly the solid electrolyte phase that covers the active metal and its interfaces with the solution and the metal (the $R_{CT}$ of the Li/Li$^+$ couple is negligible, as discussed above).

Figure 18, taken from Ref. 77, describes several models proposed for the Li electrodes in solutions, their equivalent circuit analogs, and the expected impedance spectra (presented as Nyquist plots). Assuming parallel plate geometry for the solid electrolyte interface, as well as knowledge of the surface species involved from spectroscopy (and thus their dielectric constant, which is around 5 for many surface species formed on Li, including ROCO$_2$Li, Li$_2$CO$_3$, LiF, ROLi, etc. [186]), it is possible to estimate the surface film's thickness from the electrode's capacitance (calculated from the model fitted to the spectra):

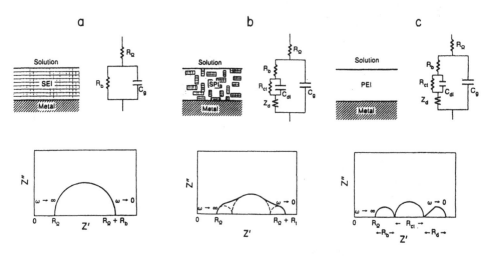

**Figure 18**  Various models proposed for the surface films that cover Li electrodes in nonaqueous solutions. The relevant equivalent circuit analog and the expected (theoretical) impedance spectrum (presented as a Nyquist plot) are also shown [77]. (a) A simple, single layer, solid electrolyte interphase (SEI); (b) solid polymer interphase (SPI). Different types of insoluble Li salt products of solution reduction processes are embedded in a polymeric matrix; (c) polymeric electrolyte interphase (PEI). The polymer matrix is porous and also contains solution. Note that the PEI and the SPI may be described by a similar equivalent analog. However, the time constants related to SPI film are expected to be poorly separated (compared with a film that behaves like a PEI) [77]. (With copyrights from The Electrochemical Society Inc., 1998.)

Film thickness $l = \dfrac{\varepsilon_0 KA}{C}$ (14)

where $\varepsilon_0$ is the dielectric constant for vacuum, $K$ is the film's dielectric constant, $A$ is the electrode's geometric area, and $C$ is its capacitance. Hence, the surface film's resistivity can be calculated as $\rho = RA/l$ ($R$ is the diameter of the semicircle in the spectra of Figure 18). However, the models of Figure 18 are too simplified. It is known that the surface films formed on Li comprise at least two parts: an inner compact layer and an outer, porous layer [21,81,234]. It is reasonable to assume that the surface films should have a multilayer structure because they are formed under a continuously changing driving force. As Li is exposed to solution, it is the most reactive and, hence, the first layers are deposited under the least selective conditions, and their constituents are expected to be at the lowest oxidation states [i.e., species such as $Li_2O$, LiX ($X$ = halide), species containing Li—C bonds, etc.]. As the surface film formation continues, the following surface layers are formed under a smaller and more selective driving force, and thus, their oxidation states and properties should differ from those of the inner shell. Finally, reduction of solution species may take place only via occasional tunneling of electrons through the surface films in points of defects in the surface species' crystal structure, etc. This process, taking place in the outer (solution) side of the surface films, is accompanied by partial dissolution of surface species and their possible secondary reactions with solution species. Such processes are expected to form surface films composed of a compact multilayer inner part and a porous outer part.

Figure 19 presents typical impedance spectra (Nyquist plots) obtained from Li electrodes freshly prepared in $LiAsF_6$ solutions of different solvents, salt concentrations, and water content [49]. The effect of aging is also demonstrated. All Nyquist plots obtained from freshly and smoothly prepared Li electrodes in solutions are similar and include a large, flat, asymmetrical semicircle and a low frequency feature that may appear as another small, flat semicircle with a Warburg-type element at the lowest frequencies. It is clear that these spectra reflect several time constants of different relaxation processes. Taking into account the expected multilayer structure of the Li-solution interphase, we have proposed the model and equivalent circuit analog shown in Figure 20 [49]. All impedance spectra obtained from hundreds of Li electrodes in different solutions could be nicely fitted with the Voigt-type analog of five $R\|C$ in series shown in Figure 20 [49,186]. In this model, the three high frequency/low capacity $R\|C$ circuits (a–c in Figure 20) represent the multilayer, compact inner part of the surface films, whereas the other two $R\|C$ circuits, related to the low frequencies (whose capacitance is indeed high, as expected for a porous film), represent the outer solution side of the Li-solution interphase.

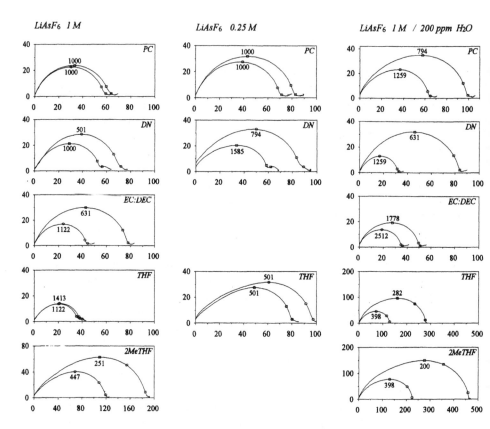

**Figure 19** Typical Nyquist plots obtained from impedance spectroscopic measurements of Li electrodes stored in different electrolyte solutions (as indicated): circles = after 3 h of storage; squares = after 6 days of storage. The first circle or square to the left of each plot related to $\omega_{max}$ (marked on the plots in Hz), and the next two (to the right) in each plot relate to 100 Hz and 10 Hz, respectively [49]. (With copyrights from The American Chemical Society, 1998.)

From the $R$ and $C$ values of the time constants a–c in the model, it was possible to estimate the thickness and resistivity of layers comprising the compact part of the surface films. The temperature dependence of these three time constants (e.g., linear Arrhenius plots for the different resistivities calculated that reflect different activation energy for $Li^+$ ion migration in each layer), as well as their dependence on the solution composition and the experimental conditions, revealed that the model has a solid physicochemical ground [48,49,186].

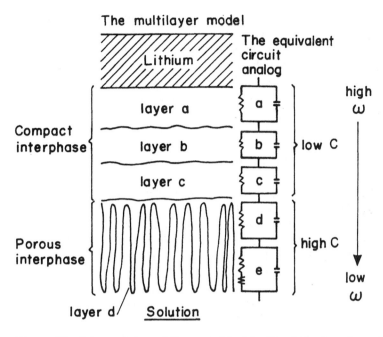

**Figure 20** Scheme of the multilayer model of the Li-solution interphase, the division of the various layers, and the corresponding equivalent circuit analog, which can be fitted very well to the experimental data [49]. (With copyrights from The American Chemical Society, 1998.)

This model and the related calculations of film thickness and resistivity values have many interesting consequences, some of which are described below [48,49,186].

1.   The average thickness of the compact part of the surface films formed on lithium electrodes in dry ethereal and alkyl carbonate solutions is around 30–50 Å.

2.   The resistivity of the layers comprising the compact part of the interphase depends strongly on their distance from the Li-film interface and the salt used. For $LiAsF_6$, $LiClO_4$, LiBr, $LiSO_3CF_3$, and $LiN(SO_2CF_3)_2$ solutions in alkyl carbonates and $LiAsF_6$ solutions in ethers (DN,THF, 2Me-THF), the resistivity of the layer close to the lithium is in the order of $0.1–0.2 \cdot 10^8 \ \Omega \cdot cm$, whereas the outers layer's (close to the porous part of the Li-solution interphase) resistivity is around $3–4 \cdot 10^8 \ \Omega \cdot cm$. This behavior is achieved during the first few hours of storage after electrode preparation and remains steady upon storage in the dry solutions for weeks.

3.   The overall thickness of the surface films formed on lithium in $LiBF_4$ and $LiPF_6$ solutions of alkyl carbonates is similar to those formed in other salt solutions, but the resistivity of the outer layers is much higher compared with the other solutions mentioned above (one order of magnitude for 1 $M$ $LiBF_4$ solutions and two orders of magnitude for $LiPF_6$ solutions, $10^9$ and $10^{10}$ $\Omega \cdot cm$, respectively). These resistivity values increase considerably upon storage. This result clearly reflects the replacement of $ROCO_2Li$ surface species by LiF in $LiPF_6$ and $LiBF_4$ solutions of alkyl carbonates. The importance of these results lies in the conclusion that it is not an increase in the surface film thickness but rather resistivity changes that lead to the high interfacial impedance of Li electrodes in $LiPF_6$ and $LiBF_4$ solutions.

4.   When the solution contains water at a concentration of 100 ppm and above, the thickness calculated for the layers comprising the compact surface films on lithium in all solutions increases upon storage, while the effect on the resistivity is not pronounced (compared with dry solutions). For wet alkyl carbonate solutions, the increase in the film thickness levels off during the first few days of storage, while in wet ethereal solutions the increase in the surface film thickness upon storage is more pronounced than in the alkyl carbonates, and its leveling off takes much longer. These changes reflect the effect of water that diffuses through the surface film and is reduced by the lithium, thus increasing the film thickness according to the parabolic law of growth [Section III.A, Eqs. (3), (4)]. In the alkyl carbonate solutions, the water diffusion is partially blocked by the formation of $Li_2CO_3$ due to reactions of $H_2O$ with $ROCO_2Li$. $Li_2CO_3$ is one of the least hygroscopic surface species formed on lithium [19], and, hence, once it is formed on the Li surface, water diffusion by hydration of surface species is difficult and slow. This is clearly reflected by the fast leveling off in the increase of the Li surface film thickness in wet alkyl carbonate solutions.

5.   When the solutions contain $CO_2$, the surface film thickness calculated is smaller than in the $CO_2$-free solutions, and the behavior is very stable upon prolonged storage. This reflects the formation of $Li_2CO_3$ on the Li surface, which is an excellent passivating agent (and which also prevents water diffusion to the Li surfaces, as discussed above). These results suggest that the high Li impedance measured in wet solutions mostly reflects growth of the surface films rather than an increase in their resistivity, while the low Li impedance measured in $CO_2$-containing solutions is due to the thin (and highly passivating) $Li_2CO_3$ thus formed.

The above discussion demonstrates the possibility of using relatively simple models (the Voigt-type analog) in describing the impedance characteristics of Li electrodes. It should be noted, however, that other models have also been proposed, such as the space-charge approach proposed by Pejovnik et al. [98–100].

The last electrochemical tool described in this section is electrochemical quartz crystal microbalance (EQCM). This technique enables mass accumulation and depletion on electrodes to be followed in conjunction with standard electrochemical measurements [252], and, thus, it may be a valuable tool for studying Li dissolution and deposition processes. The application of EQCM for the study of Li deposition-dissolution processes was demonstrated by Naoi et al. [183] and by Moshkovitz and Aurbach [184]. A key feature of these measurements is the possibility of calculating the mass per mole of electrons (m.p.e.) of the various deposited species and comparing it with their expected equivalent weight. For Li deposition and dissolution, the expected m.p.e. is 7 gr/mole. As demonstrated in Ref. 184, the deviation of the experimental m.p.e. from the expected equivalent weight (7) is diagnostic for the type of processes involved. For instance, m.p.e. >7 during deposition indicates corrosion of the Li deposits by reactions with solution species that form new surface species, while m.p.e. <7 during deposition may indicate massive dendrite formation, part of which is continuously disconnected from the bulk.

When m.p.e. values >7 are measured during Li dissolution, it indicates continuous depletion of Li by surface reactions that form soluble products. Hence, EQCM measurements may serve as a useful tool in the optimization of solutions for rechargeable Li batteries and the choice of preferred electrolyte systems for them [184].

## E.  Remarks on the Morphology of Li Electrodes

The reversibility of Li electrodes depends largely on the morphology of Li deposition and dissolution processes. As described in Section V.C, the complicated surface chemistry of Li electrodes in solutions leads to the formation of surface films which may consist of a large variety of different surface species of different chemical, physical, and electrochemical properties. Consequently, these surface films may be nonuniform and mosaic-like at the microscopic level. Such a structure induces nonuniform current distribution, since different surface species should have different Li ion conductivity. Li deposition may therefore lead to dendrite formation. These dendrites are Li deposits that grow outside the surface films and, due to primary current distribution considerations, usually continue to grow in a noodle-like structure with many bottlenecks. These may readily corrode, and these dendrites can be easily electrically disconnected from the bulk through substitution of Li metal in the bottlenecks by electrically insulating solution reduction products. This phenomenon is a major reason for irreversibility of Li anodes in most nonaqueous electrolyte solutions of interest. Until the appearance of scanning probe microscopy [253], a major tool for Li morphology studies was scanning electron microscopy (SEM), which has to be performed ex situ. Figure 21 presents several typical micrographs of Li electrodes after being cycled in different electrolyte solutions (as specified). These pictures demonstrate

**(a)**

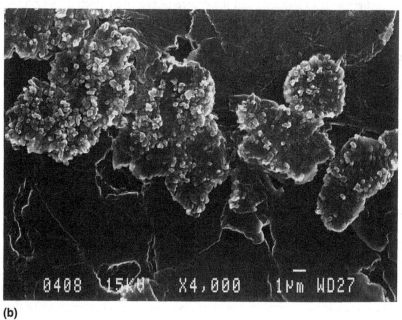

**(b)**

**Figure 21** Typical SEM micrographs that demonstrate the morphology of Li deposition processes and their dependence on the solution composition. 0.5 $M$ LiClO$_4$ solutions, Li substrates, 0.5–0.7 C/cm$^2$. A scale appears in each picture. (a) A pristine Li surface; (b) γ-butyrolactone (BL); (c) tetrahydrofuran (THF); (d) propylene carbonate (PC); (e) 1,3-

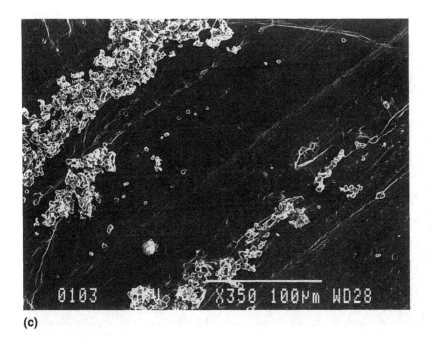

**(c)**

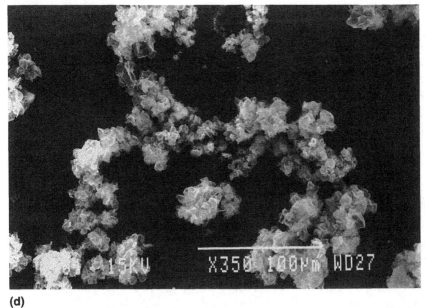

**(d)**

dioxolane (DN). The typical smooth morphology of Li deposition processes in this solvent is demonstrated: (f,g,h) typical single dendrites formed in PC, BL/O$_2$ saturated, and di-methoxyethane (DME) solution, respectively [19,144]. (With copyrights from The Electrochemical Society, Inc., 1998.)

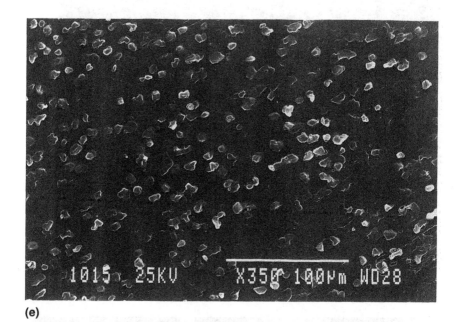

**(e)**

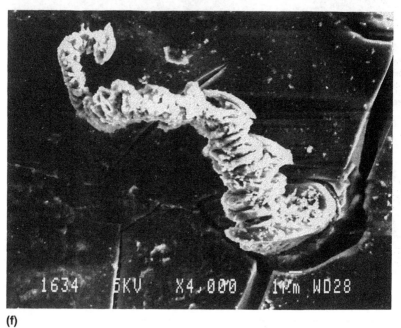

**(f)**

**Figure 21**  Continued

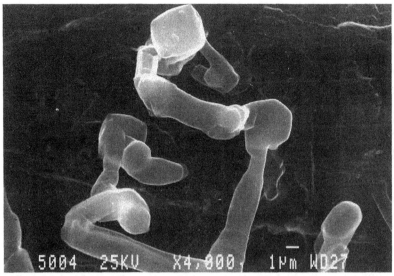

(g)

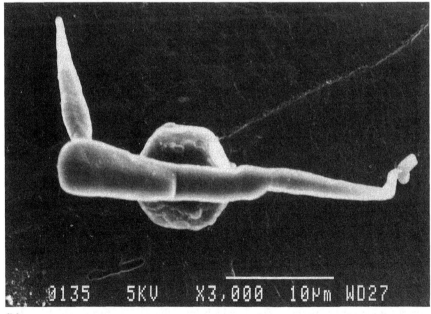

(h)

how rough the morphology of Li deposition can be, and they also feature typical Li dendrites. Several attempts have been made to observe Li morphology in situ in solutions by optical means, among them the work of Osaka et al. [254] and Yamaki et al. [255]. For instance, the latter demonstrated by in situ optical observation that dendrites can also be formed on Li deposits that are electrically disconnected from the bulk electrode (due to application of the electrical field). Scanning tunneling microscopy (STM) and atomic force microscopy (AFM) opened the door for in situ morphological studies. STM may be a problematic tool for Li electrodes since the surface films covering the electrodes are electrically insulating and may prevent any tunneling current between the STM tip and the electrode. Nevertheless, there are already reports on in situ studies of Li electrodes by STM [173]. AFM is more appropriate because it does not depend on any current passage between the tip and the electrode.

We have already reported on in situ AFM studies of Li electrodes in solutions as a function of the solvent, salt, and additives that compose the solutions [109,172]. These studies required special experimental arrangements directed toward properly isolating the systems studied from atmospheric contaminants, while still maintaining high resolution and sensitivity. Figure 22 shows AFM images of a Li electrode in a PC LiAsF$_6$ solution before and after Li deposition and then after Li dissolution. As shown in these images, this electrode has a special texture formed during its mechanical preparation. This texture was covered by the Li deposition and was then uncovered by the consecutive dissolution [172]. Figure 23 presents SEM micrographs of the same electrode after the experiments, showing its typical texture, and are of the same scale as the AFM images [172]. The comparison between the two figures clearly shows the advantage of AFM over SEM in these studies. Despite the fact that the AFM images are measured in situ, they reflect the Li morphology more clearly than do the SEM micrographs.

Figure 24 shows in situ AFM images of a Cu substrate on which Li was deposited in a PC-LiClO$_4$ solution and then dissolved [109]. These two images present a good example of nonuniform Li deposition. The consecutive dissolution leaves an appreciable percentage of Li which remains electrochemically inactive. The porous nature of the surface films which cover the electrode is also clearly reflected by the image. Hence, the application of scanning probe techniques such as AFM to Li electrochemistry provides new insights into the correlation between Li morphology and parameters such as solution composition, the substrate involved, current density, potential, etc. It should be emphasized that it is impossible to predict a priori what the morphology of Li in most of the commonly used electrolyte solutions should be. This depends on a very delicate balance which exists among the various processes which form the surface films (which control Li dissolution and deposition). This balance may be strongly influenced by contaminants at the ppm level that may differ from one experimental setup to another. There are, however, a few exceptions, some of which are as follows:

1. The use of Li alloys such as Li-Al induces very smooth Li deposition [256,257].
2. In 1,3-dioxolane/LiAsF₆/TBA solutions, Li is always deposited in a flake-like formation, which leads to the high Li cycling efficiency achieved in this system [116].
3. In solutions containing HF, Li is deposited in a unique formation—small, round, uniformly dispersed deposits [258].

In conclusion, in most polar aprotic Li salt solutions, the morphology of Li electrodes upon charge-discharge cycling is very rough. The electrodes become covered with reactive dendrites that can be readily disconnected electrically from the bulk. Nonuniform Li dissolution further enhances the fractal structure of the electrode surface. These situations were recently rationalized and well explained by theoretical approaches and calculations [259,260].

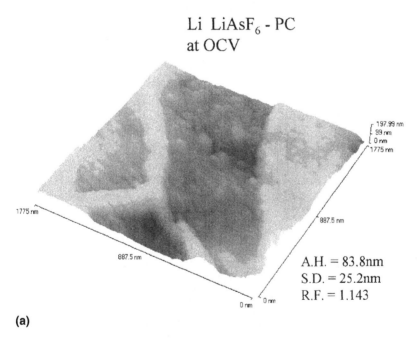

(a)

**Figure 22** AFM images of Li electrodes treated in PC/LiAsF₆ solution. (a) Li in solution at OCV condition. The pristine electrode has a unique morphology due to its preparation procedure. (b) The same electrode as in (a) after Li (0.5 C/cm²) was deposited. The pristine electrode texture disappeared; (c) the same electrode as in (a) and (b) after consecutive Li dissolution (0.5 C/cm²). Note that the pristine morphology reappeared [172]. A. H. = average height; S. D. = standard deviation; R.F. = roughness factor. (With copyrights from The Electrochemical Society Inc., 1998.)

## Li deposition 0.5C/cm²
## LiAsF₆ - PC

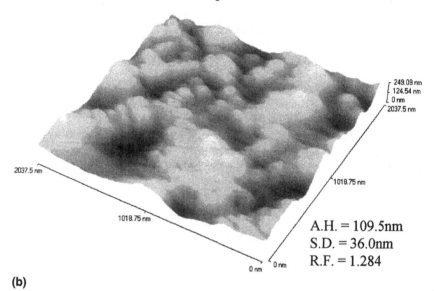

A.H. = 109.5nm
S.D. = 36.0nm
R.F. = 1.284

**(b)**

## Li disolution 0.5 C/cm²
## LiAsF₆ - PC

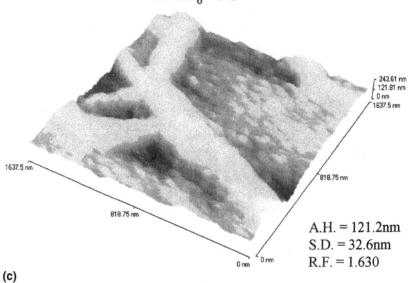

A.H. = 121.2nm
S.D. = 32.6nm
R.F. = 1.630

**(c)**

**Figure 22**  Continued

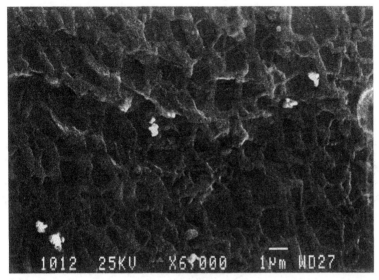

(a)

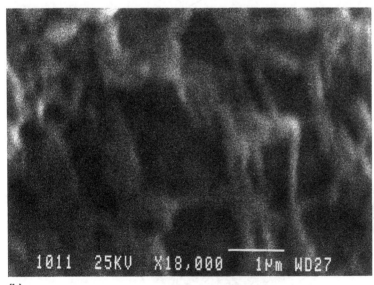

(b)

**Figure 23** SEM micrographs of the electrode imaged in Figure 22c measured from it ex situ at the end of the experiment. Two magnifications as indicated. Note that the SEM reflects the unique morphology of this electrode, but at a much lower resolution than the AFM images (although measured ex situ) [172]. (With copyrights from The Electrochemical Society Inc., 1998.)

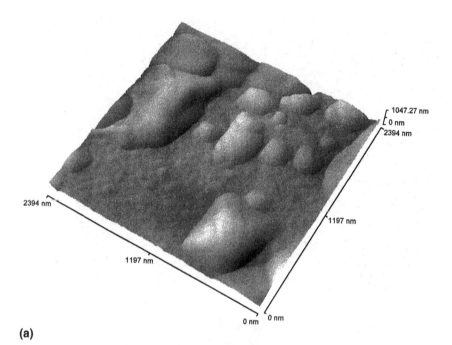

(a)

(b)

## F. Remarks on the Cycling Efficiency of Li Electrodes

A key feature of Li electrodes in electrolyte solutions, which is critical for their use as anodes in rechargeable batteries, is their cycling efficiency. In general, this term defines what percentage of Li deposited electrochemically remains electrically active and can thus be dissolved electrochemically. We can define Li cycling efficiency (Li$_{C.E.}$) per cycle as

$$E\% = \frac{Q_a}{Q_c} \cdot 100 \tag{15}$$

$Q_c$ and $Q_a$ are the charges of Li deposited and dissolved per cycle, respectively. The average Li$_{C.E.}$ is calculated as [19]

$$\bar{E}(\%) = \frac{Q - (Q_{in} - Q_{ex})/n}{Q} \cdot 100 = \frac{nQ - Q_{in} + Q_{ex}}{nQ} \cdot 100 \tag{16}$$

Where $Q$ is the charge cycled (electrochemical Li deposition-dissolution cycling), $Q_{in}$ is the charge of the Li present initially, $Q_{ex}$ is the charge of the residual Li at the end of the experiment, and $n$ is the number of cycles. In experiments of prolonged cycling in which the electrode is consumed,

$$\bar{E}(\%) = \left(1 - \frac{Q_{in}}{nQ}\right) \cdot 100 \tag{17}$$

Another useful measure for Li utility in repeated charge-discharge cycling is the figure of merit (F.O.M.), which is the ratio between the overall charge involved in the cycling ($nQ$) and the initial charge content of the electrode:

$$\text{F.O.M.} = \frac{nQ}{Q_{in}} = \frac{1}{1 - \bar{E}/100} \tag{18}$$

Since testing of practical batteries is very prolonged, it was necessary to develop quick tests for measuring Li cycling efficiency. In a typical experiment, Li is deposited electrochemically in the tested solution on a Cu or Ni substrate (thus defining $Q_{in}$), followed by charge-discharge cycling ($Q < Q_{in}$). After a pre-

---

**Figure 24** AFM images of a Cu substrate on which Li was deposited and then dissolved in PC/LiClO$_4$ solution (a,b). Note the nonuniform Li deposition in this solution, and the fact that "dead" Li remains after the dissolution step. The image in (b) reflects the porous structure of the surface films [109]. (With copyrights from The Electrochemical Society Inc., 1998.)

determined number of cycles $(n)$, the residual Li $(Q_{ex})$ is determined by electrochemical dissolution (galvanostatic experiments using a Li counterelectrode). It should be noted that $Q_{in}$ taken for the calculation of $E$ is only a fraction of the charge involved in the initial Li deposition and depends on the average Li cycling efficiency in the specific solution. Thus, $Q_{in} \approx EQ'_{in}$, where $Q'_{in}$ is the charge measured for the initial Li deposition process (in situ Li electrode preparation). As the amounts of charge $Q_{in}$ and $Q$ and the number of cycles in these experiments are higher, so the cycling efficiency measured in these experiments reflects better tests of practical batteries.

It should be noted that in practical batteries such as coin cell (parallel plate configuration) or AA, C, and D (jelly-roll configuration), there is a stack pressure on the electrodes (the Li anodes are pressed by the separator), and the ratio between the solution volume and the electrode's area is usually much lower than in laboratory testing. Both factors may considerably increase the Li cycling efficiency obtained in practical cells, compared with values measured for the same electrolyte solutions in the Li half-cell testing described above. It has already been proven that stack pressure suppresses Li dendrite formation and thus improves the uniformity of Li deposition-dissolution processes [107]. The low ratio between the solution volume and the electrode area in practical batteries decreases the detrimental effects of contaminants such as Lewis acids, water, etc., on Li passivation.

In any event, the short experiments for Li cycling efficiency measurements [19,116,144,224,230] are useful in ranking the various electrolyte solutions proposed for rechargeable Li batteries and thus serve as an important tool in the process of optimization of electrolyte solutions for Li battery systems. Table 6 presents typical Li cycling efficiency values measured in different solutions calculated in experiments in which Li was deposited on copper (5 C/cm$^2$), followed by galvanostatic cycling (1 mA/cm$^2$) at 25% DOD (10–20 cycles). The difference in the Li cycling efficiency values measured in different solutions, as demonstrated in Table 6, may be understood in light of the surface chemistry developed on Li in each solution (Section V.C), as summarized below.

1.  In general, Li cycling efficiency depends on the chemical and physical properties of the surface films (determined by a delicate balance among the various possible surface reactions, as discussed in Section V.C). Additives at the ppm level in solutions may considerably influence the Li cycling efficiency. Obvious conditions for high Li cycling efficiency are
    a.  The formation of surface species that precipitate as thin, compact, and highly passivating films.
    b.  Uniform Li deposition: the less the dendritic deposition, the higher the Li cycling efficiency.

**Table 6** Typical Li Cycling Efficiency Values Calculated in Experiments in Which Li Was Deposited on Cu (5 $C/cm^2$) Followed by Galvanostatic Cycling (1 $mA/cm^2$) at 25% DOD (10–20 cycles)

| Solvent | Salt | Additives | Li C.E. | Ref. |
|---|---|---|---|---|
| 1,3-Dioxolane | $LiClO_4$ | None | >95% | 144 |
| | $LiAsF_6$ | Tributyl amine | >96–97% | 116 |
| | $LiN(SO_2CF_3)_2$ | TBA | <50% | |
| 2Me-THF | $LiAsF_6$ | None | >95% | 140 |
| | $LiClO_4$ | None | <50% | 140 |
| THF | $LiAsF_6$ | | >90% | 140 |
| Glyme family | $LiClO_4$ | | 80% | 149 |
| (DG, EG) | $LiN(SO_2CF_3)_2$ | | <50% | 149 |
| | $LiSO_3CF_3$ | | <50% | 149 |
| $\gamma$-Butyrolactone | $LiClO_4$ | | <50% | 19 |
| | $LiAsF_6$ | | >80% | 19 |
| | $LiAsF_6$ | $O_2$ saturated | >85% | 19 |
| Methyl formate (MF) | $LiAsF_6$ | | <50% | 261 |
| Methyl formate (MF) | $LiAsF_6$ | $CO_2$ | >90% | 261 |
| MF-EC or MF-PC | $LiAsF_6$ | $CO_2$ | >90% | 261 |
| Propylene carbonate | $LiClO_4$ | | >70% | 19 |
| | $LiAsF_6$ | | >80% | 131 |
| | $LiAsF_6$ | $CO_2$ | >95% | 131 |
| | $LiAsF_6$ | Trace $H_2O$ | >90% | 131 |
| | $LiPF_6$ | | >80% | 131,48 |
| Ethers + PC | $LiAsF_6$ | | <70% | 153 |
| EC-Ethers | $LiAsF_6$ | | >70% | 153 |
| EC-Ethers | $LiAsF_6$ | $CO_2$ or trace $H_2O$ | >95% | 153,132 |
| EC-PC | $LiAsF_6$ | | >80% | 131 |
| EC-DEC | $LiAsF_6$ | | >90% | 132 |
| EC-DMC 1:1, (1:5) | $LiAsF_6$ | | <70% (>90%) | 51 |
| EC-DMC 1:1 | $LiPF_6$ ($CO_2$) | | <70% (>90%) | 51 |

    c.  Highly preferential are situations in which the surface films are flexible (e.g., containing polymeric species) and can stretch and contract according to changes in the surface of the active metal. In such cases, Li can be deposited beneath existing surface films, preventing the exposure of fresh lithium to solution phase.

  2.  Li cycling efficiency in $LiAsF_6$ solutions of all solvents is generally better than in other salt solutions [e.g., $LiClO_4$, $LiPF_6$, $LiBF_4$, $LiN(SO_2CF_3)_2$]. This is due to doping of Li with arsenic and arsenic compounds (e.g., those suggested by Koch [72–74], Irish [179], and

Aurbach [86,87]), which reduces the Li reactivity and enhances the uniformity of Li deposition processes.

3. The best cycling efficiency was obtained with $LiAsF_6$/DN solutions stabilized with tributylamine [116] or KOH and methyl furan [205]. This is due to the specific combination of surface compounds formed, which include the arsenic species, HCOOLi, ROLi, and, probably, oligomers of poly-DN with OLi and LiC bonds [159,204]. The surface films composing this combination of compounds induce highly uniform Li deposition in a flake-like morphology [144,204].

4. High cycling efficiency is also obtained in $LiAsF_6$-2Me-THF solutions containing 2-methyl furan [262]. We attribute it mostly to the effect of Li doping by arsenic compounds, which in this case is not interfered with by pronounced solvent reactions [50], since 2Me-THF is one of the least reactive toward Li of all the commonly used solvents.

5. $Li_2CO_3$ is one of the best passivating agents for Li electrodes in non-aqueous solutions. Whenever it is formed, it enhances Li cycling efficiency.

6. $Li_2CO_3$ may be formed on Li surfaces in solutions in three cases:
   a. $CO_2$-containing solutions [131,164,263].
   b. Alkyl carbonate solutions containing trace water ($H_2O$ reacts with the $ROCO_2Li$ formed by solvent reduction, and $Li_2CO_3$ is one of the products.) [131].
   c. In solvent mixtures containing EC and PC at sufficiently low concentration (e.g., with ether cosolvents). In this situation, the alkyl carbonate reduction by Li may lead to pronounced $Li_2CO_3$ formation, as explained in Ref. 191.

   In cases a and b, high Li cycling efficiencies are indeed obtained. In case c, the existence of too many surface active species in solution may have a detrimental effect, which balances the positive role of enhanced $Li_2CO_3$ formation [153].

7. Esters such as BL and MF are too reactive toward Li, and their reduction products are of relatively low passivation. Consequently, Li cycling efficiency in BL and MF solutions is very low (unless they contain $CO_2$).

It should be emphasized that Li dendrite formation *is not* prevented by surface $Li_2CO_3$ formation in $CO_2$-containing solutions. However, once the dendrites are formed, they become highly passivated, due to the $Li_2CO_3$. Thus, the corrosion of dendrites in these solutions, which is a major cause for low Li cycling efficiency, is largely prevented.

Since intensive work was devoted to the search for additives that increase

Li cycling efficiency, we conclude this section with a list of typical examples of these studies.

1. The work of Osaka et al. on the positive effect of $CO_2$ on Li cycling efficiency [164,263].
2. Covalent Assoc. devoted extensive work to the stabilization of DN solutions by species such as methyl furan and KOH [204].
3. The EIC group discovered the positive effect of methyl furan of Li cycling efficiency in ethereal solvents [262].
4. Extensive work by Yamaki and Tobishima was devoted to the effect of additives on Li cycling efficiency in ethers and alkyl carbonates [265]. This included the use of cation complexing compounds (e.g., 12 crown 4, TMEDA), reactive organic compounds (e.g., quinoneimine derivatives), and adsorption compounds (e.g., large tetraalkyl ammonium chloride substituted with aromatic rings) [265].
5. The studies of Yamaki et al. on metallic additives which modify the reactivity of Li electrodes toward solution species and thus improve Li cycling efficiency.
6. Fluorination compounds were also suggested for Li surface modifications [167].
7. Extensive work was devoted by Matsuda et al. to the study of additives for rechargeable Li battery systems [168–170]. For instance, inorganic additives such as $Mg^{2+}$, $Fe^{2+}$, and $Ga^{3+}$ ions were suggested [266]. The idea behind these additives is that their reduction on the Li surface may lead to surface alloying that can modify the reactivity of the Li toward solution species.
8. Zaban et al. studied the effect of fluorinating agents and arsenic-containing additives on Li cycling efficiency in ethereal solutions. It was found that doping Li surfaces with arsenic increases Li cycling efficiency in $LiClO_4$ solutions. This is in line with the assumption that Li cycling efficiency in $LiAsF_6$ solutions is high, due to the modification of Li with elemental arsenic and arsenic compounds [50].

It should be emphasized that the above list is only a partial one, and that much more work regarding additive effects on Li cycling efficiency has been published in the proceedings volumes related to the annual fall meetings of the Electrochemical Society and the Biannual International Meetings on Li batteries (Section V.B).

In parallel, during the past two decades, extensive efforts have been devoted to solvent effects on Li cycling efficiency. The use of solvent mixtures instead of single solvent solutions may have obvious advantages in terms of properties of the electrolyte solutions. For example, high polarity and low viscosity of the solvent media are the desired properties for obtaining high conductivity. This is indeed achieved by the use of mixtures of highly polar (and viscous) solvents such as PC, BL, or sulfolane with less polar (but less viscous) ethers [153–157]. However, the effect of solvent mixtures on Li cycling efficiency is much more problematic and unpredictable. Attempts have been made over the years to correlate Li cycling efficiency measured in the experiments described in this section to bulk properties of solvents such as oxidation potentials, donor or acceptor numbers, bonding-antibonding orbital gaps, etc. [267,268]. As discussed above, Li cycling efficiency is determined by the Li surface chemistry, which is a complicated matter that is determined by a delicate balance among competing surface reactions. This balance can be dominated by the presence of contaminants in the ppm level; hence, all the attempts to derive correlations and rules for Li cycling efficiency in terms of bulk solvent properties should be examined very carefully. Table 7 presents a partial list of references for typical studies of Li cycling efficiency in solvent mixtures.

## G.  Li-Alloy Electrodes

As is evident from data presented in the previous two sections, Li metal anodes may fail in rechargeable Li batteries because of their high reactivity and the fact that the volume and surface area of these electrodes change upon cycling. The protecting surface films that cover the active metal cannot accommodate the surface changes upon deposition and dissolution; thus fresh Li is continuously exposed to solution species, even in cases where detrimental dendrite formation does not occur. Thus, batteries based on Li metal anodes always have limited cycleability, no matter how good the chosen solution and the cell's engineering. Replacement of Li metal anodes by Li alloys may be one of the solutions for problems arising with the use of Li metal anodes in rechargeable batteries. It is possible to obtain highly reversible insertion of lithium into a large variety of Li alloys, as described further on. Li alloys may have the following advantages over Li metal as anodes in rechargeable Li batteries:

1. The Li reactivity is modified and reduced thus preventing active mass consumption by reactions with solution species.
2. Li alloys may form a rigid host matrix to which Li is inserted upon charging. Hence, the surface area of the anode remains stable upon cycling, as do the surface films that cover the electrode, so exposure of fresh, reactive Li to the solution is avoided.
3. Li melts at around 180°C, while many Li alloys have much higher melting points, and so they considerably extend the applicability of Li batteries for high temperature.

The obvious drawback to the use of Li alloys instead of Li metal in batteries is the lower potential, specific charge, and specific energy of Li-alloy-based battery systems, compared with Li metal batteries. The work with Li alloys as alternatives for Li metal batteries began in the early 1970s [293]. Table 8, taken from Ref. 294, presents some data on Li alloys tested as anodes in rechargeable Li battery systems. The various examples of Li alloys shown in Table 8 can be divided into three groups:

1.  Binary alloys (examples 2–10, Table 8). These alloys are not dimensionally stable, and considerable volume changes take place when the host metal is lithiated. For instance, lithiation of Zn to LiZn leads to volume expansion of about 71%; lithiation of Al to LiAl leads to about 97% volume expansion; $B \rightarrow Li_3B$ leads to 177% volume expansion; $Cd \rightarrow Li_3Cd$ leads to 268% volume expansion; $Si \rightarrow Li_4Si$ and $Sn \rightarrow Li_{4.4}Sn$ lead to 323% and 676% volume expansion, respectively [302]. Part of these alloys are hard and brittle (e.g., LiAl, LiPb), which makes them difficult to process mechanically.
2.  Dimensionally stable alloys (examples 11–13 in Table 8). In these composite alloys, the active alloy (usually a binary one) is encapsulated in another alloy that forms a rigid matrix. For instance, in example 13, an active LiCd alloy is hosted in a much less active, but rigid Li-Sn matrix. Hence, stable electrodes of relatively low volume change upon cycling can be composed, provided that the potential is kept strictly within limits in which only the active alloy reacts electrochemically, while the stable alloy remains inert.
3.  Wood alloys (examples 14–18, Table 8). This class includes soft alloys that can absorb a large amount of lithium at relatively low potentials and low voltammetric changes.

There are reports that the surface chemistry of Li alloys is indeed largely modified, compared with Li metal electrodes [303]. It appears that they are less reactive with solution species, as is expected. The morphology of Li deposition on Li alloys may also be largely modified and smooth, compared with Li deposition on Li substrates [302,304]. A critical point in the use of Li alloys as battery anodes is the lithium diffusion rates into the alloys. Typical values of Li diffusion coefficient into alloys are $\beta$-LiAl $\rightarrow 7 \cdot 16^{-9}$ cm$^2$/s [305], $Li_{4.4}Sn \rightarrow 2 \cdot 10^{-9}$ cm$^2$/s [306], LiCd and LiZn $\rightarrow 10^{10}$ cm$^2$/s [307]. It should be emphasized that it is very difficult to obtain reliable values of Li diffusion coefficient into Li alloys, and thus the above values provide only a rough approximation for diffusion rates of Li into alloys. However, it is clear that Li diffusion into Li alloys is a slow process, and thus is the rate-limiting process of these electrodes. Li deposition of rates above that of Li diffusion leads to the formation of a bulk metallic lithium layer on the alloy's surface which may be accompanied by mas-

**Table 7**  Representative List of Papers Reporting on the Effect of Solvent Mixtures and Solvent Effects on Li Cycling Efficiency

| Author | Year | Solvents | Salt | Remarks | Ref. |
|---|---|---|---|---|---|
| E. Peled et al. | 1989 | THF/DN/toluene | $LiClO_4$ | Li-S batteries | 269 |
| S. Subbarao et al. | 1990 | EC/THF/2Me-THF, EC 3-MeS/2Me-THF | $LiAsF_6$ | Li-TiS$_2$ batteries | 270 |
| Y. Matsuda | 1987 | PC/DME, DMSO/DME, PC/THF, S/DME | $LiClO_4$, $LiBF_4$, $LiPF_6$ | Li vs. Li comparative studies | 271 |
| K.M. Abraham et al. | 1982 | THF, DEE, 2Me-F, DN, DME, 2Me-THF | $LiAsF_6$ | Li-TiS$_2$ batteries | 272 |
| C.D. Desjardins et al. | 1985 | 2Me-THF, THF, 2Me-F | $LiAsF_6$ | Li vs. Li | 273 |
| Y. Matsuda et al. | 1985 | S/THF, S/DN, S/DME | $LiBF_4$, $LiPF_6$, $LiClO_4$ | Li vs. Li comparative studies | 274 |
| M. Anderman et al. | 1989 | THF/2Me-THF | $LiAsF_6$ | Li-TiS$_2$ batteries | 275 |
| K.M. Abraham et al. | 1986 | THF/2Me-THF/2-MeF | $LiAsF_6$ | Li-TiS$_2$ batteries | 276 |
| Y. Geronov et al. | 1989 | THF/2Me-THF, EC/2Me-THF, PC/2Me-THF, PC/DME | $LiAsF_6$, $LiClO_4$ | Li-LiV$_3$O$_8$ batteries | 277 |
| K.M. Abraham et al. | 1989 | THF, 2Me-THF, EC, 2-Me-F, 2-MeDN, DMM, 2-MeO THF | $LiAsF_6$ | C, AA Li-LiTiS$_2$ cells | 278 |
| S.V. Sazhin et al. | 1993 | PC, DME, BL, MA, SO$_2$, DMSO, | $LiSO_3CF_3$, $LiClO_4$, $LiBF_4$ | Li vs. Li | 279 |
| S.I. Tobishima et al. | 1983 | PC/THF, PC/DMSO, PC/DME, PC/DN | $LiClO_4$ | Li vs. Li | 280 |

| | | | | | |
|---|---|---|---|---|---|
| E.J. Frazer et al. | 1983 | PC/AN | $LiAsF_6$, $LiClO_4$ | Li vs. Li | 281 |
| M. Salomon et al. | 1990 | MF, DEC | $LiAsF_6$ | Li-$LiCoO_2$ batteries | 282 |
| E.S. Takuchi | 1991 | MF/PC | $LiAsF_6$, $LiBF_4$ | Li-$LiAgVO_x$ cells | 283 |
| W.B. Ebner et al. | 1989 | MF/DEC, MF/DMC, MF/ DMC/$CO_2$ | $LiAsF_6$ | Li-$Li_x$ $CoO_2$ cells | 284 |
| M. Salomon et al. | 1987 | DME, BL, MF, 2Me-THF | $LiAsF_6$ | Li-$V_6O_{13}$ cells | 285 |
| E. Plichta et al. | 1987 | MF, MA | $LiAsF_6$ | Li-$Li_x CoO_2$ cells | 286 |
| S.I. Tobishima et al. | 1995 | 2Me-THF, THF, EC, 4-Me-DN, DEE, PC, DME, AN, MA, MF | $LiAsF_6$ | Li vs. Li, ternary mixtures | 287 |
| R. Messina et al. | 1995 | EC/PC/DMC | $LiAsF_6$, $LiSO_3CF_3$ | Li vs. Li | 288 |
| M. Takahashi et al. | 1993 | EC/DME/1, 2-butylene carbonate | $LiSO_3CF_3$ | Li-$Li_x MnO_2$ cells | 289 |
| Y. Sakurai et al. | 1989 | PC, EC, DME, DN, 2Me-THF, diethoxyethane | $LiAsF_6$, $LiClO_4$, $LiBF_4$, $LiPF_6$ | Li-$Cu_2V_2O_7$ cells | 290 |
| S.I. Tobishima et al. | 1984 | EC/PC | $LiAsF_6$, $LiClO_4$ | Li vs. Li | 291 |
| M. Yoshio et al. | 1988 | PC/DME, PC/EC/DME | $LiPF_6$ | Li vs. Li | 292 |

PC = propylene carbonate; DN = 1,3-dioxolane; EC = ethylene carbonate; 3-MeS-3 = methyl sulfolane; DME = dimethoxy ethane; THF = tetrahydrofuran; S = sulfolane; DMSO = dimethyl sulfoxide, DEE = diethyl ether, 2-Me-F = 2 methyl furan; 2-MeDN = 2-methyl 1,3-dioxolane; MA = methylacetate; DMM = dimethoxymethane; 2-MeOTHF = 2 methoxytetrahydrofuran; BL = γ-butyrolactone; NM = nitromethane; AN = acetonitrile; MF = methyl formate; DEC = diethyl carbonate; DMC = dimethyl carbonate.

**Table 8**  Selected Data on Li Alloys Tested as Anodes in Rechargeable Li Battery Systems

| No. | Material | Theoretical energy density | | Average voltage, V (Li/Li$^+$) | Energy density relative to Li metal anode | | Ref. |
|---|---|---|---|---|---|---|---|
| | | Ah/Kg | Ah/L | | Gravimetric | Volumetric | |
| 1 | Li | 3860 | 2061 | 0 | 1 | 1 | — |
| 2 | LiAl | 790 | 1280 | 0.36 | 0.20 | 0.62 | 295 |
| 3 | Li$_{4.4}$Si | 2010 | 1749 | 0.2 | 0.54 | 0.85 | 295 |
| 4 | Li$_{4.4}$Pb | 500 | 1270 | 0.45 | 0.13 | 0.62 | 295 |
| 5 | Li$_{4.4}$Sn | 790 | 1003 | 0.5 | 0.20 | 0.49 | 295 |
| 6 | Li$_y$Sn (0.7 < y < 2.53) | 350 | 444 | 0.53 | 0.09 | 0.22 | 296 |
| 7 | Li$_4$Sb | 715 | — | 0.95 | 0.19 | — | 296 |
| 8 | Li$_4$Bi | 453 | — | 0.83 | 0.12 | — | 296 |
| 9 | LiZn | 371 | — | 0.2 | 0.10 | — | 297 |
| 10 | Li$_3$Cd | 604 | — | 0.08 | 0.16 | — | 297 |
| 11 | Li-B:Li$_7$B$_6$ · 17Li | 1920 | — | 0.02 | 0.50 | — | 298 |
| 12 | β-LiAl/Cu 80/20w/o | 632 | — | 0.36 | 0.16 | — | 299 |
| 13 | Li$_3$Cd/Li$_{4.4}$Sn Cd/Sn = 1 | 285 | — | 0.80 | 0.14 | — | 300 |
| 14 | Li$_x$ Bi/Cd 60/40w/o | — | 1761 | 0.3 | — | 0.85 | 301 |
| 15 | Li$_x$ Pb/Cd 60/40w/o | — | 1489 | 0.3 | — | 0.72 | 301 |
| 16 | Li$_x$ Cd/Sn 70/30w/o | — | 1479 | 0.3 | — | 0.72 | 301 |
| 17 | Li$_x$ Bi/Cd/Pb 50/30/20w/o | — | 1963 | 0.3 | — | 0.95 | 301 |
| 18 | Li$_x$ Pb/Cd/Sn 60/30/10w/o | — | 1720 | 0.3 | — | 0.84 | 301 |

*Source:* From Ref. 294.

sive dendrite formation. In any event, diffusion coefficients of the above orders of magnitude may allow Li deposition-dissolution of current densities in the mA/cm$^2$ region without Li metal bulk formation [294], which is acceptable for many applications of rechargeable Li batteries. Another point of interest would be the potential profiles of lithiation processes into Li alloys. It appears that many lithium insertion processes into binary alloys occurs at constant potential. Thus, galvanostatic lithiation of alloys such as Li$_x$Zn, Li$_x$Pb, Li$_x$Cd, Li$_x$Sn, Li$_x$Bi, etc., occurs at potential plateaus [308–310]. (Also see useful tables in Ref. 308.)

Among the various Li alloys mentioned, Li$_x$Al and Li$_x$Sn have been given special attention. Cycling of LiAl foil electrodes in liquid electrolytes has been reported [311–313] as well as in polymer electrolyte-based cells [314]. In general, the cyclability of the LiAl electrode depends on the current density and the amount of charge cycled, similar to lithium electrodes. In addition, a capacity decay is observed during cycling, and the cycle life is likewise restricted to 100–200 cycles. The limited cycling capability is related to the mechanical degradation of the electrode due to the volume changes associated with the lithium insertion processes [314,315].

It was found in these experiments that corrosion of the negative electrode is less pronounced compared with lithium. This is partly because the alloys are less reactive and partly because a protective surface film is formed on the electrode surface [311,314,315].

The Li tin systems were also studied extensively. Five voltage plateaus are observed during galvanostatic lithiation of tin. The largest one, from $x = 0.8 \rightarrow$ 2 in Li$_x$Sn, is the most favorable from the kinetic point of view (relatively high diffusion coefficient which allows high rates of lithiation without formation of Li metallic bulk Li phase [309]). Besenhard et al. [316] showed that it is possible to obtain tin-based Li alloys which behave as highly reversible Li anodes of extended cycle life by designing a specific phase of a porous microstructure. The host matrix, either pure tin or multiphase Sn/Ag or Sn/Sb alloys, is obtained by means of electrochemical deposition from aqueous solutions. The composition, porosity and particle size of these matrices can be controlled by the bath composition and the plating condition. More than 200 charge-discharge cycles at efficiencies close to 100% could be obtained with lithiated Sn/Ag and Sn/Sb alloys ($\approx 3:1$ molar ratio) up to a Li content of 1.7 moles of Li per 1 mole of Sn at practical current densities (0.25 mA/cm$^2$) in LiClO$_4$/PC solutions [316]. This achievement is very significant in light of the fact that cycling efficiency for Li metal anodes in PC solutions is very poor. The high cycling efficiency obtained for the Li-Sn-Ag or Li-Sn-Sb anodes should be attributed to the fact that lithium is inserted reversibly into a rigid matrix, and the repeated insertion-depletion of Li is accompanied by very little volume change in the matrix. Hence, the electrodes remain protected by stable surface films.

The last point in this section relates to in situ formation of composites containing Li-Sn alloys as a reversible anode material from tin oxides [317]. The Fuji Photo Film Co. recently announced the development of Li batteries based on tin oxide anodes of very high capacities [318]. Further work by Dahn et al. [319], Idota, et al. [320], and analysis by Huggins [308] led to the conclusion that during initial charging of $SnO_2$ (polarization to low potentials in a Li salt solution), $SnO_2$ is reduced to $Li_2O$ and $Li_xSn$ alloys. Hence, a composite anode is formed which contains a rigid matrix of $Li_2O$ and $Li_xSn$ phases. It appears that a similar situation may occur with carbonaceous anodes containing silica, as prepared by Dahn and Wilson [321]. These materials could be lithiated at high capacity, a great part of which is reversible; therefore, they are promising anodes for Li ion batteries. The initial lithiation of these materials probably forms $Li_xSi$ alloys encapsulated in the carbon matrix, which can insert-deplete lithium reversibly at high capacity.

In conclusion, this new class of composites, in which microparticles of Li alloys are encapsulated in rigid matrices such as oxides or carbon (which do not interfere with the reversible insertion of Li and alloying), seems to be highly promising as anodes for rechargeable, high energy density Li batteries.

## H.  Li-Carbon Electrodes

### 1.  Classification of Li-Carbon Electrodes

A wide variety of carbonaceous materials can intercalate or insert lithium reversibly and thus may be candidates for anodes for lithium ion batteries. In recent years, many types of carbons have been tested as alternative anodes for rechargeable lithium batteries, part of which have found use as anodes in practical, commercial lithium ion batteries. The most straightforward way of classifying these electrodes is according to the type of the carbon, which determines their capacity and basic electrochemical behavior. The major types of carbons tested in recent years as anode materials for Li ion batteries are listed below:

1.  *Graphite*. Several types of graphite were investigated, including HOPG (highly oriented pyrolytic graphite), synthetic graphite flakes, and natural graphite (usually appearing as flakes). HOPG does not have any practical importance, but it is highly significant as a model compound for basic studies of the lithiation process into graphite [322–323]. Synthetic and natural graphite flakes (average size 5–50 μm) may be used as practical anode materials [324–330]. They intercalate reversibly with lithium up to a stoichiometry of $LiC_6$ (theoretical capacity of 372 mAh/g [331]). These materials are characterized by a highly ordered, layered structure of graphene planes between which lithium is intercalated.

2.  *Graphite fibers*. Carbonization of several types of polymers (heat treatment at temperatures >2500°C) produces fibers of graphitic structure that can also intercalate with lithium up to a stoichiometry close to $LiC_6$ [332–337].

3. *Disordered soft carbons.* These carbons are characterized by disordered structure. They can be graphitized when heated to temperatures >2000°C. Typical examples are the petroleum cokes [338–340], which may insert lithium reversibly at a capacity lower than that of graphite. Hence, the intercalation stoichiometry is $Li_xC_6$ ($x < 1$).

4. *Disordered hard carbons.* There is a wide variety of disordered carbons produced by carbonization of organic precursors such as food products (coffee beans [341], sugar [342]), cotton [343], epoxy resins [344–345], phenolic resins [346–349], various types of polymers (e.g., PVC, PVDF, PPS, ENR) [350], poly(*p*-phenylene) [351], and others. The mechanisms of lithium insertion into these carbons is not intercalation but insertion or adsorption. These carbons are produced by heat treatment of the organic precursors at temperatures around 1000°C and below, and may thus contain C—H bonds [352]. Part of the lithium insertion into them was found to be connected with this hydrogen [353]. These disordered carbons may insert Li at higher capacities than graphite does, up to a stoichiometry of $LiC_3$ [354] and $LiC_2$ [355].

Graphitic carbons intercalate lithium at potentials between 0.3–0. V versus $Li/Li^+$ [324–330], whereas disordered carbons insert lithium at a wider potential range, 1.5–0. V versus $Li/Li^+$ [338–351]. Lithium intercalation in graphite is a staged process, and thus the voltage profile (in galvanostatic processes) is characterized by plateaus. In contrast, the voltage profile of lithium insertion into disordered carbons is a sloping one.

As discussed in the next section, lithiated carbon electrodes are covered with surface films that influence and, in some cases, determine their electrochemical behavior (in terms of stability and reversibility). They are formed during the first intercalation process of the pristine materials, and their formation involves an irreversible consumption of charge that depends on the surface area of the carbons. This irreversible loss of capacity during the first intercalation/deintercalation cycle is common to all carbonaceous materials. However, several hard, disordered carbons exhibit additional irreversibility during the first cycle, in addition to that related to surface reactions with solution species and film formation. This additional irreversibility relates to consumption of lithium at sites of the disordered carbon, from which it cannot be electrochemically removed [346–351].

The various types of carbons described differ from each other in their stability upon cycling. The behavior of graphitic carbons is highly sensitive to the solution composition. The stability of lithiated graphite anodes depends strongly on the nature of the surface films that cover them [87]. The more turbostratic disorder in the graphites (e.g., as exists in graphitic carbon fibers), the higher is its stability upon prolonged intercalation/deintercalation cycling. Disordered soft carbons such as coke exhibit much higher reversibility and prolonged cycle life than graphitic carbons. In addition, the reversibility of lithiated coke anodes is much less dependent on the solution composition than is that of lithiated graphite.

## 2. Surface Chemistry of Lithiated Carbons and Passivation Phenomena

As discussed in Chapter 4, which deals with the electrochemistry of nonactive metal electrodes in nonaqueous Li salt solutions, most of the solvents, salt anions, and contaminants relevant to Li batteries are reduced in the presence of Li ions at potentials much higher than 0. V versus Li/Li$^+$, to insoluble Li organic and/or inorganic salts. These precipitate as surface films on the electrodes. As also discussed in a previous chapter, the basic chemical composition of the surface films formed on nonactive electrodes, when polarized to potentials between 2 and 0. V (Li/Li$^+$) in the commonly used Li salt solutions, is very similar to that of the surface films formed on lithium in the same solutions [86,87]. It should be noted that the above similarity between lithium and nonactive metal electrodes polarized to low potentials is mostly in the types of surface species formed. The relative amounts of the various compounds in the surface films, the physical structure of the surface layers, and their 3D chemical profile may differ considerably when comparing Li and nonactive metal electrodes. The surface chemistry of the carbons polarized to low potentials is similar to that developed on noble metal electrodes at low potentials [87]. The onset of reduction of solution species such as alkyl carbonate solvents, $H_2O$, $CO_2$, $O_2$, $AsF_6^-$, $PF_6^-$, and $N(SO_2CF_3)_2^-$ is between 2 and 1.5 V (Li/Li$^+$). Massive reduction of alkyl carbonate solvents and the above anions occurs at potentials between 1.5 and 1 V (Li/Li$^+$) and appears as plateaus in the chronopotentiograms of carbon electrodes polarized galvanostatically to low potentials. These reduction processes precipitate surface films whose composition has been investigated extensively by several groups. We rigorously studied the surface chemistry of graphite electrodes using FTIR spectroscopy, XPS, and EDAX in the following systems:

1. Alkyl carbonates: PC [356–357], EC [51,357,358,359], DMC [51,358], DEC [359], EMC [360], and PMC [361]
2. Ethers: DME [356], THF [191,356], 2Me-THF [356], and 1,3-dioxolane [357]
3. *Salts*: LiAsF$_6$ [356–561], LiClO$_4$ [356], LiBF$_4$ [51,59], LiPF$_6$ [51,359–361], and LiN(SO$_2$CF$_3$)$_2$ [51]
4. *Additives*: CO$_2$ [356–359], H$_2$O[358], and SO$_2$ [362]

Table 3, which presents the various surface compounds formed on Li electrodes in the various solutions, together with reaction schemes 1–10, describes well the basic surface chemistry developed on the carbons. Similar results concerning the surface chemistry developed on carbons in alkyl carbonate mixtures have also been obtained by others [363–365]. Hence, carbon electrodes are also solid electrolyte interface (SEI) electrodes, similar to lithium; i.e., the overall insertion process of Li into the carbons requires the necessary step of Li ion migration through the surface layers. This is clearly reflected by impedance measurements of lithiated carbon electrodes. Impedance spectra of Li-C anodes always include well-separated time constants related to $Li^+$ migration through surface films (appearing as a depressed, high frequency semicircle in the Nyquist plots)[151,191,359]. The performance of Li-C electrodes as anodes in rechargeable battery systems largely depends on the structure of these surface species, their passivating properties, and resistivity. For instance, it appears that certain reactive solution components such as EC, $CO_2$, and $SO_2$ all react on the carbon surface, at potentials higher than that of usual Li-C intercalation potentials (1.5–1 V versus $Li/Li^+$), to form highly passivating surface films of low resistivity.

There are two undesirable processes that these surface layers should prevent:

1.  Further reactions between the lithiated carbon and solution species (irreversible charge consumption)
2.  Percolation of solution species solvating the Li ions into the carbon structure, which may lead to its partial exfoliation [366]

## 3. Graphite Electrodes

Due to its layered structure, graphite is the carbonaceous material most sensitive to the detrimental effects of interactions with solution species. Co-intercalation of solution species such as solvent molecules together with the Li ions may lead to exfoliation and destruction of their structure, as indeed happens in solutions of PC, γ-BL, THF and other solvents [356–358]. Lithiated graphite electrodes behave reversibly only in solutions in which highly passivating, stable surface layers are formed on the pristine material before any Li intercalation takes place.

Figure 25 illustrates three cases of lithiated-graphite electrodes in solutions [87]. (Typical chronopotentiograms are also presented.)

1.  Passivating surface films are formed, and thus Li intercalation is highly reversible. The initial irreversible capacity loss due to the surface reactions is about 10–30% of the reversible capacity. This behavior is typical of EC-based solutions and solutions containing $CO_2$ (e.g., $MF/CO_2/LiAsF_6$).

2.  The surface films formed are not sufficiently passivating, and their formation involves partial exfoliation of the graphite. Thick and resistive surface films are formed. The irreversible capacity loss is pronounced, and the electrode cannot reach all the Li-graphite intercalation stages (MF, DMC, and ether solutions).

3.  The electrode is deactivated during massive reduction of solution species when polarized to low potentials (as happens with PC solutions) [356–357]. The detrimental effect of co-intercalation of solvent molecules in this case is proven by the fact that, in PC solutions containing crown ethers (which solvate

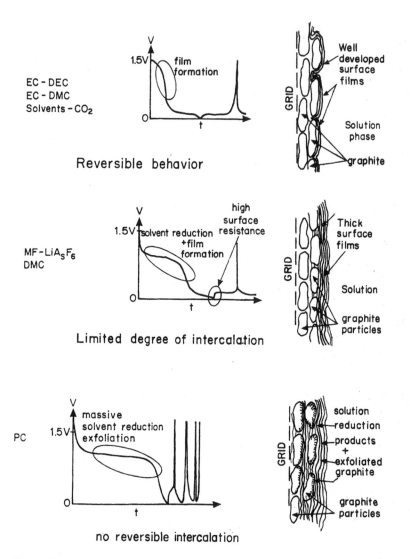

**Figure 25** Typical chronoamperograms and schematic view of the structure of lithiated graphite electrodes in three classes of electrolyte solutions (as indicated). 1. Reversible behavior; 2. partially reversible behavior, low capacity ($x < 1$ in $Li_xC_6$; 3. irreversible behavior, the electrode is deactivated and partially exfoliated before reaching intercalation stages. Note that in reality the graphite particles are usually flakes, the electrode's structure is porous and the surface films are also formed inside the electrode among the particles, and thus have a porous structure [87]. (With copyrights from Elsevier Science Ltd., 1998.)

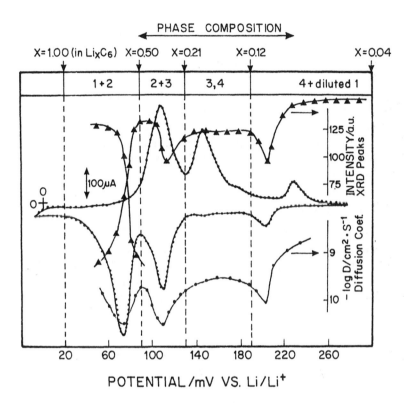

**Figure 26** A comparison of the chemical diffusion coefficient of Li into graphite (calculated from PITT), the intensity of the major XRD peaks (e.g., 002, 004) during intercalation (vs. $E$), and a completed slow scan rate cycling voltammogram of a thin (10 μm) composite graphite electrode (KS-6 from Lonza) in EC-DMC/LiAsF$_6$ solution. Note that as the electrode is thinner and the particles are more oriented (with their basal planes parallel to the current collector), the scan rate is slower and the CV peaks are sharper and better resolved. The various phases (intercalation stages) are indicated [87]. (With copyrights from Elsevier Science Ltd., 1998.)

the Li ions instead of the solvent molecules), lithiated graphite electrodes behave reversibly [367]. The basic electroanalytical behavior of reversible lithiated graphite electrodes is demonstrated in Figure 26. The figure shows a typical, slow-scan-rate cyclic voltammogram (SSCV), the variations of log $D$ (the chemical diffusion coefficient of Li ion into the graphite), with the potential calculated from PITT measurements [368], and the intensity of characteristic XRD peaks of the various Li-C intercalation stages as a function of the potential measured during in situ XRD experiments [369]. The SSCV clearly shows phase transitions during Li intercalation into graphite (marked in Figure 26), appearing as redox pairs of peaks, corresponding to the plateaus that characterize the chronopotentiograms of these electrodes (upon galvanostatic cycling). The log $D$ versus $E$ function is nonmonotonous with minima at the SSCV peak potentials. The unique electroanalytical behavior of these electrodes is dealt with in Refs. 368 and 370.

There is a possibility of modifying graphite by partial, controlled oxidation, which forms a porous structure, and thus the capacity can be increased [371]. In addition, partial oxidation may form desirable surface species (e.g., C—OH, COOH), which, upon reduction, may behave as a passivating, protective solid electrolyte interphase [372].

## 4. Disordered Carbons

Figure 27 shows a typical chronopotentiogram of the first lithiation-delithiation cycle of a petroleum coke electrode [357]. It demonstrates the irreversible capacity loss due to the carbon's surface reactions (plateau around 1 V versus Li/Li$^+$), the sloping potential profile, and the fact that the maximal reversible capacity is less than that of graphite. However, its structural disorder makes this electrode

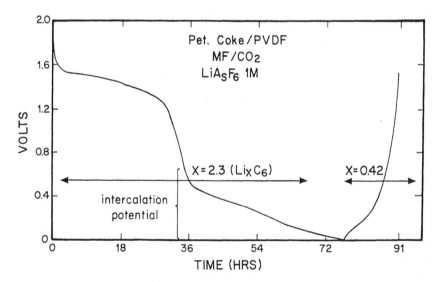

**Figure 27** Typical chronopotentiometric profile of a petroleum coke (4% PVDF) electrode. The first charge-discharge cycle of a cell with the petroleum coke electrode and Li counterelectrode loaded with MF-LiAsF$_6$ 1 $M$ (6 atm, CO$_2$), galvanostatic operation, C/33 h, 0.5 mA/cm$^2$ [357]. (With copyrights from The Electrochemical Society Inc., 1998.)

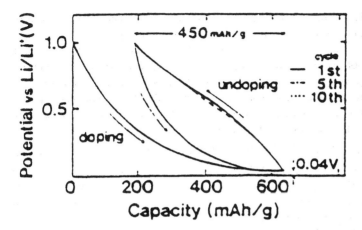

**Figure 28** Typical chronopotentiometric profile of a disordered carbon charged-discharged in Li salt-alkyl carbonate solution. The carbon was a Li doped polyacenic semiconductor (PAS) obtained by heat treatment of phenolformaldehyde resin [348]. Note the high capacity, the sloping potential profile and the hysteresis between the charge and discharge routes. (With copyrights from Elsevier Science Ltd., 1998.)

much less sensitive than graphite to detrimental processes due to interactions with solution species. For instance, these electrodes could be cycled in PC solutions (although at lower capacity than in EC-based solutions). Consequently, the cycle life of these electrodes (Li-petroleum coke) is much more prolonged than that of lithiated graphite electrodes. Figure 28 shows chronopotentiograms of lithiated, hard, disordered carbon of high capacity [348]. Several key features of the behavior of these electrodes should be noted.

1.  The potential profile includes a sloping part (1.5–0.5 V versus Li/Li$^+$) and a flat part at potentials close to 0. V versus Li/Li$^+$.
2.  The reversible capacity is higher than that of graphite. A pronounced part of the reversible capacity relates to the low potentials. This is an important anode characteristic for Li ion batteries.
3.  The first Li insertion capacity is very pronounced, but part of it appears irreversible. Most of this irreversibility is not connected with surface film formation but with irreversible insertion of lithium. The origin of this irreversibility is not yet understood.
4.  There is a pronounced hysteresis between the voltage profiles of the insertion and deinsertion processes. This hysteresis is unique to these hard-disordered carbons and is not seen with graphitic materials and soft disordered carbons.

All of the above features are connected with the fact that the interaction of these carbons with lithium does not involve intercalation but adsorption-type processes.

We should acknowledge the intensive work that is currently being devoted to the understanding of the Li insertion mechanisms into these carbonaceous materials [373–380]. Especially important are studies that utilize neutron diffraction techniques [381]. These relatively new materials are very promising as high capacity, stable, low potential anodes for a new generation of high energy density Li ion batteries. Table 9 summarizes details of several types of high-capacity-disordered carbons and their precursors.

## I. Remarks on Li Electrodes in Polymeric Electrolyte Systems

### 1. Introduction

While most of this chapter is devoted to liquid electrolyte solutions, an intensive amount of work is currently devoted to R & D of polymeric electrolyte systems for Li batteries. Of particular importance are polymeric electrolyte systems that maintain high room temperature conductivity (the order of $10^{-3}$ S/cm, which is comparable to the conductivity of Li salt solutions in many polar aprotic systems) [382].

Rigorous classification of these systems and a description of their physical, chemical, and electrical properties are beyond the scope of this chapter. However, due to the importance of these systems in modern electrochemistry, in general, and Li batteries, in particular, several important points regarding Li electrodes in these systems are mentioned below.

### 2. Brief Classification

The commonly used polymeric electrolyte systems include the following types of polymers:

1. Linear polymer hosts, e.g., polyethylene oxide (PEO).
2. Comb-branched polymers, e.g., poly[2-(2-methoxyethoxy)ethylglycidyl ether] [383], metacrylate-based polymers [384].
3. Polymers with inorganic spines, such as siloxanes [—$R_2SiO$—]$_n$ [385] and polyphosphazenes —(—P$=$N$=$)$_n$ [386]. Table 10 presents some examples of the above types of polymers and summarizes some of their properties.
4. Polymers containing plasticizing salts. It appears that when the Li salts used are of large anions, e.g., $LiN(SO_2CF_3)_2$, the conductivity of poly-

**Table 9** Summary of Details on Several Types of Disordered, High Capacity Carbonaceous Materials Which Are Promising as Anodes for Novel Li Ion Batteries

| Precursor | Preparation temperature (°C) | Discharge capacity (mAh/gr) | Irreversible capacity (mAh/gr) | Potential range vs. Li/Li$^+$ (V) | Ref. |
|---|---|---|---|---|---|
| Coffee beans | 1000–1400 | 410 | 80 | 2–0 | 341 |
| Green tea | 1000–1400 | 380 | 95 | 2–0 | 341 |
| Sugar cane | 1000–1400 | 400 | 80 | 2–0 | 341 |
| Sugar | 600–800 | 650 | 170 | 1.5–0 | 342 |
| Cotton | 700–1000 | 600 | 120–140 | 2–0 | 343 |
| Epoxy resins | 900–2200 | 590 | — | 1.5–0 | 344 |
| Epoxy resins[a] | 1000 | >600 | 200–800 | 1.5–0 | 345 |
| Phenolic resins | 700 | 630 | 260 | 1.5–0 | 346 |
| Phenolic resins | 1000[b] | 560 | 200 | 1.5–0 | 346 |
| Phenol-formaldehyde resin | 600–700 | 530 | 250 | 1–0 | 348 |
| Phenol-formaldehyde resin | 700 | 438 | 250 | 1.2–0 | 349 |
| PVC[c] | 550–900[d] | 940–400 | 340–70 | 2–0 | 350 |
| PVDF[c] | 550–900[d] | 550–380 | 680–650 | 2–0 | 350 |
| PPS[c] | 800–900[d] | 670–500 | 310–250 | 2–0 | 350 |
| ENR[c] | 700–1000[d] | 650–570[e] | 680–150 | 2–0 | 350 |
| Poly(p-phenylene) | 700 | 650–450[e] | 500 | 3–0 | 351 |
| Pyrene/benzene[f] | 500–700 | 975–800 | >600 | 3–0 | 352 |
| PTCDA[g] | 550 | 660 | 400 | 2–0 | 354 |

[a] The carbon was further partially oxidized. This increased both the reversible and irreversible capacities.
[b] Pronounced temperature effect on both reversible and irreversible capacities.
[c] PVC = polyvinyl chloride, PVDF = polyvinylidene fluoride, PPS = polyphenylene sulfide, ENR = epoxy novolac resin.
[d] The lower the preparation temperatures, the higher the reversible and irreversible capacities (within the limits specified in the table).
[e] The reversible capacity depends on the cycle number, cutoff voltage, and other experimental parameters. Maximal reversible capacity around 650 mAh/gr could be obtained.
[f] The organic precursors were heat-treated while loaded on calcified clay (based on natural montmorillonite that was further modified).
[g] PTCDA = 3,4,9,10-perylenetetra-3,4,9,10-carboxylic dianhydride.

**Table 10** Properties of Some Polymers and Polymer Electrolytes

| Polymer | Repeat unit | Glass-transition temperature, $T_g$ (°C) | Melting point, $T_m$ (°C) | Typical polymer electrolyte | Conductivity (S/cm at 25°C) |
|---|---|---|---|---|---|
| Poly(ethylene oxide) | —CH₂CH₂O— | −60 | 64 | (PEO)₈-LiClO₄ | ~10⁻⁸ |
| Poly(oxymethylene-oligo-oxymethylene) (POO) | —(CH₂O)—(CH₂CH₂O—) | −66 | 13 | (PEO)₂₅-LiCF₃SO₃ | 3 · 10⁻⁵ |
| Poly(propylene oxide) (PPO) | —(CH₃)CH₂CH₂O— | −60 | amorphous | (PPO)₈-LiClO₄ | 10⁻⁸ |
| MEEP* | OCH₂CH₂OCH₂CH₂OCH₃ ⎯P=N⎯ OCH₂CH₂OCH₂CH₂OCH₃ | −83 | amorphous | (MEEP)₄-LiBF₄ | 2 × 10⁻⁵ |

* Poly[bis-(2-(2-methoxyethoxy)ethoxy)phosphazene].
*Source:* From Ref. 382.

meric electrolyte systems based on types 1–3 improves considerably [387].

5. Polymers containing plasticizing solvents. Addition of solvents such as EC or PC to the polymeric systems of the types shown in Table 10 improves their conductivity considerably. Of special importance are conducting polymeric membranes based on polyacrylonitrile (PAN) containing EC, PC, and Li salts such as $LiClO_4$, $LiSO_3CF_3$ [382].

6. Composite materials. It has been shown that composite electrolyte systems composed of polymers such as PEO, Li salts such as LiI, and dispersed micronic and submicronic alumina ($Al_2O_3$) particles provide $Li^+$ ion conducting systems of improved properties (e.g., higher conductivity, high $Li^+$ transference number, higher stability) [388–389].

## 3. Behavior of Li Electrodes in Polymeric Electrolyte Systems

The motivation for developing polymeric electrolyte systems for lithium batteries arises from the following reasons:

1. It is assumed that the polymers are much less reactive with lithium than are the commonly used solvents. Hence, reducing the reactivity of the anode toward the components of the electrolyte systems may contribute to safety and cycle life.

2. It is assumed that a solid state electrolyte system should suppress dendrite formation upon Li deposition. This is also supposed to contribute to enhanced safety and prolonged cycle life.

3. In the event of abuse cases, such as short circuit, exposure to air, or overcharge or overdischarge, it is assumed that solid state Li batteries based on polymeric electrolyte systems should be much less dangerous than liquid-based batteries. It is expected that the former battery systems would be much better protected in the above cases from thermal runaway than are the liquid-based batteries.

The electrochemical behavior of lithium electrodes in a variety of polymeric electrolyte systems was studied extensively by a number of groups, including Scrosati et al. [390–392], Panero et al. [393], Abraham et al. [394–396], Osaka et al. [397–398], Watanabe et al. [399–401], Peled et al. [402]. It is clear that there are surface reactions between the lithium and all of the polymeric systems mentioned above. It has already been clearly shown that the ether linkage is attacked by lithium, resulting in the formation of Li alkoxy species [149]. Hence, it is expected the PEO-based polymers also react with Li surfaces. Spectroelectrochemical studies of the Li-PEO system by Scherson et al. [177] provide some evidence for this possibility. Besides the polymers, the polymeric electrolyte systems contain salts with anions such as $AsF_6^-$, $SO_3CF_3^-$, $N(SO_2CF_3)_2^-$, etc., which are obviously reactive with lithium. They also contain water contamination. It is thereby expected that Li electrodes in all commonly used polymeric electrolyte solutions develop surface films, and thus should also be considered as SEI electrodes [32], similar to the cases of lithium in the liquid electrolyte solutions discussed previously in this chapter.

Impedance spectroscopic studies of Li electrodes in polymeric electrolyte systems by many groups [388–402] have provided clear evidence that Li electrodes in polymeric electrolyte systems are covered by surface films. These grow during storage and control the electrochemical behavior of these systems. This is also true for the least reactive systems based on PEO [403].

As mentioned in Section V.H.2, highly important room temperature polymeric systems are based on polyacrylonitrile (PAN) plasticized with PC and EC. These systems are obviously highly reactive with lithium. It is clear that Li, in contact with these membranes, develops surface chemistry which is dominated by the reduction of EC and PC to $ROCO_2Li$ (Table 3 and Schemes 1 and 3, Section V.C, and related discussion).

Hence, it can be concluded that the basic surface chemistry and interfacial properties of Li electrodes in most of the commonly used polymeric electrolyte systems is very similar to that of lithium in liquid electrolyte solutions. Nevertheless, the use of polymers instead of liquid solvents is advantageous because the three assumptions mentioned at the beginning of this section are found to be correct. The use of solid membranes instead of liquid electrolyte solutions in Li batteries increases Li cycling efficiency because it suppresses dendrite formation [404]. It has already been shown that stack pressure in practical Li batteries suppresses dendrite formation and improves their cycle life [106,107,206]. The use of a polymeric electrolyte membrane thus has a similar positive effect on Li morphology and cycling efficiency. On the other hand, there might be contact problems between the solid electrodes and the polymeric membrane, which may leave voids in the electrode-membrane interface. However, these problems can be solved by appropriate engineering and should only be considered as technical difficulties. The R & D in this area is presently continuing in the following major directions:

1. Increasing low temperature conductivity
2. Increasing the transference number of the Li ion in the polymers
3. Increased compatibility of these systems with composite electrodes, thus making them suitable for use in Li ion batteries

## VI. MAGNESIUM ELECTRODES

## A. Introduction

There are two major interests in nonaqueous electrochemistry of magnesium:

1. Electroplating of magnesium for purposes of surface finishing, cathodic protection against corrosion, etc.

2. The use of magnesium as an anode in high energy density nonaqueous batteries

Indeed, over the years there have been continuous efforts in these directions, as well as papers describing attempts to electroplate magnesium [405–409]. There have been reports on the study of corrosion of magnesium in non-aqueous systems [410] and the behavior of the $Mg^{2+}$/MgHg couple [410–414]. The standard potential of this system in dimethylformamide was found to be $-2.47$ V versus $Ag/Ag^+$. The behavior of Mg electrodes in thionyl chloride was investigated by Peled and Straze [415]. Pletcher and Genders [416] studied the basic electrochemical behavior of the $Mg^{2+}$/Mg couple in THF and PC, using microelectrodes. A comprehensive report on the feasibility of rechargeable non-aqueous magnesium batteries was published by Gregory et al [417]. Reviews on the anodic behavior of magnesium in nonaqueous electrolytic solutions [418], plating of magnesium from organic solvents [419], and the reversibility of Mg electrodes in Grignard solutions [420] should also be acknowledged.

The following general conclusions can be drawn from the above reports [405–420]:

1. In general, the solubility of Mg salts such as $Mg(ClO_4)_2$, $Mg(SO_3CF_3)_2$, etc., in commonly used solvents (ethers, alkyl carbonates, and esters) is lower than that of Li salts.

2. Magnesium electrodes are passivated in most of the commonly used polar aprotic solvents by surface films. The origin of these surface films may be from reactions of atmospheric components ($O_2$, $H_2O$) on the active metal surface before introduction into the solution or from a result of the reduction of solution species.

3. In contrast to the case of lithium electrodes, where the surface films deposited on them are usually good $Li^+$ ion conductors, the $Mg^{2+}$ ion conductivity of the surface films formed on magnesium electrodes is very poor.

4. Consequently, Mg dissolution in a wide variety of Mg salt solutions in commonly used polar aprotic solvents requires high overvoltage. It probably proceeds via a breakdown of the surface films that allows exposure of fresh magnesium to the solution.

5. Magnesium deposition is even more difficult and, in fact, impossible in most of the commonly used polar aprotic solvents containing "normal" Mg salts [e.g., $Mg(ClO_4)_2$, $Mg(SO_3CF_3)_2$, etc.].

6. It is possible to obtain highly reversible behavior of the magnesium electrodes in ethereal solutions of Grignard reagents (e.g., RMgCl, RMgBr, R = alkyl, aryl) [416,417,420]. Mg can be deposited smoothly and uniformly at high plating efficiency in these solutions.

7. While the surface films formed on Mg electrodes are usually imperme-

able to $Mg^{2+}$ ions, they are permeable to $Li^+$ ions. Hence, Li can be plated on Mg substrates from Li salt solutions.

8. In contrast to the case of lithium, where extensive efforts have been devoted to the study of interfacial phenomena, surface film composition, etc., very little is yet known about the composition and structure of the surface films covering magnesium electrodes in nonaqueous media. In addition, the mechanisms of Mg dissolution and plating in the Grignard solutions are unclear.

There are reports on the reversible behavior of Mg electrodes in high temperature molten salts such as $MgCl_2$ (740°C) [421], $MgCl_2$-NaCl (700–800°C) [422], and $MgCl_2$-$MgF_2$ (700°C) [423]. It was impossible to obtain solutions of $MgCl_2$, at reasonable concentrations in room temperature, or molten salts, such as derivatives of imidazolium chloride (EMIC) with $AlCl_3$, in the same way that Li salt solutions are obtained with these systems [227–229], due to the low solubility of the $MgCl_2$ in EMIC-$AlCl_3$ at room temperature [424]. However, it was possible to obtain $MgCl_2$ solution in EMIC-$AlCl_3$, with electrochemical activity of $Mg^{2+}$ ions at elevated temperature (80°C) [425].

## B. Remarks on Mg Dissolution and Deposition Processes in Polar Aprotic Systems

From the extensive work of Lossius and Emmenegger [419], it appears that $MgCl_2$ and $Mg(CF_3SO_3)_2$ in dimethylformamide (DMF), dimethylacetamide (DMA), diethylacetamide, $\gamma$-butyrolactone ($\gamma$-BL), or binary mixtures of these have acceptable solubilities (0.1 $M$) and conductivities (>1 mS/cm). Another salt, Mg-acetylacetonate, which is very soluble, showed low conductivity. A 70:30 vol.% mixture of DMF and acetonitrile (AN) gave the highest solubility observed for $Mg(CF_3SO_3)_2$—1.0 $M$. A 50:50 DMA:$\gamma$-BL mixture gave the highest solubility for $MgCl_2$—1.5 $M$. A particularly strong effect of mixed solvent was observed for 1,2-dimethoxyethane (DME) and AN, where the solubility of $Mg(CF_3SO_3)_2$ or $MgCl_2$ increased 50–100 times compared with each solvent alone.

It was further discovered by these authors [419] that Mg dissolves with low overpotential and high dissolution efficiency in many systems with $Mg(CF_3SO_3)_2$ solute. In solutions with DMF and DMA, a $Mg^{2+}$/Mg reduction process was identified at potentials $< -3.0$ V (versus ferrocene/ferricenic picrate, FC). There was a 1.5 V gap between the dissolution and the reduction onset. In this range, the Mg surface appeared to be stable with no chemical reactions on its surface.

In none of the systems mentioned did the $Mg^{2+}$/Mg reduction lead to a measurable Mg deposition. It has been suggested that transient $Mg^0$ is too reactive for these solvents and undergoes a reaction with the solvent instead of forming a metal layer on the electrode. The addition of supporting electrolytes increased the passivation of the Mg dissolution (in the order $NBu_4PF_6 < LiCF_3SO_3 < NBu_4CF_3SO_3 < NBu_4Cl = LiCl$). In addition, the $NBu_4^+$ ion is reduced at potential close to that of the $Mg^{2+}/Mg$ couple. The presence of $Cl^-$ was found to be destructive to Mg dissolution and $Mg^{2+}/Mg$ reduction.

From the work of Lossius and Emmenegger [419], it is clear that a condition for reversible behavior of Mg electrodes is the lack of surface films on them, because the passivating layers usually formed on Mg electrodes are not $Mg^{2+}$ conductive. Thus, electrodeposition does not occur in most of the commonly used solvents, such as alkyl carbonates, ethers and esters, when only simple $Mg^{2+}$ species are present in solution [416].

The electrochemical behavior of an Mg electrode in thionyl chloride/$Mg(AlCl_4)_2$ solutions was investigated extensively by Meitav and Peled [426]. The Mg electrode in this electrolyte system is covered by $MgCl_2$, which forms a bilayered surface film: a compact one close to the metal and a porous one at the film-solution interface. This surface film determines the electrochemical behavior of these systems and can only conduct $Cl^-$ ions, and thus the mobility of $Mg^{2+}$ through it is practically zero. Thus, Mg deposition does not occur in this system, and Mg dissolution at a reasonable rate occurs via a breakdown and repair mechanism. Since the active metal is thermodynamically unstable in thionyl chloride when a fresh metal is exposed to solution, it reacts readily with the solvent to form this film.

The Mg electrode was found to be highly reversible in solutions containing Grignard reagents. Most of the commonly used polar aprotic solvents, such as alkyl carbonates, esters, and acetonitrile, are too electrophilic for Grignard reagents (which are strong nucleophiles) and thus react with them readily. The only solvents in which stable Grignard solutions can be prepared are ethers. Highly reversible behavior of Mg electrodes, which includes deposition of metallic magnesium at high coulombic efficiency in THF containing Grignard salts, was reported by a number of authors [405–408,416,417,420]. Typical Grignard reagents tested in this respect were RMgCl (with R = methyl, ethyl or butyl) [417] and $CH_3CH_2MgBr$ [416,420].

It was discovered that the addition of Lewis acids such as organoboron compounds to RMgX (X = halide) solutions in ethers improves the quality of Mg deposition in these systems [427–428]. A similar improvement in the quality of Mg deposition was observed by Gregory et al. when $AlCl_3$ was added to THF solutions of RMgCl [417]. Another interesting discovery by Gregory et al. [417] was the possibility of obtaining reversible behavior of Mg electrodes in THF solutions containing magnesium borate salts of the $Mg(BR_4)_2$ type, where R is butyl or phenyl. It was explained by these authors [417] that the reversibility of the Mg electrodes in these borate solutions should be attributed to the fact that the charge characteristics of the magnesium in these borate salts is very similar to those of Mg in the Grignard reagents (i.e., only a partial ionic nature).

The reactions that organomagnesium halides undergo to permit Mg dissolution and plating are complex and far from being fully understood. However, from what is known to date, the main reactions of importance can be summarized as follows: alkyl magnesium halides exist in ethers as equilibrium mixtures of various compounds, including dialkyl magnesium and magnesium halides. The most important equilibrium is the Schlenk equilibrium [429]:

$$2RMgX \rightleftharpoons R_2Mg + MgX_2$$
$$K_{eq} = 0.18 \quad (25°C, R = ethyl, X = Cl) \text{ [430]} \tag{19}$$

The primary ionization equilibrium is the autoionization of RMgX:

$$2RMgX \rightarrow RMg^+ + RMgX_2^- \tag{20}$$

Further ionization of RMgX can be promoted by the addition of Lewis acids such as $AlCl_3$:

$$RMgX + AlX_3 \rightarrow RMg^+ + AlX_4^- \tag{21}$$

The primary species that react at the cathode to deposit Mg metal are probably $RMg^+$. The Mg anodes can dissolve by a number of reactions:

$$Mg^0 + R_2Mg \rightarrow 2RMg^+ + 2e^- \tag{22}$$

$$Mg^0 + 2RMgX_2^- \rightarrow 2MgX_2 + R_2Mg + 2e^- \tag{23}$$

$$Mg^0 + 2AlX_4^- \rightarrow Mg(AlX_4)_2 + 2e^- \tag{24}$$

The magnesium aluminum halide tends to dissociate in ethereal solvents to form magnesium and aluminum halides:

$$Mg(AlX_4)_2 \rightarrow MgX_2 + 2AlX_3 \tag{25}$$

In conclusion, there are still many unanswered questions regarding Mg deposition-dissolution processes in these systems that call for further work in this area. It is not at all clear whether Mg deposition in Grignard/ether systems occurs under surface film free conditions or whether there are mechanisms of ion transport (e.g., $RMg^+$ or $Mg^{2+}$) through surface films on Mg electrodes. It is not clear whether the native surface films on Mg electrodes dissolve in the Grignard solutions, allowing the interaction of a fresh Mg surface with the solution species (which may explain the high reversibility of the Mg electrode), or whether, even in this case, Mg dissolution occurs via the breakdown and repair of surface films (due to reaction of the active metal with solution species).

## C.  Remarks on Attempts to Develop Secondary Nonaqueous Mg Batteries

As mentioned in Section VI.A an important driving force in the study of the behavior of Mg electrodes in nonaqueous systems has been the challenge to develop secondary Mg batteries. These batteries may be cheaper, more environmentally friendly, and safer than Li batteries. In addition, it has been assumed that the wide experience gained in nonaqueous electrochemistry, polar aprotic electrolyte solutions, and active metal electrochemistry during the three-decade process of Li battery development could facilitate the R & D of other nonaqueous batteries based on active metal anodes such as magnesium.

As shown in Section VI.B it is possible to obtain reversible behavior of Mg anodes in RMgX and $Mg(BR_4)_2$ solutions in ethers. The electrochemical window of part of the Grignard reagent solutions and the Mg borate-based systems seems to be wide enough to allow the construction of 1.5–2 V nonaqueous Mg batteries (based on ethereal RMgX or $Mg(BR_4)_2$ solutions), provided that appropriate cathodes for these systems exist. During recent years, several groups have investigated the possibility of obtaining reversible intercalation of $Mg^{2+}$ into host materials such as transition metal oxides and sulfides, in a similar manner as in the case of lithium. Chemical interaction of Mg into host materials such as $TiS_2$ and $MO_x$ (M = V, W, Mo and Mn) was studied by Bruce et al [431–432]. The maximal magnesium content in $TiS_2$, $V_6O_{13}$, and $\gamma$-$MnO_2$ was found to be 0.25, 0.48, and 0.32 mole fraction, respectively. Novak et al. investigated the electrochemical intercalation of Mg into $VO_x$ and $MVO_x$ bronzes (M = Li, Na, K, Ca) [425,433–434]. It was found that reversible intercalation of Mg into the $VO_x$ and $NaVO_x$ (e.g., $V_3O_8$) could be obtained in $Mg(ClO_4)_2$ solutions in acetonitrile and PC at RT, and in $MgCl_2/AlCl_3/EMIC$ at 80°C. The Mg insertion and deinsertion potentials are around 2.5 and 3–3.5 V versus $Li/Li^+$ for the above systems. Initial capacities around 150 Ah/g for Mg insertion into $NaV_3O_8$ could be achieved [433]. It was found that the possibility of inserting Mg into these host materials requires the presence of water in their lattice (hydration shell) [433].

So far, it seems that the cycle life obtained for these Mg insertion cathodes is rather low. For instance, Novak et al. [433,434] showed that cathodes such as $Mg_xV_3O_8$ could be cycled reversibly more than 50 times, but the capacity decreases to about 50% of its initial capacity ($\approx$150 mAh/gr) after 20 cycles. However, it is clear that this work presents an important breakthrough, as it shows the feasibility of R & D of insertion cathodes for secondary Mg batteries. Further promising work in this field was demonstrated recently by Sanchez and Pereira-Ramos [435], who showed that Mg can be inserted reversibly, by electrochemical means, into the cation-deficient oxide mix $Mn_{2.15}Co_{0.37}$(up to 0.23 Mg per mole oxide) from $PC/Mg(ClO_4)_2$ solutions at potentials around 2.9 V versus $Li/Li^+$.

Finally, the most promising results on the way to the development of secondary Mg batteries were demonstrated by Gregory et al. [417], who studied the electrochemical insertion of Mg into a number of host materials of the $MO_x$ (M = Co, Mn, Pb, Ru, V, W), $MS_2$ (M = T, V, Zr), and $MB_2$ (M = Mo, Ti, Zr) types. The most promising host material was found to be $Co_3O_4$; i.e., into which Mg could be inserted at potentials around 2.3 V ($Mg/Mg^{2+}$), with a capacity around 220 mhA/gr (corresponding to 0.8 mole fraction of Mg per mole oxide). They were able to show cycling data of a cell composed of an Mg anode, a $Co_3O_4$ cathode, and $THF/DME/MgB(Bu_2Ph_2)_2$ electrolyte solutions [417]. While only limited cycling was demonstrated for this test system, these results further show that development of a secondary Mg nonaqueous battery is feasible.

## VII.  CALCIUM ELECTRODES

### A.  Introduction

Extensive work has been devoted to the study of the electrochemistry of calcium electrodes in inorganic polar aprotic solvents such as $SOCl_2$ (TC), $SO_2$, and $SO_2Cl_2$ in connection with R & D of nonaqueous calcium batteries [436–447]. Following the success in the development and commercialization of high energy density primary $Li-SOCl_2$ and $Li-SO_2$ batteries [207], a natural follow-up step has been an attempt to also develop Ca-TC batteries [436–447]. Replacing Li by calcium in these batteries means a considerable loss in energy density, due to the higher equivalent weight and redox potential of Ca, compared with Li [20 versus 7 gram eq./mole, 0.5–1 versus 0. V $(Li/Li^+)$]. However, the use of Ca instead of Li is expected to improve considerably the safety features of the TC- or $SO_2$-based batteries. In contrast to the many papers on Ca electrodes in inorganic solvents to date, only a few papers have been published on the behavior of calcium electrodes in organic polar aprotic solvents [448–450]. It is interesting that so little work has been devoted to calcium electrochemistry in organic solvents, compared to Mg, although both belong to the same family of metals. This can be explained as follows.

Due to the difference in the atomic structure of Mg and Ca, it is impossible to form Grignard-type reagents with Ca (all the bonds of this metal are ionic in nature). Consequently, it is impossible to regard Ca as a potential anode material in rechargeable batteries, and hence, the interest in Ca/organic electrolyte systems is limited.

### B.  Behavior of Calcium Electrodes in $SOCl_2$ (TC) Solutions

Calcium thionyl chloride systems have been investigated by a number of researchers as a safer substitute for $Li-SO_2$ and $Li-SOCl_2$ batteries. It appears that as a battery system, Ca-TC cells have an open circuit potential around 3.2 V and practical operating voltages varying between 2.5 and 3 V, depending on the current density. A number of comprehensive papers describing these systems have been published by Staniewicz [446], Binder [447], and Peled [436–445].

The commonly used electrolyte in the Ca-TC system is $Ca(AlCl_4)_2$. TC is reduced on the calcium surface as follows [447]:

$$2Ca + 2SOCl_2 \rightarrow 2CaCl_2 + SO_2 + S \tag{26}$$

Even when native CaO films cover the Ca, they may be replaced by $CaCl_2$ surface species according to the following equation:

$$CaO + SOCl_2 \rightarrow CaCl_2 + SO_2 \tag{27}$$

In addition, CaO dissolves in solutions containing an excess of $AlCl_3$, as follows [28]:

$$CaO + 2AlCl_3 + SOCl_2 \rightarrow Ca(AlCl_4)_2 + SO_2 \tag{28}$$

Hence, surface films composed mostly of $CaCl_2$ cover calcium electrodes in T. As is usual for SEI-type electrodes [32], the films grow and reach a steady thickness during storage. It is assumed that the major ionic conductor in these films is $Cl^-$, as they are totally blocking for Ca deposition [32]. Their resistivity was measured by transient techniques and impedance spectroscopy [28], and is around $10^9 \; \Omega \cdot cm$ for a fresh Ca electrode and $10^{10} \; \Omega \cdot cm$ for an aged Ca electrode (500–1000 h in a TC-$Ca(AlCl_4)_2$ solution) [28]. The thickness of the surface films is estimated as several tens of Å [32].

Anodic polarization of Ca electrodes in TC leads to current passage and dissolution of the active metal at high efficiency. As expected for SEI electrodes, a Tafel-like behavior connects the current and the overpotential applied [see Eqs. (5)–(11) in Section V.C.3]. It is assumed that upon anodic polarization the anions ($Cl^-$) migrate from the surface film's solution interface to the surface film's metal interface. Two processes can thus occur:

1. Dissolution of cations left at the outer (solution side) interface of the surface film into the solution
2. Transfer of anions from the solutions into empty lattice defects in the surface films

In the former process, the surface film thickness remains constant, while in the second one the surface layer grows as the process proceeds. These two processes are accompanied by the breakdown and repair of the surface films. Hence, the passivation of the active metal is basically retained during discharge at the expense of some loss of the active metal, due to the surface reactions that repair the film [32].

These processes lead to a complicated morphology of the active metal surface and the formation of multilayer surface films whose outer solution side is porous and nonuniform (as confirmed by SEM studies [446]). Discharge curves of Ca-TC prototype cells, which are very flat [445], prove that the basic mechanism of transport through the surface film during discharge is the first one suggested above; i.e., the $Cl^-$ migrates from the outer side of the surface films to the film-metal interface, $Ca^{2+}$ ions are liberated to the solution, and the surface film retains its steady state thickness.

It appears from the papers cited above that the Ca-TC battery system that contains $Ca(AlCl_4)_2$ as a single electrolyte is inferior to Li-TC batteries due to severe corrosion problems of calcium in this medium [446]. There are reports on successful attempts to improve the performance of Ca-TC systems by modifying the surface films covering the active metal in solutions. The goals are to improve passivation and the transport properties of these surface films. Using strontium and barium salts [i.e., $Sr(AlCl_4)_2$ or $Ba(AlCl_4)_2$] instead of $Ca(AlCl_4)_2$ considerably improves the stability of the calcium electrodes during prolonged storage. In the former case, the surface films are dominated by $SrCl_2$ precipitation, whereas in the second case $BaCl_2$ is only a minor constituent in the surface films (which comprise mostly $CaCl_2$) [445]. Further improvement of these cells involves the addition of $SO_2$ as a cosolvent, and modification of the carbon cathode on which TC is reduced [447].

In summarizing the status of these R & D efforts, it can be said that Ca-TC cells are promising primary battery systems and, in several respects, appear superior to commercial Li-TC or Li-$SO_2$ cells. These include the volumetric energy density, which can be higher for Ca-TC cells than for Li-$SO_2$ cells by 30% [451]. In addition, the safety features of Ca-TC cells are better than those of Li-TC or Li-$SO_2$ batteries. While the melting point of Li (180°C) limits their use at temperatures well below this point, Ca-TC cells withstand heating up to 600°C and short circuiting at this temperature [452]. The state of the art of these studies includes tests on the performance and safety features of Ca-TC cells sized from 3.5 A · h to 7000 A · h [453].

## C. Behavior of Calcium Electrodes in Organic Solvents

As stated in Section VII.A, very few papers describe the behavior of calcium electrodes in organic solvents. Tsuchida et al. [448] reported on studies of Ca-$MnO_2$ primary cells composed of $Ca(ClO_4)_2$ solutions in various solvents including DMF, AN, PC, and BL. It appears from their studies that AN/$Ca(ClO_4)_2$ is the most preferred electrolyte solution of the highest conductivity ($>10^{-2}$ $\Omega^{-1}$ $cm^{-1}$ at RT and salt concentration between 0.6 and 0.8 $M$). In this solution, the overpotential required for Ca dissolution was the lowest (of the four solvents tested). It was possible to discharge Ca electrodes in AN/$Ca(ClO_4)_2$ solutions at current densities $>1$ mA/$cm^2$ (which are quite practical values for multipurpose batteries of a jelly-roll configuration), with an overvoltage below 0.2 V versus Ca/$Ca^{2+}$. For comparison, in PC/$Ca(ClO_4)_2$ solutions, the discharge of Ca electrodes at current densities above 0.1 mA/$cm^2$ required an overvoltage $>1$ V versus Ca/$Ca^{2+}$. It appears from this work [448] that $MnO_2$ can insert calcium at potentials varying between 1 and 2.5 V versus Ca/$Ca^{2+}$. Consequently, a discharge curve of a $Ca-MnO_2$ cathode slopes, in contrast to the flat discharging curves of $Li-MnO_2$ battery systems [454]. A rigorous study of the correlation between surface chemistry, morphology, and electrochemical behavior of calcium electrodes in PC, THF, BL and AN containing various salts [e.g., $Ca(ClO_4)_2$, $Ca(BF_4)_2$, $LiAsF_6$, and tetrabutyl ammonium $ClO_4$ and $BF_4$ salts] is described in Ref. 349. The behavior of calcium electrodes in MF solutions is described in Ref. 4.

The results of these studies can be summarized as follows:

1. The native surface films can be composed of CaO, $Ca(OH)_2$ and $CaCO_3$. These films are replaced by solvent and salt anion reduction products during storage of Ca electrodes in solutions.
2. The surface chemistry of Ca in $\gamma$-BL and MF is very similar to that of Li in these solvents. In $\gamma$-BL, the surface species include Ca butyrate, derivatives of $\gamma$-hydroxy butyrate, and a cyclic $\beta$-keto ester calcium salt (see Table 3). In MF, calcium formate is the major surface species.
3. In PC, the Ca surfaces are covered with $CaCO_3$, an alkoxide species, and $Ca(OH)_2$ (due to the presence of trace water).
4. In THF, the surface films comprised $Ca(OH)_2$ (trace water) and alkoxy species.
5. In AN, the Ca surfaces are covered by AN condensation products.
6. The $BF_4^-$ anion is stable toward calcium, while $ClO_4^-$ is reduced to form surface $CaCl_2$ species.
7. The voltammetric behavior of calcium electrodes is controlled by the surface chemistry described in 1–6. In $Ca(ClO_4)_2$ solutions, the electrodes are strongly passivated, due to the formation of $CaCl_2$. Hence, high overvoltage ($>1$ V versus $Ca/Ca^{2+}$) is required in order to drive any anodic process of calcium in the solvents/$Ca(ClO_4)_2$ solutions. In contrast, Ca electrodes dissolve at low overpotential in $BF_4^-$ salt solutions of all of the above solvents. The lowest overpotential required to obtain a massive Ca dissolution was measured in AN/TBABF₄ or AN/$Ca(BF_4)_2$ solutions (this being in line with the results in Ref. 448). As discussed in Ref. 449, the voltammetric response of Ca electrodes in these solutions reflects Ca dissolution via a breakdown and repair mechanism of the surface films.
8. It was impossible to obtain Ca deposition in any of the electrolyte systems studied because the surface films on the active metal cannot conduct $Ca^{2+}$ ions.

The results in this section show that it may be possible to optimize organic electrolyte solutions for primary Ca batteries in which Ca dissolution occurs at low overpotential, yet the electrodes remain fully passivated at open circuit conditions. The major obstacle in developing attractive, primary Ca batteries is the availability of efficient, high capacity cathodes for these batteries. However, the work presented in Ref. 448 shows that it may be possible to develop Ca insertion cathodes based on intercalation of Ca into host transition metal oxides or sulfides, as has been done for Li batteries. In any event, the larger size of the $Ca^{2+}$ ion and its double charge, compared with $Li^+$, considerably limits the number of potential oxides or sulfides that can electrochemically insert Ca ions.

## VIII.  NONAQUEOUS ALUMINUM ELECTROCHEMISTRY

### A.  Introduction

Extensive work has been devoted to aluminum electroplating in nonaqueous systems. Choosing appropriate bath compositions enables aluminum to be deposited at high efficiency and purity from nonaqueous electrolyte solutions. Comprehensive reviews on this matter have appeared recently in the literature [123,455]. This work has led to the development of a number of commercial processes for nonaqueous electroplating of aluminum. The quality of the electroplated aluminum is very similar to that of cast metal. For instance, electrodeposited aluminum can be further anodized in order to obtain hard, corrosion resistive, electrically insulating surfaces. It is also possible to electroplate Al on a wide variety of metal surfaces, including active metals (e.g., Mg, Al), nonactive metals, and steel.

It appears that in several nonaqueous systems, the difference between the potentials of the $Al^{3+}/Al$ couple and that of other couples, such as $Zn^{2+}/Zn$, $Sn^{2+}/Sn$, $Cu^{2+}/Cu$, and others is much lower than in aqueous solutions [456–458]. Consequently, electroplating of Al alloys from nonaqueous solutions is feasible and may even be advantageous over metallurgical processes of alloy formation [456,459,460]. Anodic dissolution of aluminum electrodes has also been investigated [461–465]. It appears that reversible behavior of the $Al^{3+}/Al$ couple may be obtained in a number of nonaqueous systems (see next section).

Aluminum electrochemistry differs from the electrochemistry of Li, Mg, and Ca in the following aspects:

1.  It is possible to obtain highly conductive aluminum salt solutions due to the unique properties of Al halide salts (e.g., $AlCl_3$, $AlBr_3$) which are strong Lewis acids.
2.  Aluminum is less reactive than are alkaline or alkaline earth metals, and thus its corrosion in polar aprotic solvents is much less pronounced than that of Li, Mg, Ca, etc.
3.  The acidity of aluminum halides and the relatively moderate reactivity of aluminum enables surface film free aluminum electrodes to be obtained in several types of nonaqueous electrochemical solutions, as discussed in the next section.

## B. Nonaqueous Electrolyte Solutions for Aluminum Electroplating

There are basically three classes of solutions from which aluminum can be deposited and in which the $Al^{3+}/Al$ couple is highly reversible.

1. Ethers such as diethyl ether [464–466] or THF [467], where the salts are a combination of $AlCl_3$ and $LiAlH_4$ [468] or $AlCl_3$, LiH, and $LiAlH_4$ [469].
2. Nonpolar aromatic solvents such as toluene, ethylbenzene, xylene [1,2,54–57], and substituted aniline [470–471]. The salts in these systems are usually a combination of $AlBr_3$ and KBr [1,2,54–57].
3. There are several groups of RT halo aluminate molten salts in which reversible behavior of Al electrodes is obtained:
   a. Alkylpyridinium aluminates, such as ethylpyridinium bromide (EPB)-$AlCl_3$ [472–473] and $N$-(1-butyl)pyridinium chloride (BPC)-$AlCl_3$.
   b. Alkyl imidazolium aluminates [227–229,474–476]. From among this family of molten salts, the most important one is the 1-methyl-3-ethyl imidazolium chloride (MEIC). By mixing this compound with an excess of $AlCl_3$, a low temperature ionic melt of a large electrochemical window is formed [475].
   c. Trimethyl phenylammonium chloride (TMPAC)-$AlCl_3$ [477].

In addition to these three major classes of electrolyte systems, there are also reports on the successful deposition of aluminum from $AlCl_3$:LiCl solutions in dimethyl sulfone $[(CH_3)_2SO_2]$ [478,479]. Aluminum is deposited in these systems from the complex $Al[(CH_3)_2SO_2]_3^{3+}$ [479]. The above systems differ from each other in their ionic structure, conductance mechanism, and the mechanism of Al deposition [480]. In the ethereal solutions (class 1) containing $AlCl_3$ and $LiAlH_4$, the following equilibria take place [467]:

$$2AlCl_3 + LiAlH_4 \rightarrow LiCl + AlH_4^- + Al_2Cl_5^+ \tag{29}$$

$$AlH_4^- + Al_2Cl_5^+ + Li(AlH_4) \rightleftharpoons LiCl + 4AlH_2Cl \tag{30}$$

$$LiAlH_4 + AlH_2Cl \rightleftharpoons 2AlH_3 + LiCl \tag{31}$$

The first reaction leads to the high ionic conductivity of the solutions ($10^{-3}$–$10^{-2}$ $\Omega^{-1}$ $cm^{-1}$). The following reaction paths have been suggested for the Al electrodeposition process [481]:

$$AlCl_4^- + 3e^- \rightarrow Al + 4Cl^- \tag{32}$$

$AlCl_4$ is formed by the reaction

$$LiCl + AlCl_3 \rightleftharpoons Li^+ + AlCl_4^- \tag{33}$$

The Cl$^-$ liberated during Al deposition reacts with AlCl$_3$ to produce AlCl$_4$$^-$. In aromatic hydrocarbon-based systems (class 2), AlBr$_3$ is soluble and forms dimers. Addition of KBr leads to the formation of species such as Al$_2$Br$_7$$^-$ and AlBr$_4$$^-$. The positive ions in these systems include (in addition to K$^+$) species such as AlBr$_2$(solvent)$^+$ and Al$_2$Br$_5$(solvent)$^+$ formed by the slight decomposition of the Al$_2$Br$_6$. There are also commercial aluminum deposition processes which utilize aromatic hydrocarbon solvents (e.g., toluene), in which the electrolyte is a complex based on organoaluminum compounds such as M[R$_3$AlFAlR$_3$], where M = Na, K or R$_4$N$^+$, R = alkyl (e.g., CH$_3$CH$_2$$^-$) [467,482,483].

The third system, the RT molten salt (class 3), is always a combination of organic salts R$^+$X$^-$ and aluminum halide AlX$_3$ (R$^+$ is usually a nitrogen-containing aromatic compound such as pyridinium, imidazolium, etc.). The properties of the ionic liquid are determined by the mole ratio of these components ($l/m$) according to the following reactions [455]:

$$lR^+X^- + mAlX_3 \rightleftharpoons R^+AlX_4^- + (l - m)R^+X^- \qquad (l > m) \qquad (34)$$

$$\begin{aligned} lR^+X^- + mAlX_3 &\rightleftharpoons (2l - m)R^+AlX_4^- \\ &+ (m - l)R^+Al_2X_7^- \qquad (l < m) \end{aligned} \qquad (35)$$

$$2AlX_4^- \rightleftharpoons Al_2X_7^- + X^- \qquad (36)$$

The Al$_2$X$_7$$^-$ is a strong Lewis acid, with X$^-$ as its conjugate Lewis base. Hence, at $l > m$, the melt is termed basic, while at $l < m$ it is acidic, and at $m = l$ the melts are neutral. Aluminum deposition occurs in these systems by the following reactions [455]:

$$AlCl_4^- + 3e^- \rightleftharpoons Al + 4Cl^- \qquad (32)$$

or

$$4Al_2Cl_7^- + 3e^- \rightleftharpoons Al + 7AlCl_4^- \qquad (37)$$

In all of the organic melts, the R$^+$ organic cations are reduced at higher potentials than those of the AlX$_4$$^-$ reduction. Thus, aluminum can only be deposited from the acidic melts according to the following mechanism [484]:

$$Al_2Cl_7^- \rightleftharpoons [AlCl_3 \cdot AlCl_4^-] \qquad (38)$$

$$4[AlCl_3 \cdot AlCl_4^-] + 3e^- \rightleftharpoons Al + 7AlCl_4^- \qquad (39)$$

It appears that in several organic melts the electroplated aluminum corrodes by reactions with the organic cations or impurities.

In conclusion, the common denominator of all the above systems is that a surface film free electrode is obtained. This is attributed either to the low reactivity of the solvents used (e.g., ethers, aromatic hydrocarbons) or to the acidity of the media (RT molten salts). Indeed, in more reactive solvents such as PC, the electrochemical behavior of Al is surface film controlled [463].

## C. Remarks on Nonaqueous Aluminum Batteries

The reversible behavior of aluminum anodes in a number of electrolyte solutions described above, as well as the high charge density and low redox potential of this metal, provides a strong driving force for the development of nonaqueous rechargeable aluminum batteries. There is a report by Koura and Ejiri [485] on rechargeable aluminum batteries based on an Al anode in $AlCl_3$/butyl pyridinium chloride, low temperature melt, and a polyaniline-based cathode which can be reversibly doped with $Al_2Cl_7^-$. This 1.7 V volt prototype battery has an energy density of 100 Wh/kg and could be cycled 60 times.

Another report by Koch [486] describes a rechargeable aluminum battery based on $AlCl_3$/$N$-ethylpyridinium bromide or $AlCl_3$/$N$-ethyl-$N$-methyl imidazolium chloride melts and a $TiS_2$ cathode. It has been reported that Al can intercalate reversibly into a number of sulfides, including $TiS_2$, $FeS_3$ and $TaS_2$.

In conclusion, the above reports clearly demonstrate that the high reversibility of Al electrodes in a variety of solutions can be exploited for their use as anodes in high energy density, rechargeable nonaqueous batteries. In this respect, the reports on reversible Al intercalation cathodes are also very promising.

## REFERENCES

1. A. Reger, E. Peled and E. Gileadi, *J. Electrochem. Soc. 123*, 638 (1976).
2. M. Elam, I. Bahett, E. Peled and E. Gileadi, *J. Phys. Chem. 88*, 1609 (1984).
3. D. Aurbach, M. L. Daroux, P. Faguy and E. B. Yeager, *J. Electroanal. Chem. 297*, 225 (1991).
4. D. Aurbach and Y. Ein-Eli, *Langmuir 8*, 1845 (1992).
5. V. R. Koch, *J. Electrochem. Soc. 126*, 181 (1979).
6. M. Pourbaix, *Atlas of Electrochemical Equilibria in Aqueous Solutions*, 2nd ed., Cebelcor, Brussels (1974).
7. R. T. Atanasoski, H. H. Law and C. W. Tobias, *J. Appl. Electrochem. 16*, 339 (1986).
8. J. Kruger, D. A. Shifler, J. F. Scanlon and P. J. Moran, *Russ. Elektrokhimiya* (English edition) *31*, 1004 (1995).
9. D. A. Shifler, P. J. Moran and J. Kruger, *J. Electrochem. Soc. 139*, 54 (1992).
10. K. Wang, P. N. Ross, Jr., F. Kong and F. McLarnon, *J. Electrochem. Soc. 143*, 422 (1996).
11. S. P. S. Yen, D. Shen, R. P. Vasquez, F. J. Grunthaner and R. B. Samoano, *J. Electrochem. Soc. 128*, 1434 (1981).

12. I. A. Kedrinsky, I. V. Murggin, V. E. Dmitrenko, O. E. Abolin, G. I. Sukhova and I. I. Grudyanov, *J. Power Sources 22*, 99 (1988); E. Knözinger, K.-H. Jacob, S. Singh and P. Hofmann, *Surf. Sci. 290*, 388 (1993).

13. T. Fujieda, N. Yamamoto, K. Saito, T. Ishibashi, M. Honjo, S. Koike, N. Wakabayashi and S. Higuchi, *J. Power Sources 52*, 197 (1994).

14. T. Fujieda, N. Yamamoto, K. Saito, T. Ishibashi, M. Honjo, S. Koike, N. Wakabagashi, T. Mumma, T. Taijima and Y. Matsumoto, *J. Electrochem. Soc. 142*, 1057 (1995).

15. G. Nazri, *Mat. Res. Soc. Symp. 135*, 117 (1989).

16. K. Kanamura, H. Takezawa, S. Shiraishi and Z. Takehara, *J. Electrochem. Soc. 144*, 1900 (1997).

17. K. Kanamura, H. Tamura, S. Shiraishi and Z.-I. Takehara, *J. Electrochem. Soc. 142*, 340 (1995).

18. D. Aurbach, *J. Electrochem. Soc. 136*, 1610 (1989).

19. D. Aurbach, Y. Gofer and Y. Langzam, *J. Electrochem. Soc. 136*, 3198 (1989).

20. D. Radman, The Electrochemical Society Fall Meeting, New Orleans, Oct. 1984, Extended Abstracts. The Electrochemical Society, Inc., Pennington, NJ, PV 84-2, p. 188.

21. Y. Geronov, F. Schwager and R. H. Muller, in *Proceedings of the Workshop on Lithium Nonaqueous Battery Electrochemistry*, E. B. Yeager, B. Schumm, Jr., G. Blomgren, D. R. Blankenship, V. Leger and I. Akridge (eds.), The Electrochemical Society, Inc., softbound series PV 80-7 (1980), p. 115.

22. H. Yamin and E. Peled, in *Proceedings of the Symposium on Lithium Batteries*, A. N. Dey (ed.), The Electrochemical Society, Inc., softbound series PV 84-1 (1984), p. 40.

23. A. N. Dey and C. R. Schlaikjer, in Proceedings of the 26th Power Sources Symposium, Atlantic City, NJ (1975), pp. 47–50, The Electrochemical Society, Pennington, NJ.

24. J. O. Besenhard, J. Gürtler, P. Komenda and M. Josowicz, in *Proceedings of the Symposium on Primary and Secondary Ambient Temperature Lithium Batteries*, J. P. Gabano, Z. Takehara and P. Bro (eds.), The Electrochemical Society, Inc., softbound series 88-6 (1988), p. 618.

25. S. Fujita, A. Yasuda and Y. Nishi, in *Proceedings of the Symposium on Primary and Secondary Lithium Batteries*, K. M. Abraham and M. Salomon (eds.), The Electrochemical Society, Inc., softbound series PV 91-3 (1991), p. 262.

26. Z. Takehara, *J. Power Sources 68*, 82 (1997).

27. E. B. Yeager, in *Proceedings of the Workshop on Lithium Nonaqueous Battery Electrochemistry*, E. B. Yeager, B. Schumm, Jr., G. Blomgren, D. R. Blankenship, V. Leger and I. Akridge (eds.), The Electrochemical Society, Inc., softbound series PV 80-7 (1980), p. 1.

28. A. Meitav, Ph.D. thesis, Department of Chemistry, Tel Aviv University, Israel (1982).

29. D. M. Overcash and F. C. Mathers, *Trans Electrochem. Soc. 64*, 305 (1933).

30. E. Peled, *Proceedings of the Symposium on Advanced Battery Materials, Processes*, The Electrochemical Society, Inc., softbound series PV 84-4 (1984), p. 22.

31. D. Aurbach and R. Skaletsky, *Proceedings of the Symposium in Primary and Sec-*

*ondary Lithium Batteries,* 1990. The Electrochemical Society, Inc., Vol. 91-3 (1991), pp. 429–442.

32.  E. Peled, in *Lithium Batteries*, J. P. Gabano (ed.), Academic Press, London (1982), Ch. 3, p. 43.

33.  E. Peled and H. Yamin, in *Proceedings of the 28th Power Sources Symposium*, Atlantic City, NJ (1978), The Electrochemical Society, Pennington, NJ.

34.  A. Meitav and E. Peled, *J. Electroanal. Chem. 134*, 49 (1982).

35.  E. Peled, *J. Electrochem. Soc. 126*, 2047 (1979).

36.  L. Young, *Anodic Oxide Films*, Academic Press, New York (1961).

37.  J. Frenkel, *Z. Phys. 35*, 652 (1926).

38.  N. W. Ashcroft and N. D. Mermin, *Solid State Physics*, W. B. Saunders, Philadelphia, (1976), p. 616.

39.  C. Wagner and W. Schottky, *Z. Phys. Chem. 10*, 163 (1930).

40.  D. O. Ralegh, in *Electroanalytic Chemistry 6*, A. J. Bard (ed.), Marcel Dekker, New York (1973).

41.  N. F. Mott and R. W. Gurny, in *Electronic Processes in Ionic Crystals*, Oxford University Press, London (1948).

42.  N. N. Greenword, in *Ionic Crystals Lattice Defects and Nonstoichiometry*, Chemical Publishing, New York (1970).

43.  L. Heyne, *Electrochim. Acta 15*, 7, 1251 (1970).

44.  R. J. Staniewicz, *J. Electrochem. Soc. 127*, 782 (1980).

45.  A. D. Holding, D. Pletcher and R. V. H. Jones, *Electrochim. Acta 34*, 1529 (1989).

46.  A. J. Bard and L. R. Faulkner, *Electrochemical Methods. Fundamentals and Applications*, Wiley, New York (1980), Ch. 3, p. 86.

47.  A. Zaban and D. Aurbach, *J. Electroanal. Chem. 348*, 155 (1993).

48.  D. Aurbach, A. Zaban, O. Chusid and I. Weissman, *Electrochim. Acta 39*, 51 (1994).

49.  D. Aurbach, E. Zinigrad and A. Zaban, *J. Phys. Chem. 100*, 3098 (1996).

50.  D. Aurbach, A. Zaban, Y. Gofer, O. Abramson and M. Ben-Zion, *J. Electrochem. Soc. 142*, 687 (1995).

51.  D. Aurbach, B. Markovsky, A. Schechter, Y. Ein-Eli and H. Cohen, *J. Electrochem. Soc. 143*, 3525 (1996).

52.  J. Jortner and N. R. Kestner (eds.), *Electrons in Fluids*, Springer-Verlag, Berlin (1973).

53.  I. Juchnovski, Ts. Kolev and B. Stamboliyska, *Spectrosc. Lett. 26*, 67 (1993).

54.  E. Peled and E. Gileadi, *Plating 62*, 342 (1975).

55.  E. Peled and E. Gileadi, *J. Electrochem. Soc. 123*, 15 (1976).

56.  E. Peled and E. Gileadi, *J. Electrochem. Soc. 128*, 1697 (1981).

57.  E. Peled, M. Elam and E. Gileadi, *J. Appl. Electrochem. 11*, 463 (1981).

58.  J. H. Connor, W. E. Reid and B. B. Wood, *J. Electrochem. Soc. 104*, 38 (1957); A. Brenner and J. L. Sligh, *Trans. Inst. Met. Finish 49*, 71 (1971).

59.  D. Fouchard, in *Proceedings of the 33rd Power Sources Symposium*, The Electrochemical Society, Inc., Pennington, NJ (1988); J. A. R. Stilb, *J. Power Sources 26*, 233 (1989).

60.  K. M. Abraham, T. N. Nguen, R. J. Hurd, G. L. Holleck and A. C. Macrides, in

*Proceedings of the Third International Rechargeable Battery Seminar*, Deerfield Beach, FL, March 5–7, 1990.

61.  B. Scrosati, *J. Electrochem. Soc. 139*, 2776 (1992).
62.  P. Dan, E. Mengeritsky, Y. Geronov, D. Aurbach and I. Weissman, *J. Power Sources 54*, 143 (1995).
63.  J. Broadhead and T. Skotheim, in *The 14th International Seminar on Primary and Secondary Batteries*, Fort Lauderdale, FL, March 10–13, 1997.
64.  A. N. Dey, H. C. Kuo, P. Pilliero and M. Kaliandis, *J. Electrochem. Soc. 135*, 2115 (1998).
65.  W. S. Harris, Ph.D. thesis UCRL-8431, University of California, Berkeley (1958).
66.  R. Selim and P. Bro, *J. Electrochem. Soc. 121*, 1467 (1974).
67.  R. Jasinski, in *Advances in Electrochemistry and Electrochemical Engineering*, P. Delahay and C. W. Tobias (eds.), Vol. 8, Wiley-Interscience, New York (1971).
68.  K. M. Abraham and S. B. Brummer, in *Lithium Batteries*, J. P. Gabano (ed.). Academic Press, London (1983), p. 371.
69.  A. N. Dey and E. J. Rudd, *J. Electrochem. Soc. 121*, 1294 (1974).
70.  A. N. Dey and B. P. Sullivan, *J. Electrochem. Soc. 117*, 222 (1970).
71.  A. N. Dey, *Thin Solid Films 43*, 131 (1977).
72.  V. R. Koch, J. L. Goldman and D. L. Natwing, *J. Electrochem. Soc. 129*, 1 (1982).
73.  V. R. Koch and R. H. Young, *J. Electrochem. Soc. 125*, 1371 (1978).
74.  V. R. Koch and R. H. Young, *Science 204*, 499 (1979).
75.  J, Jantsa, F. D. Dousek and J. Riha, *J. Electroanal. Chem. 38*, 445 (1972).
76.  F. P. Dousek, J. Jantsa and J. Riha, *J. Electroanal. Chem. 46*, 281 (1973).
77.  J. G. Thevenin and R. H. Muller, *J. Electrochem. Soc. 134*, 273 (1987).
78.  M. Garreau, *J. Power Sources 20*, 9–17 (1987).
79.  S. Fouache, M. Garreau and J. Thevenin, *J. Power Sources 26*, 593–596 (1989).
80.  J. P. Contour, A. Salesse, M. Froment, M. Garreau, J. Thevenin and D. Warin, *Microsc. Spectrosc. Electron. 4*, 483 (1979).
81.  Y. Geronov, F. Schwager and R. H. Muller, *J. Electrochem. Soc. 129*, 1423 (1982).
82.  F. Schwager, Y. Geronov and R. H. Muller, *J. Electrochem. Soc. 132*, 285 (1985).
83.  G. Nazri and R. H. Muller, *J. Electrochem. Soc. 132*, 1385 (1985).
84.  D. Aurbach, M. L. Daroux, P. Faguy and E. B. Yeager, *J. Electrochem. Soc. 134*, 1611 (1987).
85.  D. Aurbach, M. L. Daroux, P. Faguy and E. B. Yeager, *J. Electrochem. Soc. 135*, 1863 (1988).
86.  D. Aurbach, A. Zaban, Y. Gofer, Y. Ein-Eli, I. Weissman, O. Chusid, O. Abramson and B. Markovsky, *J. Power Sources 54*, 78 (1995).
87.  D. Aurbach, A. Zaban, Y. Ein-Eli, I. Weissman, O. Chusid, B. Markovsky, M. D. Levi, E. A. Levi, A. Schechter and E. Granot, *J. Power Sources 86*, 91 (1997).
88.  K. S. Aojula, J. D. Genders, A. D. Holding and D. Pletcher, *Electrochim. Acta 34*, 1535 (1989).
89.  M. W. Verbagree and B. I. Koch, *J. Electroanal. Chem. 367*, 123 (1994).
90.  M. Odziemkowski and D. E. Irish, *J. Electrochem. Soc. 139*, 3063 (1992).

91. M. Odziemkowski and D. E. Irish, *J. Electrochem. Soc. 140*, 1546 (1998).
92. M. Morita, S. Aoki and Y. Matsuda, *Electrochim. Acta 37*, 119 (1992).
93. N. Takami and T. Ohsaki, *J. Electrochem. Soc. 139*, 7 (1992).
94. A. V. Churikov, E. S. Nimon and A. L. Lvov, *Electrochim. Acta 42*, 179 (1997).
95. M. Ishikawa, K.-Y. Otani, M. Morita and Y. Matsuda, *Electrochim. Acta 41*, 1253 (1996).
96. S. E. Sloop and M. M. Lerner, *J. Electrochem. Soc. 143*, 3 (1996).
97. C. Liebenow and K. Lühder, *J. Appl. Electrochem. 26*, 689 (1996).
98. J. Jamnik, M. Gaberscek, A. Meden and S. Pejovnik, *J. Electrochem. Soc. 138*, 6, (1991).
99. D. Metelko, J. Jamnik and S. Pejovnik, *J. Appl. Electrochem. 22*, 638 (1992).
100. M. Gaberscek, J. Jamnik and S. Pejovnik, *J. Electrochem. Soc. 140*, 2, (1993).
101. K. Kanamura, S. Shiraishi, H. Tamura and Z.-I. Takehara, *J. Electrochem. Soc. 141*, 2379 (1994).
102. K. Kanamura, H. Tamura, S. Shiraishi, and Z.-I. Takehara, *Electrochim. Acta 40*, 913 (1995).
103. K. Kanamura, H. Tamura and Z.-I. Takehara, *J. Electroanal. Chem. 333*, 127 (1992).
104. K. Kanamura, S. Shiraishi and Z.-I. Takehara, *J. Electrochem. Soc. 141*, L108 (1994).
105. K. Kanamura, H. Tamura, S. Shiraishi, and Z.-I. Takehara, *J. Electroanal. Chem. 394*, 49 (1995).
106. T. Hirai, I. Yoshimatsu and J. Yamaki, *J. Electrochem. Soc. 141*, 611 (1994).
107. D. P. Wilkinson, H. Blom, K. Brandt and D. Wainwright, *J. Power Sources 36*, 517 (1991); D. P. Wilkinson and D. Wainwright, *J. Electroanal. Chem. 355*, 193 (1993).
108. E. Goren, O. Chusid and D. Aurbach, *J. Electrochem. Soc. 138*, L6 (1991).
109. D. Aurbach and Y. Cohen, *J. Electrochem. Soc. 143*, 3525 (1996).
110. D. Aurbach and A. Zaban, *J. Electroanal. Chem. 393*, 43 (1995).
111. M. Odziemkowski, M. Kress and D. E. Irish, *J. Electrochem. Soc. 139*, 3052 (1992).
112. M. Matsumoto, T. Ichino, J. S. Rutt and S. Nishi, *J. Electrochem. Soc. 140*, L151 (1993).
113. T. Kabata, T. Fujita, O. Kimura, T. Ohsawa, Y. Matsuda, M. Watanabe, *Polym. Adv. Tech. 4, 205* (1993).
114. J-Y. Sanchez, F. Alloin, D. Benrabah and R. Arnaud, *J. Power Sources 68*, 43 (1997).
115. B. Scrosati, in *Polymer Electrolyte Reviews 1*, J. R. MacCallum and C. A. Vincent (eds.), Elsevier Applied Science, New York (1987), Ch. 10.
116. D. Aurbach, Y. Gofer and M. Ben-Zion, *J. Power Sources 39*, 163 (1992).
117. J. Yamaki, S. Tobishima, Y. Sakurai and K. Hayashi, *J. Appl. Electrochem. 28*, 135 (1998).
118. K. M. Abraham, D. M. Pasquariello and F. J. Marin, *J. Electrochem. Soc. 133*, 661 (1986).
119. J. O. Besenhard and G. Eichinger, *J. Electroanal. Chem. 60*, 1 (1976).
120. K. M. Abraham, *J. Power Sources 7* (1981).
121. S. B. Brummer, V. R. Koch and R. D. Rauh, in *Materials for Advanced Batteries*,

D. W. Murphy, J. Broadhead and B.C.H. Steele (eds.), Plenum Press, New York (1980), p. 123.

122. M. Armand, in *Materials for Advanced Batteries*, D. W. Murphy, J. Broadhead and B. C. H. Steele, (eds.), Plenum Press, New York (1980), p. 145.

123. J. Barthel and H.-J. Gores, in *Chemistry of Nonaqueous Solutions. Current Progress*, G. Mamantov and A. I. Popov (eds.), VCH, 1004, Ch. 1, p. 1.

124. M. Hughes, N. A. Hampson and S. A. G. R. Karunathilaka, *J. Power Sources 2*, 83 (1984).

125. K. M. Abraham, *J. Power Sources 14*, 179 (1985).

126. V. S. Bagotzky and A. M. Skundin, *Russ. J. Electrochem. 31*, 308 (1995); translated from *Elektrokhimiya 31*, 342 (1995).

127. P. G. Bruce, *Phil. Trans. R. Soc. Lond. A 354*, 1577 (1996).

128. I. Epelboin, M. Froment, M. Garreau, I. Thevenin and D. Warin, *J. Electrochem. Soc. 127*, 2100 (1980).

129. R. D. Rauh, T. F. Reise and S. B. Brummer, *J. Electrochem. Soc. 125*, 186 (1978).

130. V. R. Koch and S. B. Brummer, *Electrochim. Acta 23*, 56 (1978).

131. D. Aurbach, Y. Gofer, M. Ben-Zion and P. Aped, *J. Electroanal. Chem. 339*, 451 (1992).

132. D. Aurbach, A. Zaban, A. Schechter, Y. Ein-Eli, E. Zinigrad and B. Markovsky, *J. Electrochem. Soc. 142*, 2873 (1995).

133. S. Tobishima, M. Arakawa, T. Hirai and J. Yamaki, *J. Power Sources 20*, 293 (1987).

134. S. Tobishima, J. Yamaki and T. Okada, *Electrochim. Acta 29*, 1471 (1994).

135. F. K. Shokoohi, J. M. Tarascon and D. Guyomard, *Prog. Batteries Battery Mater. 14*, 173 (1995).

136. R. S. McMillan and M. W. Juzkow, *J. Electrochem. Soc. 138*, 1566 (1991).

137. V. R. Koch, *J. Electrochem. Soc. 126*, 181 (1979).

138. V. R. Koch and J. H. Young, *J. Electrochem. Soc. 125*, 1371 (1978).

139. M. R. Bielefeld, V. A. Cipolla, C. T. Saunders, J. C. Solis, L. A. Davidson, B. Hughes, S. Steubing, G. West, J. West and C. A. Young, *Electrochim. Acta 35*, 1061 (1990).

140. D. Aurbach, Y. Malik, A. Meitav and P. Dan, *J. Electroanal. Chem. 282*, 73 (1990).

141. C. D. Desjardins, G. K. MacLean and H. Sharifian, *J. Electrochem. Soc. 136*, 345 (1989).

142. S. Subbarao, D. H. Shen, S. Dawson, F. Deligiannis, J. Taraszkiewicz and G. Halpert, *J. Power Sources 22*, 269 (1988).

143. C. D. Desjardins, T. G. Cadger, R. S. Salter, G. Donaldson and E. J. Casey, *J. Electrochem. Soc. 132*, 529 (1988).

144. O. Youngman, Y. Gofer, A. Meitav and D. Aurbach, *Electrochim. Acta 35*, 625 (1990).

145. O. Youngman, P. Dan and D. Aurbach, *Electrochim. Acta 35*, 639 (1990).

146. J. L. Goldman, L. A. Dominey and V. R. Koch, *J. Power Sources 26*, 519 (1989).

147. G. H. Newman, in *Proceedings of the Workshop on Lithium Nonaqueous Battery Electrochemistry*, E. B. Yeager, B. Schumm, Jr., G. Blomgren, D. R. Blankenship, V. Leger and I. Akridge (eds.), The Electrochemical Society, Inc., softbound series PV 80-7 (1980), p. 143.

148. P. G. Gluga, in *Proceedings of the Symposium on Power Sources Biomed. Implantable Appl. Ambient Temp. Lithium Batteries*, The Electrochemical Society, Inc., softbound series PV 80-4 (1980), p. 143.

149. D. Aurbach and E. Granot, *Electrochim. Acta 42*, 697 (1997).

150. D. Aurbach, *J. Electrochem. Soc. 136*, 2606 (1989).

151. F. P. Dousek and J. Jantsa, *J. Electroanal. Chem. 74*, 195 (1976).

152. M. Uchiyama, S. Slane, E. Plichta and M. Salomon, in *Proceedings of the Symposium on Primary and Secondary Ambient Temperature Lithium Batteries*, J. P. Gabano, Z. Takehara and P. Bro. (eds.), The Electrochemical Society, Inc., softbound series PV 88-6 (1988), p. 540.

153. D. Aurbach and Y. Gofer, *J. Electrochem. Soc. 138*, 3529 (1991).

154. S.-I. Tobishima and T. Okada, *Electrochim. Acta 30*, 1715 (1985).

155. Y. Yoshimatsu, T. Hirai and J.-I. Yamaki, *J. Electrochem. Soc. 135*, 2422 (1988).

156. J. Yamaki, *J. Power Sources 20*, 3 (1987).

157. K. Tahata, H. Sakamoto, T. Harada, I. Kishi and M. Sano, *Progr. Batteries Solar Cells 8*, 129 (1989).

158. D. Aurbach, O. Chusid and I. Weissman, *Electrochim. Acta 41*, 747 (1996).

159. D. Aurbach, I. Weissman, A. Schechter and H. Cohen, *Langmuir 12*, 3991 (1996).

160. L. P. Klemann and G. H. Newman, *J. Electrochem. Soc. 128*, 13 (1981).

161. L. P. Klemann, G. H. Newman and T. A. Whitney, in *Proceedings of the Symposium on Lithium Batteries*, H. V. Venkatasetty (ed.), The Electrochemical Society, Inc., softbound series PV 81-4 (1981), p. 189.

162. F. Kita, A. Kawakami, T. Sonoda and H. Kobayashi in the *Electrochemical Abstracts*, The Electrochemical Society Spring Meeting, Honolulu, Hawaiii, May 1993, Extended Abstracts. The Electrochemical Society, Inc., Pennington, NJ, PV 93-1, p. 115.

163. T. A. Whitney and L. P. Klemann, U.S. Patent No. 4, *104*, 450 (1978).

164. T. Osaka, T. Momma, Y. Matsumoto and Y. Uchida, *J. Electrochem. Soc. 144*, 5 (1997).

165. S. Shiraishi, K. Kanamura and Z. Takehara, *J. Appl. Electrochem. 25*, 584 (1995).

166. Y. Matsuda, *J. Power Sources 43–44*, 1 (1993).

167. A. T. Ribes, P. Beaunier, P. Willmann and D. Lemordant, *J. Power Sources 58*, 189 (1996).

168. Y. Matsuda, M. Morita and A. Aoki, in *Proceedings of the Symposium on Rechargeable Lithium Batteries*, S. Subbarao, V. R. Koch, B. B. Owens and W. H. Smyrl (eds.), The Electrochemical Society, Inc., softbound series PV 90-5 (1990), p. 67.

169. Y. Matsuda, H. Hayashida and M. Morita, *Proceedings of the Symposium on Primary and Secondary Ambient Temperature Lithium Batteries*, J. P. Gabano, Z. Takehara and P. Bro (eds.), The Electrochemical Society, Inc., softbound series PV 88-6 (1988), p. 610.

170. Y. Matsuda and M. Morita, *J. Power Sources, 26*, 579 (1989).

171. I. Yoshimatsu, T. Hirai and J. Yamaki, *J. Electrochem. Soc. 135*, 2422 (1988).

172. D. Aurbach and Y. Cohen, *J. Electrochem. Soc. 144*, 3355 (1997).

173. M. Inaba, Z. Siroma, A. Funabiki, Z. Ogumi, T. Abei, Y. Mizutani and M. Asaho, *Langmuir 12*, 1535 (1996).

174. M. Froment, M. Qarreau, J. Thevenin and D. Warin, *J. Microsc. Spectrosc. Electron. 4*, 483 (1979); M. Fromert, M. Garreau, J. Thevenin and D. Warin, *ibid., 4*, 111 (1979).
175. D. Aurbach, Y. Ein-Eli and A. Zaban, *J. Electrochem. Soc. 141*, L1 (1994).
176. D. Aurbach and O. Chusid, *J. Power Sources 68*, 463 (1997).
177. T. Ichino, B. D. Cahan and D. A. Scherson, *J. Electrochem. Soc. 138*, L59 (1991).
178. D. Aurbach, M. L. Daroux, P. F. Faguy, A. Wilkins and E. B. Yeager, The Electrochemical Society Spring Meeting, Toronto, Ont., Canada, May 1985. The Electrochemical Society, Inc., Pennington, NJ, Extended Abstracts. 85-1, p. 760.
179. M. Odziemkowski, M. Krell and D. E. Irish, *J. Electrochem. Soc. 143*, 2517 (1996).
180. D. Aurbach, M. L. Daroux, G. McDougal and E. B. Yeager. *J. Electroanal. Chem. 358*, 63 (1993).
181. V. Schily and J. Heitbaum, *Electrochim. Acta 37*, 731 (1992).
182. G. Zhuang, G. S. Chottiner and D. A. Scherson, *J. Phys. Chem. 99*, 7009 (1995).
183. K. Naoi, M. Mori and Y. Shinagawa, *J. Electrochem. Soc. 143*, 2517 (1996).
184. D. Aurbach and M. Moshkovich, *J. Electrochem. Soc. 145*, 2629 (1998).
185. J. Jamnik, M. Gaberscek, A. Meden and S. Pejovnik, *J. Electrochem. Soc. 138*, 1582 (1991).
186. D. Aurbach and A. Zaban, *J. Electroanal. Chem. 367*, 15 (1994).
187. D. Aurbach, M. L. Daroux and E. B. Yeager, Unpublished results.
188. V. W. Behrendt, G. Gatlow and M. Drager, *Z. Anorg. Allg. Chem. 397*, 237 (1973).
189. D. Aurbach and H. E. Gottlieb, *Electrochim. Acta 34*, 141 (1989).
190. D. Aurbach and A. Zaban, *J. Power Sources 54*, 289 (1995).
191. D. Aurbach, M. D. Levi, E. A. Levi and A. Schechter, *J. Phys. Chem. B 101*, 2195 (1997).
192. D. Billaud, F. X. Henry and P. Willmann, *Mater. Res. Bull. 28*, 477 (1993).
193. D. Guyomard and J. M. Tarascon, *J. Electrochem. Soc. 140*, 3071 (1993).
194. J. Yamaki, S. Tobishima, Y. Sakurai, K. Saito and K. Hayashi, *J. Appl. Electrochem. 28*, 135 (1998).
195. J. B. Kerr, *J. Electrochem. Soc. 132*, 2389 (1985).
196. K. Wang, G. H. Chottiner and D. A. Scherson, *J. Phys. Chem. 97*, 11075 (1993).
197. G. R. Zhung, K. L. Wang and P. N. Ross, *Surf. Sci. 387*, 199 (1997).
198. J. P. Gabano, *Progr. Batteries Solar Cells 8* (1989).
199. E. Santos and M. C. Giordano, *Electrochim. Acta 29*, 1327 (1984).
200. L. A. Dominey, J. L. Goldman and V. R. Koch, *Proceedings of the Symposium on Primary and Secondary Lithium Batteries*, K. M. Abraham and M. Salomon (eds.), The Electrochemical Society, Inc., softbound series PV 91-3 (1991), p. 293.
201. L. A. Dominey and J. L. Goldman, *Proceedings of the International Power Sources Symposium*, Vol. 34 (1990), p. 84.
202. F. Walsh, *Proceedings of International Power Sources Symposium*, Vol. 31 (1984), p. 125.
203. Y. Gofer, M. Ben-Zion and D. Aurbach, *Proceedings of the Symposium on High Power, Ambient Temperature Lithium Batteries*, W. D. K. Clark and G. Halpert (eds.), The Electrochemical Society, Inc., softbound series PV 92-15 (1992), p. 145.

204. D. Aurbach, Y. Ein-Eli and Y. Gofer, *Electrochim. Acta 37*, 1897 (1992).
205. D. H. Shen, S. Subbarao, F. Deligiannis, L. Dominey, V. R. Koch and J. Goldman, *Proceedings of the Symposium on Primary and Secondary Lithium Batteries*, The Electrochemical Society, Inc., softbound series PV 91-3 (1991), p. 280.
206. D. Aurbach. I. Weissman, A. Zaban, Y. Ein-Eli, E. Mengeritsky and P. Dan, *J. Electrochem. Soc. 143*, 2110 (1996).
207. C. R. Schlaikjer, in *Lithium Batteries*, J. P. Gabano (ed.), Academic Press, London (1983), Ch. 13, p. 304.
208. H. Shorab et al., *Proceedings of the International Power Sources Symposium*, Vol. 34 (1990), p. 185.
209. J. J. Auborn, K. W. French, S. I. Lieberman, V. K. Shah and A. Heller, *J. Electrochem. Soc. 120*, 1613 (1973).
210. W. K. Behl, J. A. Christopolus, M. Ramirez and S. Gilman, *J. Electrochem. Soc. 120*, 1619 (1973).
211. I. R. Hill, B. G. Anderson, M. Goledzinowski and R. J. Doré, *J. Electrochem. Soc. 142*, 10 (1995).
212. U. Schily and J. Heitbaum, *Vacuum 41*, 1736 (1990).
213. U. Schily and J. Heitbaum, *Electrochim. Acta 137*, 4, 731 (1992).
214. L. A. Dominey, in *Lithium Batteries, New Materials, Developments and Perspectives*, G. Pistoia (ed.), Elsevier, Amsterdam (1994), p. 137.
215. B. J. Carter, B. Jeffries and S. P. S. Yen, *Proceedings of the Symposium on Lithium Batteries*, A. N. Dey (ed.), The Electrochemical Society, Inc., softbound series PV 87-1 (1987), p. 145.
216. G. Nazri and R. H. Muller, *J. Electrochem. Soc. 132*, 2050 (1985).
217. O. Chusid (Youngman) and D. Aurbach, *J. Electrochem. Soc. 140*, L1 (1993).
218. H. A. Frank and D. D. Lawson, *Proceedings of the Symposium on Lithium Batteries*, H. V. Venkatasetty (ed.), The Electrochemical Society, Inc., softbound series PV 81-4 (1981), p. 364.
219. D. Aurbach and O. Chusid, *J. Electrochem. Soc. 140*, L155 (1993).
220. I. Taniguchi, in *Modern Aspects of Electrochemistry*, Vol. 20, J. O'M Bockris, R. E. White and B. E. Conway (eds.), Plenum Press, New York (1991), p. 327.
221. Y. Ein-Eli, R. Thomas and V. R. Koch, *J. Electrochem. Soc. 143*, L195 (1996); *144*, L159 (1997).
222. Lucas, *Organic Chemistry*, p. 393, American Book, New York (1953).
223. C. R. Walk, in *Lithium Batteries*, J. P. Gabano (ed.), Academic Press, London (1983), Ch. 12, p. 281.
224. S. Tobishima, M. Arakawa, T. Hirai and J. Yamaki, *J. Power Sources 28*, 449 (1989).
225. A. N. Dey and R. W. Holmes, *J. Electrochem. Soc. 126*, 1637 (1979).
226. M. Morita, F. Tachihara and Y. Matsuda, *Electrochim. Acta 32*, 299 (1987).
227. C. S. Kelley and R. T. Carlin, *J. Electrochem. Soc. 141*, 873 (1994).
228. J. Fuller, R. A. Osteryoung and R. T. Carlin, *J. Electrochem. Soc. 142*, 3632 (1995).
229. J. Fuller, R. T. Carlin and R. A. Osteryoung, *J. Electrochem. Soc. 143*, L145 (1996).
230. K. M. Abraham, J. L. Goldman and D. L. Natwig, *J. Electrochem. Soc. 129*, 2404 (1982).

231. Y. Geronov, P. Zlatilova, B. Puresheva, M. Pasquali and G. Pistoia, *J. Power Sources* 26, 585 (1989).
232. K. M. Abraham, D. M. Pasquariello and D. A. Schwartz, *J. Power Sources* 26, 585 (1989).
233. S. Tobishima, M. Arakawa, T. Hirai and J. Yamaki, *J. Power Sources* 20, 293 (1987).
234. H. Yamin, Ph.D. thesis, Department of Chemistry, Tel Aviv University, Israel (1983).
235. D. Rahner, S. Marhill and K. Sicery, *J. Power Sources* 68, 69 (1997).
236. I. A. Kedrinsky, I. V. Murygin, V. E. Dmitrenko, O. E. Abolin, G. I. Sukhova and I. I. Grudyanov, *J. Power Sources* 22, 99 (1988).
237. D. Pletcher, J. F. Rohan and A. G. Ritchie, *Electrochim. Acta* 39, 1369 (1994).
238. D. Pletcher, J. F. Rohan and A. G. Ritchie, *Electrochim. Acta* 39, 2015 (1994).
239. G. C. Farrington and J. Xu, *Solid State Ionics* 74, 125 (1994).
240. J. Xu and G. C. Farrington, *J. Electrochem. Soc.* 142, 3303 (1995).
241. M. E. Orazem, P. Agarwal and L. H. Garcia, *J. Electroanal. Chem.* 378, 51 (1994).
242. R. Xue, H. Huang, X. Huang and L. Chen, *Solid State Ionics* 74, 133 (1994).
243. T. Osaka, T. Momma, T. Tajima and Y. Matsumoto, *J. Electrochem. Soc.* 142, 4 (1995).
244. C. Fringant, A. Tranchant and R. Messina. *Electrochim. Acta* 40, 513 (1995).
245. G. Montesperelli, P. Nunziante, M. Pasquali and G. Pistoia, *Solid State Ionics* 37, 149 (1990).
246. E. M. Shembel, I. M. Maksuyta and O. S. Ksenzhek, *Élektrokhimiya* (English edition), 23, 662 (1987).
247. C. D. Desjardins and G. K. MacLean, *J. Electrochem. Soc.* 136, 2, (1989).
248. Yu. M. Povarov, L. A. Beketaeva and Vorob'eva, *Élektrokhimiya* (English edition), 19, 521 (1983).
249. E. M. Shembel, O. S. Ksenzhek and I. M. Maksyuta, *Élektrokhimiya* (English edition), 22, 4161 (1985).
250. R. V. Moshtev and B. Puresheva, *J. Electroanal. Chem.* 180, 609 (1984).
251. D. Aurbach and A. Zaban, *J. Electroanal. Chem.* 365, 41 (1994).
252. D. A. Buttry, in *Electroanalytic Chemistry*, Vol. 17, A. J. Bard (ed.), Dekker, New York (1991), p. 1.
253. D. R. Louder and B. A. Parkinson, *Anal. Chem.* 66, 84R (1994); J. Fromer, *Ang. Chem. Int. Ed. Engl.* 31, 1298 (1992).
254. T. Osaka, T. Momma, K. Nishimura and T. Tajima, *J. Electrochem. Soc.* 140, 2745 (1993).
255. M. Arakawa, S. Tobishima and J. Yamaki, The 6th International Meeting on Lithium Batteries, Münster, Germany (May 1992), Extended Abstracts, p. 7.
256. M. Garreau, J. Thevenin, D. Warin and Ph. Campion, in *Proceedings of the Workshop on Lithium Non-aqueous Battery Electrochemistry*, E. B. Yeager, B. Schumm, Jr., G. Blomgren, D. R. Blankenship, V. Leger and J. Akridge (eds.), The Electrochemical Society, Inc., softbound series PV 80-7 (1980), p. 158.
257. M. Morita and Y. Matsuda, *J. Power Sources* 26, 573 (1989).

258. K. Kanamura, S. Shiraishi and Z.-I. Takehara, *J. Electrochem. Soc. 143*, 2187 (1996).
259. H. Imoto, M. Nagamine and Y. Nishi, The Electrochemical Society Fall Meeting, Miami Beach, Florida (Oct. 1994), Extended Abstracts, The Electrochemical Society, Inc., Pennington, NJ, PV 94-2, p. 121.
260. M. Lafage, D. Windel, V. Russier and J. P. Badiali, *Electrochim. Acta 42*, 2841 (1997).
261. Y. Ein-Eli, B. Markovsky and D. Aurbach, *J. Power Sources 54*, 289 (1995).
262. K. M. Abraham, D. M. Pasquariello and B. Willstaedt, *J. Power Sources 136*, 579 (1989).
263. T. Osaka, T. Momma, Y. Matsumoto and Y. Uchida, *J. Power Sources 68*, 497 (1997).
264. S. Surampudi, D. H. Shen, C.-K. Huang, S. R. Narayanan, A. Attia, G. Halpert and E. Peled, *J. Power Sources 43–44*, 21 (1993).
265. J. Yamaki and S. Tobishima, in *Proceedings of the Symposium on Primary and Secondary Lithium Batteries*, K. M. Abraham and M. Salomon (eds.), The Electrochemical Society, Inc., softbound series PV 91-3 (1991), p. 235.
266. Y. Matsuda, M. Morita and H. Nigo, in *Proceedings of the Symposium on Primary and Secondary Lithium Batteries*, K. M. Abraham and M. Salomon (eds.), The Electrochemical Society, Inc., softbound series PV 91-3 (1991), p. 272.
267. S. Tobishima, M. Arakawa and J. Yamaki, *Electrochim. Acta 35*, 383 (1990).
268. S. Tobishima, M. Arakawa, T. Hirai and J. Yamaki, *J. Power Sources 26*, 449 (1989).
269. E. Peled, Y. Sternberg, A. Gorenshtein and Y. Lavi, *J. Electrochem. Soc. 136*, 6 (1989).
270. S. Subbarao, D. H. Shen, F. Deligiannis, C. K. Huang and G. Halpert, *J. Power Sources 29*, 579 (1990).
271. Y. Matsuda, *J. Power Sources 20*, 19 (1987).
272. K. M. Abraham, J. L. Goldman and D. L. Natwig, *J. Electrochem. Soc. 129*, 12404 (1982).
273. C. D. Desjardins, T. G. Codger, R. S. Salter, G. Donaldson and E. J. Casey, *J. Electrochem. Soc. 132*, 529 (1985).
274. Y. Matsuda, M. Morita, K. Yamada and K. Hirai, *J. Electrochem. Soc. 132*, 11 (1985).
275. M. Anderman, J. T. Lundquist, S. L. Johnson and R. T. Giovannoni, *J. Power Sources 26*, 309 (1989).
276. K. M. Abraham, D. M. Pasquariello and F. J. Martin, *J. Electrochem. Soc. 133*, 661 (1986).
277. Y. Geronov, P. Zlatilova, B. Puresheva, M. Pasquali and G. Pistoia, *J. Power Sources 26*, 585 (1989).
278. K. M. Abraham, D. M. Pasquariello and D. A. Schwartz, *J. Power Sources 26*, 247 (1989).
279. S. V. Sazhin, A. V. Gorodyskii, M. Y. Khimchenki, S. P. Kuksenko and V. V. Danilin, *J. Electroanal. Chem. 344*, 61 (1993).
280. S. Tobishima and A. Yamaji, *Electrochim. Acta 28*, 1067 (1983).

281. A. J. Parker, P. Singh and E. J. Frazer, *J. Power Sources 10* (1983).

282. S. Slane et al., *Proceedings of the International Power Sources Symposium 34*, 87 (1990).

283. D. R. Tuhovak and E. S. Takeuchi, *J. Power Sources 34*, 51–64 (1991).

284. E. Plichta, S. Slane, M. Uchiyama, M. Salomon, D. Chua, W. B. Ebner and H. W. Lin, *J. Electrochem. Soc. 136*, 7 (1989).

285. M. Uchiyama, S. Slane, E. Plichta and M. Salomon, *J. Power Sources 20*, 279 (1987).

286. E. Plichta, M. Salomon, S. Slane, M. Uchiyama, D. Chua, W. B. Ebner and H. W. Lin, *J. Power Sources 21*, 25 (1987).

287. S. Tobishima, K. Hayashi, K. Saito and J. Yamaki, *Electrochim. Acta 40*, 537 (1995).

288. M. Berhil, N. Lebrun, A. Tranchant and R. Messina, *J. Power Sources 55*, 205 (1995).

289. M. Takahashi, S. Yoshimura, I. Nakane, K. Nishio, T. Saito, M. Fujimoto, S. Narukawa, M. Hara and N. Furukawa, *J. Power Sources 43–44*, 253 (1993).

290. Y. Sakurai, S. Tobishima and J. Yamaki, *Electrochim. Acta 34*, 981 (1989).

291. S. Tobishima and A. Yamaji, *Electrochim. Acta 29*, 267 (1984).

292. M. Yoshio, H. Nakamura, K. Isono, S. Itoh and K. Holzleithner, *Progr. Batteries Solar Cells 7*, 271 (1988).

293. A. N. Dey, *J. Electrochem. Soc. 118*, 1547 (1971).

294. D. Fauteux and R. Koksbang, *J. Appl. Electrochem. 23*, 1 (1993).

295. R. W. Holmes, in *Proceedings of the Symposium on Lithium Batteries*, A. N. Dey (ed.), The Electrochemical Society, softbound series PV 84-1 (1984), p. 323.

296. J. Wang, I. D. Raistick and R. A. Huggins, *J. Electrochem. Soc. 133*, 457 (1986).

297. J. Wang, P. King and R. A. Huggins, *Solid State Ionics 20*, 185 (1986).

298. P. Sanchez, C. Belin, C. Crepy and A. de Guibert, *J. Appl. Electrochem. 19*, 421 (1989).

299. J. O. Besenhard, M. Hess and P. Komenda, *Solid State Ionics 40/41*, 525 (1990).

300. A. A. Anani, S. Crouch-Baker and R. A. Huggins, *J. Electrochem. Soc. 135*, 2103 (1988).

301. Y. Toyoguchi, J. Yamaura, T. Matsui and T. Iijima, *Prog. Batt. Sol. Cells 6*, 58 (1987).

302. J. O. Besenhardt, J. Gürtler and P. Komenda, in *Chemical Physics of Intercalation*, A. P. Legrand and S. Flandrois (eds.), NATO ASI Series, Series, Physics 172, Plenum Press, New York (1987), p. 469.

303. J. Epelboin, M. Froment, M. Garreau, J. Thevenin and D. Warin, *J. Electrochem. Soc. 127*, 2100 (1980).

304. J. O. Besenhard, H. P. Fritz, E. Wudy, K. Dietz and H. Meyer, *J. Power Sources 14*, 193 (1985).

305. T. R. Jow and C. C. Lian, *ibid.*, *129*, 1429 (1982).

306. A. Anani, S. Crouch-Baker and R. A. Huggins, in *Proceedings of the ECS Symposium on Lithium Batteries*, A. N. Dey (ed.), The Electrochemical Society, softbound series PV 84-1 (1987), p. 365.

307. A. Anani, S. Crouch-Baker and R. A. Huggins, *J. Electrochem. Soc. 134*, 3098 (1987).
308. R. A. Huggins, in *Proceedings of the Symposium on Batteries for Portable Applications and Electric Vehicles*, C. F. Holmes and A. R. Landgrebe (eds.), The Electrochemical Society, softbound series PV 97-81 (1997), p. 1.
309. J. Wang, I. D. Raistrick and R. A. Huggins, *J. Electrochem. Soc. 133*, 457 (1986).
310. J. Wang, P. King and R. A. Huggins, *Solid State Ionics 20*, 185 (1986).
311. W. C. Maskell and J. R. Owen, *ibid., 132*, 1602 (1985).
312. Y. Geronov, P. Zlatilova and R. V. Moshtev, *ibid., 12,* 155 (1985).
313. J. O. Besenhard, *J. Electroanal. Chem. 94*, 77 (1978).
314. J. R. Owen, W. C. Maskell, B. C. H. Steele, T. S. Nielsen and O. T. Sorensen, *Solid State Ionics 13*, 329 (1984).
315. A. S. Baranski, W. R. Fawcett, T. Krogulcc and M. Drogowska, *ibid., 131*, 1750 (1984).
316. J. Yang, J. O. Besenhard and M. Winter, in *Proceedings of the Symposium on Batteries for Portable Applications and Electric Vehicles*, C. F. Holmes and A. R. Landgrebe (eds.), The Electrochemical Society, softbound series PV 97-18 (1997), p. 350.
317. Y. Idto, Y. Mineo, A. Matsufujii and T. Miyasaka, *Denki Kagaku 65*, 717 (1997).
318. Fujifilm, Internet: *http://www.fujifilm.co.jp/eng/news e/nr079.html* (1996).
319. I. A. Courtney and J. R. Dahn, *J. Electrochem. Soc. 144*, 2045 (1997).
320. Y. Idota, T. Kubota, A. Matsufuji, Y. Naekawa and T. Miyasaka, *Science 276*, 1395 (1997).
321. A. M. Wilson and S. R. Dahn, *J. Electrochem. Soc. 142*, 326 (1995).
322. Q. Liu, T. Zhang, C. Bindra, J. E. Fischer and J. Y. Josefowicz, *J. Power Sources 68*, 287 (1997).
323. A. C. Chu, J. Y. Josefowicz, J. E. Fishcer and G. C. Farrington, The Electrochemical Society Fall Meeting, San Antonio, TX, Oct. 1996, The Electrochemical Society, Inc., Pennington, N. J., Extended Abstracts, 96-2, p. 1062.
324. A. H. Whitehead, K. Edstrom, N. Rao and J. R. Owen, *J. Power Sources 63*, 41 (1996).
325. H. Shi, J. Barker, M. Y. Saidi and R. Koksbang, *J. Electrochem. Soc. 143*, 3466 (1996).
326. T. D. Tran, J. H. Feikert, R. W. Pekala and K. Kinoshita, *J. Appl. Electrochem. 26*, 1611 (1996).
327. V. Manev, I. Naidenov, B. Puresheva, P. Zlatilova and G. Pistoia, *J. Power Sources 55*, 211 (1995).
328. A. Satoh, N. Takami and T. Ohsaki, *Solid State Ionics 80*, 291 (1995).
329. X. Y. Song, K. Kinoshita and T. D. Tran, *J. Electrochem. Soc. 143*, L120 (1996).
330. T. Zheng and J. R. Dahn, *Phys. Rev. B. 53*, 3061 (1996).
331. J. R. Dahn, A. K. Sleign, H. Shi, B. M. Way, W. J. Weydanz, J. N. Reimers, Q. Zhong and U. von Sacken, in *Lithium Batteries. New Materials, Developments and Perspectives*, G. Pistoia (ed.) Elsevier, Amsterdam (1994), Ch. 1, p. 1.
332. M. Morita, N. Nishimura and Y. Matsuda, *Electrochim. Acta 38*, 1721 (1993).

333.  R. Kanno, Y. Kawamoto, Y. Takeda, S. Ohashi, N. Imanishi and O. Yamamoto, *J. Electrochem. Soc. 139*, 3397 (1992).
334.  N. Imanishi, H. Kashiwagi, T. Ichikawa, Y. Takeda, O. Yamamoto and M. Inagaki, *J. Electrochem. Soc. 140*, 315 (1993).
335.  M. Kikuchi, Y. Ikezawa and T. Takamura, *J. Electroanal. Chem. 396*, 451 (1995).
336.  N. Takami, A. Satoh, M. Hara and T. Ohsaki, *J. Electrochem. Soc. 142*, 2564 (1995).
337.  M. W. Verbrugge and B. J. Koch, *J. Electrochem. Soc. 143*, 24 (1996).
338.  J. M. Tarascon and D. Guyomard, *J. Electrochem. Soc. 138*, 2864 (1991).
339.  N. Takami, A. Satoh, M. Hara and T. Ohsaki, *J. Electrochem. Soc. 142*, (1995).
340.  S. H. Ma, J. Li, X. B. Jing and F. S. Wang, *Solid State Ionics 86*, 911 (1996); N. Takami, A. Satoh, T. Ohsaki and M. Kanda, *Electrochim. Acta 42*, 2537 (1997).
341.  S.-I. Yamada, H. Imoto, K. Sekai and M. Nagamine, The Electrochemical Society Spring Meeting. Montreal, Quebec, Canada, May (1997), The Electrochemical Society, Inc., Pennington, NJ, Extended Abstracts, 97-1, p. 85.
342.  W. Xing, J. S. Xue and J. R. Dahn, *J. Electrochem. Soc. 143* 10 (1996).
343.  V. Eskenazi and E. Peled, The Electrochemical Society Fall Meeting, San Antonio, TX, Oct. (1996), The Electrochemical Society, Inc., Pennington, NJ, Extended Abstracts, 96-2, p. 1051.
344.  Y. Liu, J. S. Xue, T. Zheng and J. R. Dahn, *Carbon 34* 2, 193 (1996).
345.  J. S. Xue and J. R. Dahn, *J. Electrochem. Soc. 142*, 11 (1995).
346.  T. Zheng, Q. Zhong and J. R. Dahn, *J. Electrochem. Soc. 142*, 11 (1995).
347.  H.-Q. Xiang, S.-B. Fang and Y.-Y. Jiang, *J. Electrochem. Soc. 144*, 7 (1997).
348.  S. Yata, H. Kinoshita, M. Komori, N. Ando, T. Kashiwamura, T. Harada, K. Tanaka and T. Yamabe, *Synthetic Metals 62*, 153 (1994).
349.  B. Huang, R. Xue, G. Li, Y. Huang, H. Yan, L. Chen and F. Wang, *J. Power Sources 58*, 177 (1996).
350.  T. Zheng, Y. Liu, E. W. Fuller, S. Tseng, U. von Sacken and J. R. Dahn, *J. Electrochem. Soc. 142*, 8 (1995).
351.  M. Alamgir, Q. Zuo and K. M. Abraham, *J. Electrochem. Soc. 141*, 11 (1994).
352.  G. Sandi, R. E. Winans and K. A. Carrado, *J. Electrochem. Soc. 143*, 5 (1996).
353.  P. Papanek, M. Radosavljevic and J. E. Fischer, *Chem. Mater. 8*, 1519 (1996).
354.  M. Hara, A. Satoh, N. Takami and T. Ohsaki, *J. Phys. Chem. 99*, 16338 (1995).
355.  K. Sato, M. Noguchi, A. Demachi, N. Oki and M. Endo, *Science 264* (1994).
356.  D. Aurbach, O. Youngman Chusid, Y. Carmeli, M. Babai and Y. Ein-Eli, *J. Power Sources 43*, 47 (1993).
357.  D. Aurbach, Y. Ein-Eli, O. Chusid, M. Babai, Y. Carmeli and H. Yamin, *J. Electrochem. Soc. 141*, 603 (1994).
358.  D. Aurbach, Y. Ein-Eli, B. Markovsky, Y. Carmeli, H. Yamin and S. Luski, *Electrochim. Acta 39*, 2559 (1994).
359.  D. Aurbach, Y. Ein-Eli, B. Markovsky and A. Zaban, *J. Electrochem. Soc. 142*, 2882 (1995).
360.  D. Aurbach, A. Schechter, B. Markovsky, Y. Ein-Eli and V. Koch, *J. Electrochem. Soc. 143*, L273 (1996).

361. Y. Ein-Eli, S. F. McDevitt, B. Markovsky, A. Schechter and D. Aurbach, *J. Electrochem. Soc. 144*, L180 (1997).
362. Y. Ein-Eli, S. R. Thomas and V. R. Koch, The Electrochemical Society Fall Meeting, San Antonio, TX, Oct. (1996), The Electrochemical Society, Inc., Pennington, NJ, Extended Abstracts, 96-2, p. 1046.
363. S. Geniès, R. Yazami, J. Garden and J. C. Frison, *Synthetic Metals 93*, 77 (1998).
364. R. Imhof and P. Novak, *J. Electrochem. Soc. 145*, 1081 (1998).
365. A. Naji, J. Ghanbaja, B. Humbert, P. W. Willimann and D. Billaud, *J. Power Sources 63*, 33 (1996).
366. R. Fong, U. von Sacken and J. R. Dahn, *J. Electrochem. Soc. 137*, 2009 (1990).
367. Z. X. Shu, R. S. McMillan and J. J. Murray, *J. Electrochem. Soc. 140*, 992 (1993).
368. M. D. Levi and D. Aurbach, *J. Electroanal. Chem. 421*, 79 (1997); D. Aurbach, M. D. Levi and E. A. Levi, *J. Electroanal. Chem. 421*, 89 (1997).
369. D. Aurbach and Y. Ein-Eli, *J. Electrochem. Soc. 142*, 1746 (1995).
370. D. Aurbach and M. D. Levi, *J. Phys. Chem. B 101*, 4641 (1997); D. Aurbach and M. D. Levi, *J. Phys. Chem. B. 101*, 4630 (1997).
371. C. Menachem, D. Golodnitsky and E. Peled, in *Proceedings of the Symposium on Batteries for Portable Applications and Electric Vehicles*, C. F. Holmes and A. R. Landgrebe (eds.), The Electrochemical Society, softbound series PV 97-18 (1997), p. 95.
372. Y. Ein-Eli and V. R. Koch, *J. Electrochem. Soc. 144*, 9, 2968 (1997).
373. P. Papanek, M. Radosavljevic and J. E. Fischer, *Chemi. Mater. 8*, 1519 (1996).
374. P. Zhou, P. Papanek, R. Lee, J. E. Fischer and W. A. Kamitakahara, *J. Electrochem. Soc. 144*, 1744 (1997).
375. P. Zhou and J. E. Fischer, *Phys. Rev. B. 53*, 12643 (1996).
376. X. Qiu, Q. Liu and L. Yang, *Solid State Ionics 60*, 351 (1993).
377. T. Zheng, W. R. McKinnon and J. R. Dahn, *J. Electrochem. Soc. 143*, 2137 (1996).
378. K. Sato, M. Noguchi, A. Demachi, N. Oki and M. Endo, *Science 264*, 5158 (1994).
379. M. Endo, Y. Nishimura, T. Takahashi, K. Takcuchi and M. S. Dresselhaus, *J. Phys. Chem. Solids 57*, 725 (1996).
380. A. M. Wilson, J. N. Reimers, E. W. Fuller and J. R. Dahn, *Solid State Ionics 74*, 249 (1994).
381. P. Zhou, P. Papanek, C. Bindra, R. Lee and J. E. Fischer, *J. Power Sources 68*, 296 (1997).
382. M. Alamgir and K. M. Abraham, in *Lithium Batteries. New Materials, Developments and Perspectives*, G. Pistoia (ed.), Elsevier, Amsterdam (1994), Ch. 3, p. 93.
383. K. Motogami et al., Third International Meeting on Polymer Electrolytes, Annecy, France, June (1991).
384. D. W. Xia, S. Soltz and J. Smid, *Solid State Ionics 14*, 221 (1984).
385. K. Nagaoka, H. Naruse, I. Shinohara and M. Watanabe, *J. Polym. Sci. Polym. Lett. Ed. 22*, 659 (1984).
386. P. M. Blonsky, D. F. Shriver, P. Austin and H. R. Allcock, *J. Am. Chem. Soc. 106*, 6854 (1984).

387.  J. F. LeNest et al., Third International Meeting on Polymer Electrolytes, Annecy, France, June (1991).

388.  E. Peled, D. Golodnitsky, G. Ardel and A. Peled, in *Proceedings of the Symposium on Rechargeable Lithium and Lithium-Ion Batteries*, S. Megahed (ed.), The Electrochemical Society, softbound series PV 94-28, (1994), p. 389.

389.  E. Peled, D. Golodnitsky, E. Strauss, J. Lang and Y. Lavi, *Electrochim. Acta 43*, 1593 (1998).

390.  G. Dautzenberg, F. Croce, S. Passerini and B. Scrosati, *Chem. Mater. 6*, 1538 (1994).

391.  S. Panero, E. Spila and B. Scrosati, *J. Electrochem. Soc. 143*, L29 (1996).

392.  G. B. Appetecchi, F. Croce, G. Dautzenberg, S. Passerini and B. Scrosati, *Electrochim. Acta 39*, 2187 (1994).

393.  S. Panero, A. Clemente and E. Spila, *Solid State Ionics 86*, 1285 (1996).

394.  H. S. Choe, B. G. Carrol, D. M. Pasquariello and K. M. Abraham, *Chem. Mater. 9*, 369 (1997) and *J. Power Sources 54*, 40 (1995).

395.  Z. Jiang and K. M. Abraham, *J. Electrochem. Soc. 144*, L136 (1997).

396.  K. M. Abraham, Z. Jiang and B. Corell, *Chem. Mater. 9*, 1978 (1997).

397.  T. Osaka, T. Momma and K. Nishimura, in *Proceedings of the Symposium on New Batteries and Supercapacitors*, The Electrochemical Society, softbound series PV 93-23 (1993), p. 178.

398.  T. Osaka, T. Momma, K. Nishimura, S. Kakuda and T. Ishii, *J. Electrochem. Soc. 141*, 1994 (1994).

399.  M. Watanabe and A. Nishimoto, *Solid State Ionics 79*, 306 (1995).

400.  H. Akashi, S. L. Hsu, W. J. MacKnight, N. Ogata and M. Watanabe, *J. Electrochem. Soc. 142*, L205 (1995).

401.  M. Kono, K. Furuta, S. Mori, M. Watanabe and N. Ogata, *Polym. Adv. Tech. 4*, 85 (1993).

402.  D. Golodnitsky, G. Ardel and E. Peled, *Solid State Ionics 231*, 1 (1996).

403.  B. Scrosati, *J. Electrochem. Soc. 136*, 2774 (1989).

404.  G. Eichinger and M. Fabian, Annual Meeting of the International Society of Electrochemistry, Xiamen, China (1995). Extended Abstracts *1*, Abst. 5-01.

405.  L. W. Gaddum and H. E. French, *J. Am. Chem. Soc. 49*, 1295 (1927).

406.  W. Evans, *J. Am. Chem. Soc. 64*, 2965 (1942).

407.  J. H. Connor, W. E. Reid and G. B. Wood, *J. Electrochem. Soc. 104*, 38 (1957).

408.  A. Brenner and J. L. Singh, *Trans. Inst. Met. Finish. 49*, 71 (1971).

409.  M. M. Baizer and H. Lund (eds.), *Organic Electrosynthesis*, Marcel Dekker, New York (1983).

410.  O. R. Brown and R. McIntyre, *Electrochim. Acta 30*, 627 (1985).

411.  O. R. Brown and R. McIntyre, *Electrochim. Acta 29*, 995 (1984).

412.  W. R. Fawcett and J. S. Jaworskii, *J. Chem. Soc. Faraday Trans. 1 78*, 1971 (1982).

413.  T. Psarras and R. E. Dessy, *J. Am. Chem. Soc. 89*, 5132 (1967).

414.  J. Bittrich, R. Landsberg and W. Gaube, *Wiss. Z. Hochsch. Chem. "Carl Schorlemmer" Lenna-Merseburg 2*, 60, 449 (1959).

415.  E. Peled and H. Straze, *J. Electrochem. Soc. 124*, 1030 (1997).

416.  J. D. Genders and D. Pletcher, *J. Electroanal. Chem. 199*, 93 (1986).

417. T. D. Gregory, R. J. Hoffman and R. C. Winterton, *J. Electrochem. Soc. 137*, 3 (1990).
418. W. Vonau and F. Berthold, *J. Praktische Chemie-Chemiker 336*, 2, 140 (1994).
419. L. P. Lossius and F. Emmenegger, *Electrochim. Acta 41*, 3, 445 (1996).
420. C. Liebenow, *J. Appl. Electrochem. 27*, 221 (1997).
421. A. Kisza, J. Kazmierczak, B. Børressen, G. M. Haarberg and R. Tunold, *J. Appl. Electrochem. 25*, 940 (1995).
422. A. Kisza and J. Kazmierczak, *J. Electrochem. Soc. 144*, 5 (1997).
423. B. Børressen, G. M. Haarberg and R. Tunold, *Electrochim. Acta 42*, 1613 (1997).
424. D. Aurbach and A. Schechter, *On Non-aqueous Magnesium Electrochemistry*, in preparation (1998).
425. P. Novák, V. Shklover and R. Nesper, *Z. Phys. Chem. 185*, 51 (1994).
426. A. Meitav and E. Peled, *J. Electrochem. Soc. 128*, 825 (1981).
427. A. Brenner, *J. Electrochem. Soc. 118*, 99 (1971).
428. A. Brenner and J. L. Sligh, *Trans. Inst. Met. Finish. 49*, 71 (1971).
429. W. Schlenk and W. Schlenk Jr., *Chem. Ber. 62*, 920 (1929).
430. M. B. Smith and W. E. Becker, *Tetrahedron 23*, 4215 (1967).
431. P. G. Bruce, F. Krok, J. Nowinski, V. C. Gibson and K. Tavakkoli, *J. Mater. Chem. 1*, 705 (1991).
432. P. Lightfoot, F. Krok, J. L. Nowinski and P. G. Bruce, *J. Mater. Chem. 2*, 139 (1992).
433. P. Novák, W. Scheifele and O. Haas, *J. Power Sources 54*, 479 (1995).
434. P. Novák, W. Scheifele, F. Joho and O. Haas, *J. Electrochem. Soc. 142*, 8 (1995).
435. L. Sánchez and J.-P. Pereira-Ramos, *J. Mater. Chem. 7*, 471 (1997).
436. E. Peled, R. Cohen, A. Melman and Y. Lavi, *J. Electrochem. Soc. 139*, 1836 (1992).
437. E. Peled, R. Cohen, A. Melman and Y. Lavi in *Proceedings of the International Power Sources Symposium*, Vol. 34, 264 (1990).
438. R. Cohen, Y. Lavi and E. Peled, *J. Electrochem. Soc. 137*, 2648 (1990).
439. R. Cohen, Y. Lavi and E. Peled, *J. Electrochem. Soc. 137*, 1999 (1990).
440. E. Elster, R. Cohen and E. Peled, *J. Power Sources 26*, 42 (1989).
441. A. Meitav and E. Peled, *Electrochim. Acta 33*, 1111 (1988).
442. E. Peled, E. Elster, R. Tulman and J. Kimel, *J. Power Sources 14*, 93 (1985).
443. E. Peled, R. Tulman, A. Golan and A. Meitav, *J. Electrochem. Soc. 131*, 2314 (1984).
444. E. Peled, R. Tulman and A. Meitav, *J. Power Sources 10*, 35 (1983).
445. E. Elster, R. Cohen, M. Brand, Y. Lavi and E. Peled, *J. Electrochem. Soc. 135*, 1307 (1988).
446. R. J. Staniewicz, *J. Electrochem. Soc. 127*, 782 (1980).
447. W. L. Wade, Jr., C. W. Walker, Jr., and M. Binder, *J. Power Sources 28*, 295 (1989).
448. T. Tsuchida, M. Nakanishi, H. Takayanagi and K. Yamamoto, *Battery Mater. Symp. (Proc.) 2*, 417 (1985).
449. D. Aurbach, R. Skaletsy and Y. Gofer, *J. Electrochem. Soc. 138*, 3536 (1991).
450. F. Walsh, Final report to the NSF, Sept. (1988). Available from NTIS (U.S. Department of Commerce), Springfield, VA, PB 91-226134.

451.  E. Elster, R. Cohen, M. Brand, Y. Lavi and E. Peled, in *Proceedings of the Symposium on Primary and Secondary Ambient Temperature Li Batteries*, J. P. Gabano, Z. Takehara and P. Bro (eds.), The Electrochemical Society, softbound series PV 88-6 (1988), p. 271.
452.  R. J. Staniewicz, S. J. Kafner and R. A. Dixon, (1988), ibid., p. 279.
453.  A. Meitv and E. Peled, *J. Electrochem. Soc. 129*, 451 (1982).
454.  H. Ikeda, in *Lithium Batteries*, J. P. Gabano (ed.), Academic Press, London (1983), Ch. 8, p. 169.
455.  Y. Zhao and T. J. VanderNoot, *Electrochim. Acta 42*, 3 (1997).
456.  G. A. Capuano, R. Ducase and W. G. Davenort, *J. Appl. Electrochem. 9*, 7 (1979).
457.  E. Gileadi, in *Proceedings of the Third Symposium on Electrodes*, 1979, S. Bruckenstein, J. D. E. McIntyre, B. Miller and E. Yeager (eds.) (1980), p. 366.
458.  R. D. Blue and F. C. Mothers, *Trans. Electrochem. Soc. 99*, 234 (1952).
459.  M. Elam, E. Peled and E. Gileadi, *J. Electrochem. Soc. 130*, 585 (1983).
460.  A. Mayer, *J. Electrochem. Soc. 137*, 2806 (1990).
461.  L. Legrand, A. Tranchant and R. Messina, *Electrochim. Acta 41*, 2715 (1996).
462.  K. V. Rybakja and L. A. Bejetaeva, *J. Power Sources 42*, 377 (1993).
463.  R. T. Atanasoski, *J. Serb. Chem. Soc. 57*, 935 (1992).
464.  A. Brenneer, in *Advances in Electrochemistry and Electrochemical Engineering 5*, C. W. Tobias (ed.), Wiley-Interscience, New York (1967).
465.  R. Suchentrunk, Paper I in 68th Annual Technical Conference of the American Electroplating Society, Vol. 1 (1981).
466.  A. G. Buschow and C. H. Escola, *Plating 55*, 931 (1968).
467.  S. Birkle, The Electrochemical Society, softbound series PV 87–17 (1987), p. 369.
468.  M. Yoshio, H. Nakamura, H. Nogouchi and M. Nagamatsu, presented at the 7th International Conference on Non-aqueous Solutions, Regensburg, Vol. 1, p. E3 (1980).
469.  F. J. Schmidt and I. J. Hess, *Plating 53*, 229 (1966).
470.  T. Hisano, T. Terazawa, I. Takeuchi, S. Inohara and H. Ikeda, *Bull. Chem. Soc. Jpn. 44*, 599 (1971).
471.  T. Hisano, T. Terazawa and H. Matusi, *Chem. Lett. 4*, 213 (1973).
472.  W. H. Safranek, W. C. Schickner and C. L. Fraust, *J. Electrochem. Soc. 99*, 53 (1952).
473.  L. Simanavicius and P. Dobrovolskis, *C.A. 78*, 23137z (1973).
474.  R. A. Osteryoung and J. Robinson, *J. Am. Chem. Soc. 101*, 323 (1979); *102*, 4415 (1980); R. J. Gale and R. A. Osteryoung, *Inorg. Chem. 18*, 1603 (1979); *J. Electrochem. Soc. 127*, 2167 (1980); R. A. Osteryoung and J. Robinson, *J. Electrochem. Soc. 127*, 122 (1980).
475.  R. A. Osteryoung, in *Molten Salt Chemistry*, G. Mamantov and R. Marassi (eds.), Reidel, Dordrecht Holland (1987), p. 329.
476.  S. Takahashi, I. Saeki, I. Kazuhiko and M. Shoichiro, The 177th Meeting of the Electrochemical Society, Montreal, Quebec, Canada, May, 1990. The Electrochemical Soc. Inc., Pennington, NJ. Extended Abstract (1990), abstract no. 889.
477.  S. D. Jones and G. E. Blomgren, *J. Electrochem. Soc. 136*, 424 (1989).
478.  L. Legrand, A. Tranchant and R. Messina, *Electrochim. Acta 39*, 1427 (1994).

479. L. Legrand, A. Tranchant and R. Messina, *J. Electrochem. Soc. 141*, 378 (1994).
480. J. Robinson and R. A. Osteryoung, *J. Electrochem. Soc. 127*, 122 (1980).
481. J. Eckert and M. Galova, *Electrochim. Acta 26*, 1169 (1981).
482. H. Lehmkuhl, Dissertation, Rheinisch-Westfalische Technische Hochschule, Aachen (1954).
483. K. Ziegler and H. Lehmkuhl, *C.A. 52*, 19619d (1958),
484. P. K. Lai and M. Skyllas-Kazacos, *Electrochim. Acta 32*, 1443 (1987).
485. N. Koura and H. Ejiri, The Electrochemical Society, softbound series PV 90–17 (1990), p. 785.
486. V. R. Koch, Presented at the 3rd International Rechargeable Battery Seminar, Deerfield Beach, FL, March 7, 1990.

# 7

# Ionic Conducting Polymers for Applications in Batteries and Capacitors

**Tetsuya Osaka, Shinichi Komaba,\* and Xingjiang Liu**
*Waseda University, Shinjuku-ku, Tokyo, Japan*

## I. INTRODUCTION

Conducting polymers are classified as ''ion conductive polymers'' and ''electron conductive polymers.'' Their conductivities and those of various materials are compared in Figure 1. Electron conductive polymers appeared in 1971, as reported by Shirakawa and co-workers [1]. They synthesized conductive polyacetylene and found that it had rather high conductivity for an organic compound, $10^3$ S cm$^{-1}$ [1,2]. Since then, various conducting polymers, e.g., polypyrrole, polyaniline, polythiophene, poly($p$-phenylene), their derivatives, and their methods of synthesis were studied actively by many researchers. Generally speaking, because electron conductive polymers have $\pi$-electrons which are delocalized in conjugated systems along the whole polymer chain, their electronic conductivities are much higher than those of other polymers. For the last several decades, many kinds of devices with electron conductive polymer films have been investigated. These include electronic and display devices, chemical sensors and active materials for rechargeable batteries.

In the case of ion conductive polymers, gel polymer electrolytes which consist of a polymer matrix, organic solvents and supporting electrolyte, were introduced as novel nonaqueous electrolyte systems in electrochemical applications, such as rechargeable batteries and electric double layer capacitors [3–5]. Recently, considerable attention has been devoted to the application of gel poly-

---

\* *Present affiliation*: Iwate University, Morioka, Japan.

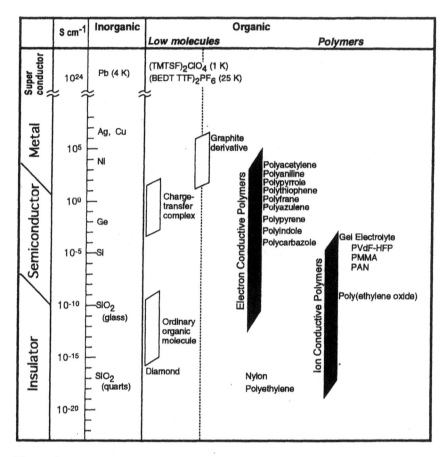

**Figure 1**   Comparison of conductivities of inorganic and organic materials.

mer electrolyte systems in Li secondary batteries. By applying the gel electrolyte to the batteries, the electrolyte solution does not leak out from the cell, and the electrolyte can be prepared as a thin film, which enables the construction of a solid state and high energy density battery. In particular, the gel electrolyte demonstrates highly ionic conductivity, about $10^{-3}$ S cm$^{-1}$, at room temperature and has sufficient mechanical strength. Typical examples of the polymers used in these matrices are poly(acrylonitrile) (PAN) [6,7], poly(methylmethacrylate) (PMMA) [5,6], and a new copolymer of vinylidene fluoride with hexafluoropropylene (PVdF-HFP) [8,9]. Based on these advances, the first reliable and practical rechargeable Li-ion plastic battery was developed during 1997. It contains a lithiated carbon material as the anode, and the PVdF-HFP copolymer type gel electrolyte [10].

In this chapter, some characteristics and applications of gel electrolyte systems are reviewed, with emphasis on their application to electrochemical devices. We also include a report on our own work in this field.

## II.  Li BATTERY APPLICATION

### A.  Li Polymer Battery Applications

Recent years have seen the accelerated development of secondary batteries with high energy density demanded for fields such as telecommunication, portable computers and electrical vehicles. The appearance of nickel metal hydride and Li ion batteries with energy density higher than 200 W h $1^{-1}$ to 300 W h $1^{-1}$ is a very important achievement. The capacity and energy density of secondary batteries can be even higher if Li metal anodes could be used as shown in Figure 2 [11]. However, the use of Li metal in secondary batteries has the problem of dendrite formation, which leads to internal electrical shorts between the anode and the cathode. This leads to serious safety problems in the use of such batteries. The use of ionic polymers as electrolyte systems in Li batteries has a chance to depress dendritic deposition of Li metal. Indeed, many research groups investigated the use of polymeric electrolyte systems in rechargeable Li metal anode batteries [12,13].

### B.  Coin Type Lithium/Polymer Battery

Solid polymer electrolyte systems, which consist of an ionically conducting polymeric matrix and a supporting electrolyte which is a Li salt, have been introduced as a new electrolyte medium for lithium batteries by Wright, Armand, and coworkers [14,15]. By using a solid polymer electrolyte in the battery, a separator is not needed, and the electrolyte medium can be fabricated as an ultrathin film, which enables an all-solid-state, high energy density lithium battery to be constructed. In the initial studies, poly(ethylene oxide) (PEO) based electrolytes were investigated [16–19]. However, it suffered from the drawback that PEO-based solid polymer electrolytes showed relatively low ionic conductivity (about $10^{-7}$ S $cm^{-1}$) compared with liquid electrolytes (about $10^{-2}$ S $cm^{-1}$) at room temperature (see Figure 1). Improvement of the ionic conductivity of such systems is mandatory for their use as battery materials. Approaches such as addition of plasticizers [20,21] and introduction of intermolecular bridge formation within the polymer matrix [22] have been tried. Especially successful were the gel electrolyte systems which consist of a polymeric matrix, supporting electrolyte (Li

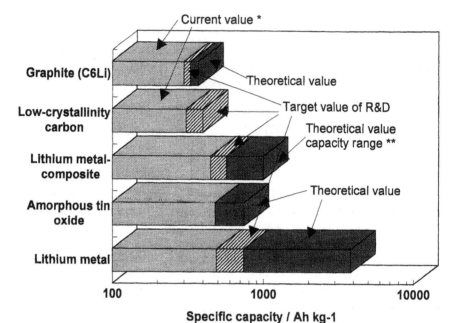

**Figure 2**   Specific capacity of anode active materials.

salt) and organic solvents. Such matrices have high ionic conductivity (about $10^{-3}$ S cm$^{-1}$ at room temperature) and sufficient mechanical strength [3–5]. Therefore, the application of gel electrolyte systems to lithium secondary batteries is expected to be suitable for low temperature performance of these batteries, all-solid-state construction, and high energy density. In this section, we describe the use of PMMA-based gel electrolyte in Li/PPy batteries and investigate their characteristics and charge-discharge performance. We compared battery systems composed of PAN-based and the PMMA-based electrolytes. The chemical composition of the PMMA and PAN gel electrolytes is as follows (numbers in parentheses are molar percentages):

PMMA-based gel    PMMA(30)-PC(19)-EC(46.5)-LiClO$_4$(4.5)
PAN-based gel       PAN(21)-PC(33)-EC(38)-LiClO$_4$(8)
                           (EC = ethylene carbonate; PC = propylene carbonate)

Both solid electrolyte systems have a transparent and elastomeric appearance, whose thickness is about 0.3 mm and whose ionic conductivities are of the

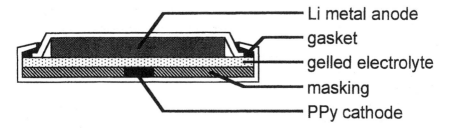

Li metal anode
gasket
gelled electrolyte
masking
PPy cathode

**Figure 3** Schematic diagram of coin-type cell.

order of $10^{-3}$ S cm$^{-1}$ [5]. Furthermore, an electropolymerized PPy film was used in the cell as a cathode material [19]. PPy was obtained by potentiostatic oxidative polymerization of pyrrole at potentials of 3.9 V versus Li/Li$^+$ in 0.2 mol dm$^{-3}$ LiClO$_4$ and 0.2 mol dm$^{-3}$ pyrrole PC solution. Figure 3 shows a schematic of the coin type cell.

Charge-discharge performance of the Li/PPy battery using PMMA gel or PAN gel as the electrolyte system was investigated by galvanostatic tests at 0.1 mA cm$^{-2}$. One slightly inclined plateau is observed in each charging and discharging curve, from which the average battery voltage, 2.9 V, is determined. Discharge capacity and coulombic efficiency estimated from these curves are shown in Figure 4. Both the PMMA and the PAN systems showed about 90–100% of coulombic efficiency through all cycles. It is obvious that the Li/PMMA gel/PPy and Li/PAN gel/PPy systems have sufficient electrochemical stability and reversibility. Furthermore, both batteries showed excellent performance with constant discharge capacities about 50–40 mC cm$^{-2}$ (discharge depth of the PPy cathode was about 90–80% using 53.6 mC cm$^{-2}$ as the maximum capacity of

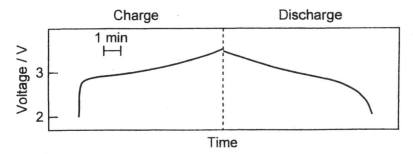

**Figure 4** Typical charge-discharge curve for Li/PMMA gel electrolyte/PPy battery. Charging and discharging current densities are 0.1 mA cm$^{-2}$.

PPy [19]) up to 500 cycles. However, when continuing the measurement, discharge capacity of both batteries gradually decreased. Especially, in Figure 5b on the PAN system, the capacity became nearly 0 after 4000 cycles were completed, and it could not be operated beyond 4000 cycles. On the contrary, the PMMA system showed no sudden decrease and more than 8000 cycles were obtained (Figure 5a).

The difference of the cycling performance between the PMMA and the PAN systems should be due to the difference in the chemical stability of the Li/ gel electrolyte interface. The interfacial stability of the Li electrodes in contact with the gel electrolytes was investigated by impedance spectroscopy. Figure 6 shows the interfacial resistance ($R_i$), which is estimated from the first semicircle

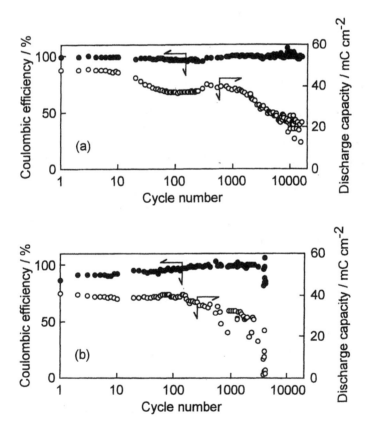

**Figure 5** Dependence of coulombic efficiency (●) and discharge capacity (○) of (a) Li/PMMA gel/PPy battery, (b) Li/PAN gel/PPy battery on number of cycles. Charging and discharging current densities are 0.1 mA cm$^{-2}$.

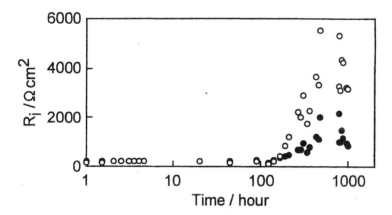

**Figure 6** Time dependence of Ri values which are estimated from the first semicircle (high frequency region) in the Cole-Cole plots of Li/PMMA gel/Li cell (●) and Li/PAN gel/Li cell (○).

in the Cole-Cole plots measured with a Li/gel electrolyte/Li, symmetrical cell. In this figure, $R_i$ is about 150–200 $\Omega$ cm$^2$, and this value remains constant during 100 h. After 100 h, $R_i$ suddenly increases with storage time for the PMMA and PAN systems. We believe that the increase of $R_i$ is due to chemical side reactions of Li and the liquid components (PC, EC) in the gel electrolyte. The side reaction products, which have poor ionic conductivity, accumulate on the interface between the Li and the gel electrolyte and form a layer which grows with storage time. As a result, the interfacial resistance to which the surface layer's resistance contributes considerably increases.

This increase in $R_i$ with storage time differs between the two systems. As shown in Figure 6, the $R_i$ value for the PAN system reaches 4000 $\Omega$ cm$^2$, while the $R_i$ value for the PMMA system remains less than 2500 $\Omega$ cm$^2$. These results indicate that the PAN system is chemically less stable than the PMMA system because of additional possible side reactions of Li and the cyano (CN) groups in the PAN polymeric matrix. This difference in the interfacial stability should be the main cause for the difference in the charge-discharge cycling performance between two batteries. The increase in the interfacial resistance leads to an increase in the total internal resistance of the cell and, thus, leads to power losses due to heat dissipation. An increase in the internal resistance of the cell reduces its practical operating voltage range. Thus, the total discharge capacity of the battery decreases with storage time and cycle number.

As mentioned above, a coin type Li polymer battery has high performance. In particular, the PMMA gel electrolyte is very stable toward lithium anodes.

This allows it to have a long cycle life of more than 8000 cycles for the Li/PMMA/PPy battery system.

## C. Metallic Lithium Anode in Poly(vinylidene fluoride) Type Gel Electrolyte

For future generations of rechargeable batteries, secondary cells using lithium metal as the anode are highly attractive candidates as power sources for portable electric devices, electric vehicles, and load leveling systems. Lithium metal demonstrates remarkable low electrochemical equivalent and the most negative redox potential of all metallic elements. There are, however, some disadvantages of the lithium metal anode compared to the carbon anode in Li-ion battery systems. The charge/discharge cycleability of lithium metal anodes is limited by the formation of dendritic deposits, which causes the isolation of active lithium metal and also may short the cell. Another reason for the poor cycleability of lithium anodes in secondary batteries is the formation of thick and resistive passivating layers whose ionic conductivity is very poor. These layers are formed by the reaction of solution components such as the solvent, the salt anion, and traces of contaminants with the highly reactive lithium metal [23,24].

It was possible to improve the interfacial properties of Li metal anodes in liquid electrolyte solutions using additives that modify the Li-surface chemistry, such as $CO_2$ [23–27] and HF [28,29]. Using PEO-based gel electrolyte systems effectively suppressed dendritic deposition of lithium [30]. In Section C we report on a very good charge-discharge performance of lithium metal anodes in PVdF-HFP gel electrolyte systems. Furthermore, addition of $CO_2$ to the PVdF-HFP gel electrolyte system considerably improves the charge/discharge characteristics [31].

Figure 7a shows charge-discharge cycling efficiency of lithium anodes in various electrolyte systems. Lithium was charged (deposited) on a nickel electrode (about 1.0 C cm$^{-2}$) and discharged (dissolved) until the voltage reached 1.0 V, repeatedly at a current density of 0.5 mA cm$^{-2}$. It should be emphasized that as depth-of-discharge and current density were decreased, efficiency and lifetime of these systems were increased; for example, several lithium/polymer battery systems showed prolonged lifetime and more than 8000 cycles with 90–100% coulombic efficiency could be obtained.

As shown in Figure 7a, the lithium anode tested in the EC/PC-based liquid solution shows relatively high efficiency only at the first cycle. In subsequent cycles, the cycling efficiency drops to around 75%. The Li cycling efficiencies measured in EC/PC are similar to those measured in the PEO gel. It is obvious that lithium anodes in the PVdF-HFP gel electrolyte system show higher cycling efficiency.

Li dendritic deposition was considerably suppressed in the PEO gel electro-

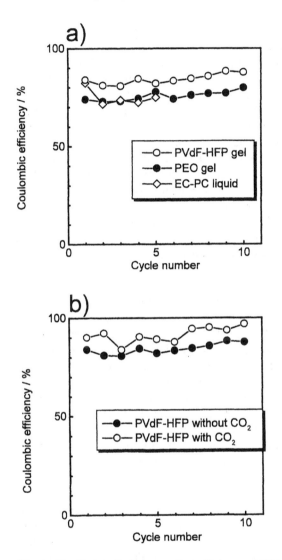

**Figure 7** Charge-discharge cycling efficiency of lithium metal anode at 0.5 mA cm$^{-2}$ (a) in EC/PC liquid, PEO gel and PVdF-HFP gel electrolyte systems without $CO_2$, and (b) in PVdF-HFP gel electrolyte system with and without $CO_2$ addition.

lytes, probably due to the immobility of the solvent molecules, as described previously [30,32]. It was thought that the Li cycling efficiencies measured in the various gel electrolytes depend on the kind of polymer matrix. There are some apparent differences between the PEO and the PVdF-HFP gel electrolytes in terms of the chemical behavior of the polymer matrix and the immobility of the solvent molecules. It is most likely that the interface between the gel electrolyte and lithium plays the most important role in the cycling characteristics. The PAN gel electrolyte shows less stability against Li metal, probably due to the cyano groups of the PAN, as discussed above. The PVdF-HFP polymer is also supposed to be reactive with lithium. It is known that poly(tetrafluoroethylene) (PTFE), which has a chemical structure similar to that of the copolymer of PVdF-HFP, reacts with lithium. (Its white color changes to black when it is in contact with Li metal.) The reaction products are probably LiF and groups with Li—C bonds. In consideration of this, we suggest that the improved performance of Li anodes in the PVdF-HFP-based electrolyte system is due to the formation of some protective layers on the lithium by the reactions of the PVdF-HFP copolymer on the lithium metal.

Figure 7b compares Li cycling efficiency in PVdF-HFP gel electrolyte systems with and without $CO_2$. It is clear that the Li cycling efficiency was increased by the $CO_2$ addition (up to 90–97%). Although a similar tendency was obtained also with PEO gel electrolyte, the Li cycling efficiency for the PVdF-HFP/$CO_2$ electrolyte system was higher than of the PEO with $CO_2$. From these results, we see that using $CO_2$ as an additive is effective not only for liquid electrolyte solutions but also for the gel electrolyte. As reported by Ishikawa et al. [33], the charge-discharge efficiency of Li anodes in an EC/dimethyl carbonate electrolyte containing $LiPF_6$ can be improved by carrying out some charge-discharge cycles at low-temperature conditions. From our results, it appears that the combination of the PVdF-HFP gel electrolyte and $CO_2$ enhances the cycling efficiency of lithium anodes more effectively.

The morphology of the lithium electrodes in the gel electrolytes is shown in Figure 8. In the case of the PEO gel electrolyte, the morphology gradually becomes rough as cycling proceeds. It is clear from the micrographs of Figure 8 that the lithium electrode tested in the PVdF-HFP gel has more uniform surface than that in the PEO gel. As explained above, this is due to the differences in the chemical behavior and the immobility of the electrolyte molecules. By the addition of carbon dioxide to the PVdF-HFP gel, the surface becomes smoother, as seen in Figures 8b and 8c. Simultaneously, impedance measurements confirmed that the interfacial resistance in the $CO_2$-containing electrolyte is lower.

---

**Figure 8**   Micrographs of lithium metal anode at the first and fifth cycle in (a) PEO gel without $CO_2$, (b) PVdF-HFP gel without $CO_2$, and (c) PVdF-HFP gel with $CO_2$.

## 1st      5th

a) PEO

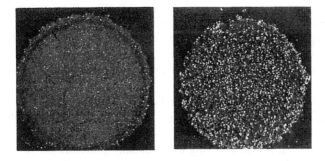

b) PVdF-HFP

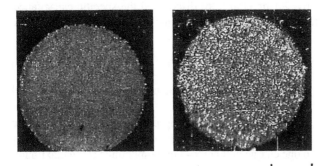

c) PVdF-HFP with $CO_2$

100 μm

(It decreases from 18 to 6 $\Omega$ cm$^2$ due to $CO_2$ addition, similar to the effect of $CO_2$ addition on the behavior of Li electrodes in liquid PC-based electrolyte solutions [14,15].)

## D.  Summary

There are possibilities for using Li metal anodes in rechargeable batteries by selecting and designing a host polymer, supporting electrolyte, plasticizer and additives in ion conducting polymeric matrices. Further analyses and discussion are required for better understanding of the characteristics of the lithium anode in the PVdF-HFP gel system. We believe that polymeric batteries with Li metal anodes have high potential for practical applications in the near future.

## III.  ELECTRIC DOUBLE LAYER CAPACITOR APPLICATIONS

### A.  Introduction

In recent years there has been increasing interest in the power capacitors, ultra-capacitors or supercapacitors based on electrochemical systems. These include electric double layer capacitor (EDLC) types based on carbon electrodes with suitable electrolyte systems, and electrochemical capacitors with pseudocapacitance [34,35].

Several types of capacitors are classified in Table 1 according to their energy storage mechanisms. Redox capacitors function by utilizing electrochemical pseudocapacitance based on fast Faradaic redox [34], insertion of protons into noble metal oxides [36–38], $p$- and $n$-doping of conducting polymers [39–41], lithium ion intercalation [42] or underpotential deposition (adsorption) processes [43]. Among the various materials for redox capacitors, proton insertion into noble metal oxides and $p$- and $n$-doping of conducting polymers have been widely investigated. Conway reported on the use of $RuO_2$ and $IrO_2$ as electrodes of redox capacitors [34]. Recently, Zheng et al. [36] discovered the specific capacitance, 760 F g$^{-1}$, of amorphous hydrous ruthenium oxide, where the process can be expressed as

$$RuO_2 \cdot xH_2O + \delta H^+ + \delta e^- \leftrightarrow RuO_{2-\delta}(OH)_\delta^-, \qquad 0 \le \delta \le 2 \qquad (1)$$

Noble metal oxides, however, are expensive and usually have poor electrical conductivity.

The process of charge storage in conducting polymers requires electronic transport through the polymer backbone. Charging commonly used conducting polymers corresponds to one charge unit per 2–3 monomer units at the maximum.

**Table 1** Classification of Capacitors According to Energy Storage Mechanism

| Type | Basis of charge or energy storage | Example |
|---|---|---|
| Regular dielectric capacitor | Electrostatic | |
| Vacuum | | — |
| Dielectric | | — |
| Electrolytic capacitor | Electrostatic | |
| Oxide-film | | Al, Ta oxide condenser |
| EDLC | Electrostatic | |
| Aqueous electrolyte | | Carbon, activated carbon electrode $H_2SO_4$ aqueous |
| Nonaqueous electrolyte | (Charge separation at | $TEABF_4$/propylenecarbonate |
| Ion-conducting polymer | double layer elec- | PAN gel electrolyte |
| Solid inorganic salt | trode interface) | $RbAg_4I_5$ |
| Redox capacitor | Faradaic charge transfer | |
| Redox oxide film | | $RuO_2 \cdot xH_2O$, $IrO_2$, NiO/Ni, $TiS_2/Li^+$ |
| Redox polymer film | (pseudocapacitance) | Polypyrrole, polythiophene, polyacene |
| Soluble redox system | | $[Fe(CN)_6]^{3-}/[Fe(CN)_6]^{4-}$ |
| Underpotential deposition | | H/Pt, Pb/Au (adsorption) |

Conducting polymers belong to a class of materials with exciting potential in the field of supercapacitors because they can be charged and discharged at high rates.

The principle of EDLC operation is very simple and is based on the well-known electrical or double layer phenomenon. The device operates within a potential range in which no Faradaic reactions take place, and thus the behavior is fully capacitive. Polarization of the electrodes in opposite directions leads to accumulation of opposite charges at the electrode-solution interfaces. The higher the electrode surface area and the polarity of the electrolyte solution and its ionic concentration, the higher are the capacity and energy density of these devices. The capacitance $C$ and the accumulated electrostatic energy $E$ stored are given by Eqs. (2) and (3), respectively [44]:

$$C = \int \frac{\varepsilon}{4\pi\delta} \cdot dS \qquad (2)$$

$$E = \frac{1}{2} CV^2 \qquad (3)$$

EDLC with very low internal resistance (no electrochemical reaction resistance), very high capacitance and extremely long cycle life (ca. $10^6$ cycles) can be composed [45]. Thus, the EDLCs have been envisaged as pulse-power sources, an especially useful complement to batteries in electric vehicles (EV) [34].

The (DOE) Department of Energy targets for EV applications are listed in Table 2 [46]. As shown in Figure 9, the EDLCs developed so far have not reached sufficient energy and power density required for EV application. Improving the energy and power density of EDLC requires development of new electrode materials and electrolyte systems as well as new cell design. It appears from our studies that using thin films of solid polymer electrolyte (SPE) or gel electrolyte in EDLC may be useful in achieving the DOE goals [47].

Ion conducting polymers may be preferable in these devices' electrolytes because of their flexibility, moldability, easy fabrication and chemical stability (for the same reasons that they have been applied to lithium secondary batteries [19,48,49]). The gel electrolyte systems, which consist of a polymeric matrix, organic solvent (plasticizer) and supporting electrolyte, show high ionic conductivity about $10^{-3}$ S cm$^{-1}$ at ambient temperature and have sufficient mechanical strength [5,7,50,51]. Therefore, the gel electrolyte systems are superior to solid polymer electrolytes and organic solvent-based electrolytes as batteries and capacitor materials for ambient temperature operation.

In this section, we review recent developments of redox capacitors with conducting polymer electrodes and EDLC with SPE or gel electrolyte systems.

**Table 2**  Near-Term and Advanced Goals for DOE Ultracapacitor Development Programs

|  | Near term | Advanced |
|---|---|---|
| Battery w/o capacitor | | |
| Weight (kg) | 500–600 | 200–300 |
| Power density (W kg$^{-1}$) | | |
|   Average | 10 | 20 |
|   Gradability | 30–50 | 110–160 |
|   Peak (acceleration) | 80 | 375–550 |
| Ultracapacitor unit | | |
|   Energy stored (Wh) | 500 | 750 |
|   Maximum Power (kW) | 50 | 80 |
|   Weight (kg) | <100 | <50 |
|   Volume (dm$^3$) | <40 | <20 |
|   Energy density (Wh kg$^{-1}$) | >5 | >15 |
|   Maximum usable power density (W kg$^{-1}$) | >500 | >1600 |
| Round-trip efficiency (%) | >90 | >90 |
| Vehicle acceleration (0–88 km h$^{-1}$)/s | <20 | <8 |

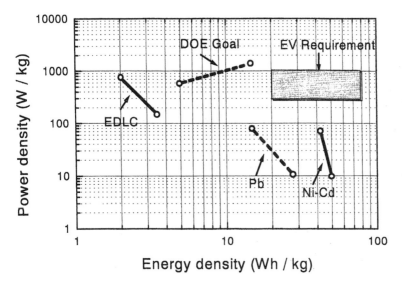

**Figure 9** Status of pulse power source for EV application.

## B. Supercapacitor Based on Electronically Conducting Polymer Electrode

Following Scrosati's suggestion [50], conducting polymers were used as active materials in supercapacitors in three different configurations: In type I the same *p*-dopable conducting polymer is used as the active material for both electrodes. In type II two different *p*-dopable conducting polymers are used as the active material in the two electrodes. In type III a conducting polymer that can be both *n*- and *p*-doped is used. Note that this type of supercapacitor is a symmetric device. In the fully discharged state both electrodes are in the undoped state, whereas in the charged state one electrode is *p*-doped and the other is *n*-doped. The main advantage of this configuration is that all the doping charge can be delivered at high voltages because of the separation of potentials related to the polymer's energy gap between the *p*- and the *n*-doping domains. Another advantage is that the charged device has both electrodes in the conducting states, thereby making type III the most promising polymer supercapacitor in terms of energy and power density. However, only a limited number of conducting polymers can be both *n*- and *p*-doped, primarily polyacetylene, polyacene (PAS), poly-*p*-phenylene and polythiophene (PTh) and its derivatives.

Rudge et al. believe that polythiophene and its derivatives are suitable materials for type III supercapacitors. Especially, the electrochemically prepared polyfluorophenylthiophene (PFPT) demonstrates good cycleability both in *p*- and *n*-doping [50]. Ren et al. have identified poly(3-(*p*-fluorophenyl)thiophene) as a promising candidate for type III supercapacitors with TEABF$_4$ in acetonitrile as the electrolyte solution [52]. Mastragostino et al. selected electrode materials prepared from dithieno(3, 4-*b*: 3′, 4′-*d*)thiophene (DTT1) and dithieno(3, 4-*b*: 2′, 3′-*d*)thiophen (DTT3). They also found that pDTT3 is a more promising and competitive candidate than pDTT1 because the characteristics of the *n*-doping process (which is the more problematic one) are much like those of the *p*-doping one [53].

There are also reports on polypyrrole (PPy) and PAS as electrode materials for supercapacitor application. Naoi et al. have been attempted to produce PPy films for supercapacitors that exchange cations instead of anions in order to increase the energy density and rate capabilities of these devices. There are supercapacitors called ''dual-mode ion switching'' in which anion and cation exchange occurs at different potentials (Figure 10) [54]. PPy is an electronically conducting polymer which is suitable for such a type of supercapacitor. The dopants that provide PPy films with dual-mode ion switching characteristics generally have both hydrophobic and hydrophilic nature. One example is anionic surfactants, and others are multivalent dopants based on large aromatic rings of benzene or naphthalene sulfonates. The capacitors with PPy electrodes exhibit both double layer capacitance and redox capacitance. Specific capacitance values higher than 400–500 F cm$^{-2}$ are possible for PPy-based supercapacitors. Moreover, Yata and Yamaguchi have developed a type III capacitor with PAS electrodes. They fabricated a cylindrical type PAS capacitor (18650 type) of improved performance in terms of high specific capacity, power and energy density [55].

## C. All-Solid-State EDLC with Ion Conducting Polymer

The application of SPE to EDLC is attractive, but, there have been few attempts in this direction for the following reasons: (1) relatively low ionic conductivity of the most relevant polymer electrolyte systems at ambient temperatures; (2) poor contact at the electrode/electrolyte interface; (3) detrimental influence of crystalline domains of the polymer electrolyte systems on their conductivity. Thus, many researchers focus on improving the ionic conductivity and increasing the electrode's capacitance by controlling the electrode/electrolyte interface. The proton conducting polymer (Nylon 6-10, 2H$_3$PO$_4$) has been used as an electrolyte of a power capacitor [56], and this polymer blend associated with activated carbon gives an electrostatic capacity similar to that obtained when using liquid electrolyte solutions. However, the energy density of a capacitor with proton conducting polymer is limited by a voltage of about 1 V (which is the maximal voltage applicable for such systems). Non-proton-conducting solid polymer electrolytes such as poly(ethylene oxide) (PEO) salt or poly(propylene oxide) (PPO)

Anode (PPy)     Cathode (PPy)

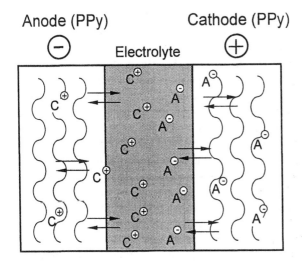

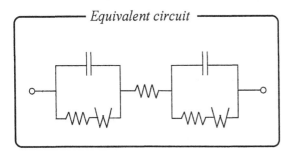

**Figure 10** Concept of supercapacitor devices based on dual-mode ion switching poly-pyrrole.

salt have a low dielectric constant (PEO: $\varepsilon = 5$) and low ionic conductivity at room temperature ($<10^{-5}$ S cm$^{-1}$) [57,58]. However, such non-proton-conducting polymers have the advantage of a large voltage region. Also, they can be fabricated as ultrathin films and thus high energy density may be obtained in devices containing them. Recently, the application of polymer electrolytes such as PEO/LiClO$_4$ SPE for electrochemical devices was attempted [59,60]. The phase diagram and ionic conductivity of PEO/LiClO$_4$ electrolyte systems have already been measured by many researchers [17,18,61]. This polymeric electrolyte consists of crystalline complexes and an amorphous phase. In fact, only the amorphous phase is conducting [62]. However, there are few recent reports on the use of these electrolyte systems for EDLC. Osaka et al. have attempted using

PEO/LiClO$_4$ SPE as an electrolyte medium for EDLC. After a survey of carbon materials as electrodes for these devices, it was found that isotropic high density graphite (HDG) is a good candidate for the electrode material for EDLC of high capacity using a solid-state electrolyte [62,63]. The effects of some properties of EDLC composed of PEO/LiClO$_4$ SPE and HDG electrodes were investigated.

Figure 11 shows the dependence of the capacitance on the concentration of the salt (LiClO$_4$). At about [EO]/[Li$^+$] = 8:1, the capacitance of HDG electrode is the highest among various molar ratios at 80°C or at room temperature [63]. As mentioned, this polymer consists of the PEO-LiClO$_4$ crystalline complex and the amorphous phase [17–19]. Only the amorphous phase provides ionic conductivity in this polymer complex. Also, the electric double layer is formed at the interface between the electrodes and the amorphous phase of the SPE. However, the content of the amorphous phase depends on the ratio of [EO]/[Li$^+$]. It appears that this SPE is mostly amorphous at [EO]/[Li$^+$] = 8:1 at room temperature. Thus, the capacitance becomes the largest at this ratio at 20°C. Although the SPE shell melts at 80°C, the LiClO$_4$ concentration dependence shows behavior similar to that at room temperature. The all-solid-state EDLC with PEO/

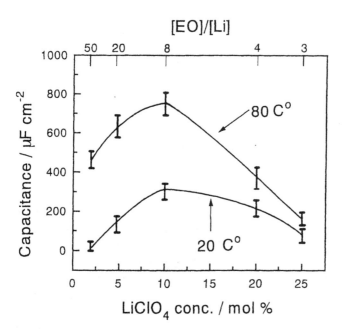

**Figure 11** Concentration dependence of capacitance of EDLC using HDG electrode with PEO/LiClO$_4$ electrolyte at 80°C and 20°C.

LiClO$_4$ SPE ([EO]/[Li$^+$] = 8:1) showed a good charge/discharge behavior over 1000 cycles between 0 and 3.0 V.

## D. EDLC with Polymer Gel Electrolyte

In addition, various gel electrolytes were investigated for EDLC as well. Poly(ethylene oxide) (PEO)-, poly(acrylonitrile) (PAN)-, poly(methylmethacrylate) (PMMA)- and poly(vinylidene flouride) (PVdF)-based gel electrolytes, which have been applied for secondary lithium batteries [5–7,50,51,64], were also studied as electrolyte systems for EDLC with carbon electrodes [65–68].

### 1. Properties of Gel Electrolytes for EDLC Application

We compared the use of PEO-, PMMA- and PAN-based gel electrolytes in EDLC [67]. The chemical composition (in molar ratio) of the gel electrolytes was polymer (PEO, PMMA, PAN)/PC/EC/LiClO$_4$ = 30:19:46.5:4.5. The PAN-based gel electrolyte membranes were prepared by slowly heating the above solution to 100–110°C and pressing it between a pair of polyimide sheets. PEO-based gel electrolyte films were obtained by casting acetonitrile solution of PEO-based mixture (12 mg/ml) on a Teflon sheet and drying it in flowing argon for two days. The PMMA-based gel electrolyte film was prepared by casting a dimethoxyethane (DME) solution of a PMMA-based mixture (11 mg/ml) in a stainless steel cell and drying it in flowing argon for more than two days. The thickness of the films was controlled within 50–500 μm by varying the quantity of the casting mixtures.

The performance of EDLCs with the above gel electrolytes was investigated using isotropic HDG electrodes and is described in Table 3. The ionic conductivities of the various gel electrolytes were around 10$^{-4}$–10$^{-3}$ S cm$^{-2}$, and they decrease in the order PAN > PEO > PMMA at ambient temperature. Capac-

**Table 3** Conductivities of Various Gel Electrolytes and Capacitances of HDG Electrode in Various Gel Electrolytes at Ambient Temperature (measured by ac impedance)

| Electrolyte | A[a] | A/PMMA | A/PAN | A/PEO |
|---|---|---|---|---|
| Conductivity(S cm$^{-1}$/10$^{-4}$) | 65 | 7.6 | 27.2 | 13.6 |
| Capacitance (mF cm$^{-2}$) | 17.6 | 15.8 | 14.6 | 9.9 |
| Stable potential range (V) | 3.5 | 3.5 | 3.5 | 2.5 |
| Mechanical strength | — | Good | Good | Poor |

[a] A = electrolyte LiClO$_4$/PC + EC.

itances approaching the value of EDLC using organic liquid electrolyte, 20 mF cm$^{-2}$, with an HDG electrode were obtained in PAN- and PMMA-based gel electrolytes. As shown in Figure 12, the EDLC with PMMA-based gel electrolyte showed good charge-discharge behavior over 10$^4$ cycles at a charging potential of 3.0 V.

## 2. Activated Carbon Fiber Cloth Electrode

When SPE was used in EDLC with activated carbon fiber cloth (ACFC) electrodes, the device did not function satisfactorily because the SPE hardly fills the micropores of the ACFC. Hence, the poor contact leads to high resistance. However, EDLC with ACFC electrodes can function if a flexible gel electrolyte is used. Matsuda et al. have succeeded in making EDLC with ACFC electrodes using various gel electrolytes based on PEO-PMMA, PAN or PVdF [65,66,68]. The model cell is illustrated in Figure 13. They investigated the effects of the parameters of the electrolyte system chosen on the ionic conductivity of the gels and the capacity of the ACFC electrodes. According to their results, the ionic conductivity increased in the order of TBAClO$_4$ < TEAClO$_4$ < TEABF$_4$, which corresponds to the ion size. The order of the magnitude of the capacitance is TEABF$_4$ > TEAPF$_6$ > TEAClO$_4$ > TBABF$_4$ > TBAClO$_4$, which may correlate with the geometric size of both the cations and the anions, as TEA$^+$ < TBA$^+$ for cations and BF$_4^-$ < PF$_6^-$, ClO$_4^-$ for anions. Using small ions resulted in an

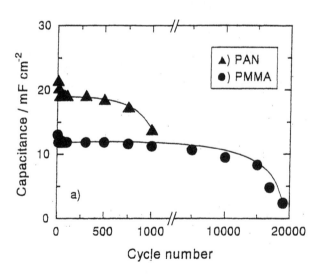

**Figure 12** Cyclability capacitance vs. cycle number for EDLC using PMMA or PAN gel electrolyte (charge-discharge current: 2 mA cm$^{-2}$).

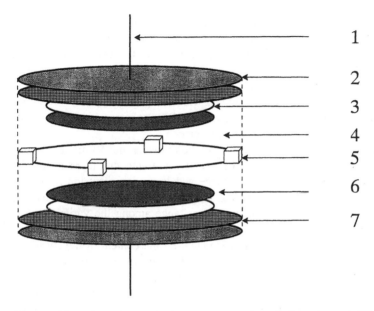

1

2

3

4

5

6

7

**Figure 13** Schematic for a model capacitor with gel electrolyte: (1) Ni wire, (2) Ni plate, (3) conducting plastic sheet, (4) gel electrolyte, (5) Teflon prop, (6) ACFC electrode, (7) Ni mesh.

increase of ionic conductivity and good adsorption of ions on the surface of the carbon electrode. Thus, the application of gel electrolyte with TEABF$_4$ to EDLC with ACFC electrodes resulted in large capacity values [66]. The highest ionic conductivity of ca. $6 \times 10^{-3}$ S cm$^{-1}$ was obtained by using the PVdF/PC/TEABF$_4$ gel electrolyte, and the highest capacitance of ca. 80 F g$^{-1}$ for ACFC electrodes was also obtained in this gel electrolyte [68].

## 3. Carbon/Gel Electrolyte Composite Electrode for EDLC Application

Carbon powder mixed with polymeric binder (PVdF, PTFE) has been widely used as anode material for lithium ion batteries and as the electrode material for EDLC with liquid electrolyte solutions. When such composite electrodes composed of carbon powder and polymer binder were used in all-solid-state EDLC, the performance was not good enough because of poor electrical contact between the electrode's active mass and the electrolyte. By having the electrolyte inside the composite electrode, the contact between the active mass in the electrode and the electrolyte can be considerably improved and hence the capacitance can be enhanced. This can be achieved by using a composite electrode made of active powder and a gel electrolyte. Such composite electrodes with gel electrolytes have been studied as battery cathodes or anodes [69–71]. Osaka et al. investigated the possibility of fabricating an all-solid-state capacitor by using composite electrodes with gel electrolyte [72,73].

The gel polymer electrolytes studied were composed of PVdF or poly(vinylidene fluoride-hexafluoropropylene) (PVdF-HFP) as base materials with the addition of $TEABF_4/EC + PC$ as the plasticizer.

Composite electrodes composed of carbon powder and PVdF gel electrolyte were prepared as electrodes of EDLC. It was found that these composite electrodes have higher specific capacitance and lower ion diffusion resistance compared to carbon electrodes with PVdF binder only. Composite electrodes with acetylene black (AB) show the highest capacitance among the various carbon electrodes because this carbon has very small particle size. The capacitance of AB composite electrodes is strongly dependent on the content of the PVdF gel electrolyte binder as shown in Figure 14. The maximum capacitance was achieved for an AB composite electrode with 30 wt.% PVdF. This high capacity is due to the good electrical contact between the composite electrode and the gel electrolyte when the percentage of the PVdF was about 30%. EDLC with the AB composite electrode and the PVdF gel electrolyte gives an excellent cycleability performance over $10^5$ cycles with very low leakage current [72].

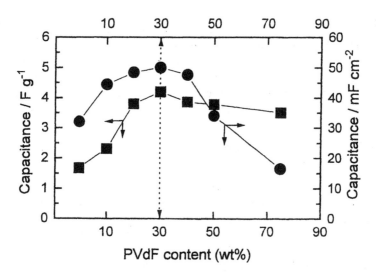

**Figure 14** Dependence of the capacitance of composite AB electrode on the PVdF content. The capacitance was calculated from impedance spectroscopic measurements.

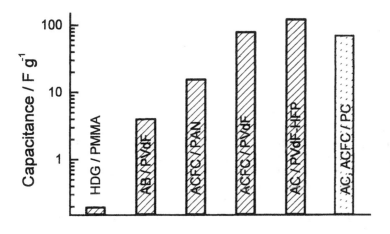

## Interface of electrode / electrolyte

**Figure 15**  Capacitances of various carbon electrodes in various gel electrolytes.

In addition, combinations of activated carbon (AC) powder with PVdF-HFP gel electrolyte as a binder (30 wt.% PVdF-HFP) were prepared and investigated as electrode material for EDLC. When activated carbon, whose specific surface area was 2500 $m^2$ $g^{-1}$ was used, a capacitance of 123 F $g^{-1}$ could be obtained. Such a capacity was higher than that of EDLC with liquid organic electrolyte solutions [73].

In the past five years, many ion conducting polymers and gel electrolytes have been investigated for EDLC application. Figure 15 shows the capacities of various carbon electrodes in SPE or gel electrolytes. The values listed in this figure do not satisfy the requirements for EV. However, it is expected that the requirements of supercapacitors for EV can be achieved by development of devices based on composite electrodes and gel electrolyte systems as described in this chapter.

## E.  Summary

Recent application of conducting polymers for electrochemical capacitors has been reviewed. The ionically conducting polymers have the potential to replace wet-type capacitors by dry-type ones. Especially important are the gel electrolyte systems whose ionic conductivity is high, and thus high capacitance can be obtained with them in EDLC. We believe that the combination of high surface area carbons and ionically conducting gel electrolyte systems is highly promising in the development of novel supercapacitors with improved performance.

## REFERENCES

1. H. Shirakawa, T. Ito, and S. Ikeda, *Polym. J.*, **2**, 231 (1971).
2. T. Ito, H. Shirakawa, and S. Ikeda, *J. Polym. Sci., Polym. Chem. Ed.*, **12**, 11 (1974).
3. G. Feuillade and P. Perche, *J. Appl. Electrochem.*, **5**, 63 (1975).
4. K. M. Abraham and M. Alamgir, *J. Electrochem. Soc.*, **137**, 1657 (1990).
5. G. B. Apetecchi, F. Croce, and B. Scrosati, *Electrochim. Acta*, **40**, 991 (1995).
6. T. Osaka, T. Momma, H. Ito, and B. Scrosati, *J. Power Sources*, **68**, 392 (1997).
7. S. Kakuda, T. Momma, T. Osaka, G. B. Apetecchi, and B. Scrosati, *J. Electrochem. Soc.*, **142**, L1 (1995).
8. J. Fuller, A. C. Breda, and R. T. Carlin, *J. Electrochem. Soc.*, **144**, L67 (1997).
9. Z. Jiang, B. Carroll, and K. M. Abraham, *Electrochim. Acta*, **42**, 2667 (1997).
10. J.-M. Tarascon, A. S. Gozdz, C. Schmutz, F. Shokoohi, and P. C. Warren, *Solid State Ionics*, **86–88**, 49 (1996).
11. B. B. Owens and T. Osaka, *J. Power Sources*, **68**, 173 (1997).
12. T. Osaka, T. Nakajima, K. Shiota, and B. B. Owens, in *Rechargeable Lithium Batteries/1989*, S. Subbarao, V. R. Koch, B. B. Owens, and W. H. Smyrl (eds.), The Electrochemical Society Proceeding Series, Pennington, NJ (1990), PV905, p. 170.
13. T. Osaka and S. Komaba, *Plastics*, **48** (7), 10 (1997).
14. D. E. Fenton, J. M. Parker, and P. V. Wright, *Polymer*, **14**, 589 (1973).
15. M. B. Armand, J. M. Chabagno, and M. J. Duclot, *Fast Ion Transport in Solids*, North-Holland, New York (1979), p. 131.
16. J. R. MacCallum and C. A. Vincent, Ed., *Polymer Electrolyte Reviews 2*, Elsevier, London (1989).
17. C. D. Robitaille and D. Fauteux, *J. Electrochem. Soc.*, **133**, 315 (1986).
18. P. Ferloni, G. Chiodelli, A. Magistris, and M. Sanesi, *Solid State Ionics*, **18,19**, 265 (1986).
19. T. Osaka, T. Momma, K. Nishimura, S. Kakuda, and T. Ishii, *J. Electrochem. Soc.*, **141**, 1994 (1994).
20. C. W. Walker, Jr., and M. Salomon, *J. Electrochem. Soc.*, **140**, 3409 (1993).
21. X. Q. Yang, H. S. Lee, L. Hauson, J. McBreen, and Y. Okamoto, *J. Power Sources*, **54**, 198 (1995).
22. M. Watanabe and A. Nishimoto, *Solid State Ionics*, **79**, 306 (1995).
23. D. Aurbach, Y. Gofer, M. B.-Zion, and P. Aped, *J. Electroanal. Chem.*, **339**, 451 (1992).
24. D. Aurbach and A. Zaban, *J. Electroanal. Chem.*, **348**, 155 (1993).
25. T. Osaka, T. Momma, T. Tajima, and Y. Matsumoto, *J. Electrochem. Soc.*, **142**, 1057 (1995).
26. T. Osaka, T. Momma, Y. Matsumoto, and Y. Uchida, *J. Electrochem. Soc.*, **144**, 1709 (1997).

27. T. Osaka, T. Momma, Y. Matsumoto, and Y. Uchida, *J. Power Sources*, **68**, 497 (1997).
28. S. Shiraishi, K. Kanamura, and Z. Takehara, *J. Appl. Electrochem.*, **25**, 584 (1994).
29. K. Kanamura, S. Shiraishi, and Z. Takehara, *J. Electrochem. Soc.*, **143**, 2187 (1996).
30. T. Osaka, T. Homma, T. Momma, and H. Yarimizu, *J. Electroanal. Chem.*, **421**, 153 (1997).
31. T. Osaka, S. Komaba, Y. Uchida, M. Kitahara, T. Momma, and N. Eda, *Electrochem. Solid State Lett.*, **2**, 215 (1999).
32. Y. Uchida, S. Komaba, T. Osaka, and N. Eda, in *Batteries for Portable Applications and Electric Vehicles*, C. F. Holmes and A. R. Landgrebe (eds.), The Electrochemical Society Proceeding Series, Pennington, NJ, **PV 97-18**, pp. 70–76 (1997).
33. M. Ishikawa, Y. Takai, M. Morita, and Y. Matsuda, *J. Electrochem. Soc.*, **144**, L90 (1997).
34. B. E. Conway, *J. Electrochem. Soc.*, **138**, 1539 (1991).
35. B. E. Conway, in *Proc. Symp. Electrochemical Capacitors*, F. M. Delnick and M. Tomkiewicz (eds.), The Electrochemical Society (1995), p. 15.
36. J. P. Zheng, P. J. Cygan and T. R. Jow, *J. Electrochem. Soc.*, **142**, 2699 (1995).
37. K. C. Liu, M. A. Anderson, *J. Electrochem. Soc.*, **143**, 124 (1996).
38. B. E. Conway and J. Mozota, *Electrochim. Acta*, **28**, 1 (1983).
39. C. Azbizzani, M. Mastagostino, and L. Meneghello, *Electrochim. Acta*, **40**, 2223 (1995).
40. D. Ofer, R. M. Crooks, and M. S. Wrighton, *J. Am. Chem. Soc.*, **112**, 7869 (1990).
41. A. Rudge, J. Davey, I. Raistrick, S. Gottesfeld, and J. P. Ferraris, *J. Power Sources*, **47**, 89 (1994).
42. S. Passerini, J. J. Ressler, D. B. Owens, and W. H. Smyrl, in *Proc. Symp. Electrochemical Capacitors*, F. M. Delnick and M. Tomkiewicz (eds.), The Electrochemical Society, (1995), p. 86.
43. B. E. Conway and B. V. Tilak, *Adv. Catal.*, **38**, 1 (1992).
44. A. Nishino, A. Yoshida, I. Tanahashi, T. Tajima, M. Yamashita, T. Muranaka, and H. Yoneda, *Nat. Tech. Rep.*, **31**, 318 (1985).
45. T. C. Murphy and W. E. Kramer, *Proc. Fourth Int. Sem. on Double Layer Capacitor and Similar Energy Storage Devices*, Florida Educational Seminars, December (1994).
46. *Proc. 3rd Int. Sem. on Double Layer Capacitor and Similar Energy Storage Devices*, Florida, December (1993).
47. T. Momma, S. Komaba, and T. Osaka, *Denki Kagaku*, **65**, 115 (1997).
48. J. R. MacCallum and C. A. Vincent, Eds., *Polymer Electrolyte Reviews 2*, Elsevier, London (1989).
49. S. Takeoka, H. Ohno, and E. Tsuchida, *Polym. Adv. Technol.*, **4**, 67 (1992).
50. F. Croce, F. Gerace, G. Dautzenberg, S. Passerini, G. B. Apetecchi, and B. Scrosati, *Electrochim. Acta*, **39**, 2187 (1994).
51. A. Reiche, T. Steurich, B. Sandner, P. Lobitz, and G. Fleischer, *Electrochim. Acta*, **40**, 2153 (1995).
52. X. Ren, S. Gottesfeld, and J. P. Ferraris, in *Electrochemical Supercapacitors / 1995*, B. M. Barnett et al. (eds.), PV95-29, The Electrochemical Society Proceedings, Pennington, NJ (1995), p. 138.

53. M. Mastragostino, C. Arbizzani, M. G. Cerroni, and R. Paraventi, in *Proc. Symp. on Electrochemical Capacitors II*, F. M. Delnick, D. Ingersoll, X. Andrieu, and K. Naoi (eds.), PV 96-25, The Electrochemical Society Proceedings, Pennington, NJ (1997), p. 109.

54. K. Naoi, Y. Oura, and H. Tsujimoto, in *Proc. Symp. on Electrochemical Capacitors II*, F. M. Delnick, D. Ingersoll, X. Andrieu, and K. Naoi (eds.), PV 96-25, The Electrochemical Society Proceedings, Pennington, NJ (1997), p. 120.

55. S. Yata and S. Yamaguchi, *Electron. Mater. Jpn.*, **11**, 50 (1997).

56. J-C. Lass gues, J. Grondin, T. Becker, L. Servant, and M. Hernandez, *Solid State Ionics*, **77**, 311 (1995).

57. K. M. Abraham and M. Alamgir, *Solid State Ionics*, **70/71**, 21 (1994).

58. W. Wieczorek, Z. Florjanczyk, and J. R. Stevens, *Electrochim. Acta*, **40**, 2251 (1995).

59. K. C. Liu and M. A. Anderson, *J. Electrochem. Soc.*, **143**, 124 (1996).

60. T. Osaka, T. Nakajima, K. Shiota, and T. Momma, *J. Electrochem. Soc.*, **138**, 2853 (1991).

61. M. J. Hudson and C. A. C. Sequeira, *J. Electrochem. Soc.*, **142**, 4013 (1995).

62. X. Liu, T. Momma, and T. Osaka, *Chem. Lett.*, 625 (1996).

63. X. Liu and T. Osaka, *J. Electrochem. Soc.*, **143**, 3982 (1996).

64. C. Schmutz, J. M. Tarascon, A. S. Gozdz, P. C. Warren, and F. K. Shokoohi, *Electrochem. Soc. Proc.*, **28**, 330 (1994).

65. M. Ishikawa, M. Morita, M. Ihara, and Y. Matsuda, *J. Electrochim. Soc.*, **141**, 1730 (1994).

66. M. Ishikawa, M. Ihara, M. Morita, and Y. Matsuda, *Electrochim. Acta*, **40**, 2217 (1995).

67. X. Liu and T. Osaka, *J. Electrochem. Soc.*, **144**, 3066 (1997).

68. M. Ishikawa, M. Morita, and Y. Matsuda, in *Proc. Symp. on Electrochemical Capacitors II*, F. M. Delnick, D. Ingersoll, X. Andrieu and K. Naoi (eds.), PV 96-25, The Electrochemical Society Proceedings, Pennington, NJ (1997), p. 325.

69. V. Haddadi-asl, M. Kazacos, and M. Skyllas-kazacos, *J. Appl. Electrochem.*, **25**, 29 (1995).

70. M. Y. Saidi, R. Koksbang, E. S. Saidi, and J. Barker, *J. Electrochim. Acta*, **42**, 1181 (1997).

71. K. M. Abraham and Z. Jiang, *J. Electrochem. Soc.*, **143**, 1 (1996).

72. T. Osaka, X. Liu, and M. Nojima, *J. Power Sources*, 70, (1998).

73. X. Liu, M. Nojima, T. Momma, and T. Osaka, *Proc. 65th Conf. Electrochem. Soc. Jpn.*, Tokyo (1998), 3F31.

# 8
# Nonaqueous Batteries

**Larry A. Dominey**
*OMG Americas, Inc., Cleveland, Ohio*

## I. INTRODUCTION

Wherever energy is needed in mobile or remote circumstances, batteries are the likely option. Water-based electrolyte battery systems supported these requirements through the first half of the 1900s. However, the window of electrochemical stability of water-based electrolytes, including overpotentials, cannot extend much beyond 2.1 V, this being so for the lead-acid system.

Since power is a function of the product of current ($i$) and voltage ($V$), $P = iE$, it was logical to search for higher voltage electrochemical couples which could only be stably interfaced with more electrochemically robust nonaqueous solvents. In the late 1950s work began in earnest to understand the behavior of electrochemical couples with expanded voltage windows, particularly in the laboratory of Charles Tobias at the University of California at Berkeley [1]. Since that early work, the field has continued to grow by way of different electrolyte systems, both organic-solvent- and inorganic-solvent-based, a multitude of cathode materials with many different fascinating structural features which govern their voltage profiles, and a variety of strongly reducing negative electrode materials as well. However, Li metal and related Li-based compounds have dominated the negative electrode research far beyond any other substance due to its extremely low reduction potential and light atomic weight. Metallic Li, Li alloys, and lithiated carbons are the key negative electrode materials for study and commercialization, particularly for small batteries.

A few words are in order regarding how battery performance is often quantified. Two of the most frequently employed properties are the gravimetric energy density and the volumetric energy density. These simply translate as energy/unit mass, e.g., (watt-hours)/kg or energy/unit volume, e.g., (watt-hours)/liter. Due to the high voltages attainable from nonaqueous couples, based on either gravimetric or volumetric measures, such systems can offer significantly more energy density than most any aqueous system.

Lead references to more comprehensive review articles are included here [2–4] and in most sections, as I have endeavored to provide a concise introduction to the broad array of technologies commercially available and under development, and to introduce and explain certain key chemical and electrochemical principles underlying this field.

*Caveats and Conventions.* A comment on descriptive conventions is in order here at the beginning. Electrochemical couples will be briefly written as "negative electrode (anode during discharge)/positive electrode." So by example a "$C/LiCoO_2$" cell is composed of a carbon negative electrode which can donate Li on discharge and a lithium cobalt dioxide positive electrode which takes up the Li, via ions diffusing through electrolyte and electrons through external circuit. The electrolyte may also be specified as a third entry between the two electrodes, e.g., $Li/LiBF_4$-PC-DME/$MnO_2$. Reference voltages are referenced to metallic Li. Abbreviations and acronyms are collected in Section XII.

The use of the word *electrolyte* is common in the battery industry to mean salt + solvent. We will use that convention here, realizing that, rigorously, such as in the more academic references, electrolyte refers to the salt alone. The salt that dissociates to provide the ionic conductivity of the electrolyte is formally known as the supporting electrolyte or solute.

As another matter of convention, a single-use, i.e., nonrechargeable, battery is known also as a primary battery. A rechargeable battery is also known as a secondary battery.

Also, due to the somewhat proprietary nature of much of this technology, the described components and chemistries are based on what can be best surmised from presentations at technical meetings, product literature and other publicly available information, and first-hand experience. There is substantial fine detail to many of these systems and many subtleties remain the secret domain of the battery experts.

## II. ADVANTAGES OF NONAQUEOUS BATTERIES

Many of the advantages of nonaqueous battery systems apply to higher value-added/extended performance situations. Appropriate use environments include high quality consumer products, extreme temperature conditions (both low and high temperatures), high energy density demand and numerous uniquely demanding military and even espionage oriented environments. The specific battery properties which enable such advantages include most importantly:

1.  The substantially lower freezing point of many nonaqueous electrolyte systems relative to aqueous systems enables reasonably good low temperature performance, sometimes down to $-40°C$.
2.  Wider voltage windows of many nonaqueous solvents as compared with water which enable the use of strong reducing metals such as Li, Na, and, in principle, Mg, Al, and Ca.
3.  Lighter negative electrode materials per unit charge (Li in particular, also Mg and Al), therefore higher gravimetric and volumetric energy densities.
4.  Lower self-discharge rates, especially in primary metal anode systems such as $Li/MnO_2$ and $Li/SO_2$.
5.  The use of solvent-free electrolytes, i.e., high temperature ionic liquids (molten salt) electrolytes, which only become ionically conductive upon melting when heated to high temperatures. Another variant is the use of low rate, room temperature solid electrolytes.

## III.  DISADVANTAGES OF NONAQUEOUS BATTERIES

1.  In comparison with the highly conductive water-based electrolytes containing high concentrations of KOH or sulfuric acid, the organic-solvent-based electrolytes are usually orders of magnitude less conductive. The organic solvents have significantly less ability to solubilize and dissociate supporting electrolytes. Lower salt concentrations and reduced extents of salt dissociation translate to orders of magnitude lower ionic conductivity and consequently lower rate capability.

2.  It is noteworthy, however, that the conductivities of a few inorganic-solvent-based electrolyte systems are of the same order as water systems. In particular, very high currents can be supported by LiBr in 30% acetonitrile/70% sulfur dioxide, which has the highest conductivity of any nonaqueous electrolyte, $5 \times 10^{-2}$ $(ohm-cm)^{-1}$ at 25°C.

3.  The solvents and salts comprising organic electrolytes cost more than common aqueous systems. The separator materials are often more costly also.

4.  Toxicity and flammability of the nonaqueous solvents are also more of a concern since there is no solvent known which is safer than water. This issue

is not of much importance to the end user under normal circumstances, but requires substantial additional precautions during the manufacturing process where solvent vapors must be controlled and special battery design considerations are required.

5.   In the case of Li metal anode batteries, care must be taken to avoid temperatures which exceed the relatively low melting point of Li (180°C) as the molten metal is far more reactive than the solid, due to the instability of the passivating film on a hot molten liquid surface. This problem is sometimes circumvented in primary cells by the use of higher melting Li alloys, but pays the cost of a lower operating voltage.

6.   Finally, the reactivity of Li metal and its related low potential materials is extreme. With the exception of noble gases such as He and Ar, Li has been found to reduce almost all other known substances. Thus, ideally, it must be handled only in highly purified atmospheres of He or Ar. Under practical manufacturing conditions it can be briefly exposed to environments of dry air when the relative humidity is below 1%.


## IV.   APPLICATIONS

Nonaqueous batteries can be designed to meet an almost unlimited variety of uses. This results from the great flexibility of choice regarding solvents, active materials, i.e., positive and negative electrode materials, and supporting electrolytes.

Numerous consumer and military applications are known. In the military arena many applications exist for primary Li batteries. The diversity of uses can be surveyed from this list of 1991 Gulf War applications [5]: SINGARS (single channel ground to air radio signal), Scrambler (communications) units, night vision goggles, laptop computers, memory backup, chemical agent monitors, field radios, helicopter pilot helmet filtration systems and survival radios, to name only some. This last device was responsible for saving the life of a fighter pilot downed inside Iraq. Another widely used military battery type is the thermal battery which operates at elevated temperatures, usually delivering large amounts of power.

In the consumer sector uses for primary nonaqueous batteries include ''button'' or ''coin'' cells for watches, calculators and other small devices with relatively low power requirements. Larger cylindrical and prismatic rechargeable consumer-oriented nonaqueous batteries are essentially all based on a Li or, most frequently, ''Li-ion'' negative electrode technology and are increasingly widely used in many portable electronic devices.

## V. SUMMARY HISTORY OF COMMERCIAL SMALL CELL NONAQUEOUS BATTERIES

Here is a short chronological overview of some of the nonaqueous concepts that have reached the marketplace.

1969: LI/SOCl$_2$ primary battery described by Gabano at SAFT. It is developed for military applications in the early 1970s.

1973: Matsushita introduces Li/(CF)$_x$ primary batteries. They were first proposed in 1970 [6].

1976: Sanyo introduces Li/MnO$_2$ primary batteries.

1977: Exxon introduces rechargeable Li/TiS$_2$ with LiAsF$_6$–2-MeTHF electrolyte (1.9 V). This was the first demonstrated rechargeable technology with reasonably good cycle life. Coin cells were introduced, but the technology did not receive significant market acceptance.

1989: Moli Energy introduces rechargeable Li/MoS$_2$ with $1M$ LiAsF$_6$ 1:1 EC/PC electrolyte (1.6 V). Spiral-wound AA size cells were introduced, but safety problems, possibly attributable to internal Li dendrite shorting failures, were not satisfactorily managed. The product was discontinued after cells in portable phones began venting with flame.

1990: Moli Energy introduces rechargeable Li/MnO$_2$ (3.0 V). Reengineered higher voltage technology made a brief entrance into the marketplace, but was upstaged by the introduction of the first Li-ion cells which had several-fold higher cycle life and improved safety.

1991: Sony introduces rechargeable nongraphite C/LiCoO$_2$ with 1 $M$ LiPF$_6$ 1:1 PC/DEC electrolyte (3.7 V). This was the first commercial Li-ion system and harbinger of numerous subsequent variations. Concurrent with the demonstration of the first anode/cathode dual intercalation system, this technology also proved that LiPF$_6$ could be purified and stabilized as a viable supporting electrolyte. Prior to this development most companies could only obtain good cycle life with high purity LiAsF$_6$. The elimination of an arsenic-based solute removed another impediment to widespread acceptance of rechargeable Li technology.

1994: Tadiran: Li/MnO$_2$ (3.0 V), a return to rechargeable Li metal utilizing novel chemical safety concepts. Widespread acceptance has been hampered by memories of earlier rechargeable Li metal systems' limitations.

1994–present: widespread introduction of graphite/LiCoO$_2$ (3.7 V) cells. Many approaches to enhance capacity and reduce cost through changes in the negative and positive electrode active materials. For instance,

1996: Moli Energy (Japan) introduces carbon/LiMn$_2$O$_4$ (3.8 V), a slightly lower capacity Li-ion technology, with safety advantages under severe overcharge conditions. A complex balance of trade-offs regarding cost, safety and performance keeps the interest level high in the $LiMn_2O_4$ positive electrode approach.

## VI. IMPORTANCE OF THE ELECTROLYTE AND THE INTERFACES WITH ACTIVE MATERIALS

### A. Electrolyte

Batteries, by their very nature, are thermodynamically unstable devices. The key to designing a system which will store energy until needed without internally corroding lies in controlling the reactivity of the active materials with the electrolyte. The solvents, organic and inorganic, have functional groups which readily react. In the case of organic solvents, the hydrocarbon groups and nonbonding oxygen electrons can be oxidized at the positive electrode and the oxygen containing groups can be reduced at the negative electrode. Several review articles have covered the evolving developments in this area over the years [7–9].

Similarly, inorganic solvents such as thionyl chloride ($SOCl_2$), sulfuryl chloride ($SO_2Cl_2$) and sulfur dioxide are readily reduced at the Li negative electrode, but are kinetically hindered from continued reaction by formation of LiCl and related passivating films which readily form on open circuit storage. These inorganic solvents actually serve as liquid positive electrodes. More on both matters will be described shortly. Complex anions of the supporting electrolyte are usually also susceptible to decomposition reactions, always at Li electrodes and frequently at positive electrodes when voltages extend much beyond 3 V.

To not compromise the active material's kinetics, i.e., its ability to deliver charge unimpeded, the ionic conductivity of the electrolyte must not be rate limiting. The conductivities of most common electrolyte categories are collected in Table 1. Li salt concentrations in the vicinity of 1 $M$ typically provide maximum conductivity. Due to the low solvating power of most organic solvents relative to water, ions are not fully dissociated at these concentration levels in nonaqueous solvents but are present as complex mixtures of equilibrating ion pairs, solvent separated ion pairs, triple ions, ion quadrupoles, and so on. Generally speaking, as electrolyte conductivity drops much below $1 \times 10^{-3}$ (ohm-cm)$^{-1}$, internal resistance becomes prohibitively high for most applications.

### B. Interfaces

During the early years of nonaqueous battery development the lack of sensitive surface analytical techniques and the availability of only limited capability impedance spectroscopy techniques precluded understanding much about the nature of the interface between the electrolyte and active materials. It was not always

**Table 1** Conductivity Ranges of Common
Electrolytes

| Electrolyte type | Specific conductivity $(ohm\text{-}cm)^{-1}$ |
|---|---|
| Aqueous | $1\text{--}5 \times 10^{-1}$ |
| Molten salt | around $10^{-1}$ |
| Inorganic | $2 \times 10^{-2}\text{--}10^{-1}$ |
| Organic | $10^{-3}\text{--}10^{-2}$ |
| Gel polymer | $10^{-5}\text{--}5 \times 10^{-3}$ |
| Neat polymer | $10^{-8}\text{--}10^{-4}$ (@ 40--100°C) |
| Inorganic solid | $10^{-9}\text{--}10^{-5}$ |

appreciated that very thin passivating films often governed the stability of the active materials through kinetic control. Since the realization of the significance of these films and the advent of numerous surface science techniques, reasonably thorough characterizations of many of these interfacial regions have occurred [10–12]. It is now understood that bulk electrolyte is separated from intimate contact with extreme high and low voltage active materials by a thin film of electrolyte decomposition products and that the existence of these films is central to the functioning of nonaqueous batteries.

A unique set of electronic and physical properties regarding ion and electron mobilities in the passivating film must exist for practical operation. The ion of the negative electrode metal must be able to readily diffuse across the film. However, the film should not support electron mobility or the film would grow indefinitely. Ideally, the film should have some flexibility to accommodate electrode shape changes caused by depletion and replenishment of active material, which must occur during the course of cell operation. And of course, the decomposition products must be insoluble in the bulk electrolyte, thereby remaining localized at the surface of the active material, or else irreversible self-discharge will readily ensue.

These requirements hold for the films at both the positive and negative electrode surfaces. Thus, these surface films frequently comprise quite complex mixtures of reaction products and their presence affects the kinetic properties of charge transfer across the interface. It is the deviation of surface film's properties from meeting this set of ideal requirements that is the single most important cause of cell failure in a large fraction of cases. When the decomposition reactions occur, a small amount of active material must also be irreversibly consumed.

A great deal of work has been devoted to understanding the reactivity of a wide variety of anions and the effects of their decomposition products on the stability of passivating films, particularly with regard to the negative electrode.

Common Li salt anions include $ClO_4^-$, $BF_4^-$, $PF_6^-$, $AsF_6^-$ and $AlCl_4^-$. A very useful alternative group of anions based on *organic* moieties possessing the most highly electron withdrawing —$SO_2CF_3$ group are now under serious investigation. Reference 9 provides a recent extensive review regarding the importance of the supporting electrolyte anion.

To summarize, the surface film must restrict electron flow from active material to bulk electrolyte. It must restrict diffusion of all solvent and anion from bulk electrolyte to the active material. Finally, it must facilitate rapid diffusion of the ion of the active material, e.g., $Li^+$, through the film in both directions.

## VII.  ELECTROCHEMICAL DEPOSITION MECHANISMS AND SUBSTRATES

### A.  Pure Metal Anodes

Metal anodes such as Li, Na, Mg, Ca, and Al do not redeposit in any highly reproducible manner. Entropic forces and current density inhomogeneities favor a redistribution, and so pure metal electrodes are most commonly found in primary batteries, where this limitation is irrelevant.

### B.  Intercalation Cathodes and Anodes

To maintain stable electrode geometry in rechargeable batteries, intercalation electrodes are most frequently employed. Such electrodes rely on an electrically conductive, mechanically stable atomically porous framework through which the active material's ion can migrate. Intercalation chemistry is a subset of the field of host-guest chemistry. In this case, the framework is the macroscopic host and the mobile ion constitutes the guest.

Positive electrode frameworks are typically either planar two-dimensional arrays or three-dimensional tunnel networks and typically based on transition metal oxide crystalline frameworks [13–15]. The framework usually consists of metal-centered oxygen-bordered octahedra or tetrahedra interconnected through the sharing of edges (two oxygen atoms) or faces (three oxygen atoms). Oxides of transition metals offer the highest voltages. Lower voltage cathodes result in materials composed of larger chalcogenide frameworks based on sulfur or selenium instead of oxygen. The lower voltages result as the larger, softer chalcogenides more effectively depolarize the transition metal ion by donating electron density, i.e., some covalency in the bonding. A variety of amorphous glasses (mostly short range order/long range disorder) of transition metal chalcogenides are also known.

Intercalation cathode materials generally exhibit electronic conductivities typical of semiconductors or poorly conductive metals. Viable intercalation compounds must conduct electrons through the host matrix and allow charge balance by allowing ready ion migration between the layers or through the channels. During discharge the Li-ion migration is driven by the electrochemical potential difference conveyed from the anode to entering electrolyte cations via the electronically conductive transition $d$ orbital manifold, as electron mobility is always higher than ion mobility.

Negative electrode intercalation frameworks are usually based on various allotropes of carbon, most commonly high temperature formed highly ordered graphite, low temperature formed disordered petroleum cokes, and numerous intermediate modifications [16–19]. Again, the diffusion of Li ions between the electronically conductive graphite planes is the mechanism for ion storage. On charging, Li ions enter and diffuse through the carbon host and electrons enter from the current collector, at higher diffusion rates, thereby driving the equilibration.

Intercalation negative electrodes provide less energy density but better rechargeability than the pure metal, since occupation in the framework by the ions affords significantly lower metal atom packing densities than the metallic elemental state. Caution must be exercised to not exceed the coulombic capacity of the negative electrode host lattice or plating of the pure metal at low voltage will occur. Should this happen the increased risks associated with highly reactive metallic deposition products present themselves.

*Lithium-Ion Technology.* This rechargeable technology [20,21] is based on coupling a graphitic or partially graphitic carbon "host" negative electrode material which can take up Li ions in the discharge mode with a metal oxide positive host which accepts Li ions. When both the positive and negative electrodes are intercalation compounds, the full cell design is sometimes colloquially termed a "rocking chair" battery because the ions rock between the two electrodes.

The most commonly encountered Li-based rechargeable nonaqueous systems rely on circa 4 V intercalation positive electrodes such as $LiCoO_2$ or lambda phase $Mn_2O_4$ prepared as $LiMn_2O_4$, a spinel structure. Less frequently encountered are $V_2O_5$, $V_6O_{13}$, and, even least frequently, lower voltage second and third row transition metal chalcogenides such as $MoS_2$, $TiS_2$, and $NbSe_3$.

Li-ion rechargeables are presently finding widespread utility in portable electronic devices such as laptop computers, camcorders and cellular telephones.

## C. Alloyed Negative Electrodes

A third type of negative electrode is based on an alloy of the negative electrode active material with one or more other metals [22]. Such anodes offer only limited rechargeability as the alloys lose their dimensional stability over long term cycling due to particle fracture induced by large volume changes encountered during uptake and loss of active metal. The use of alloy anodes also imparts a reduction in energy density relative to the use of the pure metal since the electrochemical reduction potential is diluted and volumetric and gravimetric energy densities are diluted.

Alloyed negatives are most advantageously employed in primary applications, either at room temperature or when electrode stability at higher temperatures is required. While Li metal melts at relatively low temperature, the melting point of some Li/Al and Li/Si alloys are several hundred degrees higher than that of pure Li metal. The use of these high melting alloys allows the construction of molten salt electrolyte–based high temperature, high rate nonaqueous batteries which retain much of the energy of Li metal primary batteries. Such batteries are important in many specialized military applications. When designed for room temperature use, alloy anode/liquid electrolyte designs enhance battery safety when exposed to thermal abuse conditions.

## VIII.  PRIMARY BATTERIES

Table 2 lists the important properties of the most strongly reducing metals which could, in principle, find advantageous use in nonaqueous systems. Most nonaqueous primary batteries rely on a metallic Li negative electrode because it is the lightest metal and possesses the most extreme reduction potential. The other metals find less widespread application mainly due to lower voltages, less flexibility regarding electrolyte options, and lower solid state ionic diffusivities. Many highly conductive nonaqueous Li salt solutions are known, which allow much flexibility in cell designs using this anode material. Well-designed Li metal anode cells have very low self-discharge and provide useful shelf lives of many years. A variety of practical configurations are known. Three of the most popular follow.

**Table 2**   Selected Properties of Strongly Reducing Metals

| Metal | Standard potential at 25°C (V) | Atomic weight (g) | Density (g/cc) | Melting point (°C) | Valence | Gravimetric capacity (Ah/g) | Volumetric capacity (Ah/cc) |
|-------|------|------|------|------|------|------|------|
| Li | −3.05 | 6.94 | 0.54 | 180 | 1 | 3.86 | 2.08 |
| Na | −2.7 | 23.0 | 0.97 | 97.8 | 1 | 1.16 | 1.12 |
| Mg | −2.4 | 24.3 | 1.74 | 650 | 2 | 2.20 | 3.8 |
| Al | −1.7 | 26.9 | 2.70 | 659 | 3 | 2.98 | 8.1 |
| Ca | −2.87 | 40.1 | 1.54 | 839 | 2 | 1.34 | 2.06 |

1. Liquid cathodes: These contain a soluble Li-ion-based supporting electrolyte which also serves as the cathode active material. They find a variety of military and, to a lesser extent, consumer applications, e.g., powering computer memory backup. They are becoming more widely used where light size and good energy densities are required [23,24]. The low freezing points of the two most common liquid cathodes, $SO_2$ (fp $-73°C$) and $SOCl_2$ (fp $-105°C$), afford unprecedented high quality performance at temperatures down to $-40°C$ or lower when used in conjunction with an organic solvent diluent such as acetonitrile. Their reactions and voltages are

   $$2Li + 2SO_2 \rightarrow 2Li_2S_2O_4 \quad (3.1 \text{ V})$$
   $$4Li + 2SOCl_2 \rightarrow 4LiCl + S + SO_2 \quad (3.65 \text{ V})$$

2. Liquid electrolyte/solid cathode: These usually involve a spirally wound or parallel plate configuration to maximize electrode surface area and ion diffusion. Some popular couples include 3 V $Li/MnO_2$, 3 V $Li/CF_x$ and 1.5 V $Li/FeS_2$. Common organic solvents with freezing points below that of water make these designs also superior to conventional aqueous primaries for low temperature requirements such as wintertime outdoor photography and recreational activities.

3. Low rate solid-electrolyte-based cells: For instance, $Li/I_2$ cells used primarily in implantable medical devices are well established. Another example is the developmental all-solid-state Li metal/phosphorous oxynitride (PON)/intercalation cathode cells conceived for use in microelectronic circuits. The PON is a glassy ceramic electrolyte which is stable to over 5 V [25].

## A. Selected Conventional Room Temperature Primary Cell Designs

Some popular liquid and solid cathode electrochemical couples are here described.

### 1. Liquid Cathode (Catholyte) Cells

The three most common liquid cathodes are $SO_2$, $SOCl_2$ and $SO_2Cl_2$. Due to the presence of a liquid positive electrode, all three cathode designs can deliver outstanding performance at temperatures as low as $-40°C$. As more chlorine is added to the inorganic sulfur-oxygen molecule the voltage of discharge increases. So $V(SO_2) < V(SOCl_2) < V(SO_2Cl_2)$.

The lower voltage $SO_2$ system was the first into the market. Operating at 2.7 to 2.9 V the $Li/SO_2$ cell is one of the highest power density cells known, capable of practically delivering 240 Wh/kg (400 Wh/L). The Li/SOCl$_2$ system operates at 3.5 V and, depending on design considerations, can offer in practice between 320 and 480 Wh/kg (700–950 Wh/L). The Li/SO$_2$Cl$_2$ offers the highest operating voltage and, when appropriately designed, the highest energy density of the three—3.9 V with over 450 Wh/kg (900 Wh/L).

## 2. Solid Cathode Cells

*Li/CF$_x$*

The lithium/carbon monofluoride cell possesses very high theoretical (2260 Wh/kg) and practical energy densities [26]. The cathode material is made by the reaction of carbon with fluorine. The overall reaction is

$$x\text{Li} + (\text{CF})_x \rightarrow x\text{LiF} + x\text{C} \qquad (3.2 \text{ V})$$

Of interest in this system is the generation of electronically conductive carbon in the cathode during discharge. This phenomenon leads to a reduction in cell impedance during operation. Conventional Li salts and organic solvents such as Li triflate (LiCF$_3$SO$_3$) in PC/DME are often used as the electrolyte.

*Li/MnO$_2$*

The lithium/manganese dioxide system is also very popular and along with Li/CF$_x$, was an early entry to market in the 1970s. The overall reaction is

$$\text{Li} + \text{Mn}^{(\text{IV})}\text{O}_2 \rightarrow \text{LiMn}^{(\text{III})}\text{O}_2 \qquad (3.0 \text{ V})$$

Similar electrolytes to Li/CF$_x$ cell can be used. Numerous coin-cell, spiral-wound and bobbin configurations are available, as well as three-cell-in-series prismatic 9 V [27]. As with most other Li metal anode primary systems, the very low self-discharge rate affords many years of shelf life in standby usage. Therefore, applications such as smoke detectors are very popular for this system.

*Li/I$_2$-P2VP*

The lithium/iodine-poly-(2-vinyl pyridine) system has some unique properties. Its major application is in implantable medical devices such as cardiac pacemakers, which operate at a thermostatted 37°C [28]. The reaction is

$$\text{Li} + \tfrac{1}{2}\text{I}_2 \rightarrow \text{LiI} \qquad (2.8 \text{ V})$$

The poly-(2-vinyl pyridine) is present in the cathode to solubilize the iodine via a complexation facilitating its diffusion. On fabrication, the cathode sheet, an iodine complex in a mixture with excess elemental iodine, is pressed against the Li metal anode. On contact a thin passivating, but Li-ion conductive, film of LiI is formed. It fails to grow further until the circuit is closed. Because the conductivity of $Li^+$ in LiI is appreciable, $5 \times 10^{-7}$ (ohm-cm)$^{-1}$ at 37°C, the cell supports a current density of $1-2$ microamps/cm$^2$, although internal impedance increases during discharge as the electrolyte layer increases.

## B. Calcium Batteries

Substantial effort has been devoted to developing primary batteries using a room temperature metallic Ca anode [29–31]. Like Li metal anode cells, Ca cells exhibit very low self-discharge rates. One advantage of a Ca cell is the 838°C melting point of Ca, which eliminates the high reactivity hazard associated with breakdown of passivating electrolyte interfaces, as is known with Li and Na when cell temperatures reach the anode melting point. A second potential advantage lies in the enormous overpotentials associated with attempts to plate Ca on a variety of metal and carbon substrates including Ca itself. This feature provides a measure of protection in multicell batteries when a weak unit of a series string of cells is forced into overcharge, i.e., reversal. Plating of reactive metal and formation of hazardous internal shorts is much more unlikely.

For these reasons Ca-metal-anode-based primary batteries have been well studied, but the presumed advantages have not yet led to any significant commercial products. Corrosion is significantly higher in Ca system than Li systems, although this problem may be improved by the addition of alternative solutes based on Ba and Sr. Ca-based electrolytes are significantly less conductive than Li electrolytes so sustainable current densities of Ca batteries are less than Li batteries.

## C. Reserve and Thermal Batteries

The principle of a reserve battery is to restrict the ionic conductivity between the negative and positive electrodes within the cell by precluding the introduction, or at least the activity, of the electrolyte until activation is required. This approach affords batteries with the longest possible shelf lives, decades or longer under optimal conditions. Reserve batteries can be aqueous or nonaqueous.

Thermal batteries are a special class of reserve batteries which take advantage of the long-term stability intrinsic to many interfaces when both the active material and the electrolyte are solid state, at least until activation. The stability is attributable to the very low diffusion coefficients of electrolyte ions in the solid state for the chosen electrolyte systems. These batteries are fully assembled with electrolyte present, but the electrolyte remains a solid nonconductor until it is melted by rapid heating from a pyrotechnic heat source.

Historically, a Ca metal negative electrode was used with a fusible salt electrolyte (LiCl/KCl eutectic) and a metal oxide cathode, e.g., $K_2Cr_2O_7$ (2.8–3.3 V). Since the 1980s, Li alloy negatives have gained popularity and supplanted a number of Ca anode thermal cell uses. The alloys, e.g., LiAl, Li/Si, remain solid to hundreds of degrees above the 180°C melting point of pure Li and offer the advantage of much higher rate capability due to the higher conductivity of $Li^+$ -based electrolytes. An established system here is Li-Al/LiCl-KCl/FeS$_2$ (1.6–2.1 V).

Examples of recent Gulf War uses of thermal batteries included the following missile systems. Using a Li/Si negative were the Patriot and Stinger ground to airs and the Sparrow air to air. The Tomahawk Cruise used a Ca chromate, although new versions use Li/Al. The following all used Li/Al: Hell Fire antitank, TOW (tube launched optical wire guided), AMRAAM, Copperhead antitank, and Maverick air to ground.

## IX. SECONDARY BATTERIES

The often higher costs of solvents, salts, and fabrication methods of nonaqueous systems are weaker deterrents in secondary applications than primary ones since their cost is amortized over many uses. Higher costs indicate they are used to best advantage in relatively short-term, power hungry applications where portability or high product cost are considerations. At present, laptops, cellular phones and camcorders are obvious examples.

Two key requirements must be met to produce a high quality rechargeable battery.

1. Faradaic efficiencies must approach 100%; i.e., irreversible side reactions must be minimized to maximize the cycle life.
2. Electrode dimensional stability must be preserved; i.e., ideally, each ion should redeposit where it came from.

## A. Faradaic (Coulombic) Efficiency

It normally must be well in excess of 99% to achieve the hundreds of recharges required in most applications. When this condition cannot be met, high cycle life is still achievable by packaging more than a one-to-one equivalent of the less efficient active material, typically the anode metal. However, this approach entails a substantial loss in energy density. The theoretical loss of capacity as a function of the limiting Faradic efficiency can be calculated by exponentiation of the theoretical efficiency. By example, to calculate $n$, the number of cycles obtainable, to an arbitrarily chosen 75% remaining of initial capacity from a cell which loses 1% of its capacity per recharge, solve

$$(0.99)^n = 0.75$$

Solving yields $n = 29$ cycles. Recasting the equation with a coulombic efficiency of 99.9% instead of 99%, $(0.999)^n = 0.75$, yields $n = 288$ cycles. Clearly, impressive efficiencies are demonstrated in modern Li-ion and other systems which can deliver over 500 cycles. Similarly, determination of the average capacity loss over a steadily decreasing number of cycles permits the calculation of the coulombic efficiency. Losses of less than 0.1% per cycle are required for high cycle life consumer applications.

## B. Electrode Dimensional Stability

A recurring theme in rechargeable technologies is this second point—the problem of electrode shape change which occurs over the course of many recharges. This problem is not unique to nonaqueous rechargeables. It occurs almost any time a metallic negative electrode is present and is a result of nonuniform transport processes and of the tendency of crystals to grow at different rates along different crystal faces. A thorough understanding of this phenomenon has yet to be agreed upon, but it is clear that different factors contribute to the effect.

Many factors produce inhomogeneous current densities in real batteries. The most important are

1.  Uneven spacing between negative and positive
2.  Variations in stack pressure
3.  Temperature differences which affect electrolyte conductivity and electrode kinetics
4.  Edge effects
5.  Accumulation of electrolyte decomposition products

1.  *Electrode spacing variations.* The ideal two electrode configuration is two perfectly flat electrodes with exactly constant separation distance throughout. This condition is most readily maintained when both electrodes are based on intercalation chemistry. Metallic anodes often undergo severe morphology changes upon repeated cycling. To the extent that constant separation distance is not maintained, current will preferentially flow across the areas in closest contact. Shorter ion diffusion paths in the electrolyte result in lower $iR$ drop and enhanced ion transport between these regions.

As regions of closer contact grow ever closer upon cycling, active material accumulates locally. When the growth of metallic negative electrode active material toward the positive cannot be minimized, cycle life will be very short. The failure mode is internal shorting, as these filamentary growths, known as dendrites, establish unwanted electronic contact within the cell, dropping the cell's open and closed circuit voltages to uselessly low values.

In the more spectacular conclusion to the formation of these interelectrode shorts, the sudden flow of electrons through the shunt produces an explosion, or at least what is technically termed a deflagration. The high current can heat the dendrite to very high temperatures, well above the autoignition temperature of the solvent, causing catastrophic results. Many Li metal anode rechargeable cells have exploded as a result of this phenomenon over the years.

The problem of dendrite growth is well known in the aqueous-based electroplating and electrowinning industries as well where leveling agents or brighteners are added to the plating solutions to minimize this effect. For instance, positively charged proteins are advantageously added to Cu electrowinning solutions to control dendrite growth. The presumed mechanism is that as dendrites form they bear an excess of negative charge which attracts the proteins, thereby covering the tip of the dendrite and shutting down its growth. Whereas the degree and morphology of Li dendrites are known to be very sensitive to the nature of the electrolyte, it is not clear that anyone has yet developed a truly adequate solution to control Li plating morphology. An even less satisfactory situation exists in the case of larger alkali metals.

One notable recent innovative approach [32] to improving the safety of rechargeable Li metal cells has been put forward by Tadiran Ltd. (Rehovot, Israel). It involves employing the solvent 1,3-dioxolane in conjunction with a "Lewis acid anion, $MF_6^-$." This solvent is well known to readily polymerize when initiated by high temperatures or by voltages above approximately 3.0 V versus $Li/Li^+$. Upon exposure to the heat produced in the vicinity of the internal short, this solvent rapidly polymerizes, thereby halting ion flow, which is required to balance charge transport and sustain the short.

Use of Lewis acid anions, which disproportionate on heating to yield neutral Lewis acid polymerization initiators, improves the efficiency of this safety strategy. The most common Lewis acid anions and their associated elevated temperature disproportionation products are shown in the table. Since these anions exhibit significant dissociation equilibria at temperatures less than $100°C$, they are not as useful for applications demanding long-term elevated temperature operation.

| Lewis acid anion | | Lewis acid | | Salt (ppt) |
|---|---|---|---|---|
| $Li^+BF_4^-$ | $\leftrightarrow$ | $BF_3$ | $+$ | LiF |
| $Li^+PF_6^-$ | $\leftrightarrow$ | $PF_5$ | $+$ | LiF |
| $Li^+AsF_6^-$ | $\leftrightarrow$ | $AsF_5$ | $+$ | LiF |

These disproportionations are thermodynamically driven by the very high lattice energy (stability) of LiF and its insolubility in nonaqueous solvents. The insolubility of LiF drives the reaction far to the right, as rationalized by LeChatelier's principle. Larger alkali cations do not undergo such facile dissociation, presumably because the lower MF crystal lattice energies reduce the spontaneity and rate of the anion decomposition.

2. *Stack pressure variations.* This issue relates back to electrode spacing variations. It is another means by which uneven electrode spacings are generated.

3. *Temperature variations.* Essentially all kinetic phenomena are temperature dependent: ion diffusion (in both electrolyte and active materials), electron transfer, desolvation, adsorption, etc. Additionally, thermodynamic equilibrium constants are temperature dependent, so any temperature variations within the cell will produce uneven plating and stripping, electrode shape change effects and uneven utilization again leading to compromised performance.

4. *Edge effects.* This is a particularly complex matter which is no doubt controlled by several factors. Ions diffuse and migrate down electrochemical gradients controlled by forces favoring a geometry perpendicular to the active material's interface with the electrolyte. The driving force for this orthogonal shedding and plating of ions is the electric field, its "lines of force" being always at right angles to the source of charge. At the edge of any planar electrode of finite thickness, the thin side, typically only a few mils or tens of microns thick in the case of small consumer spiral-wound cells, will shed active material concurrent with the exchange of ions across the parallel interface. Ions derived from the thin edges migrate across the interface and tend to deposit on the much larger planar region since this is the path of shortest diffusion. This deposition of edge ions occurs along with the ions derived from the parallel face, producing an accumulation of active material at the perimeter of the electrode planes and a depletion within the narrow electrode edges. Again, electrode shape change occurs and internal shorts frequently form along the perimeters of electrodes. This is a serious problem primarily with metallic electrodes and is minimally, if at all, observed in the case of intercalation electrodes.

5. *Accumulation of electrolyte decomposition products.* This phenomenon leads to blocking of ion flow near nonconductive precipitates which reduce the ability to extract capacity from microregions. Current densities build up in the remaining regions and enhance other nonideal phenomena as discussed above.

The above-mentioned factors are probably not the only reasons for unstable electrode dimensional stability, and other phenomena have been proposed and confirmed to varying degrees of certainty. Included are osmotic pumping [33,34] and nonuniform current efficiencies during charging and discharging [35]. Each electrode system's unique chemistry provides a further set of mechanisms to generate dimensional instability. An excellent discussion of several of these phenomena is available in Ref. 36.

## C.  A Sampler of Nonaqueous Secondary Battery Types

### 1.  Lithium

Rechargeable Li metal anode–based batteries have probably received more developmental effort than any other nonaqueous category for reasons already discussed. The extreme reactivity of Li metal has made the commercialization of such a rechargeable battery very risky. The problem lies in the severe difficulties in redepositing the Li metal with smooth morphology. As fresh Li metal readily reacts with every known organic solvent, mossy, dendritic, highly porous Li surfaces result. The replated metal is much less dense than the virgin Li metal and the combination of high surface area and expanded volume induced shorting problems have relegated the Li metal negative electrode to a backwater status in most laboratories.

### 2.  Sodium

Metallic sodium negative electrode batteries have been extensively researched over the years as well. The even lower melting point of Na (97.8°C), compared with Li (180.5°C), and its higher equivalent weight make it a less attractive negative electrode active material in principle. Practically speaking, it is also a very difficult metal to tame. Na passivating films with liquid electrolytes are less stable than those which form on Li. Internal shorting from Na dendrites formed on charge remains a challenge in organic liquid and polymer electrolyte systems.

   An alternative, well-studied approach to develop Na metal anode rechargeable cells is to isolate the Na electrode from the cathode via a ceramic Na ion conductor, i.e., a solid electrolyte. Such batteries are operated above the melting point of Na, thereby enhancing electrode kinetics and solid state diffusion of Na ions through the special alumina ceramic. The most commonly employed material is beta double prime alumina, which has channels permitting the facile diffusion of Na ions.

   The two most commonly investigated cathodes for the Na electrode are elemental S and $NiCl_2$ [37–39]. The redox reactions are

$$2Na + xS \rightarrow Na_2S_x \quad (E° = 2.08 \text{ V})$$
$$2Na + NiCl_2 \leftrightarrow Ni + 2NaCl \quad (E° = 2.58 \text{ V})$$

The $NiCl_2$ positive electrode technology is presently under field testing and the Na/S technology has been set aside. The $Na/NiCl_2$ system offers several potential advantages compared with Na/S:

1.  The system operates at approximately 75°C lower temperature in the vicinity of 300°C for $Na/NiCl_2$.
2.  The $Na/NiCl_2$ cell requires incorporation of an ionically conductive

secondary electrolyte, $NaAlCl_4$, in the $NiCl_2$ solid state particulate positive electrode to conduct Na ions to the ceramic electrolyte and negative electrode. In the event of a mechanical failure of the Na beta double prime alumina the molten Na preferentially reacts with the much less exothermically reactive, diffusive secondary electrolyte, $NaAlCl_4$. That is, the molten $NaAlCl_4$ tends to react with the molten Na, minimizing the energy of subsequent reaction of the Na with the less mobile $NiCl_2$ particles.

3. The molten S cathode is more corrosive.
4. There is reduced risk of solid electrolyte mechanical failure since the lower operating temperature places reduced thermal coefficient of expansion stresses on the ceramic.

## X. POLYMER ELECTROLYTES

For almost 20 years a great deal of research has focused on seeking advantageous properties by incorporating polymers into the electrolyte and active material matrices of rechargeable Li batteries. This approach may offer significant safety advantages. Ideally, such an approach will afford a battery with no volatile solvents and, therefore, a very high safety margin in the event of internal shorting. Polymer electrolyte batteries based on both Li metal anodes and lithiated carbon intercalation electrodes are under widespread investigation [40].

### A. Pure (Neat) Polymer Electrolyte Li Batteries

Solvent-free polymer-electrolyte-based batteries are still developmental products. A great deal has been learned about the mechanisms of ion conductivity in polymers since the discovery of the phenomenon by Feuillade et al. in 1973 [41], and numerous books have been written on the subject. In most cases, mobility of the polymer backbone is required to facilitate cation transport. The polymer, acting as the solvent, is locally free to undergo thermal vibrational and translational motion. Associated cations are dependent on these backbone fluctuations to permit their diffusion down concentration and electrochemical gradients. The necessity of polymer backbone mobility implies that noncrystalline, i.e., amorphous, polymers will afford the most highly conductive media. Crystalline polymers studied to date cannot support ion fluxes adequate for commercial applications. Unfortunately, even the fluxes sustainable by amorphous polymers discovered to date are of marginal value at room temperature. Neat polymer electrolytes, such as those based on poly(ethyleneoxide) (PEO), are only capable of providing viable current densities at elevated temperatures, e.g., >60°C.

As with liquid, organic-solvent-based electrolytes, it is the oxygens in the polymer molecules which coordinate to the cation. Thus, PEO is the prototypical polymer solvent. This results from its ready availability and the fact that its ($-OCH_2CH_2-$) repeat units have a very high density of oxygen atoms capable of solvating the Li ion. Unfortunately, PEO is predominantly crystalline at room temperature. On heating, it undergoes a liquid phase transition at 60°C and its ionic conductivity increases by several orders of magnitude. A loss of mechanical rigidity is associated with this melting and external pressures, and more facile penetration of dendrite growths enhance the likelihood of internal shorting.

Numerous amorphous room temperature polymer electrolytes are also known. A variant of PEO which interspaces methylene moieties with ethylene moieties is also amorphous at room temperature. Its repeat structure is ($-O-CH_2-O-C_2H_4-$). Another intensively studied Li-ion conductive amorphous polymer is known as MEEP. This acronym stands for methoxy-ethoxy-ethoxy-phosphazene. The polymer structure is a repeating ($-N=PR_2-$) phosphazene unit with two alkoxy chains dangling from the phosphorous atoms, i.e., ($-N=P-(O-C_2H_4-O-C_2H_4-O-CH_3)_2-$).

One challenge to date is the limited oxidative electrochemical stability of the ether linkages in these polymers. The fact that ethers oxidize at lower voltages than more highly oxygenated solvents such as carbonates, e.g., PC, EC, DEC, DMC, creates a challenge to use these ether-based polymers with Co-, Ni-, and Mn-based 4 V lithiated cathodes. Some aspect of the oxidized ether degradation products does not favor a beneficial passivating film. Present approaches restrict ether polymers to applications demanding no more than 3.3 V per cell.

While the successful commercial demonstration of a pure polymer electrolyte Li battery has not yet occurred, developmental batteries have been demonstrated. An ongoing joint venture between the 3M Company and the Canadian utility Hydro-Quebec has recently produced a Li metal anode/intercalation cathode electric vehicle (EV)-oriented battery which operates at 60–80°C. Approximately 600 cycles to 80% depth of discharge have been demonstrated by using an approximate threefold molar excess of Li metal foil to cathode active material with $LiN(SO_2CF_3)_2$ solute and a $VO_x$ cathode.

In this system, the relatively low conductivity of the neat polymer electrolyte is managed by good engineering practices. The electrodes are very thin, and the positive and negative are wound in a slightly offset configuration to allow continuous metallization of the top of one and the bottom of the other electrode's edge. These metallized edges are welded in their entirety from beginning to end of each strip to the case leads, maximizing current collection efficiency and reducing the likelihood of shorting.

## B. Gelled Polymer Electrolyte Li Batteries

The apparent safety advantage of a high flash point "solvent" which polymers can offer has motivated much study of liquid electrolyte systems which are diluted with polymers. These electrolytes can still have reasonably good dimensional stability and improved safety under abuse conditions. These combinations of polymer + solvent + salt are called gel polymer electrolytes.

Some recipes produce gel polymer electrolytes which are sufficiently conductive to be used in cell designs which operate at or below room temperature. They exhibit dimensional stability adequate to obviate the need for a rigid metal container. Lighter containers such as tough metallized plastic films are finding limited applications and are approaching the goal of an all-plastic battery. Such cells are referred to by some developers as pouch cells. Since all polymers are permeable to vapors and gases, a very thin metal layer is spray-coated onto the container substrate to isolate volatile cell solvents from the external environment, and to isolate detrimental ambient gases such as water vapor, oxygen, nitrogen and so forth from the cell interior. These container materials were originally developed for use in the pharmaceutical industry for product protection. The elimination of the stainless steel or aluminum can offers significantly higher practical gravimetric energy densities and is a major driver for continued developmental efforts on small cell gel polymer battery design. Large scale manufacturability and market acceptance remain to be demonstrated.

## XI.  FUTURE DIRECTIONS

Presently, a tremendous level of effort continues in the Li-ion arena to develop higher capacity cathodes and inexpensive electrolytes with wider voltage windows (>4.2 V) and temperature windows. Li-ion technology is new, and many approaches to push the extremes of service capability are under study. Materials development efforts are high regarding advanced electrolytes (ultrahigh conductivity, high $Li^+$ transport number, nonflammable), separator materials (tailored shutdown temperatures, high void volume), soluble redox-shuttle overcharge protection mechanisms and new active materials [42].

A major effort continues around the world to develop high energy, long cycle life rechargeable batteries for EV applications. Not only nonaqueous batteries, but many aqueous battery and fuel cell ideas are in development, often in the form of hybrid EV concepts, i.e., in conjunction with reduced size internal combustion engines.

No one has commercialized the fluorine positive electrode, possibly the most powerful oxidant (cathode) known, though there has been considerable effort.

One thing about the future is certain. Considering that the first commercial nonaqueous battery products have been developed only within the past generation, it is likely that more systems will be entering the marketplace and that nonaqueous battery technology will continue to offer fertile ground to electrochemists and materials scientists for many years to come.

## ABBREVIATIONS

$(CF)_x$   Fluorinated graphite-poly-(carbon monofluoride)
DEC       Diethyl carbonate
DMC       Dimethyl carbonate
EC        Ethylene carbonate
PC        Propylene carbonate
$SOCl_2$   Thionyl chloride
$SO_2Cl_2$  Sulfuryl chloride

## REFERENCES

1.  W. S. Harris, Ph.D. thesis, University of California, Berkeley (1958).
2.  J. P. Gabano, Ed., *Lithium Batteries*, Academic Press, New York (1983).
3.  G. Pistoia, Ed., *Lithium Batteries: New Materials, Developments and Perspectives*, Elsevier, New York (1994).
4.  D. Linden, Ed., *Handbook of Batteries*, 2nd edition., McGraw-Hill, New York (1994).
5.  J. J. McLaughlin, Cyprus Foote internal memo, Jan. 23, 1991.
6.  N. Watanabe and M. Fukuda, U.S. Patents 3,536,532 and 3,700,502 (1970).
7.  G. Blomgren, "Properties, Structure and Conductivity of Organic and Inorganic Electrolytes for Lithium Battery Systems," in Ref. 2.
8.  G. Blomgren, *J. Power Sources*, **14**, 39–44 (1985).
9.  L. A. Dominey, "Current State of the Art on Lithium Battery Electrolytes," pp. 137–165 in Ref. 3.
10. D. Aurbach, *Electrochimica Acta*, **39**, 51–71 (1994).
11. G. Nazri and R. H. Muller, *J. Electrochem. Soc.*, **132**, 1385–1387 (1985).
12. J. R. Hoenigman and R. G. Keil, "An XPS Study of Lithium Exposure to Low Levels of $O_2$, $H_2O$, CO, $CO_2$ and $SO_2$," in *LITHIUM: Current Applications in Science, Medicine and Technology*, 233–255, Ricardo O. Bach, Ed., Wiley, New York (1985).
13. T. Hewston and B. L. Chamberland, *J. Phys. Chem. Solids*, **48**, 97–108 (1987).
14. K. West, B. Zachau-Christiansen, T. Jacobsen and S. Skaarup, "Lithium Intercalation in Oxides: EMF Related to Structure and Chemistry," in *Solid State Ionics III*, 39–47, MRS Symposium Proceedings, Vol. 293, G. Nazri, J-M Tarascon and M. Armand, Eds., Materials Research Society, Pittsburgh, PA (1993).
15. W. R. McKinnon and R. R. Haering, "Physical Mechanisms of Intercalation," in *Modern Aspects of Electrochemistry No. 15*, 235–304, B. Conway, Ed., Plenum Press, New York (1983).
16. J. R. Dahn, A. K. Sleigh, H. Shi, B. M. Way, W. J. Weydanz, J. N. Reimers, Q. Zhong and U. von Sacken, "Carbons and Graphites as Substitutes for Lithium Anode," in Ref. 3.
17. W. Xing, J. S. Xue, T. Zheng, A. Gibaud, and J. R. Dahn, *J. Electrochem. Soc.*, **143**, 3482–3491 (1996).

18. N. Bartlett and B. W. McQuillan, "Graphite Chemistry," in *Intercalation Chemistry*, 19–47, M. S. Whittingham, Ed., Academic Press, New York (1982).
19. L. B. Ebert, *Ann. Rev. Mat. Sci.*, **6**, 181–211 (1976).
20. T. Nagura, in *Proc. 3rd Int. Battery Seminar*, Ansum Enterprises/Florida Educational Seminars, Inc., Boca Raton, FL (1990).
21. W. Ebner, D. Fouchard and L. Xie, *Solid State Ionics*, **69**, 238–256 (1994).
22. J. O. Besenhard, J. Yang and M. Winter, *J. Power Sources*, **68**, 87–90 (1997).
23. W. R. Cieslak and D. E. Weigand, in *Proc. 36th Power Sources Conf.*, 361–364, U.S. Army Research Laboratory, Ft. Monmouth, NJ (1994).
24. R. J. Staniewicz, in *Proc. 7th Int. Meeting on Li Batteries*, 106–110, H. F. Gibbard, Ed., Duracell, Inc., Needham, MA (1994).
25. X. Yu, J. B. Bates, G. E. Jellison, Jr. and F. X. Hart, *J. Electrochem. Soc.*, **144**, 524 (1997).
26. T. Nakajima, R. Hagiwara, K. Moriya and N. Watanabe, *J. Electrochem. Soc.*, **133**, 1761–1766 (1986).
27. H. Ikeda, S. Narukawa and S. Nakaido, in *Proc. 29th Power Sources Conf.*, Electrochemical Society, Pennington, NJ (1980).
28. K. R. Brennen and D. F. Untereker, in *Proc. Symp. on Biomedical Inplantable Applications and Ambient Temperature Li Batteries*, B. B. Owens and N. Margalit, Eds., Proc. Vol. 80-4, Electrochemical Society, Pennington, NJ (1984).
29. R. J. Staniewicz, *J. Electrochem. Soc.*, **127**, 782 (1980).
30. E. Peled, in *Proc. Symp. on Advances in Battery Materials and Processes*, Proc. Vol. 84-4 p. 22, Electrochemical Society, Pennington, NJ (1984).
31. E. Peled, R. Tulman, A. Golan and A. Meitav, *J. Electrochem. Soc.*, **131**, 2314–2316 (1984).
32. P. Dan, E. Mengeretski, Y. Geronov and D. Aurbach, in *Proc. 7th Int. Meet. on Li Batteries*, Boston, May 15–20, 1994, Duracell, Needham, MA (1994).
33. K. W. Choi, D. N. Bennion, and J. Newman, *J. Electrochem. Soc.*, **123**, 1616–1627 (1976).
34. K. W. Choi, D. Hamby, D. N. Bennion, and J. Newman, *J. Electrochem. Soc.*, **123**, 1628–1637 (1976).
35. J. McBreen, *J. Electrochem Soc.*, **119**, 1620–1628 (1972).
36. F. R. McLarnon and E. J. Cairns, *J. Electrochem. Soc.*, **138**, 645–664 (1991).
37. R. C. Galloway, *J. Electrochem. Soc.*, **134**, 256–257 (1987).
38. *ZEBRA High Energy Batteries* in *Electric & Hybrid Vehicle Technology '97*, UK & International Press, Surrey, UK (1997).
39. J. Sudworth and R. Tilley, *The Sodium/Sulfur Battery*, Chapman and Hall, London (1985).
40. F. M. Gray, *Solid Polymer Electrolytes Fundamental and Technological Applications*, VCH, Cambridge, UK (1991).
41. G. Feuillade, B. Cheneaux and P. Perche, French Patent No. 2,230,091 (1973), G. Feuillade and P. Perche, *J. Appl. Electrochem.*, **5**, 63 (1975).
42. K. M. Abraham, *Electrochim. Acta*, **38**, 1233–1248 (1993).

# 9
# Electrochemistry of Molten Salts

**Ioan Galasiu and Rodica Galasiu**
*Roumanian Academy, Institute of Physical Chemistry, Bucharest, Roumania*

**Jomar Thonstad**
*Norwegian University of Science and Technology, Trondheim, Norway*

## INTRODUCTION

Molten salts are favored electrolytes in cases where aqueous solutions cannot be used because the decomposition voltage of water is lower than that of the salt in question. Although high overvoltages for hydrogen and oxygen evolution allow us to extend the use of aqueous electrolytes somewhat beyond the limits set by thermodynamics, there are many substances which cannot be electrowon or plated from aqueous solutions.

Molten salts are sometimes denoted high temperature ionic liquids, but both terms need a further definition to clarify what they encompass. First of all there are several classes of high temperature ionic liquid which are not commonly classified as molten salts, such as oxides, sulfides, etc. We are then left with the halide melts and ionic melts with complex anions like sulfates, nitrates, carbonates, etc. For use as electrolytes for electrowinning the latter group is not stable enough (anodic decomposition of the anion), whereas they may be used for electroplating, and molten carbonates have a well-known use in fuel cells. With the exception of carbonates, we are then left with the halide melts as molten salts for all-round applications. They can be electrolyzed as pure salts or in mixtures, serving as electrolyte (solvent) for electrowinning, electrodeposition, and in batteries. The product may be a metal, often in the liquid form, or a compound or a gas.

Pure halides are high melting compounds, with melting points mostly above 600°C, and even for mixtures of molten salts the operating temperature is usually in the range 500–1000°C. However, there are exceptions to that rule. One important and widely studied group of low-melting salts is a group of melts rich in aluminum chloride, e.g., $AlCl_3$-NaCl, which have liquidus temperatures down to 113°C. By adding organic salts, so-called room temperature molten salts are obtained. Another example of a low melting salt mixture is KF2HF, melting at around 100°C, which serves as electrolyte for the production of fluorine.

The present chapter treats the fundamental properties of molten salts as well as the more important applications of the typical high melting halide melts and in one case halide-oxide melts as well as the low melting $AlCl_3$-NaCl system and room temperature molten salts. Industrially the high melting salts gained their importance late in the last century, when the industrial processes for aluminum and magnesium were developed, and this is still the most important application of molten salts today. The dominating process tonnage-wise is the electrowinning of aluminum, with a yearly production of close to 20 millions tonnes of aluminum. To produce this amount, some 250,000 tonnes of fluoride electrolyte is being kept in the molten state (~960°C) at all times. Other metals produced by molten salt electrolysis are magnesium, calcium and the alkali metals, but the quantities are much smaller than for aluminum.

Interesting potential applications of molten salts are electroplating and electrorefining of refractory metals and rare earth metals. Electrowinning of titanium has been tested on a pilot scale. Electrodeposition of refractory compounds like $TiB_2$ has also been demonstrated. Due to space limitations these more exotic applications of molten salts will not be treated here. However, short chapters on molten salt batteries and fuel cells are included.

## I.  PHYSICOCHEMICAL PROPERTIES OF MOLTEN ALKALI HALIDES

### A.  Molten Salt Structure

The liquid state is intermediate between the gaseous and solid states. It presents neither the structural regularity of solid crystals nor the typical disorder of gases. For that reason, theoretical studies of the liquid phase are based on the methods and results of the research of real gases (van der Waals) but also on the theory of disordered solids, like the results of X-ray experiments which show local ordering in liquids.

Molten salts constitute a category of liquids which is called ionic liquids or molten electrolytes. These liquids have some characteristics which are different from that of liquids at room temperature. Molten salt studies are very important for understanding of the liquid state because molten salts consist of ions, and the principal forces between particles are coulombic interactions. The existence of coulombic interactions in molten salts are demonstrated by very high melting and boiling points, surface tensions and electrical conductivities, in comparison with these properties of other liquids. Other properties of the pure ionic liquids are of the same order of magnitude as for nonpolar liquids, although the ionic liquids exist only at high temperatures. These properties are density, viscosity, refractive index, compressibility, vapor pressure, heat of vaporization, heat of fusion, heat capacity, etc.

A theory of molten salts must be able to predict thermodynamic and transport properties. The theory must also give a relation between state properties, pressure, temperature and volume. Such theories have been obtained by several methods.

The first method comprises theories based on a statistical description of ionic liquids in a very rigorous way using radial distribution functions obtained from X-ray diffraction. These are theories based on the principle of corresponding states.

The second group of theories is based on a general approximation and consists of the use of a molten salt model to obtain a partition function from the molecular motion. This group includes the following theories: the hole theory, the theory of significant structures and other structural models. The theories of the first group are mathematically more difficult but lead to good results for the molten salt structure.

## 1. Structural Models

Structural models for molten salts have been proposed by several authors: Bockris [1–3], Stillinger [4], Zarzycki [5], Janz [6], Kleppa [7], Blander [8], and others. These structural models are based on the older theories of the liquid state, which were applied for molecular liquids, liquified gases, molten metals, etc. Some of these models will be treated in the following.

Zarzycki [5] and Levy et al. [9] made X-ray diffraction studies of some simple ionic liquids. These studies showed the existence of an ordering at short distance after melting, changes of interionic distances with a shortening of the distance between the oppositely charged ions, i.e., the nearest neighbors, and an increase for ions of the same charge (the next nearest neighbors) and a lowering of the number of nearest neighbors (coordination number) around a central ion.

Likewise, measurements of the density of molten salts show a volume increase of 10–25% at melting. These results lead to the conclusion that the simplest model for the ionic liquid is derived from the crystalline lattice. Hence, in the liquid, some elements of the crystalline lattice are maintained, and between these some holes are produced. The holes (larger than Schottky defects) explain the volume increase by melting and also the short range ordering. This model, which is called the quasi-lattice model, was elaborated by Frenkel [10] and used by Bockris [11] for the calculation of transport phenomena in ionic liquids. According to this model the enthalpies of fusion for ionic liquids and the activation energies for diffusion were calculated.

*The hole model* for molecular liquids was elaborated by Furth [12], who supposed that the free volume of a liquid is not distributed uniformly between its molecules like in crystals, but is concentrated like some holes which can disappear in one place and appear in another place. These holes are in permanent motion, so that the situation is different from the jumps of the holes in a crystal. The appearance and disappearance of the holes in a liquid are a result of the fluctuations connected with thermal movements. These holes in liquids have no definite shape and size; they can increase or decrease spontaneously. Furth [12] tried to calculate a large number of properties of the liquids: viscosity, compressibility, thermal expansion, thermal conductivity, but the results were not successful. However, Furth obtained a precise result of the calculation of the volume change by melting and the entropy of melting.

Bockris et al. [2,3,13] applied the hole model for molten salts (Figure 1). Thus, an ionic liquid is constituted of spherical cations and anions, each of them being a harmonic oscillator through which a number of holes is distributed randomly. The volume of holes is approximately equal to the volume change by melting. Each ion is trapped in its own cell and it can hop into a hole, when this hole appears in a neighboring position. The authors calculated in good agreement with experimental data, heat capacities, coefficients of compressibility and thermal expansion for a large number of molten salts. Bockris [3] deduced from the calculation of the activation energy for viscosity that the greater part of this activation energy is needed for the hole formation.

The most important deficiency of this theory is the fact that the notion of holes has not an exact definition. Also, the hole model does not contain any

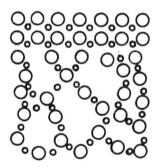

**Figure 1**   The hole model.

indication about the influence of the electrostatic forces between the ions of a molten salt upon its properties, in contrast to molecular liquids.

*The theory of "significant structure"* was developed by Eyring et al. [14] for molecular liquids. Afterward, that theory was applied for the study of some pure molten salts and salt mixtures (Figure 2) [15–17]. It was noticed an increase of the volume at melting with approximately 20%. X-ray spectrum analyses of molten salts show that the increase of volume is not associated with a corresponding increase of the distance between the nearest neighbors. The theory of "significant structures" assumes that the volume increase at melting is due to two types of defects: holes and dislocations. The holes may be of the Schottky and Frenkel types. The defects do not modify the distance between the nearest neighbors, but reduce the medium coordination number, in agreement with experimental observations [9].

In the solid phase, each ion is symmetrically surrounded by a number of ions (the nearest neighbors) that influence the central ion, which has harmonic vibrations around its equilibrium position in the lattice. At melting, according to the X-ray spectra, the long range ordering is destroyed. The liquid consists of "significant structures" described by molecules with "solid-like" translational motions and molecules which move in empty spaces with "gas-like" degrees of freedom. That theory has produced a partition function which reasonably describes molecular and ionic liquids [2,14,16]. A molecule can at one time be solid-like for in the next moment to be gas-like. That is because a molecule can vibrate around an equilibrium position and then suddenly hop into a hole in the neighborhood getting a gas-like behavior.

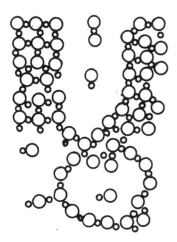

**Figure 2**   Model of significant structures.

Carlson [16] used this theory for molten alkali halides and obtained values in good agreement with experimental data for the following properties: melting and boiling temperatures, volume change at melting, entropy variation by melting and boiling. The theory has several deficiencies. Thus, there is not an objective criterion for deciding if a particle is gas-like or solid-like at a given time. The gas-like particles are not free enough to move in holes like the particles of an ideal gas, as the premises of the theory presumed.

In mixtures of molten salts it is necessary to take into account the fact that the interaction forces are very strong and that the nearest neighbors of the cations are anions and the nearest neighbors of the anions are always cations. Therefore, random distributions of anions and cations cannot be conceived. On the basis of these considerations Temkin [18] proposed a model for ideal mixtures of molten salts which assumes the existence of two interlocking sublattices, one of cations and the other of anions. In the case of mixing of two salts, the cations mix on the cation sublattice and the anions mix on the anion sublattice.

An ideal mixture of molten salts is a mixture for which the heat of mixing, energy of mixing and variation of volume of mixing have the value zero. Certainly, in practice there are no such ideal mixtures. The Gibbs energy of mixing per mole of an ideal mixture is given by

$$\Delta G^{\text{id}} = \Delta G^{\circ} + RT(\Sigma\, x_i \ln x_i + \Sigma\, x_j \ln x_j) \tag{1}$$

where the $x$'s are the ion fractions of the components: $i$, cations; $j$, anions. The chemical potential ($\mu$) of any component $ij$ in an ideal mixture is

$$\mu_{ij}^{\text{id}} = \mu_{ij}^0 + RT \ln x_i x_j \tag{2}$$

The activity $a$ of a component is defined by

$$\mu_{ij}^{\text{id}} = \mu_{ij}^0 + RT \ln a_{ij} \tag{3}$$

Consequently, for an ideal mixture,

$$a_{ij} = x_i x_j \tag{4}$$

In the case of a real mixture,

$$a_{ij} = x_i x_j \gamma_{ij} \tag{5}$$

where $\gamma$ is an activity coefficient, which is a measure of the deviation from ideal behavior. The deviations from ideal behavior are positive when $\gamma > 1$ and negative when $\gamma < 1$.

The excess chemical potential of a component $ij$ is

$$\mu_{ij}^E = \mu_{ij} - \mu_{ij}^{\text{id}} = RT \ln \gamma_{ij} \tag{6}$$

and the excess Gibbs energy is

$$\Delta G^E = \Delta G - \Delta G^{id} = RT \sum x \ln \gamma \tag{7}$$

where $\mu_{ij}^E$ is a partial molar quantity of the $ij$ component, while $\Delta G^E$ corresponds to an integral quantity.

The deviations from ideal behavior can be calculated for all thermodynamic and transport functions of mixtures of molten salts, and these reflect the phenomena which take place at mixing: ion associations, formation of complex ions, etc.

The enthalpy of mixing of some mixtures of molten salts with monovalent ions could be calculated in terms of the quasi-chemical theory given by Guggenheim [19] as follows:

$$H_m = x_1 x_2 \left( 1 - x_1 x_2 \frac{\lambda_1}{zRT} + \cdots \right) \tag{8}$$

where $z$ is a coordination number and $\lambda$ is an energy parameter related to the exchange of next nearest neighbors in the mixing process. Førland [20] calculated the $\lambda$ parameter on the basis of the coulombic interactions of cation-cation repulsion. According to Førland's theory, the mixing follows the scheme given in Figure 3.

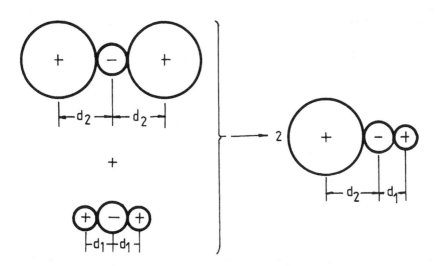

**Figure 3** Model for Førland's calculation of the mixing energy of molten salts [20].

Blander [8] assumed that the dissociation reactions could take place in a different way:

$$MX \rightleftharpoons M^+ + X^- \tag{9}$$

$$2MX \rightleftharpoons M_2X^+ + X^- \tag{10}$$

$$3MX \rightleftharpoons M_2X^+ + MX_2^- \tag{11}$$

Among these three reactions, (10) and (11) are energetically favored. Hence there is a tendency to form some ion associations by the dissociation of a simple molten salt.

## 2.  Complex Ions

The formation of complex ions is an important problem for the study of the structure and properties of molten salts. Several physicochemical measurements give evidence of the presence of complex ions in melts. The most direct methods are the spectroscopic methods which obtain absorption, vibration and nuclear magnetic resonance spectra. Also, the formation of complex ions can be demonstrated, without establishing the quantitative formula of the complexes, by the variation of various physicochemical properties with the composition. These properties are electrical conductivity, viscosity, molecular refraction, diffusion and thermodynamic properties like molar volume, compressibility, heat of mixing, thermodynamic activity, surface tension.

Bredig [21] presumed that in a mixture of molten salts the variation of the interaction parameter $\lambda$ [see Eq. (8)] makes evident the formation of complex ions. Thus, in the $KCl-CdCl_2$ and $RbCl-MgCl_2$ systems it is observed variations of the $\lambda$ parameter and with the composition of the melt, and the minimum points of those variations correspond to the composition of complex ions. However, the most convincing data on the formation of complex ions can be obtained from measurements of the enthalpy of mixing [22].

*The structure of $Na_3AlF_6-Al_2O_3$ melts* is very complicated, but very important from an industrial point of view. (See Section II. A, "Aluminum".)

Grjotheim [23] assumed that the most probable dissociation schemes of cryolite melts are

$$2AlF_6^{3-} = Al_2F_{11}^{5-} + F^- \tag{12}$$

$$AlF_6^{3-} = AlF_5^{2-} + F^- \tag{13}$$

$$AlF_6^{3-} = AlF_4^- + 2F^- \tag{14}$$

$$AlF_6^{3-} = Al^{3+} + 6F^- \tag{15}$$

Reaction (14) with formation of $AlF_4^-$ was supported by cryoscopic data [24,25], density measurements and Raman spectroscopy [26], and enthalpies of mixing [27]. Other authors [24,25,28,29] showed also the existence of the complex

$AlF_5^{2-}$. Using Raman spectroscopy data, Tixhon et al. [30] found the variation of the concentration of several ions in melt as a function of the $AlF_3$ mole fraction. This is shown in Figure 4.

Melts containing $Al_2O_3$ have more complicated structures [31–38]. At low oxide concentrations complexes of the type $Al_2OF_y^{4-y}$ have been suggested to explain the experimental results, and for higher oxide concentrations complexes of the type $Al_2O_2F_y^{2-y}$ were proposed. Førland and Ratkje [31] deduced from cryoscopic data that at low concentrations of oxide Al—O—Al bonds would form, while at higher oxide concentrations bonds of the type

would form. Gilbert [32] observed that the Raman data indicated the formation of complexes of the type $Al_2O_xF_y^{6-2x-y}$.

Sterten [33] used activity data and calculated the concentration of complex ions in cryolitic melts saturated with alumina and the distribution of anions as a function of the molar cryolite ratio ($NaF/AlF_3$), as shown in Figure 5. Julsrud [34] and Kvande [35] suggested the existence of some ions of $Al_2OF_8^{4-}$ for electrolytes with cryolitic ratio = 3, while the $Al_2O_2F_4^{2-}$ ions were in majority in

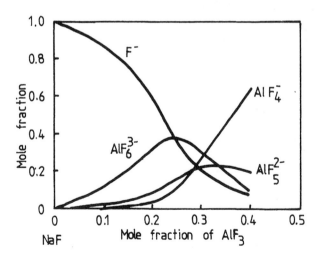

**Figure 4** The distribution of anions [30] in the NaF-$AlF_3$ molten mixtures versus the $AlF_3$ mole fraction at 1020°C.

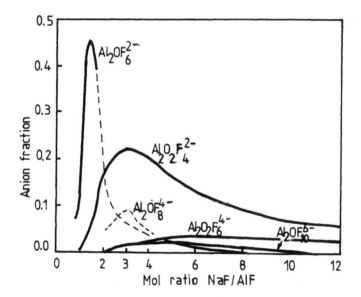

**Figure 5** Calculated [33] anion fractions of $=$Al—O—F complexes as a function of NaF/AlF$_3$ mol ratio, for melts saturated with alumina at 1012°C.

Al$_2$O$_3$-rich melts. Recently, Dewing and Thonstad [36] made a careful analysis of literature data and deduced that in Al$_2$O$_3$-rich electrolytes, the formed complex has in its composition two oxygen atoms, while for dilute electrolytes of alumina the complex has only one oxygen atom.

Gilbert et al. [37] and Olsen [38] showed that by alumina dissolution in molten cryolite the following reactions occur as a function of increasing Al$_2$O$_3$ concentration:

$$Al_2O_3 + 4AlF_5^{2-} = 3Al_2OF_6^{2-} + 2F^- \tag{16}$$

$$Al_2O_3 + AlF_5^{2-} + F^- = \frac{3}{2}Al_2O_2F_4^{2-} \tag{17}$$

$$2Al_2O_3 + Al_2O_2F_4^{2-} + 4F^- = 2Al_3O_4F_4^{3-} \tag{18}$$

and the following equilibria exist:

$$2Al_2OF_6^{2-} + 2F^- = Al_2O_2F_4^{2-} + 2AlF_5^{2-} \tag{19}$$

$$2Al_2O_2F_4^{2-} + F^- = Al_3O_4F_4^{3-} + AlF_5^{2-} \tag{20}$$

The same authors [37,38] found that in molten electrolytes with MgF$_2$ or CaF$_2$ added the following reactions take place:

$$MgF_2 + nF^- + (2 - n)AlF_5^{2-} = MgF_{2+n}(AlF_5)_{2-n}^{(4-n)-} \tag{21}$$

$$CaF_2 + (3 - n)AlF_5^{2-} = CaF_n(AlF_5)_{3-n}^{(4-n)-} + (2 - n)F^- \tag{22}$$

## B.  Electrical Conductivity of Molten Salt

### 1.  Transport Numbers and Ionic Mobilities in Molten Salts

The transport number of an ion, $t_j$, represents the fraction of the total current, $i_j$, carried by that ion as it migrates in a molten salt under the influence of an external electric field:

$$t_j = \frac{i_j}{\sum i_k} \tag{23}$$

The transport number can also be expressed in terms of the fraction of the concentration, $c_j$, the charge, $z_j$, and ionic mobility, $u_j$, of each ion in the electrolyte,

$$t_j = \frac{c_j z_j u_j}{\sum c_k z_k u_k} \tag{24}$$

For a monovalent molten salt MX the transport number for the ionic species can be reflected by the ionic mobilities:

$$t_M = \frac{u_M}{u_M + u_X}, \qquad t_X = \frac{u_X}{u_M + u_X} \tag{25}$$

These equations show that the sum of the transport numbers for the ions in a molten salt or a mixture of molten salts is always unity.

In an aqueous solution the ionic movement can be referred to the water molecules, which are unaffected by the applied electric field. But in a simple molten salt it is difficult to define the transport number because both ions are expected to move under the influence of the applied electric field. However, for a simple molten salt one can define an external transport number which is measured relative to a porous plug placed between the anode and cathode compartments of an electrolytic cell. Therefore, the ionic movement is related to a hypothetical exterior reference point of the molten salt.

The situation is less arbitrary in molten salt mixtures. In that case, the migration of two or more different ions can be defined relative to another ion, often a common ion, in order to avoid the arbitrary reference to a porous plug. That transport number is designated as internal. In a binary mixture of molten salts MX + NX, the internal mobilities of the cations $M^+$ and $N^+$ are referred to the mobility of $X^-$ and are given by the expressions [39]

$$u_{M'} = u_M + u_X, \qquad u_{N'} = u_N + u_X \tag{26}$$

The internal transport numbers of the cations $M^+$ and $N^+$ respectively are given by

$$t_{M'} = \frac{c_M u_{M'}}{c_M u_{M'} + c_N u_{N'}}, \qquad t_{N'} = \frac{c_N u_{N'}}{c_M u_{M'} + c_N u_{N'}} \tag{27}$$

Several articles and monographs [39–48] on transport numbers in molten salts have been published.

## 2. Electrical Conductivity of Simple Molten Salts

Molten salts are a special class of liquids because they consist of positive and negative ions without being in a dielectric medium like a neutral solution. The relatively high electrical conductivity of molten salts demonstrates the predominant ionic character of these liquids [41,42,44,48].

The electrical conductivity of molten salts can be expressed in two ways: equivalent conductivity $\Lambda$ (ohm$^{-1}$ cm$^2$ equiv$^{-1}$) and specific conductivity $\kappa$ (ohm$^{-1}$ cm$^{-1}$), and between these terms there is the relation

$$\Lambda = \kappa \left( \frac{\text{equivalent weight}}{\rho} \right) \tag{28}$$

where $\rho$ is the density.

Equivalent conductivities (and ionic mobilities) of the melts are similar to that of aqueous solutions. Very high specific conductivities are typical for molten salts, as seen in Table 1 [49]. The reason for this is the fact that molten salts are very concentrated solutions (for example, the concentration of molten LiF is about 65 molar; the concentration of molten KCl is about 20 molar, etc.). The electrical conductivities of various molten salts cannot be compared at constant temperature because of their different melting points. Therefore, in Table 1 the values of conductivities were selected at 50° above the melting point of each salt.

In the following text the general term *electrical conductivity* means the property of a molten salt to conduct the electric current, and when necessary the specific and equivalent conductivities will be specified.

Biltz and Klemm [50] first proposed a classification of molten chlorides according to their equivalent electrical conductivity, using Mendeleev's periodic table (Table 2). It is known that the value of the electrical conductivity is a measure of the ionic character of molten salts. In this case, the molten chlorides of metals from the groups I and II of the periodic table are good ionic conductors, while the molten chlorides of metals from groups VI and VII are nonconducting, having a covalent character. For the molten chlorides of metals from the other groups the conductivity values are intermediate. If a substance is strongly polar,

**Table 1**  Equivalent and Specific Electrical Conductivities ($\Lambda$ and $\kappa$) of Some Molten Alkali Halides at 50° above Their Melting Point

| Salt | $\Lambda$ (ohm$^{-1}$ cm$^2$ equiv$^{-1}$) | $\kappa$ (ohm$^{-1}$ cm$^{-1}$) |
|---|---|---|
| LiCl | 170.7 | 5.9 |
| NaCl | 142 | 3.7 |
| KCl | 113 | 2.3 |
| RbCl | 90 | 1.6 |
| CsCl | 78 | 1.3 |
| LiBr | 172 | 4.9 |
| NaBr | 136 | 3.0 |
| KBr | 100 | 1.7 |
| RbBr | 75 | 1.2 |
| CsBr | 64 | 0.9 |
| LiI | 174 | 4.0 |
| NaI | 134 | 2.4 |
| KI | 96 | 1.4 |
| RbI | 70 | 0.9 |
| CsI | 66 | 0.7 |

*Source*: From Ref. 49.

**Table 2**  Equivalent Electrical Conductivity of Molten Chlorides as a Function of the Position of the Element in the Periodic System, at the Melting Point

| | | | | | |
|---|---|---|---|---|---|
| HCl $10^{-6}$ | | | | | |
| LiCl 166 | BeCl$_2$ 0.086 | BCl$_2$ 0 | CCl$_4$ 0 | | |
| NaCl 133.5 | MgCl$_2$ 28.8 | AlCl$_3$ $15 \cdot 10^{-6}$ | SiCl$_4$ 0 | PCl$_5$ 0 | |
| KCl 103.5 | CaCl$_2$ 51.9 | ScCl$_3$ 15.0 | TiCl$_4$ 0 | VCl$_4$ 0 | |
| RbCl 78.2 | SrCl$_2$ 55.7 | YCl$_3$ 9.5 | ZrCl$_4$ | NbCl$_5$ $2 \cdot 10^{-7}$ | MoCl$_5$ $1.8 \cdot 10^{-6}$ |
| CsCl 66.7 | BaCl$_2$ 64.6 | LaCl$_3$ 29.0 | HfCl$_4$ | TaCl$_5$ $3 \cdot 10^{-7}$ | WCl$_6$ $2 \cdot 10^{-6}$ |
| | | | ThCl$_4$ 16 | | UCl$_4$ 0.34 |

[a] *Source*: From Ref. 50.

it will be completely dissociated in the molten state and its electrical conductivity will be high. If a substance has a molecular network in the solid state, the melt will have few ions and its electrical conductivity will be limited, like the molten halides of metals of high valence.

Redkin [51] calculated the specific conductivity of molten halides using the equation

$$\kappa = A \, \exp\left[\frac{-B - D/V}{T}\right] \exp\left(\frac{C}{V}\right) \tag{29}$$

where

$$\kappa = \text{specific conductivity (ohm}^{-1} \text{ cm}^{-1})$$
$$V = \text{molar volume (cm}^3 \text{ mol}^{-1})$$
$$T = \text{temperature (K)}$$
$$A, B, C, D = \text{empirical coefficients}$$

The relationship is valid for molten chlorides of the alkali metals, the alkaline earth metals and La, $Fe^{2+}$, $Mn^{2+}$. The author [51] proposed the following classification of molten chlorides:

1. Ideal molten salts, which are high conductivity melts. By substitution of some cation with another cation the specific conductivity in this group varies according to the change in interionic distance. Therefore, all these salts (pure salts and mixtures) have similar structures.
2. Molecular and ionic-molecular salts, which are nonconducting or low conductivity melts.

Using that classification, Redkin modified Biltz and Klemm's table (see Table 3). That classification was confirmed by Nakamura and Itoh [52], who found that the specific conductivity of molten alkali chlorides increases monotonically with decreasing cation radius, as shown in Figure 6. In the case of chlorides of alkaline earth metals there is a break at $CaCl_2$, and the specific conductivity decreases dramatically when going to $Mg^{2+}$ and $Be^{2+}$. This break may be attributed to complex formation among the component ions in those molten salts [53]. This means that in the molten alkali halides there are free ions and no complex formation, a fact confirmed by Raman spectroscopy [52].

Shabanov et al. [54,55] found that the electrical conductivity of the molten alkali chlorides and their binary mixtures is dependent on the strength of the applied electrical field. Figure 7 illustrates the increase of the equivalent electrical conductivity of molten sodium chloride with the strength of the applied electric field, reaching a limiting value of $\Lambda^\circ$ at $E^\circ \approx 10^6$ V/m. This phenomenon was observed at several temperatures. The increase of the electrical conductivity with

**Table 3**  Redkin's Classification in Good Ionic Conductors and Nonconductors as a Function of the Position of the Metal in the Periodic System

| HCl | | | | | |
|---|---|---|---|---|---|
| $10^{-6}$ | | | | | |
| LiCl | BeCl$_2$ | BCl$_2$ | CCl$_4$ | | |
| 166 | 0.086 | 0 | 0 | | |
| NaCl | MgCl$_2$ | AlCl$_3$ | SiCl$_4$ | PCl$_5$ | |
| 133.5 | 28.8 | $15 \cdot 10^{-6}$ | 0 | 0 | |
| KCl | CaCl$_2$ | ScCl$_3$ | TiCl$_4$ | VCl$_4$ | |
| 103.5 | 51.9 | 15.0 | 0 | 0 | |
| RbCl | SrCl$_2$ | YCl$_3$ | ZrCl$_4$ | NbCl$_5$ | MoCl$_5$ |
| 78.2 | 55.7 | 9.5 | | $2 \cdot 10^{-7}$ | $1.8 \cdot 10^{-6}$ |
| CsCl | BaCl$_2$ | LaCl$_3$ | HfCl$_4$ | TaCl$_5$ | WCl$_6$ |
| 66.7 | 64.6 | 29.0 | | $3 \cdot 10^{-7}$ | $2 \cdot 10^{-6}$ |
| | | | ThCl$_4$ | | UCl$_4$ |
| | | | 16 | | 0.34 |

<sup>a</sup>*Source*: From Ref. 50.

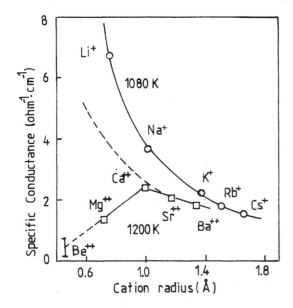

**Figure 6**  Specific conductance κ vs. ionic radii [52] for molten alkali chlorides MCl (M = Li, Na, K, Rb, Cs) at 1080 K and for molten alkaline earth chlorides MCl$_2$ (M = Be, Mg, Ca, Sr, Ba) at 1200 K.

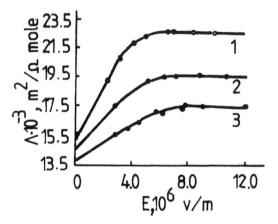

**Figure 7** Dependence of the equivalent electrical conductivity of molten NaCl on the strength of the applied electric field [54,55] at 1223 (1), 1049 (2) and 1003 K (3).

the strength of the applied electric field can be due to an increase of the mobility and/or concentration of charge carries. The concentration of charge carriers can increase due to simplification of the melt structure by the decomposition of ionic clusters. The mobility of the charge carriers depends not only on their sizes and weights. At high applied fields, the ions migrate with higher velocities and the surrounding ionic atmosphere from the oppositely charged particles is not fully established. According to a model of the structure of molten alkali halides due to Smirnov[56], ion complexes of the type $MX_n^{(n-1)-}$ and free $M^+$ ions of the second coordination sphere may exist as structural units in the melt.

The number of anions gathered around the central complex-forming ion can differ and depend on temperature and on the applied electric field. In a molten salt we may have the following equilibria [57]:

$$\frac{1}{4}MX_4^{3-} + \frac{3}{4}M^+ \rightleftharpoons \frac{1}{3}MX_3^{2-}$$

$$+ \frac{2}{3}M^+ \rightleftharpoons \frac{1}{2}MX_2^- + \frac{1}{2}M^+ \rightleftharpoons M^+ + X^- \quad (30)$$

With increasing temperature the equilibria are being shifted to the right, therefore $n$ (in $MX_n^{(n-1)-}$) decreases with the complete dissociation of the salt into simple ions. In Shabanov's opinion [54] the saturation of the electrical conductivity observed at high applied potentials is due to this complete dissociation of the salt into simple ions.

For the variation of the electrical conductivity with temperature an Arrhenius type equation is used:

$$\Lambda = Ae^{-E_\Lambda/RT} \tag{31}$$

where $E_\Lambda$ is the activation energy for conductivity and $A$ is a constant. Another equation for the temperature dependence of the electrical conductivity is [49]

$$\kappa = a + bT + cT^2 + dT^3 \tag{32}$$

where $a$, $b$, $c$, $d$ are constants. Generally it is supposed that molten salts have a structure similar to that of the respective crystals; therefore the electrical conductivity should be of the same type. In a melt, the majority of the particles have an oscillating movement around some equilibrium positions, and only a small part of the particles with a higher energy migrates to other equilibrium positions. With an applied electric field this migration of the particles takes place in the direction of the field, and at a high applied field the number of free ions increases.

It is also known that ionic substances acquire high electrical conductivity by fusion. The explanation of that fact could be the following. In the solid state there are few current conducting ions, but in the molten state the number of these ions increases dramatically. Also the kinetic energy of the ions increases (proportional to $RT$), the ionic interactions decrease and the ionic mobilities increase. At fusion, the free volume increases; consequently the ionic mobility increases and the ionic associations are broken with increasing temperature. According to this mechanism the activation energy for conductivity is the minimum energy required for an ion to make a translation movement. That increases with increasing cationic radius for a given anion and increases with the anionic radius for a given cation.

The equivalent electrical conductivity ($\Lambda$) depends on the viscosity of the liquid ($\eta$) according to the rule given in 1906 by Walden [58] for electrolyte solutions:

$$\Lambda\eta = \text{const} \tag{33}$$

Viscosity is an internal force of friction which acts oppositely to the flowing fluid. Walden's rule is also applicable to molten salts. Shabanov [54] applied that rule for the limiting equivalent electrical conductivities (see Figure 7):

$$\Lambda^0\eta = (\Lambda_+{}^0 + \Lambda_-{}^0)\eta = \text{const} \tag{34}$$

He found that for each ion, the product $\Lambda_i{}^0\eta$ changes very little, becoming constant within the limits of error of the measurements.

The Stokes-Einstein law can be applied for molten salts:

$$\Lambda_i^0 = \frac{|z_i|e^2 N_A}{6\pi\eta r_i} = \frac{|z_i|F^2}{6\pi N_A \eta r_i} \tag{35}$$

where

$N_A$ = Avogadro's number
$e$ = electron charge
$r_i$ = ionic radius

Expressing $\Lambda^0$ in m²/ohm mole and $\eta$ in Pa · s the above relation may be rewritten as

$$\Lambda_i^0 \eta = \frac{|z_i| 0.82 \cdot 10^{-5}}{r_i} \tag{36}$$

For a univalent electrolyte,

$$(\Lambda_+^0 + \Lambda_-^0)\eta = 0.82 \cdot 10^{-5}\left(\frac{1}{r_+} + \frac{1}{r_-}\right) \tag{37}$$

where $r_+$ and $r_-$ are cationic and anionic radii respectively.

Figure 8 shows a plot of $\Lambda_i^0 \eta$ versus $1/r$ for many ions [54]. This dependence is a straight line which goes through the origin of the coordinate axes. Table 4 shows that the product $\Lambda_i^0 \eta r$ is an almost constant quantity for all ions [54].

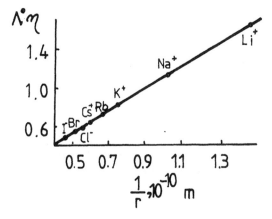

**Figure 8** Dependence of the product of the limiting electrical ionic conductivity and the viscosity of a salt on the inverse of the cation or anion radii [54].

**Table 4** The Product $\Lambda_i^0 \eta r_i$ for Different Ions

| Ion | Li$^+$ | Na$^+$ | K$^+$ | Rb$^+$ | Cs$^+$ | Cl$^-$ | Br$^-$ | I$^-$ |
|---|---|---|---|---|---|---|---|---|
| $\Lambda_i^0 \eta r_i$ | 1.13 | 1.11 | 1.10 | 1.09 | 1.07 | 1.09 | 1.11 | 1.12 |

*Source*: From Ref. 54.

For a molten alkali halide MX, the Newtonian equations for M$^+$ and X$^-$ in an applied field $E$ are [52]

$$m_M \left( \frac{dV_M}{dt} \right) = q_M E - \left( \frac{S_M}{C_M} \right) V_M \tag{38}$$

$$m_X \left( \frac{dV_X}{dt} \right) = q_X E - \left( \frac{S_X}{C_X} \right) V_X \tag{39}$$

where

$m_M, m_X$ = masses of the ions
$q_M, q_X$ = ionic charges
$S_M, S_X$ = effective cross-sectional areas
$V_M, V_X$ = average drift velocities of the ions
$C_M, C_X$ = constants that depend on temperature and composition of the molten systems
$t$ = time

In the stationary state where $dV_M/dt = dV_X/dt = 0$, Eqs. (38) and (39) become

$$V_M = \left( \frac{C_M q_M}{S_M} \right) E = u_M E \tag{40}$$

$$V_X = \left( \frac{C_X q_X}{S_X} \right) E = u_X E \tag{41}$$

where $u$ are the mobilities of the ions. If the charge carrier concentrations $N_M$ and $N_X$ are introduced, the electrical current density $i$ and the total specific conductivity ($\kappa$) at steady state conditions are

$$i = N_M q_M V_M + N_X q_X V_X = \kappa_{MX} E \tag{42}$$

where

$$\kappa_{MX} = \frac{C_M N_M q_M^2}{S_M} + \frac{C_X N_X q_X^2}{S_X} = q_M N_M u_M + q_X N_X u_X \tag{43}$$

If we introduce

$$N_M = N_X = N, \quad q_M = q_X = q, \quad S_M = ar_M^n, \quad S_X = ar_X^n$$

$$\kappa_{MX} = \frac{Nq^2}{a}\left(\frac{C_M}{r_M^n} + \frac{C_X}{r_X^n}\right) \tag{44}$$

In the special case when $C_M/r_M^n \gg C_X/r_X^n$, i.e., the cations are the principal charge carriers, the following linear relationship is obtained:

$$\kappa_{MX} = \left(\frac{Nq^2 C_M}{a}\right)\frac{1}{r_M^2} \tag{45}$$

where $n$ is set equal to 2. Figure 9 presents plots of $\kappa_{MX}$ verses $r^2$ for the alkali halides [52]. There are linear relationships with the exception of the fluorides. The deviations from the linear relationship for KF, RbF and CsF are related to

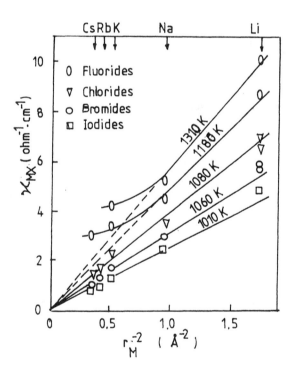

**Figure 9** Plots of specific conductivity $\kappa_{MX}$ vs. reciprocal cross-sectional area $r_M^{-2}$ of the alkali ion $M^+$ for the molten alkali halides MX (M = Li, Na, K, Rb, Cs) with the common halide ion $X^-$ at the temperatures indicated [52].

the fact that the $F^-$ anion is small and more mobile than the cation in molten KF, RbF and CsF.

The Nernst-Einstein relation shows the dependence between the self-diffusion coefficient $D_i$ and the equivalent conductivity $\Lambda$ of molten salts:

$$\Lambda = \frac{F^2}{RT} \sum D_i \tag{46}$$

There are numerous data in the literature [59] which demonstrate that for molten halides the equivalent conductivity calculated by means of the Nernst-Einstein relation is significantly higher than the directly measured conductivity value. This is due to the fact that the structural entities of molten salts make unequal contributions to diffusion and electrical conductivity.

## 3. Electrical Conductivity of Molten Salt Mixtures

The electrical conductivity of molten salts has been used to elucidate the structure of the salts. Generally the variation of that property was measured as a function of composition and temperature. The isotherm of the ideal equivalent conductivity of a binary system is additive, and it can be calculated by the relationship

$$\Lambda_m = x_1 \Lambda_1 + x_2 \Lambda_2 \tag{47}$$

where $\Lambda_1$ and $\Lambda_2$ are the conductivities of the two salts and $x_1$, $x_2$ are the molar fractions. Deviations from the ideal isotherm represent modifications in the structure of these systems and displacement of the constituent ions due to ionic interactions during mixing.

It is noticed that the law of additive equivalent electrical conductivity does not apply for simple molten salt mixtures, like the NaCl-KCl system. All other properties indicate this system to be a very simple mixture. However, the isotherm of equivalent electrical conductivity shows a maximum negative deviation from additivity at the 50-50 composition, and that indicates the presence of ionic interactions. Generally, if the deviations are not large, they can be due to the presence of ionic interactions. Large deviations indicate the presence of ionic complexes.

Markov et al. [60,61] proposed an equation for the equivalent electrical conductivity of simple binary molten salt mixtures. In binary systems ($M_1X + M_2X$ or $MX_1 + MX_2$) there is the possibility of the following ionic arrangements: $M_1X - M_1X$; $M_2X - M_2X$; $M_1X - M_2X$. The probabilities of forming the combinations $M_1X - M_1X$; $M_2X - M_2X$ and $M_1X - M_2X$ are proportional to $x_1^2$, $x_2^2$ and $2x_1x_2$, respectively, where $x_1$ and $x_2$ are the molar fractions of the two salts. For monovalent molten salts, the equivalent electrical conductivity of a mixture of these salts, $\Lambda_m$, can be written as

$$\Lambda_m = x_1^2 A_1 e^{-E_{1,1}/RT} + x_2^2 A_2 e^{-E_{2,2}/RT} + 2x_1 x_2 A_{1,2} e^{-E_{1,2}/RT} \tag{48}$$

where $E_{1,1}$ is the activation energy for electrical conductivity of the first salt, $E_{2,2}$ the same for the second salt, and $E_{1,2}$ is the activation energy for the combination $M_1X - M_2X$, the rest of the notations are known.

Since $\Lambda_1 = A_1 e^{-E_{1,1}/RT}$ and $\Lambda_2 = A_2 e^{-E_{2,2}/RT}$, Eq. (48) can be rewritten as

$$\Lambda_m = x_1^2 \Lambda_1 + x_2^2 \Lambda_2 + 2x_1 x_2 A_{1,2} e^{-E_{1,2}/RT} \tag{49}$$

The electrical conductivity of the combination $M_1X - M_2X$ is determined by the component of the lowest conductivity. Therefore, we may set $A_{1,2} e^{-E_{1,2}/RT} \approx \Lambda_1$ if the components are arranged so that $\Lambda_1 < \Lambda_2$. Modifying the last equation gives

$$\Lambda_m = x_1^2 \Lambda_1 + x_2^2 \Lambda_2 + 2x_1 x_2 \Lambda_1 \tag{50}$$

This equation is verified only for systems which are close to being ideal mixtures. The value of the deviations from that equation can be a classification criterion for binary molten salt mixtures. If the deviations are not higher than 2–3%, the mixtures are ideal according to Markov's formula. Deviations higher than 5% indicate complex processes taking place during mixing of the simple molten salts of the types: modification of the degree of ionic dissociation, modification of the screening of cations by the anions, ionic polarization or complex formation. Generally, the deviations of the equivalent electrical conductivity from the additive conductivity for molten salts according to Markov increase with the complexity of the phase diagrams from solid solutions to systems of chemical interactions. The formation of complexes in a melt changes its ionic structure. The initial smaller ions are replaced by bigger complex ions of a lower mobility, and that means a lower conductivity and a higher viscosity. Accordingly, a minimum appears on the conductivity isotherm and a maximum on the viscosity isotherm. Therefore, it is believed that the composition of the complex ions can be obtained from the location of the minimum conductivity.

Bloom et al. [62] showed that the composition of complex ions in melts can differ from that of minimum conductivity, because the complex ions are in equilibrium with the simple ions, so that the maximum negative deviation from the additive conductivity would not correspond to the stoichiometry of the complex ions. Other researchers [63,64] found that the stoichiometry of complex ions is influenced by the maximum value of the activation energy. In any case, the electrical conductivity data must be correlated with other physicochemical data, such as phase diagrams, Raman spectra, minimum thermodynamic activities, to obtain the composition of complex ions.

Until now there has not been any general theory which explains the experimental results of electrical conductivity in molten salts. Some attempts at conductivity calculations following structure models of liquids have been made, but the results are not satisfactory.

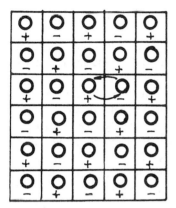

**Figure 10** Mechanism of ionic movements in a molten salt according to the cellular model [65].

According to the cellular model (Fig. 10) the motion of an ion involves a particle rotating in a liquid network [65]. For that, the ions of the same charge must be in close positions, a condition which requires a high activation energy, 120–170 kJ/mol, while the experimental values are 10 times lower.

Oldekop [66] considered that each ion in its motion repels the ions of the same charge and attracts ions of the opposite charge. Each ion is surrounded (as in the Debye-Hückel theory) by a screen of spatial charge. In the case of a charge distribution of spherical symmetry there is no resultant force which acts upon the ion under consideration. If the ion moves in an external field, a charge screen remains behind it due to the necessary time of relaxation. Therefore, a retarding force acts on the ion proportional to its migration speed in the electrical field. The electrical conductivity can be obtained from the migration speed. The difference between the Debye-Hückel and Oldekop theories is that the relaxation time in molten salts depends on the thermal motion of the ions. The results of theoretical calculations reproduce only the order of magnitude of the experimental data.

## 4. Electrical Conductivity of Metal—Molten Salt Mixtures

Generally, pure molten salts are transparent and clear. If a piece of metal is introduced into a pure molten salt, a so-called metal fog is observed streaming from the metal into the melt and coloring it. The color intensity usually increases with increasing concentration of dissolved metal. In most cases the properties of the pure molten salt change when a metal dissolves in it, and a true solution is formed. For example, the solubility is higher if the metals dissolve in their own salts, and several binary systems show a complete miscibility between metal and salt above a certain critical temperature. The metal solubility usually increases with temperature.

Many reviews of the properties of metal-molten salts have been published, of which can be mentioned Bredig [67], Cubicciotti [68], Ukshe and Bukun [69], Corbett [70], Warren [71,72], Haarberg and Thonstad [73].

Bredig [67] considered two categories of metal-molten salt mixtures: metallic and nonmetallic solutions. In metallic solutions the metal dissolves without interacting strongly with the melt. Metal ions and partially free electrons are formed. The electrical conductivity of these mixtures increases strongly due to the presence of very mobile partially free electrons. Therefore an electronic conductivity appears in these melts. In nonmetallic solutions the metal reacts with the melt under the formation of subvalent ions or subvalent compounds. The electrical conductivity of these mixtures depends only to a small extent on the concentration of dissolved metal. The variation of properties of the metal-molten salt mixtures shows a continuous change from the metallic solutions to the nonmetallic if the temperature is sufficiently high.

As in the case of molten salt mixtures, the specific conductivity of the metal-molten salt mixtures can be treated as the sum of the specific conductivity of the molten salt and that of the dissolved metal:

$$\kappa_{MX-M} = \kappa_{MX} + \kappa_{M} \tag{51}$$

Likewise, the equivalent conductivity is expressed in terms of individual equivalent conductivities:

$$\Lambda_{MX-M} = x_{MX}\Lambda_{MX} + x_{M}\Lambda_{M} \tag{52}$$

where $x_{MX}$ and $x_{M}$ are the equivalent fractions of molten salt and dissolved metal, respectively.

Several examples of studies of the solubility of metals in molten salts are given in the following. Bredig et al. [67] measured electrical conductivities of the binary alkali systems in the salt-rich region. The specific conductivity of the mixtures was found to increase dramatically by addition of metal in all these systems. The potassium systems have a larger increase of the conductivity than the sodium systems. Some results are given in Figure 11 [74]. The equivalent conductivity of dissolved metal is higher for potassium systems than for sodium systems. The authors [74] explained these results by the fact that at higher metal concentrations in the melt, dimers like $Na_2$ and $K_2$ are formed, the $Na_2$ dimers being more stable.

In molten salt systems, the polarizability of the anions increases with increasing size of the anion, and high polarizability may facilitate electron transport. Warren et al. [75] showed that the electron mobility is constant in the NaBr-Na systems at concentrations below 5.5 mol% Na.

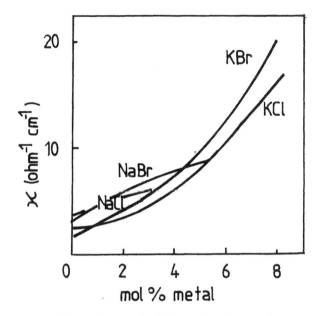

**Figure 11** Specific conductivity [74] of mixtures of sodium-sodium halides and potassium-potassium halides. KCl: 820°C; KBr: 760°C; NaCl: 850°C; NaBr: 805°C.

Several models based on the electronic properties of mixtures of metals and molten salts have been proposed, i.e., the localized electron model, the free electron model and the band model. A model which gives a good description of the properties of alkali metal–alkali halide mixtures at low metal concentrations is the model of trapped electrons or the so-called model of F-centers [76,77]. An F-center may be regarded as a localized state, and the electron is then trapped in a cavity with octahedral coordination with the neighboring cations. On average, the F-center may be considered as an $M_6^{5+}$ species.

F′-centers (i.e., two electrons trapped in the same anion vacancy) and M-centers (two electrons occupying two adjacent anion vacancies, i.e., two adjacent F-centers) can also be present. Symons [78] observed by electron spin resonance spectroscopy that in solid alkali halide crystals doped with metal the F-centers are the most stable, while F′-centers and M-centers may be formed at higher concentrations of trapped electrons. Durham and Greenwood [79] proposed that the dissolved metal dissociated into metal cations and nearly free electrons scattered in the conduction band.

Several models for the region of low concentrations of metal were proposed by other authors. Rice [80] suggested that the increase of the electrical conductivity of dissolved metal in its own salt is due to random walk of the electron between atoms and metal ions, namely a thermally activated hopping model. The calculated electron mobility agreed with experimental data.

Emi and Bockris [81] assumed that the electrons are transferred from the metal atom to a neighboring ion by hopping followed by tunneling. For the NaCl-Na and KCl-K systems at low metal concentrations, the authors supposed that the electrons were well localized but that delocalization occurred at higher metal concentrations, particularly for the KCl-K system. The calculated mobilities were in agreement with experimental data.

## C. Thermodynamics of Electrode Processes in Molten Salts

### 1. Electrodes in Molten Salts

Generally, the criteria for the choice of electrode materials in molten salts are the same as in aqueous solutions. However, some additional criteria are necessary for each particular case, due to the high working temperature. Both solid and liquid electrodes have been utilized in molten salts, and, as in aqueous solutions, these must satisfy the requirements of long-term stability, large contact surface area with the electrolyte, ease and cheapness of fabrication.

The mercury electrode has been used by researches in melts up to about 250°C, especially for nitrate and chloroaluminate melts [82,83]. Liquid lead (m.p. 327.5°C) is another liquid metal [84] used for molten halides. Liquid electrodes of bismuth (m.p. 271°C), indium (m.p. 157°C) and thallium (m.p. 303°C) have also been, used but to a limited extent [85].

The most used liquid metal pool electrode is the aluminum cathode (m.p. 660°C) in the Hall-Héroult aluminum extraction cell. Alkali metals and alkaline-earth metals are also used as liquid cathodes in their molten salt extraction processes.

A serious problem with these categories of electrodes in molten salts is the dissolution of the metal into melts having the same cation, and thus the melts may have a significant electronic conduction. (See Section I.B.4, "Electrical Conductivity of Metal-Molten Salt Mixtures".)

Probably the most common solid electrode is platinum, although it dissolves anodically in some melts, for example in halides. The choice of gold and silver [86] is also frequently made. Graphite is very often used because it is cheap and can be obtained in a wide range of sizes and qualities. These electrodes can be used over long periods of time, and they have a wide electrochemical stability, both anodic and cathodic. Vanadium and molybdenum are also used in appropriate systems. Studies for the use of some inert anodes made of semiconducting ceramics have been made, especially for aluminum electrolysis [87].

## 2.  Reference Electrodes in Molten Salts

In spite of large experimental difficulties in molten salts, many reference electrodes are used for the determination of thermodynamic activity and for kinetic measurements [88–90].

There is no fundamental difference between the two half-cells or electrodes in a cell for measuring emf (electromotive force), especially in molten salts. However, it is usual to designate one of the electrodes as reference electrode if it is used for the measurement of an emf series. In many cases a diaphragm is used to separate the two half-cells.

Two emf series [91,92] are given in Table 5, measured in the cells M/ $M_xCl_y$, LiCl, and KCl/AgCl, LiCl-KCl/Ag.

### Metal-Metal Ion Electrodes

These electrodes are frequently used and consist of a metal in equilibrium with a melt containing ions of the same metal. These electrodes are easy to prepare and are usually highly reproducible because the exchange current density is high at the interface metal-metal ion in the melt.

**Table 5**   Standard Reduction Potentials (V) on the $Ag/Ag^+$ (mole fraction) Scale[a]

| Solvent melt Temperature | LiCl-KCl 450°C | NaCl-KCl 700°C |
|---|---|---|
| $Au/Au^+$ | +0.948 | |
| $Cl/Cl^-$ | +0.853 | +0.845 |
| $Cu^+/Cu^{2+}$ | +0.682 | +0.596 |
| $Cr^{2+}/Cr^{3+}$ | +0.006 | +0.246 |
| $Ag/Ag^+$ | 0.000 | 0.000 |
| $Ni/Ni^{2+}$ | −0.158 | −0.140 (at 670°) |
| $Cu/Cu^+$ | −0.214 | −0.256 |
| $Co/Co^{2+}$ | −0.354 | −0.319 |
| $Pb/Pb^{2+}$ | −0.464 | −0.390 |
| $Fe/Fe^{2+}$ | −0.534 | −0.520 |
| $Cr/Cr^{2+}$ | −0.788 | −0.758 |
| $Zn/Zn^{2+}$ | −0.929 | −0.860 |
| $Mn/Mn^{2+}$ | −1.212 | −1.206 |
| $Li/Li^+$ | −2.773 | |

[a] Standard reduction potential means the Gibbs energy of the electrode reaction expressed as a potential (in volts).
*Source*: From Refs. 91 and 92.

## Electrodes of Sintered Oxides

Oxide electrodes have been used as reference electrodes in cryolite-alumina melts for the measurement of the oxygen-ion concentration. Thus, Rolin et al. [93–95] used galvanic cells of the following types:

$$Al/Na_3AlF_6 + Al_2O_3/SnO_2 \tag{53}$$

$$W/Na_3AlF_6 + Al_2O_3/SnO_2 \tag{54}$$

$$Al/Na_3AlF_6 + Al_2O_3/Cr_2O_3 \tag{55}$$

and found that the emf of these cells varied with the alumina concentration. The proposed cell reaction of cell (53) is

$$3SnO_2 + 4Al = 3Sn + 2Al_2O_3 \tag{56}$$

## Gas-Ion Electrodes

In principle, a gas electrode consists of an inert material immersed into a melt and flushed with the respective gas. In the case of oxygen electrodes, the inert material is a noble metal (Pt, Au, etc.), and in the case of halogen electrodes it is graphite, and the respective gas is adsorbed on the surface of the electrode. The gas molecules are in equilibrium with the ions of the same gas in the melt.

## 3. Calculation of Thermodynamic Functions from EMF Measurements

The potential of a gas electrode $X_2/X^-$ results from the equilibrium between the electrons of an inert conductor, gas molecules ($X_2$) dissolved in the electrolyte or adsorbed on the electrode and $X^-$ ions in the electrolyte. If we presume the simultaneous existence of chlorine atoms and molecules adsorbed on the graphite electrode, the behavior of the chlorine electrode results from an equilibrium between chlorine on the electrode and in the electrolyte, according to the reaction

$$2Cl^- - 2e \rightleftharpoons 2Cl_{(ads)} \rightleftharpoons Cl_{2(ads)} \rightleftharpoons Cl_{2(gas)} \tag{57}$$

The overvoltage of discharge of the chlorine ion is low, and the electrode potential $E$ is given by the Nernst equation:

$$E = E^\circ - \frac{RT}{F} \ln \frac{a_{Cl^-}}{(P_{Cl_2})^{1/2}} \tag{58}$$

The activity of chloride ions in a pure molten chloride being constant, Eq. (58) could be written

$$E = E° + \frac{RT}{2F} \ln P_{Cl_2} \tag{59}$$

At standard state, when the chlorine pressure is 1 atm, $E = E°$.

Oxygen reference electrodes are useful for the study of the thermodynamic and electrochemical properties of oxyanionic melts [96,97].

Halogen electrodes in molten salts are being used especially for the study of molten halides. In 1930 Hildebrand and Salstrom [98–100] were the first who applied a chlorine reference electrode in molten salts. Up to the present, other authors used and improved this electrode [101–105]. The reversible bromine/bromide electrode was set up for the first time by Hildebrand and Salstrom [98] and improved by Murgulescu et al. [106,107]. In the following an example is shown of how to obtain and use a reversible gas electrode in molten salts for the determination of thermodynamic activity.

The reversible iodine/molten silver iodide electrode was used for the first time by Sternberg, Adorian and Galasiu [108]. To obtain this electrode it was necessary to construct an electrochemical cell which maintained iodine in the gaseous state from the moment it was generated until it was removed from the cell. The cell, which is shown in Figure 12, was constructed of heat resistant

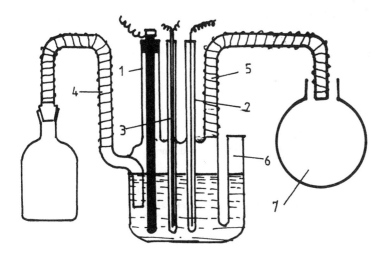

**Figure 12**  Cell scheme for the use of the iodine electrode in molten AgI [108]: 1, graphite electrode; 2, silver electrode; 3, thermocouple Pt10%Rh-Pt; 4, electrically heated tube, through which iodine vapors come from the generator; 5, heated exit tube for iodine vapor; 6, tube for introducing iodine into the cell and removing it; 7, iodine collector.

glass. The cathode consisted of a silver wire, while the positive electrode was a rod of spectral graphite. To obtain a reproducible emf it was necessary to bubble gaseous iodine through the cell for about 20 h or to make pre-electrolysis for about 5–6 h. The emf of the iodine electrode was constant and reproducible within $\pm 1.5$ mV.

The temperature dependence of the cell

$$Ag/AgI_{(l)}/I_{2(g,\ 1\ atm)}C \tag{60}$$

can be expressed by the equation

$$E^\circ = 0.550 - 0.250 \cdot 10^{-3}(T - 873) \tag{61}$$

The enthalpy, entropy, and Gibbs energy of formation for the reaction

$$Ag_{(s)} + \frac{1}{2}I_{2(g)} = AgI_{(l)} \tag{62}$$

are given in Table 6, at the melting point (557°C) of AgI. Table 6 also shows the numerical values of the same thermodynamic functions at 557°C calculated from available thermodynamic data, taken from Landolt-Börnstein [109] and Kubaschewski and Evans [110]. It may be noted that there is good agreement between the results calculated from thermodynamic data and those found from emf measurements.

The iodine electrode was also used by Sternberg, Adorian and Galasiu [111] to study the cell

$$Ag_{(s)}/AgI_{(l)} + KI_{(l)}/I_{2(g,\ 1\ atm)}C \tag{63}$$
$$\quad x_1 \qquad\qquad x_2$$

**Table 6** Thermodynamic Functions for the Formation Reaction (62) of AgI at 557°C Based on EMF Measurements and Tabulated Data

| Thermodynamic functions | Based on emf measurements[a] | Based on thermo-dynamic data[b] |
|---|---|---|
| $\Delta H^0_{830}$ (cal/mol) | −17,689 | −17,200 |
| $\Delta S^0_{830}$ (cal/mol dgr) | −5.76 | −4.57 |
| $\Delta G^\circ_{830}$ (cal/mol) | −12,917 | −13,407 |
| $E^\circ$ (volt) | 0.560 | 0.581 |

[a] From Ref. 108.
[b] From Refs. 109 and 110.

**Table 7** Measured EMF and Calculated Thermodynamic Values for AgI + KI Mixtures

| $x_{AgI}$ | Emf (V) | $a_{AgI}$ | $\gamma_{AgI}$ | $\bar{G}_1^E$ (cal/mole) | $x_{KI}$ | $\gamma_{KI}$ | $\bar{G}_1^E$ (cal/mole) |
|---|---|---|---|---|---|---|---|
| 1 | $0.550-0.250 \times 10^{-3}(T-873)$ | 1 | 1 | 0 | 0.1 | 0.380 | $-1678$ |
| 0.9 | $0.559-0.245 \times 10^{-3}(T-873)$ | 0.887 | 0.985 | $-25$ | 0.2 | 0.576 | $-959$ |
| 0.8 | $0.573-0.235 \times 10^{-3}(T-873)$ | 0.736 | 0.920 | $-144$ | 0.3 | 0.720 | $-571$ |
| 0.7 | $0.5885-0.225 \times 10^{-3}(T-873)$ | 0.599 | 0.856 | $-270$ | 0.4 | 0.811 | $-364$ |
| 0.6 | $0.605-0.210 \times 10^{-3}(T-873)$ | 0.481 | 0.802 | $-384$ | 0.5 | 0.855 | $-271$ |
| 0.5 | $0.622-0.190 \times 10^{-3}(T-873)$ | 0.384 | 0.767 | $-460$ | 0.6 | 0.901 | $-181$ |
| 0.4 | $0.645-0.160 \times 10^{-3}(T-873)$ | 0.282 | 0.706 | $-604$ | 0.7 | 0.950 | $-88$ |
| 0.3 | $0.672-0.140 \times 10^{-3}(T-873)$ | 0.197 | 0.657 | $-729$ | 0.8 | 0.990 | $-17$ |
| 0.2 | — | — | — | — | 0.9 | 0.998 | $-3$ |
| 0.1 | — | — | — | — | 1.0 | 1 | 0 |
| 0 | — | — | — | — | | | |

*Source:* From Ref. 111.

in which the electrolyte was a mixture with a common anion ($x_1$ and $x_2$ represent the molar fractions of the components). The thermodynamic activity of AgI was calculated on the basis of Nernst's equation:

$$E = E° - \frac{RT}{nF} \ln a_1 \tag{64}$$

and the partial molar excess Gibbs energy by means of the equation

$$\overline{G_1}^E = RT \ln \gamma_1 \tag{65}$$

where $\gamma$ is the activity coefficient.

The activities, the activity coefficients and the excess partial molar Gibbs energy of this mixture given in Table 7 were calculated by means of the emf data. It may be noted that this mixture, like the AgCl + KCl [103] and AgBr + KBr [98] mixtures, shows negative deviations from ideal behavior. In Figure 13 the values of the partial molar excess Gibbs energies of the three mixtures as a function of $x_2^2$ are given. It is seen that these mixtures show negative deviations from Hildebrand's relation for regular solutions [112,113]:

$$\overline{G_1}^E = Bx_2^2 \tag{66}$$

The fact that the negative deviation of the AgI + KI mixture is slightly larger than those of the chloride and bromide mixtures may be explained by the

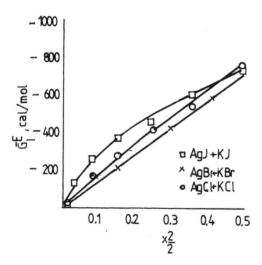

**Figure 13** Variation of $\overline{G_1}$ with $x_2$ for the mixtures [111] AgI + KI, AgBr + KBr, AgCl + KCl.

**Table 8** Activities, Activity Coefficients and Excess Partial Molar Gibbs Energies for AgI in AgI + AgBr and AgI + AgCl Mixtures at 600°C

| $\chi_{AgI}$ | AgI + AgCl | | | AgI + AgBr | | |
|---|---|---|---|---|---|---|
| | $a_{AgI}$ | $\gamma_{AgI}$ | $\overline{G}_1^E$ (cal/mole) | $a_{AgI}$ | $\gamma_{AgI}$ | $\overline{G}_1^E$ (cal/mole) |
| 1 | 1 | 1 | 0 | 1 | 1 | 0 |
| 0.8 | 0.802 | 1.002 | 4 | 0.811 | 1.051 | 84 |
| 0.6 | 0.628 | 1.046 | 78 | 0.662 | 1.103 | 170 |
| 0.5 | 0.528 | 1.056 | 95 | 0.587 | 1.174 | 278 |
| 0.4 | 0.432 | 1.081 | 153 | 0.501 | 1.251 | 389 |

*Source*: From Ref. 115.

tendency of the former to form complexes; the phase diagram shows that a compound is formed with an incongruent melting point for $\chi_{AgI} = 0.67$, whereas AgCl + KCl and AgBr + KBr yield simple eutectic systems on solidification [114]. This behavior is explained by the fact that the iodide ion is more easily polarizable than the chloride and bromide ions.

Using the same iodine electrode, Sternberg et al. [115] studied the cells

$$Ag_{(s)}/AgI_{(l)} + AgBr_{(l)}/I_{2(g,1\,atm)} C$$
$$x_1 \qquad\qquad x_2 \tag{67}$$

and

$$Ag_{(s)}/AgI_{(l)} + AgCl_{(l)}/I_{2(g,1\,atm)} C$$
$$x_1 \qquad\qquad x_2 \tag{68}$$

in which the electrolytes were mixtures with a common cation. The activity coefficients and the excess partial molar Gibbs energies for AgI in the two mixtures are given in Table 8. The experimental results show that these mixtures with a common cation display positive deviations from ideal behavior just as the AgCl + AgBr mixture does [105]. In these mixtures where the common ion is only slightly polarizable, the negative component of the heat of mixing takes small values and the positive component arising from both van der Waals and coulombian forces is prevalent.

## D. Kinetics of Electrode Processes in Molten Salts

### 1. The Electrical Double Layer

In principle, molten salts differ from aqueous solutions due to the absence of a polar solvent (water). For this reason, the structure of molten salts is different, consisting of positive and negative ions. Likewise, the structure of the molten salt/metal interface is different from that of the aqueous solution/metal interface.

Selective adsorption of ions on an electrode surface contributes to the formation of an electrical double layer. This layer is due to the electrostatic forces of adsorption between the charge of the electrode and the ions of opposite charge present in the solution. The electrical double layer generates a difference in potential and forms an electric condenser as a first approximation.

While in aqueous solutions an electrical double layer is formed, in molten salts an electrical multilayer is formed because of their structure [116]. The Russian school, particularly Dogonadze and Chizmadzhev [117], presented the first theoretical description of this layer. These authors used a binary distribution function to interpret X-ray diffraction measurements in molten salts. By means of these results they tried to describe the structure of the melt at the interface and concluded that the charge distribution was characterized by an attenuated oscillation of the charge distribution in the melt.

The general features of the capacitance of the electrical double layer can be summarized as follows:

The capacitance-potential curves have parabolic shapes with a minimum at $E_q = 0$ (i.e., at the potential of zero charge).
The potential of zero charge is the same for bromides, iodides and chlorides.
$C_{min}$ (at $E_q = 0$) increases in the order $Cl^- < Br^- < I^-$.
The zero-charge potential depends strongly on the nature of the cation.
$C_{min}$ increases with temperature.

Important work in the field of the capacitance of the electrical double layer in molten salts was presented by Ukshe and Bukun [118,119], Inman et al. [120–124], and Berge and Tunold [125].

## 2. Classification of Electrode Reactions in Molten Salts

A large number of electrochemical reactions in molten salts have been studied. In the case of electrolysis of a simple molten salt, the reduction of the metal ion takes place at the cathode; for example, for sodium deposition,

$$Na^+ + e = Na \tag{69}$$

In the case of mixtures of molten salt, usually only one cation is reduced during electrolysis; for example, in the electrolysis of a $PbCl_2$ + $NaCl$ mixture, the cathode reaction is

$$Pb^{2+} + 2e = Pb \tag{70}$$

If there are cations with multiple valence steps, the process may be very complicated, because insoluble or soluble intermediate compounds may be formed.

Some of these compounds can be oxidized by the anodic products and the current efficiency decreases. A very complicated case is the reduction of Al—O—F anions to produce aluminum (see Section II.A). Usually, the charge transfer overvoltage is low in the case of the cathodic reactions.

In the field of anodic reactions, the oxidation of some anions involves the formation of gaseous products; for example,

$$2Cl^- = Cl_2 + 2e \tag{71}$$

Another type of reaction is the oxidation of complex ions, for example in the case of aluminum electrolysis. These reactions are rather complicated and occur in several steps. During the first step, the discharge of oxygen ions takes place; the oxygen atoms formed are adsorbed on the surface of the carbon anode and molecules of $CO_2$ are then obtained. These molecules of $CO_2$ can react with the anodic carbon, and a certain proportion of CO may appear. All these gases form bubbles which escape. Usually, the anodic processes have a high overvoltage.

Other anodic reactions are the discharge of oxygen, sulfur, etc., on a metallic anode with the formation of thin layers of oxides, sulfides, etc.

## 3. Overvoltage in Molten Salts

In the case of a current passing through a cell which represents a deviation from the equilibrium state, we can write the following Butler-Volmer relation between current and voltage for a single electrode:

$$i = i_0 \left[ \exp \frac{(1 - \alpha)zF\eta}{RT} - \exp \frac{-\alpha zF\eta}{RT} \right] \tag{72}$$

where $i_0$ is the exchange current density, $\alpha$ is the transfer coefficient and $z$ is the number of electrons participating. For a large overvoltage, Eq. (72) can be simplified as follows for an anodic current:

$$i_+ = i_0 \exp \frac{\alpha zF\eta}{RT} \tag{73}$$

For a cathodic current,

$$i_- = i_0 \exp \frac{-(1 - \alpha)zF\eta}{RT} \tag{74}$$

If Eqs. (73) and (74) are transformed to a logarithmic form, the simple "Tafel equation" is obtained:

$$\eta = a + b \log i \tag{75}$$

where the constants $a$ and $b$ can be determined from experimental data:

$$b = \frac{2.303RT}{\alpha F}, \qquad a = -b \log i_0 \qquad (76)$$

From these relations also, the constant $\alpha$ and the exchange current density can be determined.

The Tafel equation implies that the overvoltage is a measure of the thermodynamic irreversibility of the electrode reaction, and it is associated with the slow step of the process. We distinguish some types of overvoltage depending on the type of slow reaction.

*Charge transfer overvoltage* is due to the fact that the electron transfer between ions and electrode occurs at a limited rate. This type of overvoltage can be influenced by the use of some "activation catalyst."

*Concentration overvoltage* is caused by slow diffusion of the electroactive species. In many cases ions which react at the electrode are not the most mobile ions which transport the current. For example, in the case of electrolysis of cryolite-alumina melts, the $Na^+$ ions transport the current, but at the cathode the Al(III) ions are discharged. In the melt, aluminum form complexes with oxygen and fluoride ions and it is transferred by diffusion.

*Reaction overvoltage* is due to secondary reactions of, e.g., the electrolysis products (for example, the association of gaseous atoms into molecules, the escape of gas bubbles, the formation of a crystalline lattice in the case of metal deposition, also called crystallization overvoltage).

The kinetics and the mechanism of electrode processes are of fundamental importance for electrolytic processes. In melts, there are generally two types of electrode processes:

1. The very fast metal-metal ion electrode processes, for which the exchange current density is very high. At steady state the overall rates of those electrode processes are controlled by the rates of mass transfer of the electroactive components to and from the electrode-melt interface.

2. Rather slow electrode processes (especially in the case of gas electrodes) which have low exchange current densities. At steady state, the overall rates are generally determined by the rates of charge transfer and/or of secondary chemical reactions at the electrode-melt interface.

The most important feature of fast electrode processes is that they usually involve the discharge of an adsorbed layer of the electroactive ions. These adsorption processes are complicated by the fact that in molten salts the electroactive ions are usually complex ions. In this case it is possible to presume [126] that the rates of charge transfer for different electrode processes are determined by the distance at which the surface excess of the electroactive ions is located from the metal-melt interface. Thus, if the electroactive ions are located close to the electrode surface, the discharge of the ions is effectively instantaneous and the electrode will always behave as an *ideally reversible electrode*. In this case, the exchange current density is very high. At the other extreme, if the surface excess of electroactive species is located relatively far from the electrode surface, the rate of charge transfer is low and the electrode behaves as an *ideally polarizable electrode*. All practical electrodes are between these extreme cases.

An example of the study of the kinetics of anodic processes in molten salts is given in the following. Sternberg and Galasiu [127–131] studied the kinetics of the evolution of halogens on active carbon electrodes in molten eutectic mixtures of AgX + KX (X = Cl, Br, I). Active carbon was used because the quantity of the adsorbed halogen is higher on this type of carbon than on other types (spectral carbon, graphite, etc.).

The Tafel slopes are given in Table 9. The first slope ($b_1$) is the lower one and it occurs at more positive potentials; the second ($b_2$) has higher values and occurs at slightly positive potentials, i.e., at lower coverages ($\theta$). Two slopes were not observed at all temperatures. Table 9 also contains corresponding coverages and the charge transfer coefficient, $\alpha$:

$$\alpha = \frac{RT/F}{b} \tag{77}$$

It is evident that $\alpha$ is strongly dependent on the nature of the halogen, on the potential, on $\theta$ and on temperature. The surface occupied by the halogen was calculated from the quantity of halogen adsorbed on the electrode and the surface area occupied by an halogen atom. This surface was divided by the surface of the carbon measured by the BET method and the degree of coverage was obtained. The exchange current densities ($i_0$) and the coverage ($\theta$) [130,131] are given in Table 10. Tunold et al. [132–134] found exchange currents of the same order of magnitude.

To evaluate the kinetics of fast reactions, transient techniques are being used, such as voltammetry, potential step amperometry and ac impedance spectroscopy.

Thus, Thonstad et al. [135,136] used the relaxation method with galvanostatic perturbation and electrochemical impedance spectroscopy to study the kinetics of the Al(III)/Al electrode reaction in cryolite-alumina melts and found that the exchange current densities were of the order of 5–15 A cm$^{-2}$. The general electrode reaction scheme may be written

$$AlF_4^- + 2e = AlF_{2\,(ads)}^- + 2F^- \tag{78}$$

$$AlF_{2\,(ads)}^- + e = Al + 2F^- \tag{79}$$

**Table 9** Experimental Tafel Slopes, Charge Transfer Coefficients and Surface Coverages for Halogen-Active Carbon Electrodes at Various Temperatures and Coverages

| $t(°C)$ | 300 | 400 | 500 | 600 | 700 |
|---|---|---|---|---|---|
| Chlorine | | | | | |
| $b_1$ | | 250 | 242 | 355 | 400 |
| $\alpha$ | | 0.53 | 0.63 | 0.49 | 0.48 |
| $\theta$ | | 0.35–0.54 | 0.47–0.60 | 0.60–0.72 | 0.55–5.66 |
| $b_2$ | | 401 | 431 | 562 | 755 |
| $\alpha$ | | 0.33 | 0.35 | 0.31 | 0.26 |
| $\theta$ | | 0.17–0.35 | 0.25–0.47 | 0.27–0.60 | 0.25–0.56 |
| Bromine | | | | | |
| $b_1$ | | 99 | 180 | 170 | |
| $\alpha$ | | 1.4 | 0.85 | 1.06 | |
| $\theta$ | | 0.70–1.0 | 0.67–0.80 | 0.84–0.91 | |
| $b_2$ | | | | 466 | 451 |
| $\alpha$ | | | | 0.37 | 0.42 |
| $\theta$ | | | | 0.57–0.84 | 0.53–0.85 |
| Iodine | | | | | |
| $b_1$ | 80 | 82 | 144 | | |
| $\alpha$ | 1.5 | 1.6 | 1.06 | | |
| $\theta$ | 0.80–1.70 | 0.73–0.97 | 0.66–0.81 | | |
| $b_2$ | | | | 301 | 370 |
| $\alpha$ | | | | 0.57 | 0.52 |
| $\theta$ | | | | 0.46–0.73 | 0.44–0.67 |

*Source*: From Ref. 131.

**Table 10** Exchange Current Densities and Coverage ($\theta$) for Halogen Electrodes in Molten Salts

| | Chlorine | | | Bromine | | | Iodine | |
|---|---|---|---|---|---|---|---|---|
| $t$ (°C) | $i_0$ (mA cm$^{-2}$) | $\theta$ | $t$ (°C) | $i_0$ (mA cm$^{-2}$) | $\theta$ | $t$ (°C) | $i_0$ (mA cm$^{-2}$) | $\theta$ |
| | | | | | | 300 | 170 | 1.7 |
| 400 | 18.6 | 0.54 | 400 | 15.5 | 1.0 | 400 | 63 | 0.97 |
| 500 | 20.5 | 0.60 | 500 | 15.5 | 0.8 | 500 | 41 | 0.81 |
| 600 | 23.0 | 0.72 | 600 | 25.0 | 0.91 | 600 | 29 | 0.73 |
| 700 | 24.0 | 0.66 | 700 | 17.5 | 0.85 | 700 | 30 | 0.67 |

*Source*: From Ref. 131.

The kinetics of the liquid magnesium electrode in pure molten $MgCl_2$ was determined by Kisza et al. [137,138]. The authors showed that the reduction of the Mg(II) ion to metallic Mg is a two-step process. The slow step is the reduction of a Mg(II) species to a Mg(I) species with an exchange current density of about 2.5 A $cm^{-2}$ at 750°C. The reduction of the Mg(I) species to metallic magnesium is a very fast step with an exchange current density of about 25 A $cm^{-2}$ at the same temperature. The charge transfer process could be preceded by a slow dissociation of a polynuclear magnesium complex. This topic is also treated in Section II.B.

The sodium electrode [139] in molten sodium chloride is very reversible with an exchange current density that varies from about 100 A $cm^{-2}$ at 820°C to about 200 A $cm^{-2}$ at 920°C.

The kinetics of the electrodeposition and electrocrystallization of titanium were studied in alkali chloride melts by Haarberg et al. [140]. The cathodic reduction Ti(III)/Ti(II) was very irreversible, and the Ti(II)/Ti reduction was found to be quasi-reversible, as shown in the voltammogram in Figure 14. In LiCl-KCl

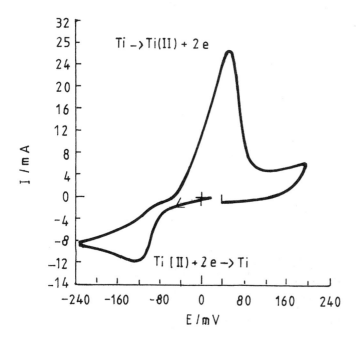

**Figure 14** Cyclic voltammogram on a W electrode showing the reduction and reoxidation of titanium [140]. KCl-LiCl eutectic; $t = 455$°C; $[TiCl_2] = 1.0 \times 10^{-4}$ mol $cm^{-3}$; sweep rate = 10 mV $s^{-1}$ [140].

melts coherent deposits were obtained for current densities of 60–120 mA cm$^{-2}$, while higher current densities gave dendritic and spongy deposits.

## II. METAL PRODUCTION

### A. Aluminum

Aluminum is the most abundant metal in the earth's crust, 7.3 wt% of it consists of aluminum. Due to its chemical reactivity, aluminum is never found in nature in elemental form, but in compounds in the oxidized state (bauxite, feldspars, sulfates, etc.). Aluminum has an atomic weight of 27, a valency 3, and its affinity for oxygen is so strong that significant energy is required for its production. For this reason aluminum is produced only by electrolysis of a molten solution of alumina in cryolite. Paul Heroult and Charles Martin Hall were the inventors of the electrolytic process for the production of aluminum in 1886 [141]. Apart from the technological progress made since, the electrolytic aluminum process remains basically the same as it was more than 100 years ago. An electrolysis cell for aluminum production is shown in Figure 15. The electrolyte is molten cryolite ($Na_3AlF_6$) in which alumina and various salts are dissolved; the anodes are made from carbon. The anode process consists of oxygen discharge and combustion of carbon, giving the overall cell reaction

$$\frac{1}{2}Al_2O_3 + \frac{3}{4}C = Al + \frac{3}{4}CO_2 \tag{80}$$

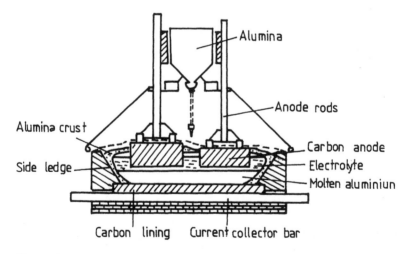

**Figure 15**   Hall-Heroult electrolysis cell using prebake anode technology.

For this reason the consumable anodes must be replaced periodically. The cathode consists of a molten aluminum layer on the bottom of the cell, and the anode-cathode distance is 4–5 cm. Alumina is periodically added to the cell in the proportion that it is consumed by electrolysis. The electrode processes during aluminum electrolysis are very complex [141] and a proper understanding of these processes is important because of the economic implications: energy and carbon consumption, cell control, pollution of the environment, etc.

The electrolytic process for the production of aluminum is briefly discussed in the following section.

## 1. The Electrolyte

The electrolyte composition used in the Hall-Heroult process has remained almost the same from the beginning of the process in 1886. Initially, in industrial cells, a neutral or basic electrolyte was used, but all plants are now using acid electrolytes. A neutral electrolyte means pure cryolite $Na_3AlF_6$ (3 NaF · $AlF_3$). The cryolite ratio (CR) means the molar ratio of $NaF/AlF_3$; if the cryolite ratio is 3 the electrolyte is neutral. When there is an excess of NaF in the electrolyte, the CR is more than 3 and the electrolyte is basic; when there is an excess of $AlF_3$, the CR is less than 3 and the electrolyte is acid. At present the common electrolyte composition is cryolite with additions of 6–12% $AlF_3$, 3–5% $CaF_2$, 2–5% $Al_2O_3$ and in some cases 0–4% LiF and 0–4% $MgF_2$. The working temperature is 950–970°C. Cryolite is used as solvent for alumina because it has high chemical stability and a high solubility for alumina.

### *The Melting Temperature*

The phase diagram [142] of the system $NaF-AlF_3$ is shown in Figure 16. At atmospheric pressure the presence of two compounds is clearly indicated in this system, viz. cryolite $Na_3AlF_6$, which melts congruently, and chiolite $Na_5Al_3F_{14}$, which melts incongruently [143–146]. The congruently melting cryolite, $Na_3AlF_6$, divides the $NaF-AlF_3$ system into two partial systems. The quasi-binary system $NaF-Na_3AlF_6$ is a simple eutectic system, the eutectic point being located at 12.8–13.8 mol% $AlF_3$ and 888°C. The partial system $Na_3AlF_6-AlF_3$ contains the compounds $Na_5Al_3F_{14}$, which melts incongruently, and $NaAlF_4$ (metastable).

Different salt additions to the electrolyte improve its physicochemical properties: melting temperature, electrical conductivity, density, interfacial tension, etc. The general trend is to use low melting electrolytes to obtain higher current efficiencies.

The cryoscopic constant of the cryolite is 25.31; therefore, for each mole of foreign ions added to 1 kg cryolite, the cryolite liquidus temperature decreases

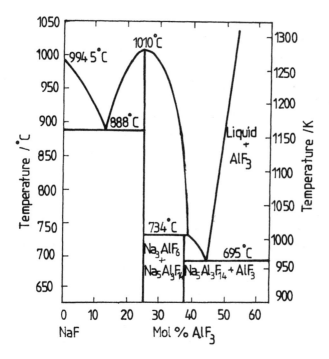

**Figure 16**  Phase diagram [142] of the system NaF-AlF$_3$.

with 25.31°C. In the case of compounds forming complex ions which affect the total number of ions in the electrolyte, there are discrepancies from that rule.

Solheim et al. [147] found an equation for the liquidus temperature as a function of the composition (wt%) in the system Na$_3$AlF$_6$-AlF$_3$-Al$_2$O$_3$-LiF-CaF$_2$-MgF$_2$-KF:

$$t = 1011 + 0.50[\text{AlF}_3] - 0.13[\text{AlF}_3]^{2.2} - \frac{3.45[\text{CaF}_2]}{1 + 0.0173[\text{CaF}_2]}$$

$$+ 0.124[\text{CaF}_2][\text{AlF}_3] - 0.00542([\text{CaF}_2][\text{AlF}_3])^{1.5}$$

$$- \frac{7.93[\text{Al}_2\text{O}_3]}{1 + 0.0936[\text{Al}_2\text{O}_3] - 0.0017[\text{Al}_2\text{O}_3]^2 - 0.0023[\text{AlF}_3][\text{Al}_2\text{O}_3]} \qquad (81)$$

$$- \frac{8.90[\text{LiF}]}{1 + 0.0047[\text{LiF}] + 0.0010[\text{AlF}_3]^2} - 3.95[\text{MgF}_2] - 3.95[\text{KF}]$$

where $t$ is the temperature in °C and the brackets denote wt% of the components. This complex equation shows that the liquidus temperature (temperature of primary crystallization of the electrolyte) is not a simple function of the concentration of the compounds, because of ionic interaction in the melt. Other authors [148,149] found similar equations for this dependence.

*The alumina solubility* in the molten electrolyte represents an important problem for the aluminum industry. The highest solubility of alumina is in pure cryolite (13%). The alumina solubility decreases when changing the cryolitic ratio and adding various substances. According to the literature [145,150–152], in the system $Na_3AlF_6$-$Al_2O_3$ the eutectic point appears at 960–963°C and 10.0–11.5 wt% $Al_2O_3$. Generally, additives lower the alumina solubility in molten cryolite [153,154], as shown in Figure 17, except for KF, which increases it slightly [149]. Also, all additives lower the dissolution rate of alumina [153, 154].

The dissolution rate of alumina depends on its properties: specific surface, α-alumina content, particle size, the content of volatile components, the way of addition, etc. Corundum, or α-alumina, is the crystal modification of alumina which crystallizes in a compact lattice and it is the stable phase at high temperature.

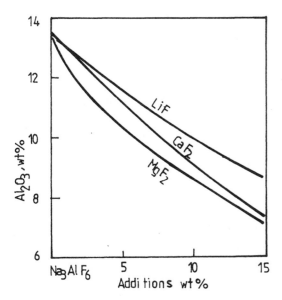

**Figure 17** Influence of additives on the solubility of alumina [153,154] in cryolite melts at 1010°C.

Bagshaw et al. [155] noticed that a low dissolution time is achieved by using alumina which release volatiles in the bath. Generally as the contact surface of alumina-electrolyte increases, the dissolution time of alumina decreases [156], but also when the $\alpha$-alumina content increases the dissolution rate of alumina decreases [157–160].

The heat of dissolution of alumina in molten cryolite, $\Delta H_{dis}$, varies with the alumina concentration, an average value [161] being $\Delta H_{dis} = 125,000$ J $\cdot$ mol$^{-1}$. When a large quantity of alumina is added to the bath, the heat needed for its heating from room temperature to the electrolyte temperature and the heat of dissolution are taken from the bath. For this reason, localized freezing occurs, with the formation of aggregates which tend to drop to the bottom of the cell. In this manner a sludge is formed [161].

*The electrolyte density* is very important in order to obtain a good separation between melt and metal and to avoid the risk of mixing bath and metal. The density of molten cryolite has been measured by several authors [162–167]. According to Edwards et al. [164] the temperature dependence of the density of cryolite is given by

$$\rho_{Na_3AlF_6} = (3.032 - 0.937 \cdot 10^{-3} \, t) \text{ g cm}^{-3} \tag{82}$$

and the temperature dependence of the density of the aluminum (99.75% pure) is [168]

$$\rho_{Al} = (2.561 - 0.272 \cdot 10^{-3} \, t) \text{ g cm}^{-3} \tag{83}$$

The variation of the density of molten cryolite with various additives [141] at 1000°C is shown in Figure 18.

Paucirova et al. [169] studied the density of NaF-AlF$_3$ mixtures and found that in the concentration range 0–20% mole% AlF$_3$ complex AlF$_6^{3-}$ anions are probably formed. These complex anions are then partially dissociated into AlF$_4^-$ and F$^-$ anions. At higher AlF$_3$ concentrations the presence of other complex anions, probably Al$_3$F$_{14}^{5-}$, is to be considered. Kvande and Rorvik [170] derived an equation for the calculation of the melt density as a function of temperature and electrolyte composition:

$$d = 2.64 - 0.0008t + 0.18(\text{wt. ratio NaF/AlF}_3) - 0.008(\text{wt}\% \text{ Al}_2\text{O}_3) \\ + 0.005(\text{wt}\% \text{ CaF}_2) + 0.008(\text{wt}\% \text{ MgF}_2) - 0.004(\text{wt}\% \text{ LiF}) \tag{84}$$

where $t$ is the temperature in °C.

*The electrical conductivity* of the electrolyte is influenced principally by two factors: the concentration of additives and the temperature. The influence of

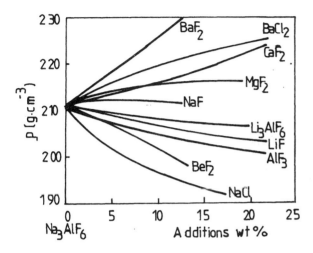

**Figure 18**   Isothermal densities [141] of binary cryolite melts at 1000°C.

additives [141] is presented in Figure 19. The conductivity of a multicomponent electrolyte can be described by an equation [171] of the type

$$\kappa(\text{S cm}^{-1}) = 7.22 \exp\left(\frac{-1204.3}{T}\right) - 2.53[\text{Al}_2\text{O}_3] - 1.66[\text{AlF}_3]$$

$$- 0.76[\text{CaF}_2] - 0.206[\text{KF}] + 0.97[\text{Li}_3\text{AlF}_6]$$
$$- 1.07[\text{MgF}_2] - 1.80[\text{Al}_2\text{O}_3][\text{CaF}_2] \qquad (85)$$
$$- 2.59[\text{Al}_2\text{O}_3][\text{MgF}_2] - 0.942[\text{AlF}_3][\text{Li}_3\text{AlF}_6]$$

where $T$ represents the temperature in K and the brackets represent the concentration of additives in mole fractions. Other authors [172,173] found similar equations for that dependence. The electrical conductivity of the electrolyte in industrial cells is also influenced by other factors, for example the presence of gas bubbles [174–176], carbon dust [177,178] and alumina particles [141], which act like insulators and decrease the electrical conductivity, while dissolved aluminum increases it [179].

*The interfacial tension* at the metal-electrolyte boundary influences the dissolution of aluminum in the bath [180] and the current and energy efficiency [181]. Thus, a high interfacial tension at the metal-electrolyte boundary reduces the rate of dissolution of aluminum in the bath and increases the current efficiency

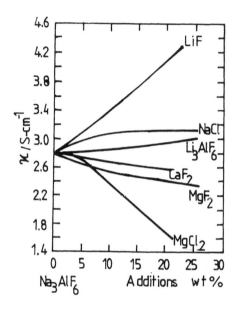

**Figure 19**  Specific conductivity [141] of $Na_3AlF_6$-$MA_x$ mixtures at 1000°C.

[182–185], especially in the case of a quiescent electrolyte. The effect of various additives on the interfacial tension at the metal-electrolyte phase boundary [186] is shown in Figure 20 [141]. In Figure 21 [182–184,187] is shown a marked decrease in the interfacial tension with increasing $NaF/AlF_3$ ratio. This effect could be due to the adsorption of Na atoms on the metal surface, generated by the reaction [182]

$$Al_{(l)} + 3NaF_{(l)} = 3Na_{(dis)} + AlF_{3(dis)} \tag{86}$$

It is known that the predominant carriers of current in the electrolyte are sodium ions, and therefore the cryolite ratio is higher at the metal-electrolyte interface than that in the bulk. From this consideration the equilibrium of reaction (86) will be displaced to the right, and the sodium concentration at the interface will be higher than in the bulk of the electrolyte.

*Solubility of Aluminum in the Electrolyte*

When a piece of aluminum is added to clear, transparent molten cryolite, foglike streamers spread out from the metal and they soon render the melt completely opaque [188–195]. It is clear that the metal fog is not a stable chemical phase in equilibrium with the electrolyte, since it dissipates when it rises from the molten

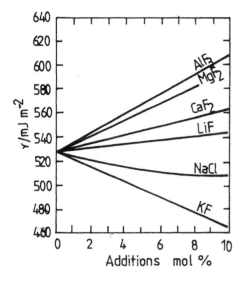

**Figure 20** Influence of different additives on the interfacial tension [141] at the electrolyte/Al boundary at 1000°C.

aluminum. Truly dissolved aluminum has been shown to exist both in the form of monovalent aluminum species, probably bound to one more fluoride ions in the melt, and as a sodium-containing species, whose identity is still uncertain [196]. These amounts of dissolved metal in the electrolyte can combine with anodic products and cause a loss in current efficiency:

$$Al + \frac{3}{2}CO_2 = \frac{1}{2}Al_2O_3 + \frac{3}{2}CO \tag{87}$$

The solubility of dissolved metal in the electrolyte at 1000°C is in the range 0.05–0.1 wt%, the amount of sodium being higher than that of aluminum [195].

## 2. The Anode Process

The anode process on carbon anodes during electrolysis of cryolite-alumina melts probably constitutes the most complex electrode process known in molten salts. The principal questions that have been raised are

1. What is the primary anode product: $CO_2$, CO or both?
2. What is the magnitude of the overvoltage?
3. What is the rate-determining step in the anode reaction?

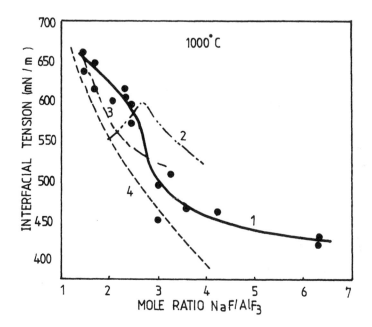

**Figure 21**  Interfacial tension of aluminum in cryolite melts at 1273 K: 1, Utigard and Toguri [183]; 2, Gherasimov and Belyaev [184]; 3, Zemchuzina and Belyaev [187]; 4, Dewing and Desclaux [182].

There are two possible cell reactions:

$$Al_2O_3 + \frac{3}{2}C = 2Al + \frac{3}{2}CO_2 \tag{88}$$

$$Al_2O_3 + 3C = 2Al + 3CO \tag{89}$$

Dissolved alumina in cryolite may form the following complex ions [197–199]: $Al_2OF_6^{2-}$, $Al_2OF_8^{4-}$, $Al_2O_2F_4^{2-}$. Discharge of fluoride ions is less probable, except during the anode effect (see below). Reactions (88) and (89) or a combination of these are then the only possible overall reactions. Reaction (89) is favored thermodynamically, the standard emfs at 1010°C being −1.161 V for (88) and −1.020 V for (89), respectively [141]. The Boudouard reaction,

$$CO_2 + C = 2CO \tag{90}$$

is displaced far to the right at 1010°C, the equilibrium composition corresponding to 99.5% CO [141]. However, the presence of considerable overvoltage, being around 0.5–0.6 V at normal current density, invalidates such thermodynamic considerations.

Reaction (89) requires twice as much carbon as reaction (88) for the same quantity of electricity. One might then expect that the problem of the primary anode product could be easily solved by determining the gas composition and the carbon consumption.

Data from industrial cells show that the exit gas contains 90–60% $CO_2$ and 10–40% CO. The carbon consumption usually ranges from 400 to 550 kg/ton Al, while the theoretical amount at 95% current efficiency is 350 kg/ton Al for reaction (88) and 700 kg/ton Al for reaction (89). The question is then whether the extra consumption, compared to the theoretical value for reaction (88), is due to simultaneous primary formation of $CO_2$ and CO or if it is due to the Boudouard reaction between gaseous $CO_2$ and the sides and the interior parts of the anode [141], or carbon dust [200] dispersed in the electrolyte, or air burning of the anode. In any case, it is known [200,201] that in the laboratory a high content of CO is formed at low cds ($<0.05$ A cm$^{-2}$) and almost pure $CO_2$ is formed at 1 A cm$^{-2}$.

According to the preceding treatment, aluminum electrolysis takes place with an anodic overvoltage of about 0.5–0.6 V. The difference between the polarization potential $E_i$ and the reversible anode potential is the anodic overvoltage:

$$\eta = E_i - E_{rev} \tag{91}$$

and it is usually presented as Tafel plots:

$$\eta = a + b \log i \tag{92}$$

Generally the Tafel equation is obeyed for charge-transfer-controlled processes and, under certain conditions, for reaction control and diffusion control [202].

Recently, Kisza et al. [203] found that the total electrode reaction can be interpreted by a two-step two-electron charge transfer process with an intermediate adsorption:

$$1.\ Al_2OF_6^{2-} + C = C_xO_{(ads)} + 2AlF_3 + 2e^- \tag{93}$$

$$2.\ Al_2O_2F_6^{4-} + C_xO_{(ads)} = CO_2 + Al_2OF_6^{2-} + 2e^- \tag{94}$$

With increasing alumina content the rate-determining step is shifted from the second step to the first step. The concentration dependence on the alumina content allows the understanding of the otherwise strange variation of the Tafel lines with the alumina content.

Grjotheim [23] suggested a reaction scheme for the anode reaction with formation of C=O double bonds at the surface or that the first discharged oxygen formed a bridge between two carbons. Thonstad [200] also suggested formation of an oxygen bridge between two adjacent carbons or possible peroxy-type bonds.

## 3.  The Anode Effect

The anode effect is observed on carbon anodes in cryolite-alumina melts at low concentrations of alumina. That phenomenon can be described as an appearance of a gas film on the anode surface, the anode being isolated from the electrolyte, and the current transport between the anode and the melt is inhibited. The anode effect appears at a certain current density, the so-called critical current density, which is attained before the normal anode reaction is superceded by the anode effect. During the anode effect a sudden increase of the cell voltage and temperature near the anode takes place. Principally, the phenomenon is due to the depletion of the alumina content near the anode surface. As a result, the anode process changes and fluoride is discharged instead of oxygen, forming the $CF_4$ and $C_2F_6$ gases. Likewise, the occurrence of the anode effect depends on the properties of the carbon anode. Galasiu [204,205] showed that the critical current density for the occurrence of the anode effect decreased in proportion to the porosity of the anodes.

The following reaction takes place during the anode effect:

$$\frac{4}{3}\,Na_3AlF_6 + C = \frac{4}{3}\,Al + CF_4 + 4NaF \tag{95}$$

and a similar reaction, forming $C_2F_6$, also occurs. A typical gas composition during the anode effect can be 20% $CF_4$, 1% $C_2F_6$, 60% $CO$, 20% $CO_2$.

Once an anode effect has occurred it is terminated by feeding alumina to the cell and by stirring the electrolyte. Stirring can be achieved by moving the anode, blowing air into the electrolyte or by the use of wooden poles.

## 4.  The Cathode Process

The cathode process has received much less attention than the anode process, probably because the reactions occurring at the cathode have been considered to be simple. However, the cathode reactions are important not only because of cathodic overvoltage but also because they may be connected with the formation of dissolved metal and thereby with the current efficiency of the process.

The ionic species present in the electrolyte appear to be $Na^+$, $AlF_6^{3-}$, $AlF_4^-$, $F^-$ and certain Al—O—F complexes (see above). The fact that sodium is present as free ions, whereas aluminum is bound in complexes, and the fact that the sodium ion is the carrier of current, have led many authors to the assumption that sodium is the primary discharge product at the cathode. However, the thermodynamic data favor primary aluminum deposition on aluminum in cryolite melts.

The question of which of the two metals, aluminum and sodium, is more noble in the electrolyte can be discussed in terms of reaction (86). At the temperature of electrolysis cell, the equilibrium reaction (86) is displaced to the left except in very NaF-rich melts. The discharge potential of $Na^+$ at a sodium pressure of 1 atm in molten cryolite was estimated to be about 250 mV negative compared to that of aluminum. Kubik et al. [206] and Silny et al. [207] found that the difference between the potential steps ascribed to aluminum and sodium was 220 mV. These data point in favor of primary aluminum discharge. Primary sodium discharge is conceivable only if the aluminum discharge steps are retarded or if the electrolyte near the cathode becomes strongly enriched with NaF due to the current transport process.

Piontelli and co-workers [208,209] reported cathodic overvoltage studies on aluminum and found the charge transfer overvoltage to be low, i.e., a few mV. The low charge transfer overvoltage was taken as an indication of primary aluminum deposition. A persistent residual overvoltage of 0.4 V at a current density of 1 A $cm^{-2}$ was also noticed in laboratory experiments. That residual overvoltage was assumed mainly to be caused by accumulation of NaF near the cathode. This residual overvoltage may be lower in industrial cells due to the convection in the bath. Thonstad and Rolseth [210] assessed the cathodic overvoltage in industrial cells to be about 100 mV.

Thonstad et al. [211,212] found that the general cathode reaction scheme may be written as two charge transfer reactions:

$$AlF_4^- + 2e^- = AlF_{2(ads)}^- + 2F^- \tag{96}$$

$$AlF_{2(ads)}^- + e^- = Al + 2F^- \tag{97}$$

The exchange current density for the first reaction ranges from 2 A $cm^{-2}$ for 2 wt% $Al_2O_3$ to 10 A $cm^{-2}$ for 12 wt% $Al_2O_3$ at 1000°C. A dramatic change of the activation energy for this charge transfer step is observed between pure molten cryolite ($E_A = 530$ kJ $mol^{-1}$) and alumina containing melts ($E_A = 86$ kJ $mol^{-1}$). The exchange current density of the second charge transfer step, which is independent of the alumina concentration, is much higher (14.7 A $cm^{-2}$ at 1000°C) and its activation energy $E_A = 183$ kJ $mol^{-1}$.

In laboratory experiments [213] the equilibrium concentration of Na in aluminum in equilibrium with cryolite at 1020°C is about 170 ppm. For an industrial electrolyte with 11 wt% $AlF_3$ and 960°C it is 60 ppm Na [214]. The average sodium content of aluminum of "older" prebake cells is low, 40–80 ppm, corresponding to a high interfacial stirring and low current efficiency (87–90%). In spite of the decrease of the cryolite ratio, the average sodium content of aluminum of modernized prebake cells is higher (90–160 ppm) after modification of the magnetic design [215], corresponding to reduced interfacial stirring, a larger concentration gradient at the interface and a higher current efficiency (93–96%).

## 5. Current Efficiency

The principal reason for the loss in current efficiency (CE) in aluminum electrolysis is the metal reoxidation by the anode gas, according to reaction (87).

Many studies have been made of the metal reoxidation reaction to evaluate the reaction mechanism and the rate-determining step. Gjerstad and Welch [216] found that the slow step is the transport of dissolved metal if the convection is very low. Thonstad [200] showed that the slow step is metal dissolution in melts agitated with $CO_2$. Generally, in industrial cells, the mass transport is very important, even if it is not the rate-determining step. It is known from laboratory studies [217] that the overall rate of the metal reoxidation reaction decreases with decreasing metal solubility. That could be obtained by using electrolytes with low cryolite ratio and low electrolysis temperature (additives of $CaF_2$, $MgF_2$ or LiF).

Generally, the current efficiency decreases with increasing temperature because the rate of reoxidation of aluminum increases. The average temperature coefficient is 0.19%/°C, which corresponds to a current efficiency increase of 1% for a temperature decrease of 5–6°C [218–220]. However, in order to maintain stable operation the working temperature must be 5–10° above the liquidus temperature of the electrolyte.

The current efficiency increases with increasing anode-cathode distance, because for a longer anode-cathode distance there is less convection, less interaction between gas bubbles and the metal, and more stable conditions. It was found that the current efficiency is independent of the anode-cathode distance above a certain distance. However, the ohmic drop in the electrolyte increases with the interpolar distance, so the cell voltage increases. These two factors influence the energy efficiency in an opposite manner. Furthermore, the heat balance of the cell sets limits to the variation of the anode-cathode distance.

In the laboratory as well as in industrial cells it was found that the current efficiency increases with increasing excess $AlF_3$. The increase is of 1% for an $AlF_3$ excess of 1.7–2%, and it is due to the influence of $AlF_3$ on the metal solubility and on the electrolyte temperature [221,222].

Lillebuen and Mellerud [223] developed a model for the current efficiency as a function of various parameters. The variation with the alumina concentration showed a curve with a minimum at about 4% $Al_2O_3$. The authors showed that the reoxidation reaction rates may be limited by gas dissolution at low alumina contents. The presence of rather large gas bubbles reduces the gas-bath interfacial area to such a degree that the reoxidation reaction rates are being reduced and the current efficiency increased, in spite of the fact that metal solubility is increased at lower alumina concentrations [195,196,223]. The current efficiency increases for alumina concentrations over 4% due to the decrease of the metal solubility in the electrolyte [195].

By mass spectrometry analysis of the escaped gas from an industrial cell, Leroy et al. [224] found that current efficiency increases strongly when the alumina content of the bath decreases (by about 2% for 1% variation of alumina content). On the other hand, Paulsen et al. [225] found that the current efficiency increases with increasing alumina content in the electrolyte of industrial cell. The influence of the alumina concentration on current efficiency thus needs to be clarified.

## 6. Energy Consumption, Energy Efficiency

The energy consumption in aluminium electrolysis, $W$, is calculated [226] as a function of the cell voltage, $V_{cell}$, and the current efficiency, $x(x = CE/100)$, by the equation

$$W = \frac{2.98 V_{cell}}{x} \tag{98}$$

where the constant 2.98 results from Faraday's law. The actual normal consumption is 13–16 kWh/kg Al. While the industrial average of the energy consumption has been claimed to approach 15.4 kWh/kg Al in 1982, Jarrett's [227] data indicate that by the year 2000 the industrial average may be 14.3 kWh/kg Al.

The energy efficiency (EE) can be calculated as a function of the current efficiency ($x$) and cell voltage ($V_{cell}$) as follows [228]:

$$EE = \frac{0.481 + 1.65x}{V_{cell}} \tag{99}$$

An energy efficiency of 48.8% is obtained for a current efficiency of 95% ($x = 0.95$) and a cell voltage of 4.2 V.

The theoretical energy requirement for the electrochemical reaction in aluminum electrolysis is 6.34 kWh/kg Al (at a current efficiency of 100%). The rest of the energy output is in the form of heat loss from the cell body; it varies largely from 5.5 to 9.0 kWh/kg Al depending on the type of cell.

Energy consumptions lower than 10 kWh/kg Al will require fundamental changes in the aluminum industrial technology, such as the use of bipolar electrodes or the use of aluminum chloride electrolysis [229].

## 7. New Technologies for Aluminum Electrolysis

*Inert Electrodes*

The main objective for the development of inert electrodes for aluminum cells is to save energy in the form of electricity and carbon anodes. Additional advantages should be environmental benefits, to save labor, tighter hooding, possible heat recovery. The inert cathode should be a solid body wetted by a film of aluminum. Electrochemically it would then act as an aluminum electrode in the same way as the liquid aluminum pool in Hall-Heroult cells. The inert anode would be an oxygen-evolving electrode in contrast to the $CO_2$-evolving carbon anode. The cell reactions and the corresponding emf and anodic overvoltage data [230] are given in Table 11. In the table the net voltage requirement is 0.6 V higher with inert anodes. On an energy basis ($\Delta H$) the difference is 1 V. But the difference in cell voltage can be more than offset by lowering the anode-cathode distance from 4.5 to 2 cm [231].

*Inert Anodes*

The working surface of an oxygen-evolving inert anode must consist of an oxide or a mixture of oxides. The following properties are desired: electronic conductivity, low solubility in the electrolyte, chemical resistance to electrolyte and oxygen, and thermal stability. Advantages are saves carbon, stable surface, needs little attendance, a tighter hooding is possible, no emissions of $CO_2$, $SO_2$, $CF_4$, etc. Disadvantages are higher emf, contamination of the produced metal by anode corrosion products, high cost. There are some literature reviews published in this field [231–234].

In the work of the Alusiusse Company, $SnO_2$-based materials were tested extensively [235]. Also other researchers have tested $SnO_2$-based anodes for aluminum electrolysis. So, Liu and Thonstad [230] studied the overvoltage on the $SnO_2$-based inert anodes and the corrosion of anodes in the electrolyte [236,237]. Galasiu et al. [238–240] studied the influence of the tin oxide anode composition on its electrical conductivity and the influence of the manufacturing conditions on the anode properties in a laboratory cell and in a pilot cell [241]. The authors found a method for increasing the thermal shock resistance of the anodes by adding presintered material of the anode composition [242]. Likewise, Galasiu et al. [243,244] proposed the use of ZnO-based ceramics for inert anodes for aluminum electrolysis.

The Aluminum Company of America, currently referred to as Alcoa, tested a new and interesting concept of using a cermet as the base material [245]. It was composed of a mixture of Ni—Fe—O with excess Ni or Cu. The recommended composition was $NiFe_2O_4$ + 18% NiO + 17% Cu. The excess metal provided good electrical conductivity and a simple connection to the current lead. The surface layer of the anode was completely oxidized, but this outer oxide layer could be kept fairly thin.

**Table 11** Cell Reactions in Aluminum Electrolysis When Using Inert and Carbon Anodes and the Corresponding EMF and Anodic Overvoltage ($\eta$) Data at 1000°C

| Anode | Cell reactions | −Emf (V) | $\eta_{an}$(V) |
|-------|----------------|----------|----------------|
| Inert | $Al_2O_3 = 2Al + \frac{3}{2} O_2$ | 2.20 | 0.10 |
| Carbon | $Al_2O_3 + \frac{3}{2} C = 2Al + \frac{3}{2} CO_2$ | 1.20 | 0.50 |

*Source*: Data from Refs. 141 and 232.

Since all these materials have a certain solubility in the electrolyte, some contamination of the metal will occur. The Eltech Systems Corporation announced [246,247] a method to protect the substrate anode material by forming a Ce(IV) oxyfluoride layer on the anode. The protective layer is maintained by adding a Ce(III) compound to the electrolyte. However, cerium contaminates the aluminum metal so that it must be removed and recirculated to the electrolyte.

*Inert Cathodes*

An inert cathode must satisfy a number of requirements: electronic conductivity, wettable by aluminum, low solubility and mechanical strength. The work on inert cathodes is focused on one particular material, namely titanium diboride ($TiB_2$) or composites thereof. Problems encountered with parts made of pure $TiB_2$ are brittleness, propagation of cracks and low thermal shock resistance. The high price of massive $TiB_2$ is also a problem. $TiB_2$-carbon composites have been tested extensively in China [248–250], by Reynolds Metals Company, USA [251,252], and by the Comalco Company, Australia. The Moltech Company markets a so-called TINOR coating which consists of $TiB_2$ powder with a colloidal alumina binder [253–255].

The inert electrode development work may need as much as 10–15 years before commercialization can be undertaken. If these efforts are met with success, we may be faced with a new cell design, radically different from the present Hall-Heroult cells. The goals may be

Energy consumption of 9–11 kWh/kg Al
Anode-cathode distance of 1–2 cm
Energy recovery systems

In this way it is possible to use bipolar cells, which would allow for a very compact cell design with high capacity. A patented bipolar cell design is shown in Figure 22.

*Aluminum Chloride Electrolysis*

Due to the high energy consumption and the high investment and operating cost of the Hall-Heroult process, several other ways of making aluminum have been studied over the years. Some features of a process based on electrolysis of aluminum chloride are treated in the following [230].

In 1972, The Aluminum Company of America (Alcoa) announced a large-scale project to develop a commercial electrolytic aluminum chloride process [256]. Based on extensive laboratory studies, a demonstration plant (15,000 tons per year) was built and operated for several years. Alcoa used commercial grade low calcined alumina as input material for the chlorination step, where it was reacted with chlorine and carbon to form gaseous $AlCl_3$ and $CO_2$ and CO. The

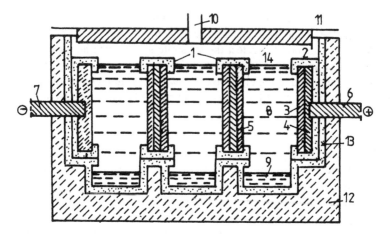

**Figure 22**  Bipolar cell for aluminum electrolysis due to Alusuisse [236]. The vertical bipolar electrodes consist of: 3, an anode layer, a ceramic oxide; 4, an intermediate conducting layer; 5, a cathode layer, e.g., $TiB_2$.

anhydrous $AlCl_3$ was dissolved in a LiCl-NaCl melt and electrolyzed between graphite electrodes to form liquid aluminum and chlorine gas:

$$AlCl_3 = Al + \frac{3}{2}Cl_2 \tag{100}$$

The electrodes were arranged as a stack of bipolar electrodes as shown in Figure 23. Aluminum formed at the lower electrodes in each individual cell flows concurrently together with the chlorine gas to the central vertical shaft, where the metal sinks to the bottom and the gas rises to the top. The gas movement promotes the necessary circulation of the electrolyte. Due to the compact bipolar arrangement and the short interpolar distance (10–20 mm) the electrical energy consumption was as low as 9.5 kWh/kg Al.

Several electrolyte compositions were suggested, but it appears that the simple LiCl-NaCl system is the best choice. Aluminum chloride, which is readily soluble in alkali halide salts, is a covalent compound, but it is ionized upon dissolution, forming anionic complexes like $AlCl_4^-$ [257] (See Section IV. A. 1).

The standard Gibbs energy for the cell reaction can be expressed by the equation

$$E° = -2.26 + 0.408 \cdot 10^{-3}T \tag{101}$$

yielding $-1.86$ V at 700°C [258]. Correction for the $AlCl_3$ activity at 5 wt% $AlCl_3$ [259] yields $E_{rev} = -2.16$ V. The anodic overvoltage on graphite has not

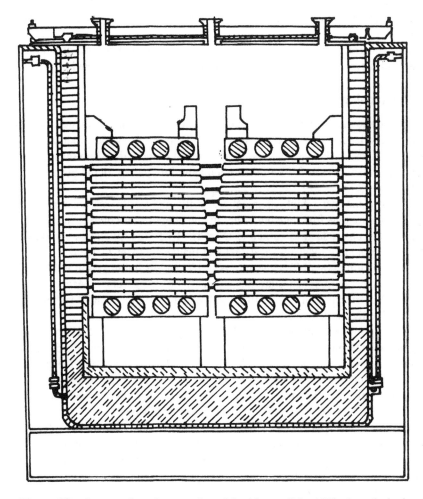

**Figure 23** Cross section of one version of the Alcoa cell for AlCl₃ electrolysis showing a stack of bipolar graphite electrodes. U.S. Patent 413, 3727, 1979.

been determined in these melts. On vitreous carbon in $AlCl_3$-rich melts it is of the order of 0.15 V at 1 A cm⁻², and it appears to be activation controlled [260].

Few data are available on current efficiency in aluminum chloride electrolysis. The CE in the Alcoa process is assumed to be 90% [259]. A major part of the losses are due to current bypassing the bipolar electrodes [261]. Losses due to reoxidation, i.e., the reverse of reaction (100) should then be low. Gas bubbles and metal droplets do not seem to react spontaneously upon contact in the melt.

It appears that the reacting species are dissolved chlorine and possibly dissolved metal.

Development work on the Alcoa process encountered several difficulties, notably in the chlorination step, and the project has been abandoned.

## B. Magnesium

One million tonnes of magnesium [262] are found in a cubic kilometer of seawater, while in one year all the magnesium plants in the world produce about 300,000 tonnes. In the earth's crust there is much more magnesium than in seawater (seawater contains 3.7% of the total magnesium present in the earth's crust [263]).

Magnesium has a high chemical reactivity, and for this reason it is never found in nature in its free state but rather in compounds—as chloride in seawater and in salt deposits such as carnallite ($MgCl_2 \cdot KCl \cdot 6H_2O$) and as carbonates in the ores dolomite ($MgCO_3 \cdot CaCO_3$) and magnesite ($MgCO_3$), etc.

The melting point of magnesium is 651°C and its boiling point is 1120°C. At room temperature, magnesium has a density of 1.74 g $\cdot$ cm$^{-3}$. Magnesium is 4.5 times lighter than iron and 1.6 times lighter than aluminum, and it is therefore an attractive lightweight material.

Two principal methods of producing magnesium metal [264] are being used: metallothermic reaction and electrolysis of molten salts. Two-thirds of the magnesium production is obtained by the electrolytic process. On the market there are few producers of magnesium. Therefore, process research and development work have been conducted by these producers, and the results have often been kept secret. Hence, there are few publications in this field, except for patents.

The standard potential of magnesium chloride of $-2.54$V at 700°C is highly electronegative, and because of this the electrolytic recovery of magnesium from aqueous media is impossible. Magnesium can be obtained only by molten salt electrolysis. Furthermore, its high affinity for oxygen (magnesium oxide is very stable, $\Delta G_f^0 = -493$ kJ/mol MgO at 1000 K) [265] requires that any nonaqueous electrolyte used for the recovery of magnesium metal be free of oxygen.

Theoretically, it is possible to obtain magnesium metal by electrolysis of pure $MgCl_2$. However, this salt has a high melting point, a high volatility and a low electrical conductivity. Furthermore, the solubility of metallic magnesium in molten $MgCl_2$ is high enough to cause a low current efficiency. All these disadvantages can be eliminated by using a mixture of molten chlorides with $MgCl_2$ dissolved in.

The extraction of Mg by electrolysis consists of two steps:

1. Preparation of $MgCl_2$ cell feed
2. Electrolysis

All extraction processes are followed by refining and casting.

The two main routes to the production of dehydrated magnesium chloride cell feed are

1. Chlorination of magnesia (MgO) or magnesite ($MgCO_3$) in the presence of carbon or carbon monoxide
2. Dehydration of aqueous magnesium chloride solutions or hydrous carnallite ($MgCl_2 \cdot KCl \cdot 6H_2O$)

## 1. Chlorination of Magnesia and Magnesite

Two principal processes are used industrially to chlorinate magnesia and magnesite:

1. The I.G. Farben process presently in operation in several countries [266]
2. The ManCan process under installation in Alberta, Canada [266]

### *The I.G. Farben Process*

In the process used by Norsk Hydro, magnesium hydroxide extracted from seawater with the aid of calcined dolomite is mixed with charcoal and magnesium chloride brine and is heated to 1000–1200°C in the presence of chlorine produced during subsequent electrolysis of magnesium chloride. The main reactions are [266]

$$2MgO + C + Cl_2 \rightarrow 2MgCl_2 + CO_2 \tag{102}$$

$$Cl_2 + C + H_2O \rightarrow 2HCl + CO \tag{103}$$

$$MgO + 2HCl \rightarrow MgCl_2 + H_2O \tag{104}$$

The reaction mechanism is very complex.

### *The ManCan Process*

In the ManCan process natural magnesite is crushed and screened. That is introduced into a reactor at 1000°C, where it reacts with chlorine produced during $MgCl_2$ electrolysis and CO obtained from a gas generator. The main reactions are [266]

$$MgCO_{3(s)} + CO_{(g)} + Cl_{2(g)} \rightarrow MgCl_{2(l)} + 2CO_{2(g)} \tag{105}$$

$$CO_{2(g)} + O_{2(g)} + 3C_{(s)} \rightarrow 4CO_{(g)} \tag{106}$$

*Dehydration of Aqueous Magnesium Chloride Solutions [267]*

Brines containing 33–34% $MgCl_2$ may be derived as a by-product from the potassium industry or produced by dissolving magnesium-bearing minerals in hydrochloric acid. Naturally occurring dilute brines are concentrated by solar evaporation or conventional dehydration processes. To avoid hydrolysis above 200°C, the final dehydration is performed in an atmosphere of hydrogen chloride.

## 2. Physicochemical Properties of the Electrolytes

*The melting temperature (liquidus temperature)* should be lower than the melting point of magnesium if the metal is to be obtained in the solid state. Thus, the working temperature is low and the electrical energy consumption is also low. Furthermore, the metal losses through evaporation, oxidation and dissolution in the melt are minimized. On the other hand, an electrolyte with a very low melting temperature penetrates easily the refractory cell liner and destroys it. For these reasons, the metal and the electrolyte should have a liquidus temperature very close to the melting point of magnesium [268].

*The density* of the electrolyte is a very important factor in cell operation. The electrolysis produces liquid metallic magnesium. It rises to the surface of the electrolyte because of its lower density. Therefore, the electrolyte density should be sufficiently high so as to promote the separation of the molten magnesium. Table 12 reports some properties of different electrolytes [266]. It is seen from this data that the role of $CaCl_2$ in the electrolyte is to increase the density. If an electrolyte without $CaCl_2$ is used, the concentration of $MgCl_2$ should be rather high. Otherwise the densities of magnesium metal and electrolyte get so close that their separation is unacceptably slow. Likewise, addition of $BaCl_2$ can be used to increase the density of the electrolyte [263,269]. It is also shown in Table 12 that the densities of the lithium-potassium chloride electrolyte and lithium-sodium chloride electrolyte are lower than the density of liquid magnesium. In this case, the metal can be collected at the bottom of the cell [269,270]. However, such cells have not found commercial application for a number of reasons. First, LiCl is an expensive component, with a strong affinity for water, and complete dehydration is very difficult and energy intensive. There are also difficulties in separating the metal because of the sludge which sinks to the cell bottom together with the magnesium.

*Electrical conductivity* is a very important property of the electrolyte. The voltage drop in the electrolyte depends on the electrical conductivity of the melt. Generally, a high current efficiency is obtained at a high current density, but a high energy efficiency is obtained at a low current density. That is so because the cell voltage is low at low current density, and therefore the energy efficiency is high. Also, the quantity of the electrolysis products increases with the current

**Table 12** Composition and Properties of Electrolytes (at 700°C)

| Electrolyte | Composition (wt%) | mp (°C) | Density (kg m$^{-3}$) | Conductivity (S m$^{-1}$) |
|---|---|---|---|---|
| Potassium | 5–12 MgCl$_2$<br>70–80 KCl<br>12–16 NaCl | 650 | 1600 | 183 |
| Sodium-potassium | 10 MgCl$_2$<br>50 NaCl<br>40 KCl | 625 | 1625 | 200 |
| Sodium-calcium | 8–16 MgCl$_2$<br>30–40 CaCl$_2$<br>35–45 NaCl<br>0–10 KCl | 575 | 1780 | 200 |
| Lithium-potassium | 10 MgCl$_2$<br>70 LiCl<br>20 KCl | 550 | 1500 | 420 |
| Lithium-sodium | 10 MgCl$_2$<br>70 LiCl<br>20 NaCl | 560 | 1521 | 488 |
| Sodium-barium | 10 MgCl$_2$<br>20 BaCl$_2$<br>50 NaCl<br>20 KCl | 686 | 1800 | 217 |
| Magnesium | | 649 | 1580 | |

*Source*: From Ref. 266.

density, but the rate of the recombination reaction is not much influenced by the current density. Therefore, the current efficiency is high at high current density.

A reduction in the ohmic drop across the electrolyte can be achieved either by a decrease of the electrical resistivity of the electrolyte or by a decrease of the interelectrode distance or lower current density. Table 12 shows that sodium and lithium salts increase the electrical conductivity of the electrolyte. The electrolysis products, magnesium and chlorine, get more easily in contact with each other and recombine by a decrease of the interelectrode distance. This so-called recombination reaction,

$$Mg + Cl_2 = MgCl_2 \qquad (107)$$

lowers the current efficiency. However, the total resistance of the electrolysis cell must be such that the generation of Joule heat balances the heat losses from the cell to maintain the temperature and thus keep the electrolyte in the molten state.

*Structure of the Electrolyte*

The data published in the literature suggest that $MgCl_2$ does not exist as $Mg^{2+}$ cations and $Cl^-$ anions when dissolved in alkali chloride melts. These solvents act as ligand donors, and magnesium is found as anionic complexes in the melt.

Flood and Urnes [271] studied the freezing point depression of the binary systems $MgCl_2$-KCl, $MgCl_2$-NaCl and $MgCl_2$-RbCl. The cryoscopic data show that for low contents of $MgCl_2$, all $MgCl_2$ present in the melt is completely complexed according to the reaction

$$2M^+ + 2Cl^- + MgCl_2 = 2M^+ + MgCl_4^{2-} \qquad (108)$$

where $M^+$ is an alkali ion. Kleppa and McCarty [272] made calorimetric studies of these systems and found that the deviations from ideality of the heats of mixing are large. This fact shows that these systems are very far from being ideal because they form complexes. From these studies the authors concluded that the composition of the complex ions is $MgCl_4^{2-}$.

Neil et al. [273] suggested that the complex species formed according to Eq. (108) partly dissociate by the reaction

$$MgCl_4^{2-} = Mg^{2+} + 4Cl^- \qquad (109)$$

On the basis of emf data, obtained for the system $MgCl_2$-KCl at 800°C, the equilibrium constant of reaction (109) was calculated to be $1.8 \cdot 10^{-3}$. Such a low value of the dissociation constant is an indication of the high stability of the $MgCl_4^{2-}$ ion. The conclusion which results from these thermodynamic studies is that there are complex species present in melts of alkali chlorides containing magnesium chloride.

Spectroscopic studies also provide evidence of the presence of some complex ions in such melts [274–277]. Results obtained by Raman spectroscopy showed the presence of a residual ionic lattice composed of polynuclear aggregates of the formula $(MgCl_2)_n$. Recent Raman spectroscopic investigations by Brooker et al. [278–280] showed that $MgCl_4^{2-}$ is the predominant species in $MgCl_2$-rich melts, while other authors [281] found that the presence of small amounts of other species such as $MgCl_3^-$ cannot be excluded. An equilibrium reaction,

$$MgCl_4^{2-} = MgCl_3^- + Cl^- \qquad (110)$$

has been proposed [281] to explain the formation of $MgCl_3^-$ in these melts.

## 3.  Electrolysis

Magnesium chloride is electrolyzed in a molten mixture with alkali chlorides at 700–800°C, the main reaction being

$$MgCl_{2(l)} = Mg_{(l)} + Cl_{2(g)} \tag{111}$$

while the individual electrode reactions may be written as

cathode     $MgCl_4^{2-} + 2e = Mg + 4Cl^-$ (112)

anode       $2Cl^- = Cl_{2(g)} + 2e$ (113)

From thermodynamic data, the standard Gibbs energy of reaction (111) is

$$\Delta G^\circ = 148,620 + 6.422\, T \ln T - 0.14 \cdot 10^{-3} T^2 - 77.47T \tag{114}$$

The standard reversible decomposition potential, $E^\circ$, for reaction (111) is given by the well-known equation

$$\Delta G^\circ = -nFE^\circ \tag{115}$$

Substitution of (114) in (115) gives the temperature dependence of the standard reversible potential:

$$E^\circ_{MgCl_2} = -3.222 + 1.679 \cdot 10^{-3} T - 1.392 \\ \cdot 10^{-4} T \ln T + 3.035 \cdot 10^{-3} T^2 \tag{116}$$

Measurements of the characteristics of various electrolyte compositions and the pure salts are reported by some authors [266,267,282]. At 750°C the standard decomposition potentials of the electrolyte components are [282] $MgCl_2$: $-2.51$ V; NaCl: $-3.22$ V; KCl: $-3.27$ V; LiCl: $-3.30$ V; $CaCl_2$: $-3.33$ V; $BaCl_2$: $-3.40$ V. Codeposition of sodium or calcium will thus occur only by depletion of $MgCl_2$ ($<3\%$). This lowers the current efficiency and causes a temperature increase due to the recombination of sodium and chlorine.

The cathodic current efficiency normally ranges from 85% to 94%. Possible mechanisms of the loss of current efficiency are summarized as follows [267]:

1. Simultaneous discharge of the magnesium ions and other ions present in electrolyte
2. Redox reactions of some multivalent ions
3. Recombination reactions of the chlorine bubbles with liquid magnesium droplets
4. Dissolution of the reaction products (magnesium and chlorine) in the electrolyte and their recombination in the melt to form magnesium chloride
5. Reaction of magnesium with some impurities in the electrolyte to finally form sludge which sinks to the bottom of the cell
6. Oxidation of magnesium metal by oxygen from the air

The simultaneous operation and interrelation of the above mechanisms depend on the working conditions of each electrolytic cell. However, magnesium metal is well wetted by the electrolyte, and the chlorine bubbles will have to penetrate this encapsulating electrolyte film before reaction (item 3) can occur.

If magnesium and chlorine are dissolved in the electrolyte, the recombination reaction is

$$Mg_{(dis)} + Cl_{2(dis)} = MgCl_{2(dis)} \tag{117}$$

In this reaction the form and the extent of dissolution of magnesium and chlorine in the electrolyte are important.

Øye and co-workers [283–286] attempted to determine the rate-determining step in the recombination reaction and found that the dissolution rate of $Cl_2$ or Mg in the electrolyte is the most probable rate-determining step. The chemical solubility of magnesium in the electrolyte increases with temperature, and it was considered in terms of the reactions

$$Mg + Mg^{2+} = 2Mg^+ \tag{118}$$

$$Mg + Mg^{2+} = Mg_2^{2+} \tag{119}$$

Both reactions imply the formation of monovalent magnesium ions, which may exist as the monomer $Mg^+$ or the dimer $Mg_2^{2+}$ (the dimer is the more likely species, as shown by Van Norman and Egan [287]).

Cathodic deposition of magnesium from various chloride melts on different substrates has been studied by several authors [288–290]. In dilute solutions of Mg(II) species the cathode process has been found to be controlled by diffusion of the reactant. Alloy formation has been observed on platinum, as reported by Tunold [288] and Duan et al. [290]. The rate constant of the charge transfer process on a Mg/Ni electrode in molten $NaCl-CaCl_2-MgCl_2$ was reported by Tunold to have a value of about 0.01 cm s$^{-1}$. This author also reported underpotential deposition of a monolayer on iron electrodes, at potentials approximately 100 mV positive to the Mg deposition potential.

In pure molten $MgCl_2$ the primary step of the cathodic process probably consists of breaking up molecular associations. In pure or nearly pure $MgCl_2$ it might be written in the following way [291–295]:

$$(MgCl_2)_n \rightleftharpoons Mg^{2+} + Mg_{n-1}Cl_{2n}^{2-} \tag{120}$$

The addition of NaCl increases the availability of "free" $Mg^{2+}$. At higher concentrations of alkali chloride the tetrahedral complex is formed, which dissociates according to Eq. (109).

This dissociation process is followed by two charge transfer steps [291–293]:

$$Mg^{2+} + e \rightarrow Mg_{(ads)}^+ \tag{121}$$

$$Mg_{(ads)}^+ + e \rightarrow Mg \tag{122}$$

The first charge transfer step (reaction 121) is the rate-determining process. The exchange current density of this step increases sharply with decreasing $x_{MgCl_2}$ from 2.7 A cm$^{-2}$ in pure $MgCl_2$ to 11.6 A cm$^{-2}$ at 75 mol% $MgCl_2$ and then decreases continuously to about 3.0 A cm$^{-2}$ at 20 mol% $MgCl_2$ [295]. The second charge transfer step (reaction 122) is very rapid, and the value of the exchange current density is about 20 A cm$^{-2}$ at 725°C, increasing to 40 A cm$^{-2}$ at 850°C [290]. The very high rate of the second charge transfer step probably means that the monovalent species is very unstable and exist only at the interface.

## 4. Industrial Electrolysis Cells

One of the most important problems in industrial cells for producing magnesium metal is the separation of the electrolysis products. Chlorine is a gas and rises to the top of the electrolyte. However, the liquid magnesium produced has a density lower than that of the electrolyte, so it is also collected at the top of the cell. The problem can be solved in two ways: either through the use of physical barriers (for example a diaphragm) or by control of the electrolyte circulation patterns in the cell. The use of a diaphragm increases the cell resistance, and hence the consumption of electrical energy also increases. Control of the circulation patterns requires that the cell components be dimensionally stable. All magnesium producers have chosen the second way, but many details, particularly those related to cell design and engineering, are not public knowledge because the magnesium industry operates in a very competitive environment.

### I. G. Farben Cell

The most important feature of these cells in the hoodlike semiwall which surrounds the graphite anode, forming a chlorine collection compartment. This structure permits separation of the products of electrolysis and removal of chlorine gas from the electrolyte [266,267]. The compartments are separated by semiwalls which are made of acid chamotte or fused alumina. Figure 24 illustrates two individual cells of an I. G. Farben cell which contain a series of electrode assemblies. In one cell, the anode compartment is surrounded by two cast steel cathodes whose construction is shown in the side view of the figure. During electrolysis, chlorine bubbles are formed and rise to the electrolyte surface in the anode box. The rising chlorine bubbles cause an electrolyte motion in the way depicted in Figure 24. The moving electrolyte carries along the magnesium droplets which then rise to the surface in the quiescent cathode compartment. The moving electrolyte also acts to bring fresh electrolyte containing magnesium chloride to the interelectrode space. The electrolysis temperature is about 750°C, and the cell life is about one year, depending on the deterioration of the semiwalls.

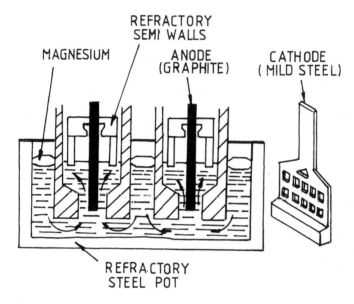

**Figure 24**   The I. G. Farben cell for anhydrous MgCl$_2$ electrolysis [266,267].

Electrolysis of hydrous cell feed (MgCl$_2$ · 1.5 H$_2$O) in the so-called Dow cell requires graphite anodes which can be lowered into the cell to compensate for graphite consumed by the electrolytic decomposition of water.

*The Norsk Hydro Cell*

This type of cell has been in operation since 1978, and it consists of a sealed, brick-lined casing, divided into two chambers for electrolysis and metal collection, as shown in Figure 25. The chlorine gas is collected above the electrolysis cell proper, while the magnesium is carried by the flowing melt into a separate chamber which runs alongside the electrolysis cell. The hollow cathodes and the metal collecting chamber can be seen in Figure 25. The energy consumption is of the order of 12–13 kWh/kg Mg and the cell life exceeds five years [282].

*The Alcan Cell*

Also in this case the operating principle of the cell is not public knowledge, and the following description is taken from a patent. The cell in Figure 26 is divided by a curtain wall into a front compartment where the metal accumulates on the heavier chloride bath and an electrolysis compartment with bipolar electrodes where chlorine is collected [296,297]. As shown in Figure 27 the metal droplets are collected from all compartments of the cell, whereby the metal separates out

Collection    Electrolysis
chamber       chamber

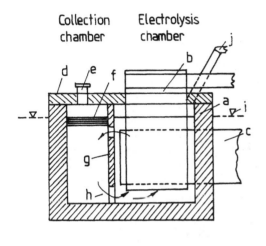

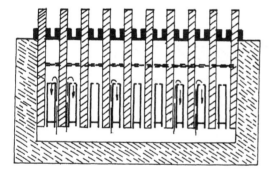

**Figure 25** Sketch of the Norsk Hydro magnesium cell [282]. Lower figure shows a longitudinal section of the cell depicting the hollow cathodes. Upper figure shows the metal collecting chamber to the left: (a) refractory material; (b) graphite anode; (c) steel cathode; (d) refractory cover; (e) metal outlet; (f) metal; (g) partition wall; (h) electrolyte flow; (i) electrolyte level; (j) chlorine outlet.

under an underflow weir alongside the cell [296,297]. The operating temperature, 660–680°C, is just above the melting point of the metal. The energy consumption of this cell when operating on high-purity $MgCl_2$ is 9.5–10.0 kWh/kg Mg [297]. The cell operates at 100 kA with a life of up to two years [297].

*Ishizuka Cell*

The Ishizuka Research Institute in Japan [298] has, beginning in 1983, developed a method using a bipolar cell for industrial magnesium production as shown in Figure 28. The bipolar electrodes are made of graphite (anode) with steel plates

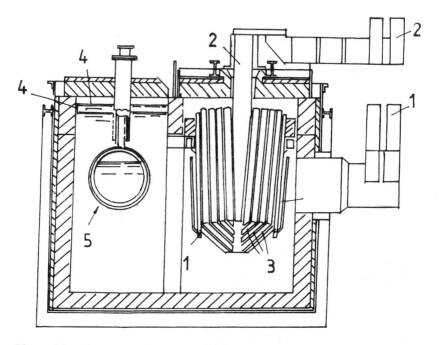

**Figure 26**   Alcan magnesium cell [296,297]; (1) cathode; (2) central anode; (3) bipolar electrode; (4) metal outlet; (5) cylinder used to adjust the electrolyte level.

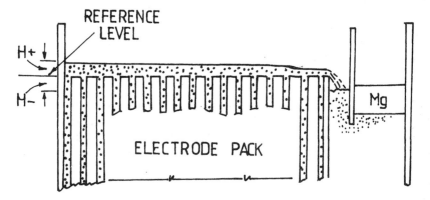

**Figure 27**   Way of collecting the Mg droplets in the Alcan cell [296,297].

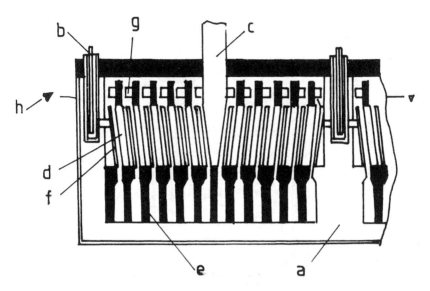

**Figure 28** Vertical section of Ishizuka cell [298]: (a) refractory material; (b) steel cathode; (c) graphite anode; (d) bipolar electrode/steel + graphite; (e) bipolar support; (f) terminal cathode; (g) electrolyte ports; (h) electrolyte level.

attached to the cathode side. The electrolyte flow produced by escaping gas carries the metal into the collecting chamber. The operating temperature of the cell is 670°C and the current is 50 kA. The quantity of magnesium produced using bipolar electrodes corresponds to a monopolar cell of 300 kA. The current density is 0.56 A m$^{-2}$, and the interpolar gap is 4 cm. At a current efficiency of 76%, the power consumption is 11 kWh/kg Mg when operated on molten magnesium chloride from titanium production by the Kroll process. The cell life is about three years.

Recently, Sharma [299], at General Motors, proposed a low cost method for the production of Al-Mg alloys, which can be used in the automobile industry and for desulfurization of steel. The unit cell is shown in Figure 29. The cathode of liquid aluminum is at the bottom of the cell where magnesium is deposited. The electrolyte is a mixture of molten chlorides in which partly dehydrated $MgCl_2$ is dissolved and the electrolysis temperature is 750°C. The same author also described another method to produce magnesium metal using an electrochemical cell with bipolar electrodes, shown in Figure 30.

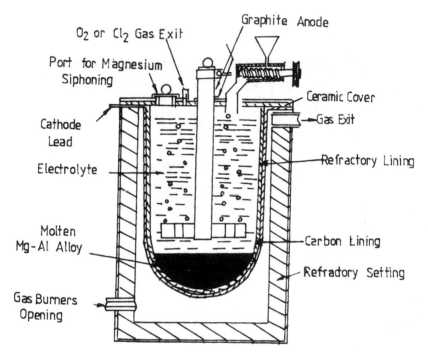

**Figure 29**  Schematic of one electrolytic cell proposed by Sharma [299].

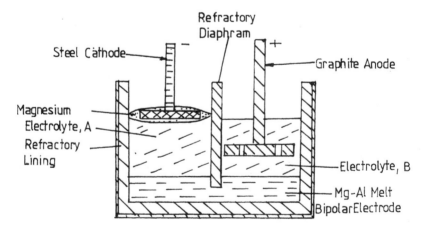

**Figure 30**  Schematic of a cell with a bipolar electrode proposed by Sharma [299].

## C. Sodium

Sodium is a strongly electronegative metal, being very reactive, and it does not occur in nature in the elemental state but always in cationic form in salts or minerals (sodium chloride, sodium sulfate, sodium carbonate, sodium borate, sodium nitrate, feldspar, kaoline, etc.).

Sodium can be obtained from its compounds by thermochemical reduction [300]. A serious disadvantage of thermochemical methods was that reverse and secondary reactions occurred.

### 1. Electrolytic Processes for Sodium Production

The first commercial process for sodium production used the electrolysis of sodium hydroxide, but it had the disadvantage that the current efficiency was less than 50%. At present this process is the only way of producing metallic sodium. The cell reaction is

$$2NaCl_{(l)} = 2Na_{(l)} + Cl_{2(g)} \tag{123}$$

The melting point of sodium chloride is 800°C [301]. At this high temperature, the chlorine gas obtained at the anode, would corrode all the cell components getting into contact with it. The sodium solubility in the molten salt is high (4.2 wt%) and so is its vapor pressure (50 kPa) [302]. Therefore an electrolyte of lower melting point must be used.

These problems have been overcome by the Downs process, which uses a 58 wt% calcium chloride–42% sodium chloride electrolyte at an operating temperature of 580–590°C. This composition represents a compromise between the melting point of the salt mixture and the sodium purity. The eutectic mixture contains 66.8 wt% $CaCl_2$ and 33.2 wt% NaCl and melts at 505°C. Increasing $CaCl_2$ content reduces the melting point of the mixture and increases the concentration of calcium in the cathode metal due to the equilibrium

$$2Na + CaCl_2 \rightleftharpoons 2NaCl + Ca \tag{124}$$

The density of the melt is 1.89–1.94 $gcm^{-3}$. During electrolysis, along with sodium metal, small quantities of calcium ($\sim$4 wt%) are deposited at the cathode. Calcium has a melting point (804°C) far higher than sodium and a low solubility in sodium (see Table 13) [302]. Because of that at the cathode a solid alloy phase Na-Ca is accumulated and this blocks the circulation of the electrolyte in the electrolysis cell and the removal of sodium from the cell. Also this solid alloy sometimes causes short-circuiting in the electrolysis cell. Sodium obtained by electrolysis is cooled to 110–120°C and filtered for the removal of calcium. The effect of temperature on the solubility of calcium in sodium [302] is shown in Table 13. At 110°C the calcium content in sodium is reduced to <0.04%.

**Table 13** Solubility of Calcium
in Sodium

| Temperature (°C) | wt% Ca |
|---|---|
| 97.5 | 0.01 |
| 135 | 0.15 |
| 160 | 0.20 |
| 170 | 0.25 |
| 210 | 0.55 |
| 400 | 4.0 |
| 600 | 4.0–5.5 |

*Source*: From Ref. 302.

Introduction of the ternary electrolyte $BaCl_2$-$CaCl_2$-$NaCl$ reduces the calcium content in sodium in proportion to the change in the molar ratio $CaCl_2/NaCl$. The melting point of the ternary eutectic of these salts is 450°C. However, the eutectic mixture is not suitable as a working electrolyte since the $CaCl_2$ content of ca. 50 wt% would result in an unacceptably high proportion of calcium in the sodium product. Following a study of the ternary system with respect to electrical conductivity, viscosity and melting point, Du Pont [303] patented the composition 46–53 wt% $BaCl_2$, 23–26 wt% $CaCl_2$, 24–28 wt% $NaCl$. The sodium produced by the electrolysis of this electrolyte contains ca. 1 wt% calcium. During electrolysis the sodium chloride content of the electrolyte and its melting point fluctuate; however, the $BaCl_2$-$CaCl_2$ weight ratio remains practically constant at 1.88. For example, the melting points of the mixture are 550°C for 26 wt% $NaCl$ and 600°C for 32 wt% $NaCl$.

The kinetics of the sodium electrode reaction in molten sodium chloride has been investigated by Kisza et al. [304]. The electrode process exhibits a mixed charge transfer-diffusion character. The electrode is very reversible with an exchange current density that varies from about 100 A $cm^{-2}$ at 820°C to about 200 A $cm^{-2}$ at 920°C.

## 2. Design and Operation of Sodium Electrolytic Cells

The scheme of an electrolytic cell is given in Figure 31. The cathodes and anodes are subdivided into four pairs of electrodes. The four cathodes are welded together, surrounding the four cylindrical graphite anodes concentrically [305–308]. The anode-cathode distance is <50 mm. A cylindrical diaphragm of iron mesh is placed between the electrodes to separate the electrolysis products (sodium and chlorine). The density of molten sodium is lower than the density of the electrolyte.

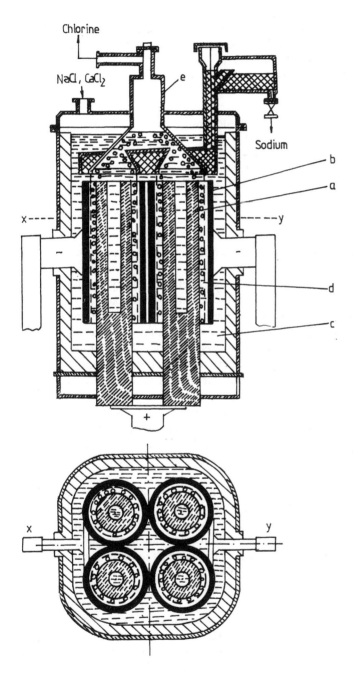

**Figure 31** Downs cell [309]: (a) anode; (b) cathode; (c) molten salt; (d) diaphragm; (e) chlorine dome.

The chlorine bubbles that form and rise at the graphite anodes improve the circulation of the melt in the anode space. Circulation of the molten catholyte is promoted by the upward flow of sodium.

Operating data for a Downs cell [309] are

| | |
|---|---|
| Operating temperature | 590–610°C |
| Cell voltage | 6.5–7 V |
| Cell current | ≤45 kA |
| Current efficiency | 80–90% |
| Consumption of electrical energy | 9.8–10.5 kWh/kg Na |
| Lifetime of the cells | 1100–1200 days |

The decomposition potential of sodium chloride is approximately 3.4 V, which represents about 50% of the cell voltage.

During the working of an electrolysis cell, the energy consumption increases because of various changes in the operating conditions. Thus, a loss of graphite from the anodes [310] takes place due to corrosion and erosion. This leads to the accumulation of precipitated material on the cell floor. Heat is produced continuously due to the internal resistance of the cell and the current passing through it. Likewise, heat is produced because of the recombination of sodium and chlorine forming sodium chloride. Excess heat is removed from the cell through the cell walls to maintain a constant temperature in the cell.

The salts for feeding (sodium chloride, calcium chloride and barium chloride) must be clean and dry [311]. Similarly, salt additions must be made as uniformly as possible to avoid salt deposits in the base of the cell. This can reduce the internal circulation of the electrolyte, cause cell overheating and reduce the current efficiency.

The current efficiency would be 100% if the recombination reaction of sodium with chlorine did not occur. Although the diaphragm prevents sodium droplets formed at the cathodes to react with the chlorine bubbles formed at the anodes, another phenomenon causes the decrease in current efficiency. Thus, the slight solubility of sodium in the melt causes the melt to become a partial electronic conductor (see Section I.B.4 "Electrical Conductivity of Metal-Molten Salt Mixtures"). This electronic conductivity and the recombination reaction of sodium and chlorine dissolved in the electrolyte decrease the current efficiency.

The consumption of electrical energy for the electrolysis is 9.8–10.5 kWh/ kg Na, depending on the lifetime of the cell. If the auxiliary equipment (ventilation, heating of equipment, etc.) is included, this increases to 11.5–12.0 kWh/ kg Na. If the energy needed to purify the raw salt and for the purification and liquefaction of the chlorine is taken into account, the total energy consumption is 12.0–13.0 kWh/kg Na.

*Alternative Process*

Ford Motor Co [312] obtained a patent for a method of producing sodium which uses a solid ceramic (β-alumina) that conducts sodium ions as a divider for a two-compartment cell. This method allows low melting sodium salts or salt mixtures to be used to produce sodium [312–315]. The cell operates at 200°C and 6 V instead of the conventional 7 V for Downs cells, making sodium at 600°C. The average current efficiency is 100% compared to 85–92% for Downs cells. The power consumption of the process at the same productivity is 20–30% lower than for the Downs cell. This process has, however, not been operated on a large scale because the $\beta\text{-Al}_2\text{O}_3$ does not have sufficient lifetime.

## III. ELECTROCHEMICAL ENERGY CONVERSION AND STORAGE IN MOLTEN SALTS

### A. Fuel Cells

The development of a modern society depends on the availability of abundant energy. The outlook for the world energy perspective has changed after the first oil crisis of 1973. Most nations were obliged to reduce their dependency on oil and to diversify their sources of primary energy. Recently, discussions on the greenhouse effect have concluded that carbon dioxide emissions cause global warming. For this reason the world is obliged to use the earth's fossil fuel resources more efficiently in order to secure a sustainable future.

The conventional generation of electrical energy from a fuel requires the use of a heat engine which converts thermal energy to mechanical energy. All heat engines operate by the Carnot cycle, and their maximum efficiency is about 40–50% (for the modern gas-fired power stations, the efficiency is about 55%).

### 1. Fuel Cell Efficiency

In an ideal electrochemical cell, the Gibbs energy of reaction is converted to electrical energy, then the intrinsic maximum energy conversion efficiency of an electrochemical cell is

$$\eta = \frac{\Delta G_T}{\Delta H_0} = 1 - \frac{T\Delta S_T}{\Delta H_0} \tag{125}$$

where $\Delta G_T$ is the Gibbs energy at $T$, the cell operating temperature, $\Delta H_0$ is the enthalpy of formation and $T\Delta S_T$ is the reversible heat produced. An electrochemical converter is not subject to the Carnot limitation. Therefore, for most electrochemical processes the intrinsic maximum energy conversion efficiency of a fuel cell is very high (>90%). The efficiency of real fuel cells compared with other generators is shown in Figure 32 [316].

Like heat engines, fuel cells can never approach maximum efficiencies. The energy used to produce electrical energy is always greater than the reversible heat, $T\Delta S$. The additional energy which is transformed into heat is given by the difference between the reversible and the real electric work output of the cell: $nF(E_{\text{rev}} - E_{\text{cell}})$, where $n$ is the number of electrons transferred in the reaction, $F$ is the Faraday constant, $E_{\text{rev}}$ is the reversible cell voltage and $E_{\text{cell}}$ is the actual cell output voltage.

The open circuit voltage (OCV) of a fuel cell is identical to the reversible cell potential, provided that the cell is in thermodynamic equilibrium. That is usually achieved, especially at high temperatures. There are some causes which

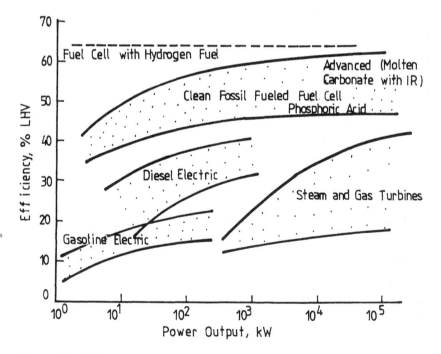

**Figure 32**  Efficiency characteristics of different technologies as a function of scale [316].

contribute to irreversible losses in a practical fuel cell, such as ohmic resistance and polarization phenomena, also called overpotential or overvoltage.

The influence of the electrode kinetics on the performance of a fuel cell is shown in Figure 33 [317]. The greatest difficulty in attaining high performance of a fuel cell, especially at low temperatures, is that many electrodes have a low electrocatalytic activity. Likewise, the reversible potential is attained only when the fuel cell does not produce current. Therefore, the efficiency of a real fuel cell is always less than the theoretical value.

The practical efficiency, $\eta_P$, of a cell under load condition is

$$\eta_P = \frac{E_{cell}}{E^\circ - T(dE^\circ/dT)} = -\frac{nFE_{cell}}{\Delta H} \tag{126}$$

where $E_{cell}$ is the operating potential and $E^0$ is the standard electrode potential. During operation of the cell $E_{cell} < E_{rev}$. Then the fuel cell voltage efficiency, or free energy efficiency, $\eta_v$, can be expressed as

$$\eta_V = \frac{E_{cell}}{E_{rev}} \tag{127}$$

$E_{rev}$ is the open circuit potential of an electrochemical cell in which an equilibrium chemical reaction takes place.

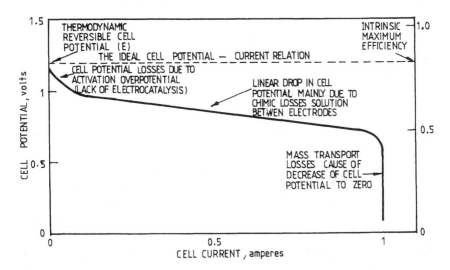

**Figure 33** Typical plot of cell potential vs. current for fuel cells illustrating regions of control by various types of overpotentials [317].

The Faradaic efficiency is defined as

$$\eta_f = \frac{I}{I_m} \tag{128}$$

where $I$ is the actual current from the fuel cell, and $I_m$ is the maximum current which can be obtained on the basis of the amount of reactants consumed, provided that the reaction is complete. The Faradaic efficiency is analogous to the current efficiency in an electrolysis cell. In most fuel cells, $\eta_f$ is less than 1 due to parallel chemical reactions or chemical reactions between the reactants.

The theoretical maximum energy capacity is given by

$$\text{Max. energy capacity} = \frac{E_{rev}F}{3600E_{wt}} \text{ kWh/kg} \tag{129}$$

where $E_{wt}$ is the equivalent weight.

The literature in the field of fuel cells is very rich. We will refer to only some books and published monographs [316–334]. In a fuel cell the reduction of the oxidant OX (usually oxygen but also other oxidants) takes place at the positive pole (cathode), and oxidation of the reductant RED (fuel) at the negative pole (anode).

The use of molten salts in fuel cells offers some advantages over aqueous or organic electrolytes. First, molten salts have a higher electrical conductivity than that of aqueous or organic electrolytes. The very high electrical conductivity of fused salts is due to the high temperature, the high concentration of ions (typically 20 molar) and the complete ionization of simple molten salts. In these systems, the absence of any solvent also increases the conductivity. Another advantage of molten salts is the high temperature of operation. Thus, the electrode reactions generally are fast and hence the overvoltage is low.

The most important advantage of molten salt electrolytes is their very high decomposition potential. Fuel cells with aqueous electrolytes have the potentials limited by the decomposition potential of water of about 1.23 V. The maximum decomposition voltage of fused salts is about 5 V, as shown in Table 14 [335]. Because of that, fuel cells with molten salts may have high potentials and high current densities. It is known that molten salts may be very corrosive. For that reason the materials of construction for molten salt fuel cells must be capable of operating at high temperature and resist the corrosive attack of the electrolytes and reactants for long periods of time.

In Table 15 are given some theoretical energy capacities [335] calculated by means of Eq. (129) and the data from Table 14. These data are based on the hypotheses that the fuel cell potential is the reversible potential, the Faradaic efficiency is 100% and the weight taken into the calculation is that of the re-

**Table 14** Calculated Reversible Potentials for Hypothetical Cells at 500°C

|     | F    | Cl   | Br   | I    | O    | S    |
|-----|------|------|------|------|------|------|
| Li  | 5.56 | 3.65 | 3.21 | 2.57 | 2.14 | —    |
| Na  | 5.12 | 3.52 | 3.16 | 2.59 | 1.62 | 1.48 |
| K   | 5.02 | 3.76 | 3.48 | 2.98 | 1.27 | —    |
| Be  | 4.41 | 2.14 | 1.58 | 0.91 | 2.70 | 1.15 |
| Mg  | 5.01 | 2.68 | 2.26 | 1.61 | 2.68 | —    |
| Ca  | 5.60 | 3.53 | 3.06 | 2.48 | 2.88 | 2.33 |
| Al  | 3.87 | —    | —    | —    | 2.48 | —    |

*Source*: From Ref. 335.

actants. Experience has shown that in a practical system about 15–30% of this theoretical energy capacity generally can be attained.

## 2. Molten Carbonate Fuel Cell (MCFC)

The first MCFC was demonstrated by Broers in 1950 [336], and the first MCFC at high pressure was built by Reiser and Schroll [337] in 1980. As shown in Figure 32, at present the MCFC it is the most efficient fuel cell, and this will be discussed in the following. The MCFC, operating at a temperature between 600 and 650°C, is generally considered a second-generation fuel cell [316,338,339]. It can be used with coal gas and even more so with natural gas as a fuel.

The electrodes are porous, favorable for gas diffusion in the reaction zone, i.e., at the contact between gaseous reactants, liquid electrolyte and electrode [340].

**Table 15** Some Theoretical Energy Capacities (kWh/kg) Calculated from the Reversible Potentials Data of Table 14 (500 °C)

|     | F    | Cl   | Br   | I    | O    | S    |
|-----|------|------|------|------|------|------|
| Li  | 5.74 | 2.31 | 0.99 | 0.57 | 3.84 | —    |
| Na  | 3.27 | 1.61 | 0.82 | 0.50 | 1.40 | 1.25 |
| K   | 2.32 | 1.35 | 0.78 | 0.52 | 0.72 | —    |
| Be  | 5.03 | 1.44 | 0.50 | 0.20 | 5.79 | 1.50 |
| Mg  | 4.31 | 1.51 | 0.66 | 0.34 | 3.56 | —    |
| Ca  | 3.84 | 1.70 | 0.82 | 0.49 | 2.75 | 1.73 |
| Al  | 3.71 | —    | —    | —    | 3.91 | —    |

*Source*: From Ref. 335.

Molten carbonates are an extremely corrosive medium for the majority of metals and alloys at temperatures of around 600°C. Therefore, the choice of materials at a reasonable cost, to be used as stable cathodes under an oxidizing atmosphere such as air or oxygen-$CO_2$ mixtures, is limited. Only semiconducting oxides are materials with such properties. In the MCFC the currently used material for the cathode is lithiated nickel oxide. Initially, the cathode is metallic nickel, but during the first period of time of cell operation, a lithiation and oxidation of the nickel occurs spontaneously in the presence of lithium carbonate under an oxidizing atmosphere at high temperature. The original structure of the material is completely changed due to these phenomena. The cathode mass acquires many very small pores which increase the contact surface area between the electrolyte and the gaseous reactant. The thickness of the NiO cathode ranges from 0.4 to 0.8 mm, and it has an electronic conductivity [338] of approximately 5 ohm$^{-1}$ cm$^{-1}$.

The MCFC anode operates under reducing atmosphere, at a potential typically 700–1000 mV more negative than that of the cathode. Many metals are stable in molten carbonates under these conditions, and several transition metals have electrocatalytic activity for hydrogen oxidation. Nickel, cobalt, copper and alloys in the form of powder or composites with oxides are usually used as anode materials. Ceramic materials are included into the anode composition to stabilize the anode structure (pore growth, shrinkage, loss of surface area) at the time of sintering. An alloy powder of Ni + 2–10 wt% Cr can be used. The initial formation of $Cr_2O_3$, followed by surface formation of $LiCrO_2$, can stabilize the anode structure.

The electrolyte has a particular structure. A mixture of $LiAlO_2$ and alkali carbonates (typically >50 vol%) is hot pressed (about 5000 psi) at temperatures slightly below the melting point of the carbonate salts. In this way, a porous matrix support material of ceramic particles ($LiAlO_2$) is formed that contains a capillary network filled with molten electrolyte. The ceramic material in the electrolyte structure represents a mechanical resistance which does not participate in the electrical or electrochemical processes. The prepared electrolyte has a thickness of 1–2 mm, and it is very difficult to produce it in large shapes.

The preferred electrolyte is a binary lithium-potassium carbonate consisting of 62 mole% $Li_2CO_3$ and 38 mole% $K_2CO_3$, with a liquidus temperature of 490°C.

Figure 34 shows an outline of a MCFC cell which uses a gaseous mixture of $H_2$ + CO [316,333,338]. The principal electrode reactions are

Coal gasification

$$C + H_2O \xrightarrow{\quad O_2 \quad} CO + H_2 \qquad (130)$$

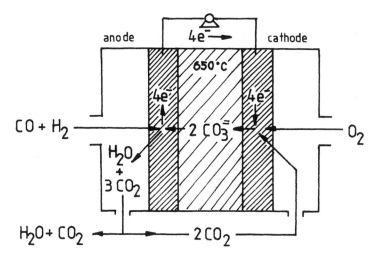

**Figure 34** Working scheme of a MCFC cell [316,333,338].

Anode

$$H_2 + CO_3^{2-} \rightarrow CO_2 + H_2O + 2e^- \tag{131}$$

$$CO + CO_3^{2-} \rightarrow 2CO_2 + 2e^- \tag{132}$$

Overall anode

$$H_2 + CO + 2CO_3^{2-} \rightarrow 3CO_2 + H_2O + 4e^- \tag{133}$$

Cathode

$$4e^- + 2CO_2 + O_2 \rightarrow 2CO_3^{2-} \tag{134}$$

Overall fuel cell

$$CO + H_2 + O_2 \xrightarrow{4F} CO_2 + H_2O \tag{135}$$

Overall power plant

$$C + O_2 \xrightarrow{4F} CO_2 \tag{136}$$

A gaseous mixture of $H_2$ and CO can be obtained by coal gasification or reforming of natural gas. That mixture reacts at the anode with molten $CO_3^{2-}$ ions to produce $CO_2$ and $H_2O$ vapors, releasing electrons to the external circuit. Also, in the anode compartment, in the presence of the metallic cell parts, at 650°C, CO present in the fuel gas reacts with $H_2O$ vapor, yielding hydrogen:

$$CO + H_2O \rightarrow CO_2 + H_2 \tag{137}$$

At the anode, hydrogen is oxidized more easily than CO, so reaction (131) is more important than reaction (132). For this reason the main electroactive species is hydrogen [316].

Carbon dioxide produced at the anode is recycled to the cathode, where it reacts with electrons and atmospheric oxygen to regenerate the carbonate ion consumed at the anode. Two mechanisms have been proposed for the anode reaction. The first is

$$2H_2 + 4M \rightleftharpoons 4MH \qquad (M = Ni) \tag{138}$$

$$2MH + 2CO_3^{2-} \rightarrow 2OH^- + 2CO_2 + 2M + 2e^- \qquad (r.d.s) \tag{139}$$

$$2MH + 2OH^- \rightleftharpoons 2H_2O + 2M + 2e^- \tag{140}$$

The second is reaction (138) with (139) followed by

$$2OH^- + CO_2 \rightleftharpoons H_2O + CO_3^{2-} \tag{141}$$

where reaction (139) is the rate-determining step (r.d.s).

The cathodic mechanism is complex and it depends on the melt composition, especially the acidity of the electrolyte, namely the cation composition. In less acidic electrolytes (see below) the peroxide mechanism prevails [341,342]:

$$O_2 + 2CO_3^{2-} \rightleftharpoons 2O_2^{2-} + 2CO_2 \tag{142}$$

$$2O_2^{2-} + 4e^- \rightarrow 4O^{2-} \qquad (r.d.s.) \tag{143}$$

$$4O^{2-} + 4CO_2 \rightleftharpoons 4CO_3^{2-} \tag{144}$$

where $O_2^{2-}$ is the peroxide ion, and $O^{2-}$ is the oxide ion.

In acidic melts, however, the superoxide mechanism is dominant [341,342]:

$$3O_2 + 2CO_3^{2-} \rightleftharpoons 4O_2^- + 2CO_2 \tag{145}$$

$$2O_2^- + 2e^- \rightarrow 2O_2^{2-} \tag{146}$$

$$2O_2^{2-} + 4e^- \rightarrow 4O^{2-} \tag{147}$$

$$4O^{2-} + 4CO_2 \rightleftharpoons 4CO_3^{2-} \tag{148}$$

where $O_2^-$ is the superoxide ion.

In these melts, the concept of acidity of Lux-Flood [343] (L-F) is applied. A Lux-Flood acid is defined as an oxide ion acceptor leading to the formation of the conjugate base:

$$\text{L-F acid} + O^{2-} \rightleftharpoons \text{base} \tag{149}$$

For example, in molten carbonates,

$$CO_2 + O^{2-} \rightleftharpoons CO_3^{2-} \tag{150}$$

The relative basicity of the commonly used alkali carbonates is [344]

$$(\text{basic}) \quad Li_2CO_3 > Na_2CO_3 > K_2CO_3 \quad (\text{acidic})$$

The cathode reaction mechanism is not well understood, and more studies are necessary to elucidate it as a function of the electrode material and electrolyte composition [345,346].

The reversible cell potential for a MCFC depends on the gas composition at the anode (partial pressures of $H_2$, $H_2O$ and $CO_2$) and at the cathode (partial pressures of $O_2$ and $CO_2$),

$$E_{rev} = E° + \frac{RT}{2F} \ln \left\{ \left[ \frac{P_{H_2} P_{O_2}^{1/2}}{P_{H_2O}} \right] \cdot \left[ \frac{P_{CO_{2,c}}}{P_{CO_{2,a}}} \right] \right\} \tag{151}$$

where the subscripts $a$ and $c$ refer to the anode and cathode gases, respectively. When the $CO_2$ partial pressures in the cathode and the anode gases are identical, the cell potential depends only on the partial pressures of $H_2$, $O_2$ and $H_2O$. The values of $E°$ (the standard cell potential) for hydrogen oxidation and for several other reactions, at 650°C are given in Table 16 [338].

From Eq. (151) it follows that a pressure increase from $P_1$ to $P_2$, causes an increase in the reversible cell voltage:

$$\Delta E = \frac{RT}{4F} \ln \left( \frac{P_2}{P_1} \right) = 46 \log \left( \frac{P_2}{P_1} \right) \tag{152}$$

Therefore, a 10-fold increase in cell pressure corresponds to an increase of 46 mV in the reversible cell potential at 650°C. Both the cell voltage and the gas solubility in the electrolyte increase with the gas pressure.

**Table 16** Thermodynamic Characteristics and Voltages of Fuel Cell Reactions at 650°C

| Reaction | $\Delta G°$ (kJ mol$^{-1}$) | $E°$ (V) |
|---|---|---|
| $H_2 + \frac{1}{2} O_2 = H_2O$ | $-196.92$ | 1.020 |
| $CO + \frac{1}{2} O_2 = CO_2$ | $-202.51$ | 1.049 |
| $CH_4 + 2O_2 = CO_2 + 2H_2O$ | $-800.89$ | 1.038 |
| $CH_4 + H_2O = CO + 3H_2$ | $-7.62$ | 0.010 |
| $CH_4 + CO_2 = 2CO + 2H_2$ | $-2.04$ | 0.003 |
| $CO + H_2O = CO_2 + H_2$ | $-5.58$ | 0.029 |
| $2CO = C + CO_2$ | $-14.62$ | 0.076 |

*Source*: From Ref. 338.

A serious difficulty is the solubility of the nickel oxide cathode in the electrolyte. The solubility of NiO depends on the $CO_2$ partial pressure, according to the equilibrium

$$NiO + CO_2 \rightleftharpoons Ni^{2+} + CO_3^{2-} \tag{153}$$

The equilibrium constant of this reaction is about $5.7 \cdot 10^{-6}$ at 600°C [316]. The NiO solubility is low in basic electrolytes or at high oxide ion concentrations ($O^{2-}$) [347–352]. Under the conditions of an operating fuel cell, the dissolution reaction which takes place is

$$NiO \rightarrow Ni^{2+} + O^{2-} \tag{154}$$

Therefore, in the presence of an excess of $O^{2-}$ ions, the solubility of NiO is suppressed. The solubility of NiO is lowered by a factor of 3 in the $Li_2CO_3$-$Na_2CO_3$ mixture compared to the $Li_2CO_3$-$K_2CO_3$ mixture. The $Ni^{2+}$ ions formed at the cathode migrate toward the anode under the influence of the electrical field and the concentration gradient. At the anode, the $Ni^{2+}$ ions are reduced and deposited, and for this reason short-circuiting of the cell can result after some time. Alternative cathode materials are $LiCoO_2$, $LiFeO_2$ and $LiMnO_3$.

Recently, many corrosion studies of the materials used in MCFC were made [353–357]. These works led to an increase of cell life up to 40,000 h.

The internal reforming molten carbonate fuel cell has a particular construction. In the anode chamber there is a catalyst for the reforming reaction of natural gas. In this case, the following reactions occur at the anode:

$$CH_4 + H_2O \rightarrow 3H_2 + CO \tag{155}$$

$$3H_2 + 2CO_3^{2-} \rightarrow 3H_2O + 3CO_2 + 6e^- \tag{156}$$

$$CO + CO_3^{2-} \rightarrow 2CO_2 + 2e^- \tag{157}$$

Overall reaction $\quad CH_4 + 4CO_3^{2-} \rightarrow 2H_2O + 5CO_2 + 8e^- \tag{158}$

The principal product from the reforming reaction, which is $H_2$, is consumed in the anode reaction. Figure 35 is a picture of a molten carbonate fuel Cell [316]. It is composed of a stack of many superposed cells, each cell having a thickness of 5 mm and an area of 1 m².

## 3. MCFC R & D Programs

In 1993 Energy Research Company (ERC) of Danbury and Torrington, CT, made a test of a 120 kW MCFC for 250 h [333]. This is a prototype stack for a 2 MW, natural-gas-fueled power plant [358]. ERC has announced that after these initial demonstrations it expects its MCFC power plant of about 100 MW to be commercialized at the beginning of the next century. In Japan, Tonen Corp., Sanyo Electric Corp. and Toyo Engineering Corp. are conducting tests on 10 kW external-reforming-type MCFC stacks [338]. Likewise, in 1993, Hitachi Ltd. [359] has

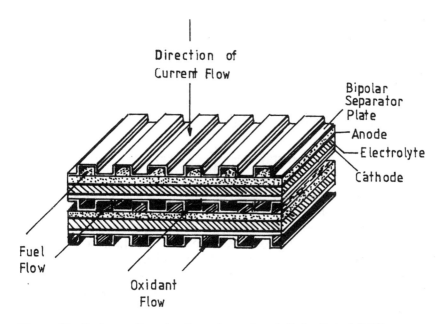

Direction of
Current Flow

Bipolar
Separator
Plate

Anode

Electrolyte

Cathode

Fuel
Flow

Oxidant
Flow

**Figure 35**   Design configuration for molten carbonate fuel cell stack [316].

obtained a stack of 100 kW and has planned the building of a plant of 1000 kW. In Europe, from 1986 to 1989, several national programs were conducted as a part of the development of MCFC. The Netherlands Energy Research Center planned to construct and test a plant [338] up to 250 kW. In Germany, Messerschmitt Bölkow Blohm GmbH has started a collaboration with ERC in the United States for a program of transfer of technology and construction of a 250 kW plant which will produce electrical energy at the end of this century [338]. In Italy, Ansaldo intends to construct a 300 kW plant with external reforming [338].

Electrical plants based on MCFC of high power (GW scale) are to be expected after some experimental difficulties have been overcome. The MCFC thus appears to have a bright future in the twenty-first century.

## B.  Batteries

Like fuel cells, batteries using molten salt electrolytes offer high performance. Molten salts have very high electrical conductivity, which permits the use of high current densities. Likewise, molten salts permit the use of highly reactive electrode materials, which cannot be used in aqueous electrolytes. For these reasons, batteries with molten salts offer very high specific energy (>100 Wh/kg). To obtain such performance it is necessary to use reactants which have a large difference in electronegativity, like the alkali metals or the alkaline-earth metals and elements from groups VIA–VIIA of the the periodic table. Also, elements having a low equivalent weight and belonging to the upper portion of the periodic table are preferable.

The theoretical specific energy of a cell can be calculated if data for the Gibbs energy of formation for the cell reactants are known:

$$\text{Theoretical specific energy} = \frac{-\Delta G}{\sum v_i M_i} \text{ (Wh/kg)} \tag{159}$$

where $\Delta G$ is the Gibbs energy change for the cell reaction, $v_i$ is the number of moles of reactant $i$ and $M_i$ is the molecular weight of the reactant $i$.

In this section we discuss three types of advanced batteries with molten salts: LiAl/LiCl-KCl/FeS, LiAl/LiCl-KCl/FeS$_2$ and Li$_4$Si/LiCl-KCl/FeS$_2$. They all derive from the initial battery developed at the Argonne National Laboratories in the United States [360–364]. In the battery type Li/LiCl-KCl/S, both reactants and the electrolyte are in the molten state. The overall cell reaction is

$$2\text{Li} + \text{S} = \text{Li}_2\text{S} \tag{160}$$

The theoretical specific energy for this battery is 2600 Wh/kg and the cell voltage is 2.2 V at 375°C [360–364]. A comparison between the performance of this battery and that of the lead acid battery (Pb/H$_2$SO$_4$/PbO$_2$) is given in Figure 36 [365].

A permanent loss of capacity takes place because of the appreciable solubility of sulfur in the electrolyte, and for that reason the cell is not stable, since a continuous self-discharge is taking place. Likewise, lithium dissolves easily in the electrolyte. In order to improve the stability of the battery a method was found to "immobilize" the lithium and sulfur at the electrodes. For the negative electrode a solid Li-Al alloy was used. This alloy has a low weight per equivalent of lithium stored, high transport rates for lithium, low cost, and it causes only a modest reduction of the cell voltage (0.3 V). The high transport rate of lithium in the Li-Al alloy permits a current density of 100 mA cm$^{-2}$ at 450°C [366]. The phase diagram for the lithium-aluminum system is given in Figure 37 [367]. The electrode works in the region of the phases $\alpha + \beta$, up to a maximum concentration of about 50% lithium.

FeS is a low-cost material which can be used as a reactant for the positive electrode. The iron is used to immobilize the sulfur and lowering the solubility of sulfur in the electrolyte. The FeS electrode has a potential of 1.6 V versus lithium. Due to these modifications, the theoretical specific energy is reduced from 2600 Wh/kg for the Li/S battery to 458 Wh/kg for the LiAl/FeS battery, but the stability of cell operation increases and the cycle life also increases from 100 cycles to several hundred cycles (for example 700 cycles and 15.000 h).

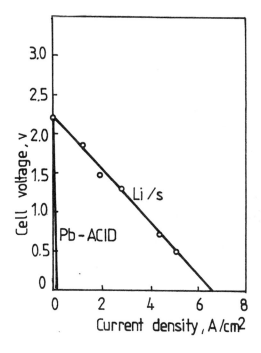

**Figure 36** Voltage vs. current density for Pb/PbO$_2$ and Li/S cells [365].

The overall cell reaction is

$$2\text{LiAl}_{(s)} + \text{FeS}_{(s)} \overset{\text{discharge}}{\underset{\text{charge}}{\rightleftharpoons}} \text{Li}_2\text{S}_{(s)} + \text{Fe}_{(s)} + \text{Al}_{(s)} \qquad (161)$$

The mechanism of the reaction is very complex and not completely understood [368–370]. These reactions take place both at the interface between FeS and the electrolyte and in the interior of the FeS granules. During battery discharge, the following reactions take place [371]:

Surface

$$6\text{LiAl} + 26\text{FeS} + 6\text{KCl} \rightarrow \text{LiK}_6\text{Fe}_{24}\text{S}_{26}\text{Cl} + 2\text{Fe} + 5\text{LiCl} + 6\text{Al} \qquad (162)$$

Interior

$$2\text{LiAl} + 2\text{FeS} \rightarrow \text{Li}_2\text{FeS}_2 + \text{Fe} + 2\text{Al} \qquad (163)$$

$$2\text{LiAl} + \text{Li}_2\text{FeS}_2 \rightarrow 2\text{Li}_2\text{S} + \text{Fe} + 2\text{Al} \qquad (164)$$

$$20\text{LiAl} + \text{LiK}_6\text{Fe}_{24}\text{S}_{26}\text{Cl} + 5\text{LiCl} \rightarrow 13\text{Li}_2\text{FeS}_2 \\ + 11\text{Fe} + 6\text{KCl} + 20\text{Al} \qquad (165)$$

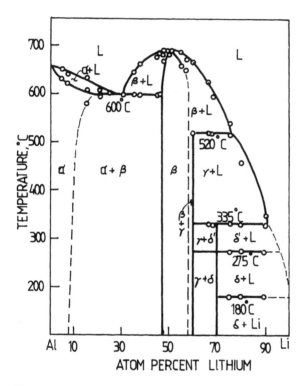

**Figure 37**   The lithium-aluminum phase diagram [367].


Surface

$$46LiAl + LiK_6Fe_{24}S_{26}Cl + 5LiCl \rightarrow 26Li_2S$$
$$+ 24Fe + 6KCl + 46Al \tag{166}$$

During charging the following reactions take place:

$$Fe + 2Al + 2Li_2S \rightarrow Li_2FeS_2 + 2LiAl \tag{167}$$

$$Fe + 2Al + Li_2FeS_2 \rightarrow 2FeS + 2LiAl \tag{168}$$

Equation (162) shows the formation of djerfischerite ($LiK_6Fe_{24}S_{26}Cl$) as the first step in the discharge process. Then this substance reacts to form $Li_2FeS_2$ [Eq. (165)] or $Li_2S$ [Eq. (166)], depending on the temperature. Due to the formation of the djerfischerite a swelling effect takes place in the FeS cathode during discharge. This effect can be avoided at high temperatures (500°C) and by the incorporation of $Cu_2S$ in the cathode. Likewise Eq. (162) shows that the djerfischerite is formed in the presence of KCl. The elimination of the potassium ion from the electrolyte by using an electrolyte like 22 mol% LiF–31 mol% LiCl–47 mol% LiBr (liquidus temperature = 430°C) is another way of avoiding the formation of djerfischerite.

Figure 38 shows discharge curves [372] for the LiAl/LiCl-KCl/FeS battery. It can be noticed that the discharge appears to take place at a single potential (1.6 V versus Li), yielding only one plateau in the voltage capacity plot.

A similar battery is

$$LiAl_{(s)}/LiCl\text{-}KCl/FeS_{2(s)} \tag{169}$$

which has a higher potential (emf = 1.76 V) and a theoretical energy density of 650 Wh/kg [371]. The overall cell reaction is

$$4LiAl_{(s)} + FeS_{2(s)} \underset{charge}{\overset{discharge}{\rightleftarrows}} 2Li_2S_{(s)} + 4Al_{(s)} + Fe_{(s)} \tag{170}$$

Again, the mechanism of the cell reactions of this battery is very complex and incompletely understood. The discharge reactions may be written as

$$3LiAl + 2FeS_2 \rightarrow Li_3Fe_2S_4 + 3Al \tag{171}$$

$$LiAl + Li_3Fe_2S_4 \rightarrow 2Li_2FeS_2 + Al \tag{172}$$

$$2LiAl + Li_2FeS_2 \rightarrow 2Li_2S + Fe + 2Al \tag{173}$$

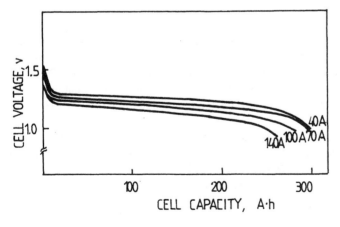

**Figure 38** Discharge curves for a LiAl/LiCl-KCl/FeS cell designed for high specific power [372].

Although the $FeS_2$-based cells have better performance than FeS cells, in practice these are not suited for storage systems because of thermal decomposition of $FeS_2$.

In the cell

$$Li_4Si/LiCl-KCl/FeS_2 \qquad (174)$$

the weight per equivalent of lithium stored is lower compared with the same weight in the $LiAl/FeS_2$ cell. The theoretical specific energy for the $Li_4Si/FeS_2$ cell is 944 Wh/kg [373–374], and the overall reaction is

$$Li_4Si + FeS_2 \rightarrow 2Li_2S + Fe + Si \qquad (175)$$

The $LiAl/FeS_2$ and $Li_4Si/FeS_2$ cells are in an earlier stage of development than the LiAl/FeS cell, because of corrosion problems at the positive electrode and problems with sulfur loss.

In recent years, batteries with bipolar electrodes, higher performance and lower material costs [375] are being developed. The higher performance was obtained by reducing the internal impedance. Thus the specific power and specific energy increase and the weight and volume of nonactive materials were reduced. For that purpose, several changes of electrolyte as well as the negative electrode were made.

During overcharge of a cell an instability was observed because of an increase of lithium activity in the negative electrode, of its dissolution in the electrolyte and therefore the self-discharge [375] increases. To avoid this phenomenon, approximately 10 mol% $Al_5Fe_2$ was added to the negative electrode. These electrodes were employed throughout 900 cycles without an appreciable capacity loss.

The use of other electrolyte compositions with a low melting point and higher electrical conductivity was also tried. The following electrolytes were tested: the eutectics LiCl-KCl, LiCl-LiBr-KBr and LiF-LiCl-LiBr and also a LiCl-rich composition 34 mol% LiCl–32.5 mol% LiBr–33.5 mol% KBr.

The LiF-LiCl-LiBr electrolyte has the highest conductivity, but due to the high liquidus the battery can work only at a temperature of 475°C, where corrosion limits the cycle life.

Use of the LiCl-LiBr-KBr eutectic allows the battery to function at 400–425°C. Also the use of LiCl-rich composition is advantageous because it can work at temperatures of 400–425°C and at the same time this composition has 25% higher electrical conductivity than that of the eutectic formed from the same salts [375], as in Figure 39.

For a very compact battery, the electrolyte is included in a "separator." This must be a porous medium, whose pores are filled with molten electrolyte. This separator must also be an electronic insulator. Till now, the best material

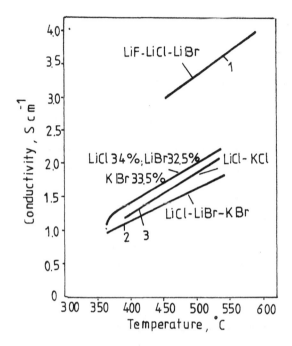

**Figure 39** Conductivities of molten salts for rechargeable lithium batteries: 1, 2, 3, eutectics.

selected for the separator is MgO. The MgO separator is fabricated as a cold-pressed plaque (at 280 MPa) from high surface area MgO and powder of electrolyte salt. Such batteries have been developed at Argonne National Laboratory [375]. They have a diameter of 17–18 cm and a thickness of 1 cm. The LiAl/FeS$_2$ battery has a specific energy of 152 Wh kg$^{-1}$, and the LiAl/FeS battery has a specific energy of 98 Wh kg$^{-1}$. Saft America Inc. [376,377] and Westinghouse Electric Corporation [378,379] have continued this research to obtain LiAl/FeS or LiAl/FeS$_2$ batteries which are capable of generating current pulses of high intensity for military applications.

The most interesting application of these batteries is found in electric vehicles. It was calculated that an electric vehicle with a battery of 200 Wh/kg, occupying 30% of the weight of the vehicle, can be expected to have an urban driving range of about 400 km between recharges. Other applications of these batteries are the storage of off-peak energy in electric utility systems, or electrical energy generated by solar- or wind-powered generators.

## IV.  LOW TEMPERATURE MOLTEN SALTS

### A.   Molten Inorganic Salts

1.   Structure and Chemical Equilibria in Molten
     Chloroaluminates

In the last few decades, molten salts which melt at low temperature have been
the subject of considerable interest, because they retain the advantages offered
by molten salts and at the same time make it possible to work at low temperatures.
The most important classes of these salts are haloaluminates, chlorocuprates,
chlorozincates, etc.

Haloaluminates have special properties from the molten salt point of view.
These salts form low melting liquids with good thermal stability, and they have
a wide range of acid-base properties as a function of composition. Likewise,
recent studies have shown the existence of some unusual dissolved species, and
this represents a vast potential field for the use of molten haloaluminates as solvents.

The study of molten haloaluminates was initiated because these melts are
interesting from a scientific point of view. Practically, these melts are used for
aluminum electroplating, as electrolytes in low temperatures batteries (see below)
and, likewise, aluminum production by electrolysis of these melts has been tried
(see Section II.A.7 ''Aluminum Chloride Electrolysis'').

The high sensibility of the low temperature molten salts toward moisture
and organic substances constitutes a great problem for the use of these salts on
a large scale.

Although the Al—X bond is predominantly ionic [380–383] (two-thirds
ionic and one-third covalent character) the pure aluminum halides do not form
ionic liquids upon melting. For example, the specific conductivity of molten
$AlCl_3$ [384] is approximately $10^{-7}$ S cm$^{-1}$ (see Section I.B ''Electrical Conductivity of Molten Salts''). This is due to the fact that, upon melting, $AlCl_3$ has a
strong tendency toward the formation of highly stable $Al_2Cl_6$ dimers. On melting,
the molar volume of $AlCl_3$ doubles, and its electrical conductivity strongly decreases [385,386].

On the other hand, the aluminum halides have a strong Lewis acid character, and because of this they can react strongly with the alkali halides and form
complex ions of the type $AlCl_4^-$. In this way the nonconducting liquid $Al_2Cl_6$
reacts with other halides, forming strongly ionic melts. According to the Lewis
definition [387] of acids and bases, an acid is defined as a substance that can
accept a pair of electrons to form a bond; a base is a substance that can donate
a pair of electrons to the formation of a covalent bond.

The vapor pressure of molten $AlCl_3$ is more than one atmosphere, but by
addition of NaCl it decreases strongly [388,389]. This fact suggests the formation

of some stable polymeric anions in the liquid such as $Al_2Cl_7^-$ in addition to the well-established monomer $AlCl_4^-$.

A common characteristic of $AlCl_3$-MCl mixtures (where M is an alkali metal) is the low melting point of the eutectic. The composition and melting points of eutectic mixtures on the $AlCl_3$-rich side of the phase diagram [390–392] are shown in Table 17.

The eutectic mixtures of the other aluminum halides with alkali halides also have low melting points (of about 100°C). The phase diagram of the $AlCl_3$-NaCl system is given in Figure 40 [390]. The phase diagrams of $AlCl_3$ mixtures with other alkali metal chlorides have similar shapes.

By changing the composition of these melts, their acid-base properties can be modified and hence the redox and the coordination chemistry in these media can be changed. The following species have been reported to exist in $AlCl_3$-NaCl melts [393–397]:

$$AlCl_3 \quad AlCl_4^- \quad Na^+$$
$$Al_2Cl_6 \quad Al_2Cl_7^- \quad Cl^-$$
$$Al_3Cl_{10}^-$$

The following set of thermodynamic equilibria were used to interrelate these species in the various melts [398–400]:

$$2AlCl_3 \rightleftharpoons Al_2Cl_6 \qquad K_0 = 2.9 \times 10^7 \qquad (176)$$

$$Cl^- + AlCl_3 \rightleftharpoons AlCl_4^- \qquad K_1 = 3.2 \times 10^{11} \qquad (177)$$

$$AlCl_4^- + AlCl_3 \rightleftharpoons Al_2Cl_7^- \qquad K_2 = 2.5 \times 10^4 \qquad (178)$$

$$Al_2Cl_7^- + AlCl_3 \rightleftharpoons Al_3Cl_{10}^- \qquad K_3 \le 10^3 \qquad (179)$$

$$2AlCl_4^- \rightleftharpoons Al_2Cl_7^- + Cl^- \qquad K_m = 1.06 \times 10^{-7} \qquad (180)$$

**Table 17**  Composition and Melting Points of the $AlCl_3$-MCl Eutectics (M = Li, Na, K, Cs)

| Eutectics | Composition (mole % $AlCl_3$) | Melting point (°C) |
|---|---|---|
| $AlCl_3$-LiCl | 57 | 110 |
| $AlCl_3$-NaCl | 60 | 113.2 |
| $AlCl_3$-KCl | 67 | 128 |
| $AlCl_3$-CsCl | 75 | 148 |

*Source*: From Refs. 390, 391 and 392.

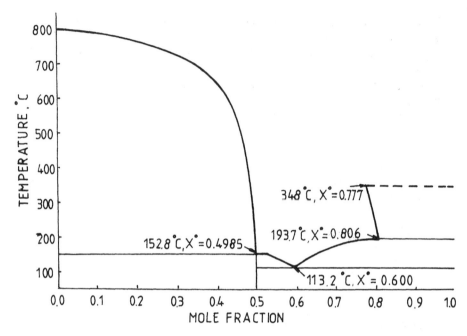

**Figure 40**  Phase diagram of the AlCl$_3$-NaCl binary system [390].

The values of the equilibrium constants are given at 175°C. Aluminum chloride in the vapor phase exists almost exclusively as the dimer Al$_2$Cl$_6$ at temperatures below 400°C [401].

The dominant acid-base equilibrium is described by Eq. (180), where Al$_2$Cl$_7^-$ is the Lewis acid, the Cl$^-$ is the base and AlCl$_4^-$ is neutral. A melt with the molar ratio AlCl$_3$/NaCl = 1/1 is neutral. With an excess of AlCl$_3$, the melt is acidic and with an excess of NaCl the melt is basic. In a neutral melt only reaction (180) takes place and it has an equilibrium constant of 1.06 × 10$^{-7}$. Therefore, in these melts there are only Na$^+$ and AlCl$_4^-$ ions present.

Reaction (180) is an autosolvolysis reaction, analogous to water autoprotolysis. The pCl$^-$ = $-$log[Cl$^-$] of a neutral melt is 3.5 at 175°C, on a mole fraction scale [400].

In acidic melts, the Al$_2$Cl$_7^-$ ions are predominant. In basic melts, pCl$^-$ is constant, in the presence of solid NaCl [401], and it is equal to 1.1 at 175°C. Likewise, in the molten chloroaluminate systems the terms "acid" and "base" denote a chloride ion acceptor and a chloride ion donor, respectively. The pCl$^-$ may be measured with either an aluminum or a chlorine electrode immersed in the melt.

**Table 18** Thermodynamic Constants for AlCl$_3$-NaCl Equilibria, 175–355°C

| Reaction (liquid states) | $\Delta H°$ (kcal mole$^{-1}$) | $\Delta S°$ (cal mole$^{-1}$ deg$^{-1}$) | $\Delta G°$ at 200°C (kcal mole$^{-1}$) |
|---|---|---|---|
| 2AlCl$_3$ = Al$_2$Cl$_6$ | −17.66 | −5.2 | −15.2 |
| NaAlCl$_4$ + AlCl$_3$ = NaAl$_2$Cl$_7$ | −13.00 | −8.9 | −8.8 |
| 2NaAlCl$_4$ = NaCl + NaAl$_2$Cl$_7$ | 12.54 | −4.1 | 14.5 |
| NaCl + AlCl$_3$ = NaAlCl$_4$ | −25.54 | −4.8 | −23.3 |

*Source*: From Ref. 399.

In the following, the apparent mole fraction of AlCl$_3$ will be denoted by N instead of the usual $x$ to emphasize that there is no free AlCl$_3$ in the melt.

The variation of enthalpy and entropy for the reactions which take place in AlCl$_3$-NaCl melts are shown in Table 18 [399]. The concentration of the species implicated in reactions (176)–(180) varies as a function of the mole fraction of the components.

The variation in concentration of the species present in AlCl$_3$-NaCl melts, at around the mole fraction N(AlCl$_3$) = 0.5, is given in Figure 41 [399]. It was noticed that in acidic AlCl$_3$-NaCl melts (rich in AlCl$_3$) some unusual ionic species

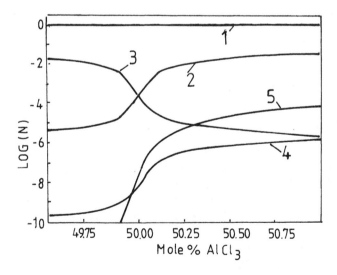

**Figure 41** Mole fractions of the various species [399] in AlCl$_3$-NaCl melts at 175°C as a function of the net AlCl$_3$/NaCl ratio at around 1/1 : 1, NaAlCl$_4$; 2, NaAl$_2$Cl$_7$; 3, NaCl; 4, AlCl$_3$; 5, Al$_2$Cl$_6$.

are stable in a low oxidation state. Likewise in these melts, metallic clusters of refractory metals such as niobium, tantalum and tungsten [402] are stable.

By addition of $ZrCl_2$ or $Na_2ZrCl_6$ to weakly acidic $AlCl_3$-NaCl melts (51–52 mole% $AlCl_3$), the formation of insoluble $ZrCl_3$ [403] was observed. In more acidic melts (60 mole% $AlCl_3$) at temperatures higher than 250°C, Gilbert et al. [403] demonstrated the formation of Zr(II) ions and of soluble Zr(III) at temperatures lower than 140°C. In acidic melts (63 mole% $AlCl_3$) the formation of the cluster $Ta_6Cl_{12}^{2+}$ was observed. It is slightly soluble in these melts and insoluble in a NaCl saturated melt [404].

The electrochemistry of sulfur has been studied by several authors [405–408]. In basic melts, coulometric and voltammetric measurements have shown the presence of $S_2Cl_2$ at 175°C [405,407]. In acidic melts (63 mole% $AlCl_3$), $SCl_3AlCl_4$ is obtained by the electrooxidation of sulfur [408]. The reduction of $SCl_3^+$ to S is obtained by passing through the intermediate species [408] $S_2^{2+}$ and $S_8^{2+}$.

By spectrophotometric and potentiometric measurements in $AlCl_3$-NaCl melts, Bjerrum [409] found the following ions and species: S(IV), S(II), $S_2^{2+}$, $S_8^{2+}$ and $S_{16}^{2+}$. In the case of selenium, in acidic chloroaluminate melts, the same author [409] discovered the existence of six different oxidation states: Se(IV), $Se_2^{2+}$, $Se_4^{2+}$, $Se_8^{2+}$, $Se_{12}^{2+}$, $Se_{16}^{2+}$. Likewise, some unusual cationic species were found, such as $Bi^+$, $Bi_5^{3+}$, $Bi_8^{2+}$, between which there are the following equilibria [410,411]:

$$6Bi^+ \rightleftharpoons Bi_5^{3+} + Bi^{3+} \tag{181}$$

$$2Bi + Bi^{3+} \rightleftharpoons 3Bi^+ \tag{182}$$

$$4Bi + Bi^{3+} \rightleftharpoons Bi_5^{3+} \tag{183}$$

$$22Bi + 2Bi^{3+} \rightleftharpoons 3Bi_8^{2+} \tag{184}$$

## 2.  Electrochemical Studies in Molten Chloroaluminates

The electrical conductivity of these melts was measured by Abraham and Elliot [412], who found for $NaAlCl_4$ an electrical conductivity of 0.46 ohm$^{-1}$ cm$^{-1}$ at 170°C. Howie and MacMillan [413] measured the electrical conductivity for the binary molten salt system $AlCl_3$-NaCl in the composition range 15–30 wt% NaCl and temperature range 155–195°C, and found the following equation:

$$\kappa = (-0.1594 + 0.207 \times 10^{-2}T)$$
$$- (-0.1475 \times 10^{-1} + 0.143 \times 10^{-3}T)W \tag{185}$$
$$+ (-0.4022 \times 10^{-3} + 0.548 \times 10^{-5}T)W^2 \text{ ohm}^{-1} \text{ cm}^{-1}$$

where $W$ is wt% NaCl.

One of the most frequently used electrodes for electrochemical studies in these melts is the aluminum electrode, because it is intended for use as the electrode for high capacity batteries at low temperatures. Holleck and Giner [414] showed that during anodic polarization of the aluminum electrode in a AlCl$_3$/KCl/NaCl melt, with the molar ratio 67/13.6/19.4, electrode passivation occurred. This anodic passivation did not occur at a constant potential, but when a certain quantity of charge was passed through the electrode. This behavior seems to show that in this melt, when a certain quantity of AlCl$_3$ is obtained by anodic polarization, AlCl$_3$ is precipitated and forms a solid passivating layer at the anode surface.

During cathodic polarization, the same authors [414] observed the formation of dendrites. The critical current, at which the formation of dendrites begins, is dependent on bath composition and temperature. The authors deduced that concentration polarization and ionic conductivity had a strong effect on dendrite formation. The exchange current density is 270 mA cm$^{-2}$ at 130°C. Generally, the aluminum electrode is highly reversible and able to support large current densities at low activation overvoltage. The polarization observed is to a large extent ohmic in nature and depends on melt composition and temperature.

Del Duca [415] studied the kinetics for dissolution and deposition of solid aluminum in molten AlCl$_3$-NaCl and AlCl$_3$-(LiCl-KCl)$_{eut}$ electrolytes over the temperature range of 175–310°C. It was noticed that the exchange current densities for dissolution varied from 2 to 20 mA cm$^{-2}$ as the concentration of AlCl$_3$ in the KCl-LiCl eutectic increased from 50 to 75 mole%. The exchange current densities for deposition varied from 10 to 40 mA cm$^{-2}$ over the same range of composition. The rate-determining steps of the reactions which take place at dissolution or deposition of aluminum are given in Table 19 [415].

Gale and Osteryoung [416] studied anodic and cathodic polarization of an aluminum electrode in the nominal 1:1 and 1:2 molar NaCl-AlCl$_3$ molten electrolytes and found that these processes were Faradaic three-electron-step re-

**Table 19** The Rate-Determining Steps for the Charge Transfer Reaction on an Aluminum Electrode, $t = 175–300$°C

| Electrolyte (mole %) | Anodic rate-determining step | Cathodic rate-determining step |
|---|---|---|
| 50AlCl$_3$-50(LiCl.KCl)$_{eut}$ | Al$^0 \to$ Al$^+$ + e$^-$ | — |
| 67AlCl$_3$-33(LiCl.KCl)$_{eut}$ | Al$^0 \to$ Al$^+$ + e$^-$ | Al$^{3+}$ + e$^- \to$ Al$^{2+}$ |
| 75AlCl$_3$-25(LiCl.KCl)$_{eut}$ | Al$^0 \to$ Al$^+$ + e$^-$ | Al$^{3+}$ + e$^- \to$ Al$^{2+}$ |
| 50AlCl$_3$-50NaCl | Al$^0 \to$ Al$^+$ + e$^-$ | Al$^{2+}$ + e$^- \to$ Al$^+$ |

*Source*: From Ref. 415.

actions. The authors [416] did not find subvalent ions to be involved in these electrochemical phenomena.

Torsi and Mamantov's works [417] constitute an example of a study of the acid-base equilibrium in $AlCl_3$-NaCl melts by means of an electrochemical cell. The authors used the following cell:

$$Al/Al^{3+}(AlCl_3\text{-}NaCl_{sat})/Pyrex/Al^{3+}(AlCl_3\text{-}NaCl_a)/Al \tag{186}$$

where $a$ refers to a known composition of the melt. For the 50:50 mole % composition of $AlCl_3$-NaCl the principal anion is $AlCl_4^-$ and the electrode reaction of interest is

$$AlCl_4^- + 3e^- = Al + 4Cl^- \tag{187}$$

Therefore, the potential ($\Delta E$) of this concentration cell is

$$\Delta E = \frac{4RT}{3F} \ln \frac{(x_{Cl^-})_{sat}}{(x_{Cl^-})_a} + \frac{RT}{3F} \ln \frac{(x_{AlCl_4^-})_a}{(x_{AlCl_4^-})_{sat}} \tag{188}$$

where $x_{Cl^-}$ and $x_{AlCl_4^-}$ are the mole fractions of $Cl^-$ and $AlCl_4^-$ in the respective melts. In acidic melts the electrode reaction can be written

$$4Al_2Cl_7^- + 3e^- = Al + 7AlCl_4^- \tag{189}$$

and the potential of the cell is

$$\Delta E = \frac{4RT}{3F} \ln \frac{(x_{Al_2Cl_7^-})_a}{(x_{Al_2Cl_7^-})_{sat}} + \frac{7RT}{3F} \ln \frac{(x_{AlCl_4^-})_{sat}}{(x_{AlCl_4^-})_a} \tag{190}$$

On the basis of the experimental data, Torsi and Mamantov [417] deduced that in acidic melts ($>$55 mol% $AlCl_3$) reaction (178) occurred along with the reaction

$$2Al_2Cl_7^- = Al_3Cl_{10}^- + AlCl_4^- \tag{191}$$

On the basis of potentiometric data, obtained by means of the cell (186), the authors [417] calculated the equilibrium constant $K$ of reaction (180) and found the values of p$K$ ($-\log K$) given in Table 20 as a function of temperature and electrolyte composition.

## 3. Batteries

One of the most important possible applications of these salts is for the use in high energy density batteries for load leveling and in electric vehicles. Molten chloroaluminates offer many advantages such as low liquidus temperatures and high electrical conductivity [418].

Aluminum is used as the negative electrode because it is quite electronegative, relatively inexpensive and has a low equivalent weight. Problems due to

**Table 20** Variation of p$K$ for Reaction
$2AlCl_4^- = Al_2Cl_7^- + Cl^-$ with Temperature
for the $AlCl_3$-MCl Systems

| Temp. (°C) | p$K$ | | | |
|---|---|---|---|---|
| | $Li^+$ | $Na^+$ | $K^+$ | $Cs^+$ |
| 175 | 4.3 | 7.1 | | |
| 250 | 4.0 | 6.3 | | |
| 300 | 3.9 | 5.7 | 7.1 | |
| 350 | 3.8 | 5.3 | 6.4 | |
| 400 | 3.8 | 5.0 | 5.8 | 7.4 |
| 450 | | | | 6.8 |

*Source*: From Ref. 417.

the formation of dendrites during charging and of the formation of a nonconducting layer of $Al_2Cl_6$ at high current densities, (during discharge) can be avoided by using some additions.

Many types of batteries have been proposed, in which the electrolyte is the molten system $AlCl_3$-NaCl. Thus, the following are known: $Al/Cl_2$ [419], $Na/Cl_2$ [420], Al/S [421], $Al/MCl_x$ and $Na/MCl_x$ [422].

Mamantov et al. [423] obtained a rechargeable cell Na/$Na^+$ ionic conductor/$SCl_3^+$ in molten $AlCl_3$-NaCl, which works in the temperature range 180–250°C and has an open circuit voltage of 4.2 V. the discharge process involves the reduction of tetravalent sulfur to the elemental state. The sodium ion conductor is $\beta''$-$Al_2O_3$. The ceramic $\beta$-$Al_2O_3$ serves a dual role in all types of liquid sodium cells acting both as an ionic electrolyte and as a separator between the liquid constituents of the electrodes. The stability of the $\beta$-$Al_2O_3$ is a crucial factor in determining the long-term performance of such cells. High energy density values and good energy efficiency have been demonstrated for this cell. The number of deep charge/discharge cycles has exceeded 400.

The reactions describing the discharge and charge of the positive electrode may be written as

$$SCl_4 + 6e^- + 6Na^+ + Al_2Cl_7^- \underset{\text{charge}}{\overset{\text{discharge}}{\rightleftharpoons}} AlSCl + 6NaCl + AlCl_4^- \quad (192)$$

Such a cell has been in continuous operation for more than nine months. The lifetime is limited because of cracks which tend to appear in the $\beta$-alumina.

Abraham and Elliot [412] used a battery of the type

$$Na_{(l)}/\beta''\text{-}Al_2O_3/NaAlCl_{4(l)}/NiS_x \quad (193)$$

The battery operates in the temperature range 170–190°C. A Na/NiS$_2$ cell, operating at 190°C, exceeded 600 deep discharge/charge cycles with practically no capacity deterioration. The overall electrode process in the cycling of the NiS$_2$ cathode is

$$NiS_2 + 4Na^+ + 4e^- + 2NaAlCl_4 \underset{charge}{\overset{discharge}{\rightleftharpoons}} Ni + 2NaAlSCl_2 + 4NaCl \quad (194)$$

A new electrolyte composition for this type of battery was found by Hjuler et al. [424]. The electrolyte has the composition LiAlCl$_4$-NaAlCl$_4$-NaAlBr$_4$-KAlCl$_4$ in the molar proportions (3:2:3:2) and the melting point is 86°C. The electrical conductivity shows linear additivity of the conductivities of the four individual salts and is 0.142 S cm$^{-1}$ at 100°C. This low melting electrolyte was employed in the rechargeable battery system

$$Al/electrolyte/Ni_2S_3 \quad (195)$$

at 100°C. The open circuit voltage of this battery is 0.82–1.0 V.

Another type of battery is the so-called Zebra cell, obtained for the first time in South Africa by Coetzer [425] at Zebra Power Systems (Pty) Limited. The development of this battery is being actively pursued in the United Kingdom for high energy density applications such as electric vehicles, load leveling and spacecraft. This type of battery has sodium as the negative electrode, and the positive electrode is made from Fe/FeCl$_2$ or Ni/NiCl$_2$.

The battery scheme is shown in Figure 42, and it has the configuration

$$Na_{(l)}/\beta-Al_2O_3/NaAlCl_{4(l)}/FeCl_2(Fe) \quad (196)$$

The overall reaction is

$$FeCl_2 + 2Na \underset{charge}{\overset{discharge}{\rightleftharpoons}} 2NaCl + Fe \quad (197)$$

The open circuit voltage is 2.35 V at 250°C. In the charged state the cathode consists of a porous Fe matrix partially chlorinated to FeCl$_2$. This matrix is impregnated with NaAlCl$_4$. During discharge, Na$^+$ ions pass through the $\beta$-Al$_2$O$_3$ and NaAlCl$_4$, and FeCl$_2$ is reduced to Fe, forming NaCl in the process. This type of positive electrode is different from other electrodes, because Fe, FeCl$_2$ and NaCl are insoluble in NaAlCl$_4$, and therefore they must be in physical contact and at the same time in contact with the current collector. To achieve this, a basic electrolyte with NaCl/AlCl$_3$ >1 is used. For a Fe matrix porosity of 85%, a specific energy value of 150–170 Wh/kg was obtained [425]. Metallic nickel is added to the iron cathode matrix to avoid the formation of Fe$^{3+}$ ions during overcharge. Thus, in the case of some overcharge, the Ni metal is chlorinated to Ni$^{2+}$ in preference to the iron being oxidized to Fe$^{3+}$. In this way, positive electrodes with capacities up to 100 Ah were obtained. The electrochemical reversibility of

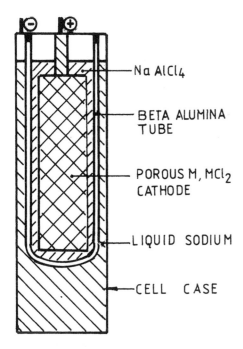

**Figure 42** Schematic of a ZEBRA cell [425].

the cell is quite good; one such cell achieved >1000 cycles, retaining 60% of its initial capacity [426]. The performance of a small 8 Ah cell at 250°C is shown in Figure 43.

Another similar battery was proposed by Galloway [427], with the configuration

$$Na_{(l)}/\beta\text{-}Al_2O_3/NaAlCl_{4(l)}/NiCl_2 \tag{198}$$

for which the cell reaction can be written as

$$2Na + NiCl_2 \underset{\text{charge}}{\overset{\text{discharge}}{\rightleftharpoons}} Ni + 2NaCl \tag{199}$$

Good results were obtained in terms of low resistance and high rates of discharge for cells up to 20 Ah in capacity. Cells were operated over the temperature range 230–400°C, where the open circuit voltage varied from 2.60 to 2.56 V.

In the lifetime of this cell a decrease of the capacity was observed due to the formation and increase of some large Ni crystals, which considerably reduced the surface area of Ni and eliminated a part of the fine porosity. This loss of

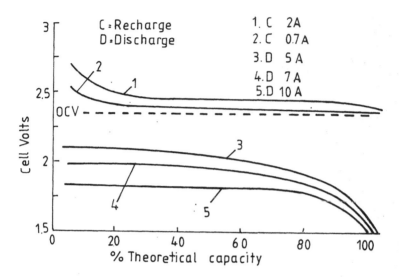

**Figure 43**   Charge-discharge cycles of a 8 Ah (9.4.1.22) cell at 250°C [425].

capacity was prevented [428] by the addition of 5 wt% elemental sulfur to the
NaAlCl$_4$. In this way there is obtained a modified high surface area and an active
metallic phase which is stable for many hundreds of cycles. Cells using these
materials have achieved a high degree of reversibility and long cell life (>2000
cycles), operating over the temperature range 200–350°C. The stability of β-
Al$_2$O$_3$ can limit the cell life.

Yao and Kummer [429] showed that the sodium ions in β-Al$_2$O$_3$ could be
replaced by other ions, especially monovalent ions. This exchange of ions is
determined by the ratio of ion activities in the melt and the diffusion coefficients
of these ions. The properties of the ceramic material (especially the electrical
resistance) and its lifetime are altered when replacing the sodium ions by other
ions. The Fe$^{2+}$ ion has an ionic radius which is close to that of Na$^+$, and therefore
the sodium is easily replaced by iron in β-Al$_2$O$_3$, especially in the presence of
an electric field when current is passing [430]. The electrical resistance of the
iron-doped regions of the ceramic is very high. It was also noticed that β″-Al$_2$O$_3$
is not affected by Fe$^{2+}$ ions. Chloroaluminate melts also damage β-Al$_2$O$_3$ irreversibly in the absence of an electric field [430]. Ni$^{2+}$ ions do not exchange significantly with Na$^+$ ions [430] at temperatures up to 400°C.

The variation of the electrical conductivity of β-Al$_2$O$_3$ with temperature
[431] is shown in Figure 44 along with the same variation for NaAlCl$_4$. It is seen
from Figure 44 that the electrical conductivity of the ceramic material suddenly

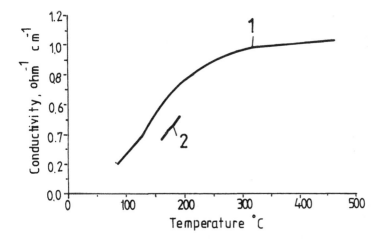

**Figure 44**   Variation of the electrical conductivities of β″-Al₂O₃ (1) and NaAlCl₄ (2) with temperature [431].

decreases as the temperature gets lower than 200°C, a fact which restricts the use of these batteries to temperatures above 250°C.

Cleaver et al. [432] measured the thermal effect of a Na/NiCl₂ cell with a capacity of 40 Ah and found that a temperature increase of 30–35° occurred at complete discharge, depending on the starting temperature; the effect is reversed upon charging.

## B.   Room Temperature Molten Salts

The high melting points of molten salts prevent their use as reaction media for many practical and fundamental applications. For example, many organic substances have low boiling points and a low thermal stability and cannot be dissolved in molten salts.

In 1948, Frank Hurley and Thomas Wier, Jr. [433–436] obtained the first haloaluminate molten salts which melted at room temperature. The authors intended to use these salts as electroplating baths. These first room temperature melts were obtained by combining aluminum chloride with certain organic halide salts. Today, the most useful room temperature molten salts consist of mixtures of aluminum chloride with 1-(1-butyl)pyridinium chloride (BPC) or 1-methyl-3-ethylimidazolium chloride (MEIC). The structures of BPC and MEIC are given

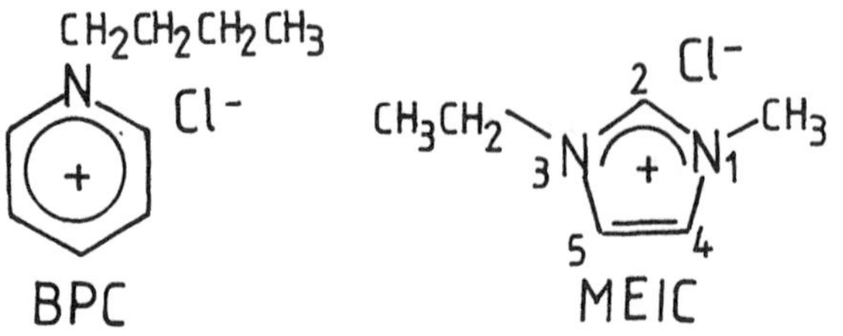

$CH_2CH_2CH_2CH_3$

BPC                          MEIC

**Figure 45** Structure of 1-(1-butyl) pyridinium chloride (BPC) and 1-methyl-3 ethyl-imidazolium chloride (MEIC).

in Figure 45. MEIC is easier to synthesize and purify than BPC, and therefore it is more in use.

Phase diagrams of the MEIC-AlCl$_3$, BPC-AlCl$_3$ and NaCl-AlCl$_3$ (partially) systems are shown in Figure 46 [437,438]. It is seen from these phase diagrams that over a large range of compositions, the MEIC-AlCl$_3$ system has melting points much lower than room temperature.

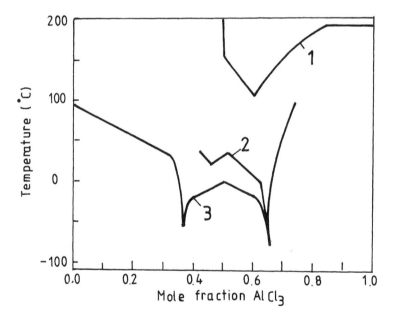

**Figure 46** Phase diagrams [437,438] for (1) NaCl-AlCl$_3$ (partial), (2) BPC-AlCl$_3$ and (3) MEIC-AlCl$_3$.

## 1. Acid-Base Equilibria

As in the case of the $AlCl_3$ system, the acidity of these molten salts can be dramaticaly modified by varying their compositions. The mixtures which include aluminum chloride in excess are acidic because this compound is a Lewis acid. The melt is considered basic at an excess of the organic salt component. In these melts the equilibrium reactions (176)–(180) are also valid, with only one difference, i.e., that the $Na^+$ ion is replaced by an organic cation.

In usual molten salts (for example the NaCl-KCl mixture) a change of composition changes the concentration of ions, whereas in $AlCl_3$-MCl mixtures (where M is an alkali metal or an organic cation) by change of composition, the acid-base properties as well as the speciation and the concentration of each anionic species are modified. The concentration variation of anionic species from the BPC-$AlCl_3$ or MEIC-$AlCl_3$ melts as a function of the melt composition is shown in Figure 47 [439]. Also in these melts, the major equilibrium describing acid-base properties is the dissociation of the tetrachloroaluminate anion [reaction (180)]. At high concentrations of $AlCl_3$, the $Al_2Cl_6$ dimer appears in the melt. Melts with $N(AlCl_3) > 0.8$ are not studied because they have high melting points and high vapor pressures.

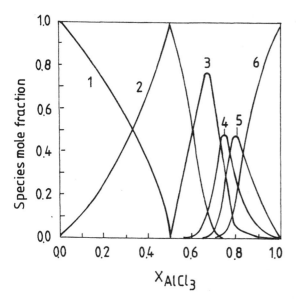

**Figure 47** Species of the melt of BPC-$AlCl_3$ or MEIC-$AlCl_3$, according to Øye et al. [439]: 1, $Cl^-$; 2, $AlCl_4^-$; 3, $Al_2Cl_7^-$; 4, $Al_3Cl_{10}^-$; 5, $Al_4Cl_{13}^-$; 6, $Al_2Cl_6$; $t = 40°C$.

**Table 21**  Values of the
Tetrachloroaluminate Ion Dissociation
Constants for the Reaction
$2AlCl_4^- \rightleftharpoons Al_2Cl_7^- + Cl^-$

| Melt composition | $\log K$ |
|---|---|
| 0.9:1 AlCl₃-BPC | −16.9 |
| 0.9:1 AlCl₃-MEIC | −16.6 |

*Source*: From Ref. 441.

The equilibrium constant of this reaction was calculated by many authors [440–442] and is given in Table 21.

Figure 48 shows the variation of the concentration of the respective species in reaction (180) in the AlCl₃-BPC system. The variation of the concentrations of the Cl⁻ and Al₂Cl₇⁻ ions is similar to the proton-hydroxyl equilibrium in aqueous solutions.

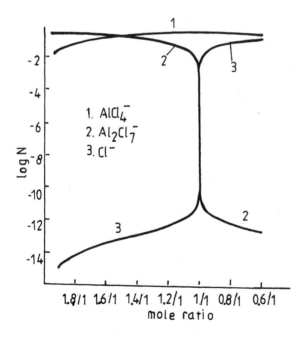

**Figure 48**  Mole fraction, N, of the major species in the melt as a function of the net AlCl₃-BPC molar ratio [440]; $t = 30°C$.

## 2. Electrochemical Properties of Room Temperature Molten Salts

*Transport Numbers*

It was assumed (without experimental data) that in melts of the $AlCl_3$-NaCl type, the alkali metal cation transports virtually all the charge passed through such a melt [443,444]. The transport numbers (see Section I.B.1 "Transport Numbers and Ionic Mobilities in Molten Salts") of anions and cations in the $AlCl_3$-MEIC melts were measured by many authors [445–447]. Although the $MEI^+$ cation is significantly larger than the $AlCl_4^-$ ion in these melts, the internal transport number of the $MEI^+$ relative to $Cl^-$ as reference is 1.00. This transport number does not depend on the composition of the melt [445–447]. The aluminum transport number is zero, a fact which shows that aluminum and chlorine are bonded together in the same chemical species, and the single chloride ions carry no charge. The external transport numbers of $MEI^+$ were found also to be independent of the composition of the melt, having an average value of 0.71. The external transport numbers of the anions depend on the composition of the melt, and the following relationships [447] exist:

$$t(AlCl_4^-) = 0.29x(AlCl_4^-);$$
$$t(Al_2Cl_7^-) = 0.29x(Al_2Cl_7^-) \quad t(Cl^-) = 0.29x(Cl^-)$$

where $x$ is the ionic fraction of the ion.

The mobilities of $AlCl_4^-$ and $Cl^-$ ions are identical in basic melts, whereas the mobilities of $AlCl_4^-$ and $Al_2Cl_7^-$ ions are identical in acidic melts. The concentration of these ions as a function of melt composition [440] is shown in Figure 48. It might be expected that the ion mobilities vary in the order $Cl^- > AlCl_4^- > Al_2Cl_7^-$, due to the different sizes. However, the fact that the ion mobilities are identical, irrespective of size of the ions, can be explained by ionic interactions in these melts. For example, the strong ionic interactions between the organic cation and the more polarizable $Cl^-$ ion in the basic melt may counteract the size advantage of this ion, while the comparatively weaker ionic interactions between the organic cation and the $AlCl_4^-$ ion may result in greater relative mobility of the latter.

*The electrical conductivity* of $AlCl_3$-MEIC melts was measured by Wilkes et al. [448] and is given in Table 22.

*Electrochemical Windows*

One of the most important properties of all solvents used for electrochemical studies is the available potential range or the "electrochemical window." The electrochemical window is the range of potentials across which it is possible to observe an electrochemical reaction of a substance dissolved in a solvent. Each

**Table 22**  Specific Conductivity of
AlCl₃-MEIC Melts

| Compn. mol% AlCl₃ | $\kappa$ ohm$^{-1}$ cm$^{-1}$ | Temp. °C |
|---|---|---|
| 44 | 0.0117 | 29.7 |
| 50 | 0.0227 | 30.9 |
| 67 | 0.0154 | 32.2 |

*Source*: From Ref. 448.

solvent has a potential range in which oxidation and reduction reactions of component ions occur.

In basic melts, at the positive potential limit the oxidation of halogen ions occurs [449]:

$$Cl^- = \frac{1}{2} Cl_2 + e^- \tag{200}$$

The positive limit decreases with increasing basicity because the chloride ion concentration increases. At the negative potential limit of basic melts the reduction of the organic cation occurs.

In acidic melts, at the positive potential limit [449], oxidation of AlCl₄⁻ ions takes place:

$$4AlCl_4^- = 2Al_2Cl_7^- + Cl_2 + 2e^- \tag{201}$$

At the negative potential limit of acidic melts [449] aluminum deposition takes place:

$$4Al_2Cl_7^- + 3e^- = Al + 7AlCl_4^- \tag{202}$$

Lipsztajn and Osteryoung [450] found that melts which contain exactly 50 mole% AlCl₃ are neutral and possess an electrochemical window that corresponds to the positive limit of an acidic melt and the negative limit of a basic melt. As Figure 49 shows, in the neutral AlCl₃-MEIC melt the potential window is about 4.4 V [451]. In a neutral AlCl₃-BPC melt, the electrochemical window is about 3.6 V [450].

Melton et al. [452] noticed that if NaCl is added to acidic AlCl₃-MEIC melts, it acts as a Lewis acid-base buffer. In these melts, NaCl reacts with Al₂Cl₇⁻ and a neutral salt results, as follows:

$$NaCl_{(s)} + Al_2Cl_7^- = 2AlCl_4^- + Na^+ \tag{203}$$

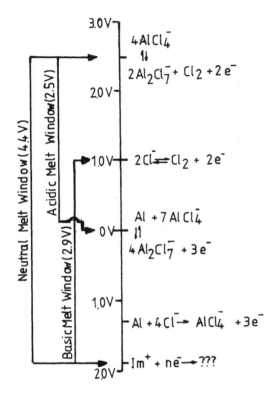

**Figure 49** Electrochemical windows for AlCl₃-MEIC molten salts [451].

On the other hand, if MEIC is added in excess to a melt buffered with NaCl it reacts with the $Na^+$ ions:

$$Na^+ + MEICl_{(s)} \rightarrow NaCl_{(s)} + MEI^+ \qquad (204)$$

Melton et al. [452] found that the electrochemical properties of buffered melts remained identical to those of neutral melts. The buffered melts maintained their neutrality, as well as the electrochemical window, even if changes were induced in the chloroacidity by the addition or electrogeneration of a Lewis acid or base.

The very wide electrochemical window of these melts appears very promising from the point of view of use as a solvent for new high energy density batteries. Although the electrochemical window of these melts is wide, it was tried in several ways to increase it. As shown above, the anodic decomposition potential is dependent on the chloroaluminate ion. But, if in the imidazole ring some electron-donating substituents are used, it should result in a more negative reduction potential for the imidazolium cation. Thus, Gifford and Palmisano [453] showed that by the use of 1,2-dimethyl-3 propylimidazolium chloride in a mixture with $AlCl_3$ at a 1:1 mole ratio, an electrochemical window of nearly 5 V was obtained at glassy carbon. A value for the acid-base equilibrium constant of $2.0 \times 10^{-15}$ was determined in these melts.

Also other electron-donating substituents can be used. Thus Reichel and Wilkes [454] observed that if 1 eq HCl was added to a mixture of 49:51 $AlCl_3$: 1 ethyl-3-methylimidazolium (EtMeim), an expansion of the electrochemical window was obtained. The mechanism of the HCl effect is unknown; however, it is supposed that in these melts there is the equilibrium

$$H_2Cl_3^- + Cl^- = 2HCl_2^- \qquad (205)$$

It is known that $AlCl_3$ is very unstable in air due to its reactivity with moisture. A way to improve the properties of these salts is to replace $AlCl_3$ with other substances. Thus, Wilkes and Zaworotko [455] prepared the following substances, which are stable toward air and water: $(EtMeim)NO_3$, m.p. $= 38°C$; $(EtMeim)NO_2$, m.p. $= 55°C$; $(EtMeim)BF_4$, m.p. $= 15°C$; $(EtMeim)MeCO_2$, m.p. $= -45°C$. These salts are hygroscopic but stable.

Koch et al. [456] prepared the following melts: 1,2-dimethyl-3-propylimidazoliumX, where $X = AsF_6^-$, $PF_6^-$, $(CF_3SO_2)_2N^-$ and $(CF_3SO_2)_3C^-$. Also these salts have low melting points, are stable in air and have an electrochemical window of about 5 V.

Suarez et al. [457] prepared and investigated the electrochemical window for the following ionic liquids: 1-*n*-butyl-3-methylimidazolium tetrafluoroborate $(BMI^+)(BF_4^-)$ and 1-butyl-3-methylimidazolium hexafluorophosphate $(BMI^+)(PF_6^-)$ on tungsten and vitreous carbon. They found the following values for the electrochemical window: 6.1 V and 5.45 V for $(BMI^+)(BF_4^-)$ and 7.1 V and 6.35 V for $(BMI^+)(PF_6^-)$, respectively.

These room temperature molten salts are promising electrolytes for industrial aluminum electroplating. There are many studies in the literature of this subject [458–461]. In acidic melts, aluminum deposition occurs according to reaction (202). In basic melts, the concentration of the $Al_2Cl_7^-$ ion is very low, and the $AlCl_4^-$ ions in these melts are not being reduced, as Carlin and Osteryoung [462] showed. Therefore, aluminum deposition is possible only in acidic melts. An underpotential deposition of aluminum [458,463] was observed on tungsten and platinum electrodes. The deposition/stripping efficiency was 100%, indicating the formation of a stable aluminum deposit [458,459]. The salient feature of the aluminum deposition is a nucleation phenomenon with diffusion-controlled growth of three-dimensional centers [458,460,461,463]. Carlin et al. [464] obtained a continuous layer of aluminum on a tungsten electrode from weakly acidic salts: $AlCl_3$/MEIC = 1/1. The average size of the nuclei formed was <0.5 μm, when the overpotential was more negative than $-0.2$ V. Deposits of high quality were observed only when the concentration of $Al_2Cl_7^-$ ions in the melt was low. In more acidic melts, i.e., at higher concentrations of $Al_2Cl_7^-$ ions, aluminum is deposited as large crystals with the formation of dendrites.

Quing Liao et al. [465] investigated aluminum deposition of high quality on copper substrates in $AlCl_3$/MEIC = 60/40 melts. The authors found that the quality of the electrodeposit was greatly enhanced by the addition of benzene as a "cosolvent." This improved the properties of the electrolyte by an increase in electrical conductivity and a decrease in viscosity.

Aluminum anodization in basic $AlCl_3$-MEIC melts was studied by Carlin and Osteryoung [462], and two different anodization processes were observed. The first step occurred in the catholic region, at a potential of $-1.1$ V, versus the aluminum electrode, and it was controlled by diffusion of chloride to the electrode surface. The authors found that the number of chlorides required to produce one $AlCl_4^-$ anion for each Al being anodized was 4. The second anodization which occurs on the anodic side of 0 V was not diffusion limited. It has not been possible to reduce the $AlCl_4^-$ anion in basic melts.

## 3. Batteries

The urgent need for new power sources for electric vehicles as well as consumer electronics has stimulated tremendous efforts in the search for new secondary batteries. These must have low cost, high energy density (>200 Wh/kg), long cycle life and an acceptable temperature range. The alkali metals are attractive negative electrodes for these batteries. This is because the alkali metals provide a high energy storage density due to their low equivalent weight and high thermodynamic reduction potentials, as demonstrated in Table 23 [451].

**Table 23** Comparison of Electricity Storage Densities of Active Metal Anodes

| Active metal | Equivalent weight (g faraday$^{-1}$) | Electrical storage density (Ah kg$^{-1}$) |
|---|---|---|
| Al | 8.99 | 2982 |
| Li | 6.94 | 3862 |
| Na | 23.0 | 1165 |
| K | 39.0 | 685 |
| Mg | 12.2 | 2206 |
| Zn | 32.7 | 820 |
| Cd | 56.2 | 479 |

Equivalent weight = (formula weight)/$n$ in the reaction $M = M^{n+} + ne^-$.
*Source*: From Ref. 451.

To use these metals, they must have a deposition potential inside the electrochemical window of the electrolyte used. Aluminum cannot be deposited in basic melts, but it is reversibly deposited and stripped in acidic electrolytes, and it has excellent cycling efficiencies [459]. Lithium appears to be the most attractive anode material, and aluminum is the second choice. Likewise sodium is attractive because of its lower cost than lithium. The $AlCl_3$-MEIC melts have a relatively high electrical conductivity at room temperature [466], 0.035 S cm$^{-1}$.

It is necessary to know the thermodynamic reduction potentials of the active metals in chloroaluminate melts. Scordilis-Kelley et al. [451,467] have studied standard reduction potentials in ambient temperature chloroaluminate melts for lithium and sodium, and they have calculated those of K, Rb, Cs. The values are, respectively, $-2.066$ V, $-2.097$ V, $-2.71$ V, $-2.77$ V and $-2.87$ V [versus Al(III)/Al in a 1.5/1.0 $AlCl_3$/MEIC reference melt].

These values show that lithium and sodium are at the negative potential limit of the electrochemical window ($-2$ V) (see Figure 49), close to the reduction potential of the imidazolium cation to neutral radical. Therefore, there is a competition between these processes with a resulting decrease in current efficiency. But Reichel and Wilkes [454], Campbell and Johnson [468], Scordilis-Kelley and Carlin [467,469] and Gray et al. [470] showed that an extension of the electrochemical window to $-2.4$ V is obtained by the addition of HCl to the $AlCl_3$-MEIC neutral melt buffered with NaCl or LiCl. Under these conditions, plating and stripping of sodium and lithium occurs at inert electrodes in room temperature chloroaluminate molten salts. The effect of HCl addition disappears quickly because of evaporation.

Fuller et al. [471] and Gray et al. [472] added thionyl chloride ($SOCl_2$), which allows sodium and lithium plating and stripping in 1-methyl-3-propyl-imidazolium chloride/$AlCl_3$ melts, because it was found that these melts have the widest stability window. Likewise, the $SOCl_2$ solute promotes high cycling efficiencies of the alkali metals in these electrolytes. The performance of the $SOCl_2$-promoted systems is improved substantially over previous studies in room temperature melts containing HCl. In addition, the low volatility of $SOCl_2$ maintains the high cycling efficiencies for extended periods of time.

Piersma et al. [473] used an addition of triethanolamine · HCl for widening of the voltage window of these melts. This substance is a stable reagent, easy to handle, and has a prolonged effect on the melt. While the concentration of the added HCl is maintained for only a few hours in these melts, the triethanolamine · HCl concentration stays high for months. The studies of these authors [473] suggest that deprotonation of one ethanolic group of triethanolamine HCl is responsible for the effect, the reagent being the best proton source.

It was noticed that in neutral MEIC-$AlCl_3$-NaCl melts the efficiency of sodium and lithium deposition is zero, because the sodium and lithium deposition potentials are not inside the electrochemical window. But if the above-mentioned reagents are added, the plating and stripping of both lithium and sodium are nearly reversible, the cycling efficiency being about 95%. The coulombic efficiency of the sodium couple (and lithium) is affected by several phenomena, for example, coreduction of metal and organic cation at negative potentials of the metal couple, and direct chemical reaction of the melt with the electrodeposited metal.

It was also shown that if graphite electrodes are used, the carbon material could reversibly accept and donate significant amounts of metal, especially lithium (Li:C = 1:6) without affecting its mechanical and electrical properties. Carbon materials can be used as electrodes in such a cell because the chemical potential of lithiated carbon material is almost identical to that of metallic lithium.

Furthermore, Carlin et al. [474] demonstrated the intercalation of both anions and cations into graphite, reversible electrodes being obtained. They used an electrolyte which contained 1-ethyl-3-methylimidazolium or 1,2-dimethyl-3-propylimidazolium as the cation and $AlCl_4^-$, $BF_4^-$, $PF_6^-$, $CF_3SO_3^-$ or $C_6H_5CO_2^-$ as the anion. Such cells could be very advantageous because the graphite electrodes are inexpensive, the battery can be assembled in the discharged state (the electrodes are formed during the first charging), and it uses only single room temperature molten salts which must be synthesized and purified.

A battery which uses room temperature molten salts as electrolyte could

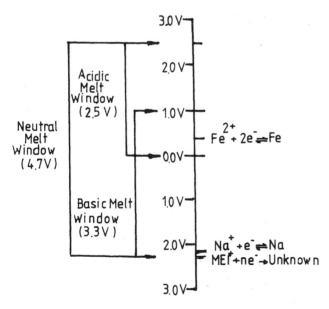

**Figure 50**   Potential of $Na/FeCl_2$ in the $MEIC-AlCl_3$ [470] electrolyte.

be similar to the Zebra battery. In Zebra batteries, which operate at about 250°C and utilize liquid sodium as the anode, a conductive ceramic, solid $\beta''$-alumina is used as separator. This prevents reaction of liquid sodium with the electrolyte. Improvements to this battery can be achieved by using a room temperature molten salt, so the sodium will remain in the solid state and the $\beta''$-alumina separator is no longer necessary.

Such a battery was proposed for the first time by Yu et al. [475] and by Gray et al. [470]. These batteries utilize the same pair of electrodes as the Zebra battery, $Na/FeCl_2$, but the electrolyte is a room temperature molten salt, $AlCl_3$-MEIC-NaCl. Yu et al. [475] used this electrolyte with the following composition: $47:45:8$ mole%. Gray et al. [470] proposed the addition of HCl to the electrolyte. The electrode reactions during discharge are

Positive electrode

$$2Na \rightleftharpoons 2Na^+ + 2e^- \tag{206}$$

Negative electrode

$$FeCl_2 + 2e^- \rightleftharpoons Fe + 2Cl^- \tag{207}$$

Overall

$$FeCl_2 + 2Na \rightleftharpoons Fe + 2NaCl \tag{208}$$

The cell potential is between $Fe^{2+}$ reduction and sodium oxidation, as seen in Figure 50.

## REFERENCES

1. Bockris J. O'M., *Modern Aspects of Electrochemistry*, Plenum Press, New York, 1959, p. 160.
2. Bloom H., Bockris J. O'M., "Structural Aspects of Ionic Liquids" in *Fused Salts*, B. R. Sundheim, ed., McGraw-Hill, New York, 1964, p. 1.
3. Bockris J. O'M., Richards N. E., *Proc. R. Soc.*, 1957, 241A, 44.
4. Stillinger F. M., "Equilibrium Theory of Pure Fused Salts" in *Molten Salts Chemistry*, M. Blander, ed., Interscience, New York, 1964, p. 1.
5. Zarzycki J., *J. Phys. Rad. (Suppl. Phys. Appl.)* 1957, 18, 65 A; *ibid.*, 1958, 19, 13 A.
6. Janz G. J., *J. Chem. Educ.*, 1962, 39, 59.
7. Kleppa O. J., *Ann. Rev. Phys. Chem.*, 1965, 16, 187.
8. Blander M., "Some Fundamental Concepts in the Chemistry of Molten Salts" in *Molten Salts*, G. Mamantov, ed., Marcel Dekker, New York, 1969, p. 1.
9. Levy H. A., Argon P. A., Bredig M. A., Danford M. D., *Ann. N. Y. Acad. Sci.*, 1960, 79, 762.
10. Frenkel J., *Kinetic Theory of Liquids*, Clarendon Press, Oxford, 1947.
11. Bockris J. O'M., Krook E. H., Richards N. E., Bloom H., *Proc. R. Soc.* (London), 1960, A 225, 558.

12. Furth R., *Proc. Cambridge Phil. Soc.*, 1941, 37, 252.
13. Bockris J. O'M., Pilla A., Barton J. L., *Rev. Chim. Acad. R.P.R.*, 1962, 7, 59.
14. Eyring H., Ree T., Hirai N., *Proc. Natl. Acad. Sci.*, 1958, 44, 683.
15. Blomgren G. E., *Ann. N.Y. Acad. Sci.*, 1960, 79, 781.
16. Carlson C. M., Eyring H., Ree T., *Proc. Natl. Acad. Sci.*, 1960, 46, 333.
17. Vilcu R., Misdolea C., *J. Chem. Phys.*, 1967, 46, 906; *ibid.* 1966, 45, 3414.
18. Temkin M., *Acta Physicochim. U.R.S.S.*, 1945, 20, 411.
19. Guggenheim E. A., *Mixtures*, Oxford University Press, London, 1952.
20. Førland T., *J. Phys. Chem.*, 1955, 59, 152.
21. Bredig M. A., "The Experimental Evidence for 'Complex Ions' in Some Molten Salt Mixtures", in *Molten Salts. Characterization and Analysis*, G. Mamantov, ed., Marcel Dekker, New York, 1969, p. 55.
22. Kleppa O. J., McCarty F. G., *J. Phys. Chem.*, 1966, 70, 1249.
23. Grjotheim K., *Contribution to the Theory of the Aluminium Electrolysis*, Det. Kgl. Norske Videnskabers Selskabs Skrifter, 1956, nr. 5, Trondheim.
24. Dewing E. W., *Proc. Electrochem. Soc.*, 1986, 86, 262.
25. Gilbert B., Materne T., *Appl. Spectrosc.*, 1990, 44, 299.
26. Gilbert B., Mamantov G., Begun G. M., *J. Chem. Phys.*, 1975, 62, 950.
27. Hong K. C., Kleppa J. O., *J. Phys. Chem.*, 1978, 82, 1596.
28. Feng N. K., Kvande H., *Acta Chem Scand.*, 1986, A 40, 622.
29. Zhon H., Herstad O., Østvold T., *Light Metals*, 1992, p. 511.
30. Tixhon E., Robert E., Gilbert B., *Appl. Spectrosc.*, 1994, 48, 1477.
31. Førland T., Ratkje S. K., *Acta Chem. Scand.*, 1973, 27, 1883.
32. Gilbert B., Mamantov G., Begun G. M., *Inorg. Nucl. Chem. Lett.*, 1976, 12, 415.
33. Sterten A., *Electrochim. Acta*, 1980, 25, 1673.
34. Julsrud S., Ph.D. thesis, Inst. of Physical Chemistry, NTH, University of Trondheim, Norway, 1983.
35. Kvande H., *Light Metals*, 1986, TMS, Warrendale, PA, p. 451.
36. Dewing E. W., Thonstad J., to be published.
37. Gilbert B., Robert E., Tixon E., Olsen J., Østvold T., *Light Metals*, 1995, W. Hale, ed., p. 181.
38. Olsen J. E., Ph.D. thesis, Inst. of Inorganic Chemistry, NTH, University of Trondheim, Norway, 1996.
39. Klemm A., "Ionic Mobilities", in *Advances in Molten Salt Chemistry*, vol. 6, G. Mamantov, C. B. Mamantov, J. Braunstein, eds., Elsevier, Amsterdam, 1987, p. 1.
40. Reddy T. B., *Electrochem. Tech.*, 1963, 1, 325.
41. Klemm A., in *Molten Salt Chemistry*, M. Blander, ed., John Wiley, New York, 1964, p. 535.
42. Sundheim B. R., in *Fused Salts*, B. R. Sundheim, ed., McGraw–Hill, New York, 1964, Chap. 3.
43. Bailey R. A., Janz G. J., in *The Chemistry of Non–aqueous Solvents*, J. J. Lagowski, ed., Academic Press, New York, 1966, Chap. 7.
44. Janz G. J., *Molten Salts Handbook*, Academic Press, New York, 1967, p. 344.
45. Moynihan C. T., in *Ionic Interactions*, vol. 1, S. Petrucci, ed., Academic Press, New York, 1971, Chap. 5.
46. Richter J., *Angew. Chem.*, 1974, 13, 438.

47. Hussey C. L., in *Molten Salt Chemistry. An Introduction and Selected Applications*, G. Mamantov, R. Marassi, eds., Riedel, Dordrecht, 1987, p. 141.

48. Delimarskii Iu. K., Markov B. F., *Electrochemistry of Fused Salts*, Sigma Press, Washington, DC, 1961, Chap. 1.

49. Janz G. J., Dampier F. W., Lakshminoraganan G. R., Lorenz P. K., Tomkins R. P. T., *Molten Salts: Vol. 1, Electrical Conductance, Density and Viscosity Data*, United States Department of Commerce, National Standard Reference Data Series, Natural Bureau of Standards 15, Washington, DC, 1968.

50. Biltz W., Klemm A., *Z. Phys. Chem.*, 1924, 110, 318.

51. Redkin A. A., Rasplavy, 1989, 3, 111; Molten Salts, Ninth International Symposium, 1994, The Electrochemical Society, Inc., Pennington, NJ, p. 211.

52. Nakamura T., Itoh M., *J. Electrochem. Soc.*, 1990, 137, 1166.

53. Sakai K., Nakamura T., Umesaki N., Iwamoto N., *Phys. Chem. Liq.*, 1984, 14, 67.

54. Shabanov O. M., *Melts*, Consultants Bureau, New York, January 1989, p. 444.

55. Shabanov O. M., Godzhiev S. M., Tagirov S. M., *Elektrokhim.*, 1973, 9, 1742; *ibid*, 1973, 9, 1828.

56. Smirnov M. V., Shabanov O. M., Hkaimenov A. P., *Elektrokhim.*, 1966, 2, 1240.

57. Smirnov M. V., *Electrode Potentials in Molten Chlorides*, Nauka, Moscow, 1973.

58. Walden P., *Z. Phys.*, 1906, 55, 207.

59. Chemla M., Lantelme F., Mehta O. P., *J. Chem. Phys. Phys. Chim. Biol.*, Numero special, 1969, 136–144.

60. Markov B. F., Sumina L., *Dokl. Akad. Nauk S.S.S.R.*, 1956, 110, 411; *Zh. Fiz. Khim.*, 1957, 31, 1767.

61. Delimarski Iu. K., Markov B. F., *Electrochimia rasplevleni͡h solei*, Metalurgizdat, Moscow, 1960.

62. Bloom H., Knaggs J. W., Molley J. J., Welch B., *Trans Faraday Soc.*, 1953, 49, 1458.

63. Harrap B. S., Heymann E., *Trans. Faraday Soc.*, 1955, 51, 259; ibid, 1955, 51, 268.

64. Bockris J. O'M., Lowe A., *Proc. R. Soc.*, 1954, A 226, 423.

65. Bockris J. O'M., Crook C. H., Bloom H., Richards N. E., *Proc. R. Soc.*, 1960, A 255, 558.

66. Oldekop W., *Z. Phys.*, 1955, 140, 181.

67. Bredig M. A., in *Molten Salt Chemistry*, M. Blander, ed., Interscience, New York, 1964, p. 367.

68. Cubicciotti D., *J. Metals*, 1953, 5, 1106.

69. Ukshe E. A., Bukun N. G., *Russ. Chem. Rev.*, 1961, 30, 90.

70. Corbett J. D., in *Fused Salts*, B. R. Sundheim, ed., McGraw-Hill, New York, 1964, p. 341.

71. Warren Jr. W. W., in *Advances in Molten Salt Chemistry*, G. Mamantov, J. Braunstein, eds., Plenum Press, New York, 1981, vol. 4, p. 1.

72. Warren Jr. W. W., in *The Metallic and Nonmetallic States of Matter*, P. P. Edwards, C. N. R. Rao, eds. Taylor and Francis, London, 1985.

73. Haarberg G. M., Thonstad J., *J. Appl. Electrochem.*, 1989, 19, 789.

74. Bronstein H. R., Bredig M. A., *J. Am. Chem. Soc.*, 1958, 80, 2077.

75. Warren Jr. W. W., Sotier S., Brennert G. F., *Phys. Rev. Lett.*, 1983, 50.
76. Kittel C., *Introduction to Solid State Physics*, Wiley, New York, 1971.
77. Feher G., *Phys. Rev.*, 1957, 105, 1122.
78. Symons M. C. R., *Chem. Soc. Rev.*, 1976, 5, 337.
79. Durham P. J., Greenwood D. A., *Phil. Mag.*, 1976, 33, 427.
80. Rice S. A., *Discuss. Faraday Soc.*, 1961, 32, 181.
81. Emi T., Bockris J. O'M., *Electrochim. Acta*, 1971, 16, 2081.
82. Inman D., Lovering D. G., Narayan R., *Trans. Faraday Soc.*, 1967, 63, 3017.
83. Nachtrieb N., Steinberg M., *J. Am. Chem. Soc.*, 1950, 72, 3558.
84. Kitazawa K., Asakura S., Fueki K., Mukaibo T., *J. Electrochem. Soc.*, 1969, 37, 45.
85. Egan J. J., Heus R., *J. Electrochem. Soc.*, 1960, 107, 824.
86. Berge T., Engseth P., Tunold R., *Acta Chem. Scand.*, 1974, A28, 1139.
87. Billehaug K., Øye H. A., *Aluminium*, 1981, 57, 146.
88. Alabyshev A. F., Lantratov M. F., Morachevskii A. G., *Reference Electrodes for Fused Salts*, English transl., A. Peiperl, Sigma Press, Washington, DC, 1965.
89. Delimarskii Yu. K., Markov B. F., *Electrochemistry of Fused Salts*, English transl., A. Peiperl, Sigma Press, Washington D. C. 1961.
90. Laity R. W., in *Reference Electrodes; Theory and Practice*, D. J. C. Ives, G. J. Janz, eds., Academic Press, New York, 1961, p. 524.
91. Inman D., The Electrode-Melt Interface at High Temperature in *Electromotive Force Measurements in High-Temperature Systems*, Proc. Symp. Nuffield Research Group, Imperial College, London, 1968, p. 163.
92. Lantelme F., Inman D., Lovering D. G., "Electrochemistry. I" in *Molten Salt Techniques*, vol. 2, R. J. Gale, D. G. Lovering eds., Plenum Press, New York, 1984, p. 137.
93. Rolin M., *Rev. Int. Hautes Temper. Refract.*, 1972, 9, 333.
94. Rolin M., Gallay J. J., *Bull. Soc. Chim. France*, 1960, pp. 2093, 2096, 2101.
95. Rolin M., Ducouret A., *Bull. Soc. Chim. France*, 1964, pp. 790, 794.
96. Janz G. J., Saegusa F., *Electrochim. Acta*, 1962, 7, 293.
97. Delimarskii Yu. K., Andreeva V. N., *Zh. Neorg. Khim.*, 1960, 5, pp. 1123, 1800, 2070.
98. Hildebrand J., Salstrom E., *J. Am. Chem. Soc.*, 1930, 52, 4641.
99. Hilldebrand J., Salstrom E., *J. Am. Chem. Soc.*, 1932, 54, 4257.
100. Salstrom E., *J. Am. Chem. Soc.*, 1933, 55, 2426.
101. Menes F., *J. Chim. Phys.*, 1965, 65, 1107.
102. Senderoff S., Mellors G., *Rev. Sci. Instr.*, 1958, 29, 151.
103. Leonardi J., Brenet J., *C. R. Acad. Sci. Paris, Ser. C*, 1965, 261, 116.
104. Murgulescu I. G., Sternberg S., *Rev. Chim. Acad. R.P.R.*, 1957, 2, 251; 1958, 3, 55.
105. Stern K., *J. Phys. Chem.*, 1956, 60, 679; 1958, 62, 385.
106. Murgulescu I. G., Marchidan D. I., *Rev. Chim. Acad. R.P.R.*, 1958, 3, 47.
107. Murgulescu I. G., Popescu L., *Rev. Roumaine Chim.*, 1972, 17, 1287.
108. Sternberg S., Adorian I., Galasiu I., *J. Chim. Phys.*, 1965, 62, 63.
109. Landolt–Börnstein, *Sechste Auflage*, II Band, 4 Teil, Springer-Verlag, Berlin, 1961, pp. 179, 474.

110. Kubaschewski O., Evans E. L., *Metallurgical Thermochemistry*, Pergamon Press, 1958.
111. Sternberg S., Adorian I., Galasiu I., *Electrochim. Acta*, 1966, 11, 385.
112. Blander M., "Thermodynamic Properties of Molten Salt Solutions" in *Molten Salt Chemistry*, M. Blander, ed., Interscience, New York, 1964, p. 127.
113. Forland T., "Thermodynamic Properties of Fused-Salt Systems" in *Fused Salts*, B. R. Sundheim, ed., McGraw-Hill, New York, 1964, p. 63.
114. Dombrovskaia N. S., Koloskova Z. A., *Izv. SFHA*, 1953, 22, 178.
115. Sternberg S., Adorian I., Galasiu I., *Rev. Roumaine Chim.*, 1966, 11, 581.
116. Esin O. A., *Zh. Fiz. Khim.*, 1956, 30, 3.
117. Dogonadze R. R., Chizmadzhev Yu. A., *Proc. Acad. Sci., SSSR, Phys. Chem. Sect.* (Engl), 1964, 157, 778.
118. Ukshe E. A., Bukun N. G., Leikis D. I., Frumkin A. N., *Electrochim. Acta*, 1964, 9, 431.
119. Bukun N. G., Ukshe E. A., *Physical Chemistry and Electrochemistry of Molten Salts and Slags*, Proceedings 3rd All-Soviet Conference, May 1966, Khimiya, Leningrad, 1968.
120. Inman D., *Electrode Reactions in Molten Salts*, in Proceedings of the Fifth International Symposium on Molten Salts, The Electrochemical Society, Pennington, NJ, 1986, p. 109.
121. Graves D., Inman D., *J. Electroanal. Chem.*, 1970, 25, 357.
122. Inman D., Lovering D. G., "Electrochemistry of Molten Salts" in *Comprehensive Treatise of Electrochemistry*, vol. 7, B. E. Conway, J. O'M. Bockris, E. Yeager, S. U. M. Khan, R. E. White, eds, Plenum Press, New York, 1983, p. 593.
123. Inman D., "Electrode Kinetics and Double Layer in Molten Salts" in *Molten Salt Chemistry. An Introduction and Selected Applications*, G. Mamantov, R. Marassi, eds., Reidel, Dordrecht, 1987, p. 271.
124. Inman D., Bowling J. E., Lovering D. G., White S. H., in *Specialist Periodical Reports—Electrochemistry*, vol. 4, Chemical Society, London, 1974.
125. Berge T., Tunold R., *Electrochim. Acta*, 1974, 19, 483.
126. Inman D., "The Electrode-Melt Interface at High Temperatures" in *Electromotive Force Measurements in High-Temperature Systems*, Proc. Symp. Nuffield Research Group, Imperial College, London, 1968, p. 163.
127. Sternberg S., Galasiu I., *Rev. Roumaine Chim.*, 1972, 17, 1697.
128. Sternberg S., Galasiu I., *Rev. Roumaine Chim.*, 1974, 19, 523.
129. Sternberg S., Galasiu I., *Rev. Roumaine Chim.*, 1974, 19, 967.
130. Sternberg S., Galasiu I., *Electrochim. Acta*, 1977, 22, 9.
131. Sternberg S., Galasiu I., Geana D., *Electrochim. Acta*, 1979, 24, 115.
132. Tunold R., Bø H. M., Paulsen K. A., Yttredal J. O., *Electrochim. Acta*, 1971, 16, 2101.
133. Tunold R., Berge T., Paulsen K. A., *Electrochim. Acta*, 1974, 19, 477.
134. Tunold R., Vatland A., *Electrochim. Acta*, 1974, 19, 375.
135. Thonstad J., Kisza A., *Molten Salt Forum*, vols. 1, 2, Trans. Tech. Publications, Aldermannsdorf, Switzerland, 1993/94, p. 195.
136. Thonstad J., Kisza A., Kazmierczak J., *J. Appl. Electrochem.*, 1996, 26, 102.

137. Kisza A., Kazmierczak J., Børresen B., Haarberg G. M., Tunold R., *J. Appl. Electrochem.*, 1995, 25, 940.
138. Børresen B., Haarberg G. M., Tunold R., Kisza A., Kazmierczak J., Proceedings of the Tenth International Symposium on Molten Salts, The Electrochemical Society 189th Meeting, Los Angeles, May 5–10, 1996, p. 1.
139. Kisza A., Kazmierczak J., Børresen B., Haarberg G. M., Tunold R., *J. Electrochem. Soc.*, 1995, 142, 1035.
140. Haarberg G. M., Rolland W., Sterten A., Thonstad J., *J. Appl. Electrochem.*, 1993, 23, 217.
141. Grjotheim K., Krohn C., Malinovsky M., Matiasovsky K., Thonstad J., *Aluminium Electrolysis, Fundamentals of the Hall-Heroult Process*, Aluminium-Verlag, Düsseldorf, 1982.
142. Holm J. L., Holm B. J., *Acta Chem. Scand.*, 1973, 27, 1410.
143. Fedotiev P. P., Ilinskii V. P., *Izv. SPB Politekn. Inst. (Proc. SPB Polytechn. Univ.)* 1912, 18, 147; *Z. Anorg. Chem.*, 1913, 80, 113; *Z. Anorg. Allgem. Chem.*, 1923, 129, 93.
144. Abramov G. A., Vetyukov M. M., Gupalo I. P., Kostyukov A. A., Lozhkin L. M., *Teoreticeskie osnovy elektrometallurghii alumina (Theoretical Principles of the Electrometallurgy of Aluminium)* Metallurghizdat, Moscow, 1953.
145. Fenerty A., Hollingshead E. A., *J. Electrochem. Soc.*, 1960, 107, 993.
146. Foster P. A., *J. Am. Ceram. Soc.*, 1970, 53, 598.
147. Solheim A., Rolseth S., Skybakmoen E., Støen L., Sterten A., Store T., in *Light Metals 1995*, J. W. Evans, ed., TMS, Warrendale, PA, 1995, p. 451.
148. Lee S. S., Lei K. S., Xu P., Brown J. J., in *Light Metals 1984*, J. R. McGees, ed., TMS, Warrendale, PA, 1984, p. 841.
149. Fernandez R., Grjotheim K., Ostvold T., in *Light Metals 1985*, H. O. Bohner, ed., TMS, Warrendale, PA, 1985, p. 501.
150. Rolin M., *Bull. Soc. Chim. France*, 1960, p. 1202.
151. Brynestad J., Grjotheim K., Gronvold F., Holm J., Urnes D., *Disc. Faraday Soc.*, 1962, 32, 90.
152. Chin D. A., Hollingshead E. A., *J. Electrochem. Soc.*, 1966, 113, 736.
153. Gerlach J., Hennig U., Tötsch H. D., *Erzmetall*, 1978, 31, 496.
154. Grjotheim K., Kvande H., Matiasovsky K., in *Light Metals 1983*, E. M. Adkins, ed., TMS, Warrendale, PA, 1983, p. 379.
155. Bagshaw A. N., Kuschel G. I., Taylor M. P., Tricklebank S. B., Welch B. J., in *Light Metals 1985*, H. O. Bohner, ed., TMS, Warrendale, PA, 1985 p. 649.
156. Jaim R. K., Tricklebank S. B., Welch B. J., Williams, D. J., in *Light Metals 1983*, E. M. Adkins, ed., TMS, Warrendale, PA, p. 609.
157. Gerlach J., Hennig U, Hern K., in *Light Metals 1974*, H. Forberg, ed., TMS, Warrendale, PA, 1974, p. 49.
158. Kuschel G. I., Welch B. J., in *Light Metals 1991*, E. Rooy, ed., TMS, Warrendale, PA, 1991, p. 299.
159. Phan-Xuan D., Castand R., Lafite M., Goret J., in *Light Metals 1975*, TMS, Warrendale, PA, 1975, p. 159.
160. Thonstad J., Nordmo F., Paulsen J. B., in *Light Metals 1971*, T. G. Edgeworth, ed., TMS, Warrendale, PA, 1971, p. 213.

161. Thonstad J., Johansen P., Kristensen E. W., in *Light Metals 1980*, C. McMinn, ed., TMS, Warrendale, PA, 1980, p. 227.
162. Matiasovsky K., Jaszova A., Malinovsky M., *Chem. Zvesti*, 1963, 17, 605.
163. Vajna A., *Alluminio*, 1950, 19, 541.
164. Edwards J. D., Taylor C. S., Cosgrove L. A., Russel A. S., *J. Electrochem. Soc.*, 1953, 100, 508.
165. Mashovets V. P., Petrov V. I., *Zh. Prikl. Khim.*, 1959, 32, 1258.
166. Matsushima T., Yoshida Y., Takahashi N., *Denki Kagaku*, 1971, 39, 102.
167. Paucirova M., Matiasovsky K., Malinovsky M., *Rev. Roumaine Chim.*, 1970, 15, 33.
168. Edwards J. D., Moormann T. A., *Chem. Met. Eng.*, 1921, 24, 61.
169. Paucirova M., Mariasovsky K., Malinovsky M., *Rev. Roumaine Chim.*, 1970, 15, 201.
170. Kvande H., Rorvik H., in *Light Metals 1985*, H. O. Bohner, ed., TMS, Warrendale, PA, 1985, p. 671.
171. Hives J., Thonstad J., Sterten A., Fellner P., *Electrochim. Acta*, 1993, 38, 2165; Proceedings of VII Aluminium Symposium, Donovaly, Slovakia, 1993, p. 115; in *Light Metals 1994*, V. Mannweiler, ed., TMS, Warrendale PA, 1994, p. 187.
172. Wang X., Peterson R. D., Tabereaux A. T., in *Light Metals 1993*, S. U. Das, ed., TMS, Warrendale, PA, 1993, p. 247.
173. Choudhary G., *J. Electrochem. Soc.*, 1973, 120, 381.
174. Solheim A., Thonstad J., in *Light Metals 1986*, W. Hale, ed., TMS, Warrendale, PA, 1986, p. 397.
175. Solheim A., Thonstad J., in *Light Metals 1987*, R. D. Zabreznik, ed., TMS, Warrendale, PA, 1987, p. 239.
176. Solheim A., Johansen S. T., Rolseth S., Thonstad J., in *Light Metals 1989*, P. G. Campbell, ed., TMS, Warrendale, PA, 1989, p. 245.
177. Foosnaes T., Naterstad T., Bruheim M., Grjotheim K., in *Light Metals 1986*, W. Hale, ed., TMS, Warrendale, PA, 1986, p. 729.
178. Galasiu I., Galasiu R., *Rev. Roumaine Chim.*, 1994, 39, 1415.
179. Oblakowski R., Pietrzyk S., Haarberg G. M., Thonstad J., Egan J. J., Proceedings of VIII Aluminium Symposium, Donovaly, Slovakia, 1995, p. 165.
180. Utigard T., Toguri J. M., in *Light Metals 1991*, E. Rooy, ed., TMS, Warrendale, PA, 1991, p. 273.
181. Grjotheim K., Welch B. J., *Aluminium Smelter Technology* Aluminium-Verlag, 1988, p. 170.
182. Dewing E. W., Desclaux P., *Metall. Trans. B*, 1977, 8 B, 555.
183. Utigard T., Toguri J. M., *Metall. Trans. B*, 1985, 16 B, 333.
184. Gherasimov A. D., Belyaev A. I., *Izv. Vyssh. Ucheb. Zavod., Tsvet. Met.*, 1958, 1, 50.
185. Utigard T., Toguri J. M., *Metall. Trans. B*, 1986, 17 B, 547.
186. Grjotheim K., Kvande H., Matiasovsky K., in *Light Metals 1983*, E. M. Adkins, ed., TMS, Warrendale, PA, 1983, p. 397.
187. Zemchuzina E. A., Belyaev A. I., *Fiz. Khim. Raspl. Solei i Shlakov*, Sverdlovsk, 1960, p. 207.

188. Qiu Zhuxian, Fan Li Man, Grjotheim K., in *Light Metals 1986*, W. Hale, ed., TMS, Warrendale, PA, 1986, p. 525.
189. Haupin W. E., McGrew W. C., in *Light Metals 1974*, H. Forberg, ed., TMS, Warrendale, PA, 1974, p. 37.
190. Zhuxian Qiu, Fan Li Man, Freng Nai Xiang, Grjotheim K., Kvande H., in *Light Metals 1987*, R. D. Zabreznik, ed., TMS, Warrendale, PA, 1987, p. 409.
191. Zhuxian Qiu, Fan Li Man, Grjotheim K., in *Light Metals 1983*, E. M. Adkins, ed., TMS, Warrendale, PA, 1983, p. 357.
192. Rolseth S., Thonstad J., in *Light Metals 1981*, G. M. Bell, ed., TMS, Warrendale, PA, 1981, p. 289.
193. Thonstad J., Rolseth S., in *Light Metals 1976*, S. R. Leavitt, ed., TMS, Warrendale, PA, 1976, p. 171.
194. Thonstad J., *Can. J. Chem.*, 1965, 43, 3429.
195. Wang Xiangwen, Peterson R. D., Richards N. E., in *Light Metals 1991*, E. Rooy, ed., TMS, Warrendale, PA, 1991, p. 323.
196. Ødegard R., Sterten A., Thonstad J., *Metall. Trans. B*, 1988, 19B, 449.
197. Sterten A. *Electrochim. Acta*, 1980, 25, 1673.
198. Sterten A., *Acta Chem. Scand.*, 1985, A 39, 241.
199. Kvande H., in *Light Metals 1986*, W. Hale, ed., TMS, Warrendale PA, 1986, p. 451.
200. Thonstad J., *J. Electrochem. Soc.*, 1964, 111, 959.
201. Ginsberg H., Wrigge H. C., *Metall.*, 1972, 26, 997.
202. Vetter K., *Electrochemical Kinetics*, Academic Press, New York, 1967.
203. Kisza A., Thonstad J., Eidet T., *Polish J. Chem.*, 1997, 71, 346.
204. Galasiu I., Galasiu R., *Rev. Roumaine Chim.*, 1994, 39, 525.
205. Galasiu I., Galasiu R., Proceedings of VIII Aluminium Symposium, Donovaly, Slovakia, 1995, p. 45.
206. Kubik C., Matiasovsky K., Malinovsky M., Zeman J., *Electrochim. Acta*, 1964, 9, 1521.
207. Silny A., Malinovsky M., Matiasovsky K., *Chem. Zvesti*, 1969, 23, 561.
208. Piontelli R., Mazza B., Pedeferri P., *Electrochim. Met.*, 1966, 1, 217; *Alluminio*, 1965, 34, 623.
209. Piontelli R., *Electrochim. Met.*, 1966, 1, 191.
210. Thonstad J., Rolseth S., Proceedings of 3rd ICSOBA Conference, Nice, 1973, p. 657.
211. Thonstad J., Kisza A., *Molten Salt Forum*, vols. 1, 2, Trans. Tech. Publications, Aldermannsdorf, Switzerland, 1993/94, p. 195.
212. Thonstad J., Kisza A., Kazmierczak J., *J. Appl. Electrochem.*, 1996, 26, 102.
213. Dewing E. W., *Met. Trans.*, 1972, 3, 495.
214. Tingle W. H., Petit J., Frank W. B., *Aluminum*, 1981, 57, 286.
215. Tabereaux A. T., in *Light Metals 1996*, W. Hale, ed., TMS, Warrendale, PA, 1996, p. 319.
216. Gjerstad S., Welch B. J., *J. Electrochem. Soc.*, 1964, 111, 976.
217. Grjotheim K., Haupin W. E., Welch B. J., in *Light Metals 1985*, H. O. Bohner, ed., TMS, Warrendale, PA, 1985, p. 679.
218. Fellner P., Grjotheim K., Matiasovsky K., Thonstad J., *Can. Met. Quart.*, 1969, 8, 245.

219. Schmitt H., *Extractive Metallurgy of Aluminium*, vol. 2, Interscience Publishers, New York, 1963.
220. Rolseth S., Müftüoglu T., Solheim A., Thonstad J., in *Light Metals 1986*, W. Hale, ed., TMS, Warrendale, PA, 1986, p. 517.
221. Kvande H., in *Light Metals 1989*, P. G. Campbell, ed., TMS, Warrendale, PA, 1989, p. 261.
222. Lewis R. A., *J. Metals*, 1967, 19, 30.
223. Lillebuen B., Mellerud T., in *Light Metals 1985*, H. O. Bohner, ed., TMS, Warrendale, PA, 1985, p. 637.
224. Leroy M. J., Pelekis T., Jolas J. M., in *Light Metals 1987*, R. D. Zabreznik, ed., TMS, Warrendale, PA, 1987, p. 291.
225. Paulsen K. A., Thonstad J., Rolseth S., Ringstad T., in *Light Metals 1993*, S. U. Das, ed., TMS, Warrendale, PA, 1993, p. 233.
226. Kvande H., *Light Metals 1991*, E. Rooy, ed., TMS, Warrendale, PA, 1991, p. 421.
227. Jarrett N., *Future Developments in the Bayer-Hall-Heroult Process*, A. R. Burkin, ed., Wiley, New York, 1987.
228. Bratland D., Grjotheim K., Krohn C., in *Light Metals 1976*, S. R. Leavitt, ed., TMS, Warrendale, PA, 1976, p. 3.
229. Thonstad J., "Aluminium Electrolysis, Electrolyte and Electrochemistry" in *Advances in Molten Salt Chemistry*, vol. 6, G. Mamantov et al., eds., Elsevier, Amsterdam, 1987, p. 73.
230. Liu Y. X., Thonstad J., *Electrochim. Acta*, 1983, 28, 113.
231. Thonstad J., *Min. Process. Extract. Metall. Rev.*, 1992, 10, 41.
232. Billehaug K., Øye H. A., *Aluminium*, 1980, 56, 642.
233. Thonstad J., *High Temp. Mater. Process.*, 1990, 9, 135.
234. Pawlek R. P., *Aluminium*, 1995, 71, 202; ibid. 1995, 71, 340; in *Light Metals 1996*, W. Hale, ed., TMS, Warrendale, PA, 1996, p. 243.
235. Alder H., U. S. Patent No. 3,960,678, June, 1976; Swiss Patent No. 14609, 1973.
236. Wang H., Thonstad J., in *Light Metals 1989*, P. G. Campbell, ed., TMS, Warrendale, PA, 1989, p. 23.
237. Xiao Haiming, Hovland R., Rolseth S., Thonstad J., in *Light Metals 1992*, E. L. Cutshall, ed., TMS, Warrendale, PA, 1992, p. 389.
238. Galasiu I., Galasiu R., Proceedings of VII Aluminium Symposium, Donovaly, Slovakia, 1993, p. 57; ibid. p. 69.
239. Galasiu I., Galasiu R., Andronescu E., Popa N., Chivu V., Proceedings of VIII Aluminium Symposium, Donovaly, Slovakia, 1995, p. 67.
240. Galasiu R., Galasiu I., Andronescu E., Proceedings of VIII Aluminium Symposium, Donovaly, Slovakia, 1995, p. 61.
241. Galasiu I., Popescu D.D., Galasiu R., Modan M., Stanciu P., Proceedings of Ninth International Symposium on Light Metals Production, Trondheim, Norway, August, 1997, J. Thonstad, ed., p. 273.
242. Galasiu I., Galasiu R., Comanescu I., Proceedings of VIII Aluminium Symposium, Donovaly, Slovakia, 1995, p. 55.
243. Galasiu I., Galasiu R., Proceedings of VIII Aluminium Symposium, Donovaly, Slovakia, 1995, p. 51.
244. Galasiu I., Galasiu R., Popa N., Chivu V., Proceedings of Ninth International Sym-

posium on Light Metals Production, Trondheim, Norway, August, 1997, J. Thonstad, ed., p. 189.

245. Ray P. S., in *Light Metals 1986*, W. Hale, ed., TMS, Warrendale, PA, 1986, p. 287.

246. Duruz J. J., Dervaz J. P., Debely P. E., U.S. Patent No. 4,614,569, 1986.

247. Duruz J. J., V. de Nora, U.S. Patent No. 4,680,094, 1987.

248. Liao Xianan, Liu Yexiang, in *Light Metals 1990*, C. M. Bickert, ed., TMS, Warrendale, PA, 1990, p. 409.

249. Liu Yexiang, Liao Xianan, Tang Fuling, Chen Zheming, in *Light Metals 1992*, E. L. Cutshall, ed., TMS, Warrendale, PA, 1992, p. 427.

250. Qiu Zhuxian, Li Qinfeng, Chen Xuesen, Wang Jing, Li Bing, Grjotheim K., Kvande H., in *Light Metals 1992*, E. L. Cutshall, ed., TMS, Warrendale, PA, 1992, p. 431.

251. Alcorn T. R., Stewart D. V., Tabereaux A. T., Jao L. A., Tucker K. W., in *Light Metals 1990*, C. M. Bickcrt, ed., TMS, Warrendale, PA, 1990, p. 413.

252. McMinn C. J., in *Light Metals 1992*, E. L. Cutshall, ed., TMS, Warrendale, PA, 1992, p. 419.

253. Sekhar J. A., de Nora V., Liu J., Duruz J. J., in *Light Metals 1996*, W. Hale, ed., TMS, Warrendale, PA, 1996, p. 271.

254. Sekhar J. A., Bello V., de Nora V., Liu J., Duruz J. J., in *Light Metals 1995*, J. W. Evans, ed., TMS, Warrendale, PA, 1995, p. 507.

255. Øye H. A., de Nora V., Duruz J. J., Johnston G., in *Light Metals 1997*, R. Huglen, ed., TMS, Warrendale, PA, 1997, p. 279.

256. Russell A., *J. Metals*, 1981, 33, 132.

257. Torsi G., Mamantov G., *Inorg. Nucl. Chem. Lett.*, 1970, 6, 553.

258. Bjørgum A., Sterten A., Sorensen V. B., Thonstad J., Tunold R., *Electrochim. Acta*, 1981, 26, 487; ibid. 1981, 26, 491.

259. Gardner H. J., Grjotheim K., Welch B. J., Proc. Int. ICSOBA Symposium on Alumina Production until 2000, Tilrany, Hungary, Oct. 1981, p. 27.

260. Berge T., Tunold R., Ytterdal S. A., Extended Abstracts, 31st Meeting of International Society of Electrochemistry, Venice, 1980, p. 151.

261. Beck T. R., Rousar I., Thonstad J., *Metall. Trans. B*, 1994, 25 B, 661.

262. Wilson C. B., Claus K. G., Earlam M. R., Hillis J. E., in *Kirk-Othmer Encyclopedia of Chemical Technology* 4th ed., vol. 14, Wiley, New York, 1995, p. 622.

263. Strelets Kh. L., *Electrolytic Production of Magnesium*, transl. J. Schmorak, Keter, Jerusalem, 1977.

264. Buck R. S., *Magnesium Products Design*, Marcel Dekker, New York, 1987.

265. Chase M. V., Curnutt J. L., Donald R. A., Syverud A. N., *JANAF Thermochemical Tables, 1978 Supplement, J. Phys. Chem. Ref. Data*, 1978, 7 (3), 793.

266. *Ullmann's Encyclopedia of Industrial Chemistry*, vol. A15, Barbara Elvers, Stephen Hawkins, Gail Schulz, eds., VCH, New York, 1990, p. 559.

267. Kipouros G. J., Sadoway D. R., "The Chemistry and Electrochemistry of Magnesium Production" in *Advances in Molten Salt Chemistry*, vol. 6, G. Mamantov et al., eds., Elsevier, 1987, p. 127.

268. Maurits A. A., *J. Appl. Chem. U.S.S.R.*, 1971, 44, 1671.

269. Høy-Petersen N., *J. Metals*, 1969, 21, 43.

270. Holliday R. D., McIntosh P., *J. Electrochem. Soc.*, 1973, 120, 858.

271. Flood H., Urnes S., *Z. Elektrochem.*, 1955, 59, 834.

272. Kleppa O. J., McCarty F. G., *J. Phys. Chem.*, 1966, 70, 1249.
273. Neil D. E., Clarck H. M., Wiswall R. H., *J. Chem. Eng. Data*, 1965, 10, 21.
274. Maroni V. A., Cairns E. J., in *Molten Salts*, G. Mamantov, ed., Marcel Dekker, New York, 1969, p. 231.
275. Maroni V. A., Hathaway E. J., Cairns E. J., *J. Phys. Chem.*, 1971, 75, 155.
276. Maroni V. A., *J. Chem. Phys.*, 1971, 55, 4789.
277. Capwell R. J., *Chem. Phys. Lett.*, 1972, 12, 443.
278. Brooker M. H., *J. Chem. Phys.*, 1975, 63, 3054.
279. Huang C. H., Brooker M. H., *Chem. Phys. Lett.*, 1976, 43, 180.
280. Brooker M. H., Huang C. H., *Can. J. Chem.*, 1980, 58, 168.
281. Laursen M. M., H. von Barner J., *J. Inorg. Nucl. Chem.*, 1979, 41, 185.
282. Thonstad J., *High Temp. Mater. Process.*, 1990, 9, 135.
283. Aarebrot E., Andersen R. E., Østvold T., Øye H. A., in *Light Metals 1977*, K. B. Higbic, ed., TMS, Warrendale, PA, 1977, p. 491.
284. Østvold T., Øye H. A., in *Light Metals 1980*, C. McMinn, ed., TMS, Warrendale, PA, 1980, p. 937.
285. Aarebrot E., Andersen R. E., Østvold T., Øye H. A., *Metall (Berlin)*, 1978, 32, 41.
286. Wypartowicz J., Østvold T., Øye H. A., *Electrochim. Acta*, 1980, 25, 151.
287. Van Norman J. D., Egan J. J., *J. Phys. Chem.*, 1963, 67, 2460.
288. Tunold R., in *Light Metals 1980*, C. McMinn, ed., TMS, Warrendale, PA, 1980, p. 949.
289. Rao G. M., *J. Electroanal. Chem.*, 1986, 16, 775.
290. Duan S., Dudley P. G., Inman D., Proc. Fifth Int. Symp. Molten Salts, The Electrochemical Society, Pennington, NJ, 1986, 86-1, 248.
291. Kisza A., Kazmierczak J., Børresen B., Haarberg G. M., Tunold R., *J. Appl. Electrochem.*, 1995, 25, 940.
292. Børresen B., Haarberg G. M., Tunold R., O. Wallevik, *J. Electrochem. Soc.*, 1993, 140(6), L99.
293. Kisza A., Kazmierczak J., Børresen B., Haarberg G. M., Tunold R., *J. Electrochem. Soc.*, 1995, 142, 1035.
294. Kisza A., *Polish J. Chem.*, 1993, 67, 885; *ibid.* 1994, 68, 613.
295. Børresen B., Haarberg G. M., Tunold R., Kisza A., Kazmierczak J., Proceedings of the Tenth International Symposium on Molten Salts, The Electrochemical Society 189th Meeting, Los Angeles, May 1996, p. 1.
296. Sivilotti O. G., U. S. Patent Nos. 4,514,269 (1985); 4,518,475 (1985); 4,560,449 (1985); 4,594,177 (1986).
297. Sivilotti O. G., in *Light Metals 1988*, R. E. Miller, ed., TMS, Warrendale, PA, 1988, p. 817.
298. Ishizuka H., U. S. Patent Nos., 4,495,037 (1985); 4,647,355 (1987).
299. Sharma R., in *Light Metals 1996*, W. Hale, ed., TMS, Warrendale, PA, 1996, p. 1113.
300. Sitting M., *Sodium, Its Manufacture, Properties and Uses*, Reinhold, New York, 1956.
301. Janz G. J., Dampier F. W., Lakshminarayanan G. R., Lorenz P. K., Tamkins R. P. T., *Molten Salts.* Volume 1: *Electrical Conductance, Density and Viscosity*

*Data*, U. S. Department of Commerce, National Bureau of Standards, Washington, DC, 1968, p. 48.

302. Cowley W. E., "The Alkali Metals" in *Molten Salt Technology*, David G. Lovering, ed., Plenum Press, New York, 1982, p. 57.
303. Loftus W. H., (E. I. Du Pont), U. S. Patent, 3,020,221 (1967).
304. Kisza A., Kazmierczak J., Børresen B., Haarberg G. M., Tunold R., *J. Electrochem. Soc.*, 1995, 142, 1035.
305. "The Alkali Metals," in *Mellor's Comprehensive Treatise on Inorganic and Theoretical Chemistry*, vol. 2, suppl. 2, part 1, Longmans, London, 1961.
306. Ethyl Corp., U.S. Patent, 2,944,950 (1957).
307. Du Pont, U.S. Patent, 3,037,927 (1962).
308. Olin Mathieson Chem. Corp., U.S. Patent, 3,085,967 (1960).
309. *Ullmann's Encyclopedia of Industrial Chemistry*, vol. A24, Barbara Elvis, Stefan Hawkins, William Russey, Gail Schultz, eds., VCH, New York, 1993, p. 277.
310. Ruppeert K. A., Degussa A. G., Company information, Frankfurt/Main, 1992.
311. Brown O. M., (Ethyl Corp), U.S. Patent, 3,245,899 (1960).
312. Kummer J. T., Weber N., (Ford Motor Co), U.S. Patent, 3,488,271 (1966).
313. Cope S. A. (Du Pont), U.S. Patent, 4,203,819 (1978).
314. Cope S. A. (Du Pont), U.S. Patent, 4,133,728 (1978).
315. Lemke C. H. (Du Pont), U.S. Patent, 4,089,770 (1977).
316. Appleby A. J., Foulkes F. R., *Fuel Cell Handbook*, Van Nostrand Reinhold, New York, 1989, p. 15.
317. Scrinivasan S., Dave B. B., Murugesamoorthi K. A., Parthasarathy A., Appleby A. J., in *Overview of Fuel Cell Technology in Fuel Cell Systems*, Leo Blumen, M. N. Mugerva, eds., Plenum Press, New York, 1993, p. 37.
318. Berger C., *Handbook of Fuel Cell Technology*, Prentice Hall, Englewood Cliffs, NJ, 1968.
319. McDougall A., *Fuel Cells*, Macmillan, London, 1976.
320. Tilak B. V., Yeo R. S., Srinivasan S., in *Comprehensive Treatise of Electrochemistry*, vol. 3, J. O'M. Bockris et al., eds., Plenum, New York, 1981, Chap. 2.
321. White R. E., Bockris J. O'M., Conway B. E., Yeager E., eds., *Comprehensive Treatise of Electrochemistry*, vol. 8, Plenum, New York, 1984.
322. Yeager E., Bockris J. O'M., Conway B. E., Sarangapani S., eds., *Comprehensive Treatise of Electrochemistry*, vol. 9, Plenum, New York, 1984.
323. Camara E. H., Wilson G. G., eds., *Fuel Cell Technology. Status and Applications*, Institute of Gas Technology, Chicago, 1982.
324. Penner S. S., ed., *Assessment of Research Needs for Advanced Fuel Cells*, U.S. Department of Energy, 1985.
325. Kinoshita K., McLarnon F. R., Cairns E. J., *Fuel Cells: A Handbook*, U.S. Department of Energy, 1988, p. 26.
326. Benjamin T. G., Camara E. H., Marianowski L. G., *Handbook of Fuel Cell Performance*, Institute of Gas Technology, U.S. Department of Energy, Washington, DC, 1980.
327. Liebhalsky H. A., Cairns E. J., *Fuel Cells and Fuel Batteries*, Wiley, New York, 1968.

328. Selman J. R., Maru H. C., in *Advances in Molten Salt Chemistry*, vol. 4, G. Mamantov, J. Braustein, eds., Plenum, New York, 1981, p. 159.
329. Selman J. R., Marianovski L. G., in *Molten Salt Technology*, D. G. Lovering, ed., Plenum, New York, 1982, p. 323.
330. Bockris J. O'M., Scrinivasan S., *Fuel Cells: Their Electrochemistry*, McGraw-Hill, New York, 1969.
331. Eisenberg M., *Thermodynamics of Electrochemical Fuel Cells, Fuel Cells*, vol. 1, Will Mitchell, ed., Academic Press, New York, 1963.
332. Kordesch K., *Brennstoffbatterien*, Innovative Energietechnik, Springer-Verlag, Wien and New York, 1984.
333. Kordesch K., Simader-Weinheim G., *Fuel Cells and Their Applications*, VCH, New York, Basel, 1996, p. 111.
334. Appleby A. J., "Characteristics of Fuel Cells Systems" in *Fuel Cell Systems*, Leo Blomen, M. N. Mugerwa, eds., Plenum Press, New York, 1993, p. 157.
335. Swinkels D. A. J., "Molten Salt Batteries and Fuel Cells" in *Advances in Molten Salt Chemistry*, vol. I, J. Braunstein, G. Mamantov, G. P. Smith, eds., Plenum Press, New York, 1971, p. 165.
336. Broers G. H. J., High Temperature Galvanic Fuel Cells, Doctoral thesis, University of Amsterdam, 1958.
337. Reiser C. A., Schroll C. R., Abstracts National Fuel Cell Seminar, Norfolk VA, 1981, p. 144.
338. Selman J. R., "Research, Development and Demonstration of Molten Carbonate Fuel Cell Systems" in *Fuel Cell Systems*, Leo Blomen, M. N. Mugerwa, eds, Plenum Press, New York, 1993, p. 345.
339. E. Barendrecht, "Electrochemistry of Fuel Cells" in *Fuel Cell Systems*, Leo Blomen, M. N. Mugerwa, eds., Plenum Press, New York, 1993, p. 73.
340. Minh N. Q., *J. Power Sources*, 1988, 24, 1.
341. Jewulski J., Suski L., *J. Appl. Electrochem.*, 1984; 14, 135.
342. Appleby A. J., Nicholson S. B., *J. Electroanal. Chem.*, 1974, 53, 105; 1977, 83, 309; 1980, 112, 71.
343. Lux H., *Z. Elecktrochem.*, 1939, 45, 303; Flood H., Forland T., *Acta Chem. Scand.*, 1974, 1, 592.
344. Selman J. R., Maru H. C., "Physical Chemistry and Electrochemistry of Alkali Carbonate Melts" in *Advances in Molten Salt Chemistry* G. Mamantov, J. Braunstein, C. B. Mamantov, eds., Plenum Press, New York, 1981, p. 160.
345. Uchida I., Nishina T., Mugicura Y., Itaya K., *J. Electroanal. Chem.*, 1986; 206, 229.
346. Uchida I., Nishina T., Mugicura Y., Itaya K., *J. Electroanal. Chem.*, 1986; 206, 241.
347. Kasai H., Suzuki A., "Dissolution and Deposition of Nickel-Oxide Cathode in MCFC," Proc. Third International Symposium on Carbonate Fuel Cell Technology, vol. 93-3, p. 240.
348. Braumgartner C. E., *J. Electrochem. Soc.*, 1984, 131, 1850.
349. Doyon J., Gilbert T., Davies G., Paetsch L., *J. Electrochem. Soc.*, 1987, 134, 3035.
350. Kunz H. R., *J. Electrochem. Soc.*, 1987, 134, 105.

351. Appleby A. J., Nicholson S. B., *J. Electroanaly. Chem.*, 1974, 53, 105; *ibid.* 1977, 83, 309; *ibid.* 1980, 112, 71.
352. Selman I. R., Lee G., Proc. 3rd Int. Symp. Carbonate Fuel Cell Technology, Honolulu, HI, May 1993, Proc. Vol. 93-3, The Electrochemical Society, Pennington, NJ, p. 309.
353. Nishina T., Yuasa K., Uchida I., Proc. 3rd Int. Symp. Carbonate Fuel Cell Technology, Honolulu, HI, May 1993, Proc. Vol. 93-3, The Electrochemical Society, Pennington, NJ, p. 264.
354. Shores D., Pischke M., Proc. 3rd Int. Symp. Carbonate Fuel Cell Technology, Honolulu, HI, May 1993, Proc. Vol. 93-3, The Electrochemical Society, Pennington, NJ, p. 214.
355. Yuh C., Johnsen R., Farooque M., Maru H., Proc. 3rd Int. Symp. Carbonate Fuel Cell Technology, Honolulu, HI, May 1993, Proc. Vol. 93-3, The Electrochemical Society, Pennington, NJ, p. 158.
356. Yokokawa H., Sakai N., Kawada T., Dokiya M., Ota K., *J. Electrochem. Soc.*, 1993, 140, 3565.
357. Yuh C., Johnsen R., Farooque M., *J. Power Sources*, 1995, 56, 1.
358. Williams M. C., Fuel Cell Seminar, Orlando, FL, Nov. 1996, Courtesy Associates, Inc., Washington, DC, p. 1.
359. Takashima S., Kahara T., Takeuchi M., Amano Y., Yoshida T., Fuel Cell Seminar, Orlando, FL, Nov. 1996, Courtesy Associates, Inc. Washington, DC, p. 532.
360. Cairns E. J., Steunenberg R. K., in *Progress in High Temperature Physics and Chemistry*, vol. 5, C. A. Rouse, ed., Pergamon Press, New York, 1973, p. 63.
361. Cairns E. J., Shimotake H., *Science*, 1969, 163, 1347.
362. Cairns E. J., Ackerman J. P., Hunt P. D., Tani B. S., presented at the Electrochemical Society Meeting, Cleveland, Oct. 1971, Abstract no. 48.
363. Cairns E. J., Shimotake H., Gay E. C., Selman J. R., presented at the International Society of Electrochemistry Meeting, Stockholm, Sweden, Sept. 1972.
364. Gay E. C., Steunenberg R. K., Battles J. E., Cairns E. J., Proceedings of the 8th IECEC, New York, 1973, p. 96.
365. Cairns E. J., "Batteries" in *Molten Salt Technology*, D. G. Lovering, ed., Plenum Press, New York, 1972, p. 287.
366. Shimotake H., Walsh W. J., Carr E. S., Bartholme L. G., in Proceedings of the 11th Intersociety Energy Conversion Engineering Conference, vol. 1, American Institute of Chemical Engineers, New York, 1976, p. 473.
367. Myles K. M., Mrazek F. C., Smaga J. A., Settle J. L., in Proceedings of the Symposium and Workshop on Advanced Battery Research and Design, March 1976, Argonne National Laboratory Report, A.N.L. 76-8, 1976, p. B-69.
368. Tomczuk Z, Preto S. K., Roche M. F., *J. Electrochem. Soc.*, 1981, 128, 760.
369. Tomczuk Z., Roche M. F., Vissers D. R., *J. Electrochem. Soc.*, 1981, 128, 2255.
370. Warin D., Tomczuk Z., Vissers D. R., *J. Electrochem. Soc.*, 1983, 130, 64.
371. Vincent C. A., Bonino F., Lazzary M., Scrosati B., *Modern Batteries. An Introduction to Electrochemical Power Sources*, Edward Arnold, London, 1984, p. 168.
372. Kaun T. D., Kilsdonk D. J., in *Lithium/Iron Sulfide Batteries for Electric Vehicle Propulsion and Other Applications*, Progress Report for October 1978–September 1980, Argonne National Laboratory Report A.N.L. 80-128/1981, p. 156.

373. Cairns E. J., "High Temperature Lithium Batteries" in *Lithium Battery Technology*, H. V. Venkatasetty, ed., Wiley, New York, 1984, p. 179.

374. Marassi R., Zamponi S., Berrettoni M., "Molten Salt Batteries" in *Molten Salt Chemistry. An Introduction and Selected Applications*, G. Mamantov, R. Marassi, eds., Reidel, Dordrecht, Boston, 1987, p. 491.

375. Kaun T. D., Nelson P. A., Redey L., Vissers D. R., Henriksen G. L., *Electrochim. Acta*, 1993, 38, 1269.

376. Briscoe J. D., Staniewicz R. J., Williams M., Proceedings of the 1992 IEEE 35th International Power Sources Symposium, Cherry Hill, NJ, June 1992, p. 294.

377. Oweis S. M., Embrey J., Willson J., Alunans P., Proceedings of the 1992 IEEE 35th International Power Sources Symposium, Cherry Hill, NJ, June 1992, p. 350.

378. Papadakis N., Barlow G., Bennett W., Schuster N., Specht S., Proceedings, of the 1992 IEEE 35th International Power Sources Symposium, Cherry Hill, NJ, June 1992, p. 285.

379. Schuster N., Seiger H. N., Proceedings of the 1992 IEEE 35th International Power Sources Symposium, Cherry Hill, NJ, June 1992, p. 354.

380. Barnes R. G., Segel S. L., *J. Chem. Phys.*, 1956, 25, 180.

381. Segel S. L., Barnes R. G., *J. Chem. Phys.*, 1956, 25, 578.

382. Casabella P. A., Bray P. J., Barnes R. G., *J. Chem. Phys.*, 1959, 30, 1393.

383. Casabella P. A., Miller N. C., *J. Chem. Phys.*, 1964, 40, 1363.

384. Boston C. R., Yosim S. J., Grantham L. F., *J. Chem. Phys.*, 1969, 51, 1669.

385. Schinke H., Sauerwald F., *Z. Anorg. Allg. Chem.*, 1956, 278, 313.

386. Semenenko K. N., Naumova T. N., *Zh. Neorg. Khim.*, 1964, 9, 1316.

387. Lewis G. N., *Valence and the Structure of Atoms and Molecules*, Chemical Catalogue, New York, 1923.

388. Grothe K., *Z. Elektrochem.*, 1950, 54, 216.

389. Dewing E. W., *J. Am. Chem. Soc.*, 1955, 77, 2369.

390. Fannin Jr., A. A., King L. A., Seegmiller D. W., Øye H. A., *J. Chem. Eng. Data.*, 1982, 27, 114.

391. Carpio R. A., Fannin Jr., A. A., Kibler F. C., King L. A., Øye H. A., *J. Chem. Eng. Data.*, 1983, 28, 34.

392. Boston C. R., "Molten Salt Chemistry of the Haloaluminates" in *Advances in Molten Salt Chemistry*, vol. 1, J. Braunstein, G. Mamantov, G. P. Smith, eds., Plenum Press, New York, 1971, p. 129.

393. Øye H. A., Gruen D. M., *Inorg. Chem.*, 1964, 3, 836.

394. Moore R. H., Morrey J. R., Vailand E. W., *J. Phys. Chem.*, 1963, 67, 744.

395. Torsi G., Mamantov G., Begun G. M., *Inorg. Nucl. Chem. Lett.*, 1970, 6, 553.

396. Tremillon B., Letisse G., *J. Electroanal. Chem.*, 1968, 17, 371.

397. Torsi G., Mamantov G., *Inorg. Chem.*, 1971, 10, 1900.

398. Fannin Jr., A. A., King L. A., Seegmiller D. W., *J. Electrochem. Soc.*, 1972, 119, 801.

399. Boxall L. G., Jones H. L., Osteryoung R. A., *J. Electrochem. Soc.*, 1973, 120, 223.

400. Chum H. Li., Osteryoung R. A., "Chemical and Electrochemical Studies in Room Temperature Aluminium-Halide-Containing Melts" in *Ionic Liquids*, D. Inman, D. G. Lovering, eds., Plenum Press, New York, 1981, p. 407.

401. Vrieland G. E., Stull D. R., *J. Chem. Eng. Data*, 1967, 12, 532.

402. Mamantov G., Gilbert B., Fung K. W., Marassi R., Rolland P., Torsi G., Bowman K. A., Brotherton D. L., McCurry L. E., Ting G., in Proceedings of the International Symposium on Molten Salts, J. P. Pemsler, ed., The Electrochemical Society, Princeton, NJ, 1976, p. 234.

403. Gilbert B., Mamantov G., Fung K. W., *Inorg. Chem.*, 1975, 14, 1802.

404. Mamantov G., Osteryoung R. A., "Acid-Base Dependent Redox Chemistry in Molten Chloroaluminates" in *Characterization of Solutes in Nonaqueous Solvents*, G. Mamantov, ed., Plenum Press, New York, 1978, p. 223.

405. Paulsen K. A., Osteryoung R. A., *J. Am. Chem. Soc.*, 1976, 98, 6866.

406. Marassi R., Mamantov G., Chambers J. Q., *Inorg. Nucl. Chem. Lett.*, 1975, 11, 245.

407. Marassi R., Mamantov G., Chambers J. Q., *J. Electrochem. Soc.*, 1976, 1123, 1128.

408. Doorenbos H. E., Evans J. C., Kagel R. O., *J. Phys. Chem.*, 1970, 74, 3385.

409. Bjerrum N. J., "Electrochemical and Spectroscopic Studies of the Chalcogens in Chloroaluminate Melts" in *Characterization of Solutes in Nonaqueous Solvents*, G. Mamantov, ed., Plenum Press, New York, 1978, p. 251.

410. Bjerrum N. J., Boston C. R., Smith G. P., *Inorg. Chem.*, 1967, 6, 1162.

411. Bjerrum N. J., Smith G. P., *Inorg. Chem.*, 1967, 6, 1968.

412. Abraham K. M., Elliot J. E., *J. Electrochem. Soc.*, 1984, 131, 2211.

413. Howie R. C., MacMillan D. W., *J. Inorg. Nucl. Chem.*, 1971, 33, 3681.

414. Holleck G. L., Giner J., *J. Electrochem. Soc.*, 1972, 119, 1161.

415. Del Duca B. S., *J. Electrochem. Soc.*, 1971, 118, 405.

416. Gale R. J., Osteryoung R. A., *J. Electrochem. Soc.*, 1974, 121, 983.

417. Torsi G., Mamantov G., *Inorg. Chem.*, 1971, 10, 1900.

418. Howie R. C., MacMillan D. W., *J. Inorg. Nucl. Chem.*, 1971, 33, 3681.

419. Brabson G. D., Fannin Jr., A. A., King L. A., Seegmiller D. W., Abstract 26, The Electrochemical Society Extended Abstracts, Chicago, IL, May 13–18, 1973, p. 61.

420. Werth J. J., U.S. Patent, 3,847,667 (1974).

421. Greenberg J., U.S. Patent, 3,573,986 (1971).

422. Sholette W. P., Klein I. S., Werth J., in Proceedings of the Symposium on Load Leveling, N. P. Yao and J. R. Selman, eds., The Electrochemical Society Softbound Proceedings Series, Princeton, NJ, 1977, p. 306.

423. Mamantov G., Marassi R., Matsunaga M., Ogata Y., Wiaux P. J., Frazer E. J., *J. Electrochem. Soc.*, 1980, 127, 2319.

424. Hjuler H. A., von Winbusch S., Berg R. W., Bjerrum N. J., *J. Electrochem. Soc.*, 1989, 136, 901.

425. Coetzer J., *J. Power Sources*, 1986, 18, 377.

426. Bones R. J., Coetzer J., Galloway R. C., Teagle D. A., *J. Electrochem. Soc.*, 1987, 134, 2379.

427. Galloway R. C., *J. Electrochem. Soc.*, 1987, 134, 256.

428. Bones R. J., Teagle D. A., Brooker S. D., Cullen F. L., *J. Electrochem. Soc.*, 1989, 136, 1274.

429. Yao Y., Kummer J. T., *J. Inorg. Nucl. Chem.*, 1967, 29, 2453.

430. Moseley P. T., Bones R. J., Teagle D. A., Bellamy B. A., Hawes R. W. M., *J. Electrochem. Soc.*, 1989, 136, 1361.

431. Ratnakumar B. V., DiStefano S., Halpert G., *J. Electrochem. Soc.*, 1990, 137, 2991.

432.   Cleaver B., Cleaver D. J., Littlewood L., Demott D. S., *J. Appl. Electrochem.*, 1995, 25, 1128.
433.   Hurley F. H., U.S. Patent 2, 446,331 (1948).
434.   Wier T. P., Hurley F. H., U.S. Patent 2, 446, 349 (1948).
435.   Hurley F. H., Wier T. P., *J. Electrochem. Soc.*, 1951, 98, 203.
436.   Hurley F. H., Wier T. P., *J. Electrochem. Soc.*, 1951, 98, 207.
437.   Hussey C. L., "Room-Temperature Molten Salt Systems" in *Advances in Molten Salt Chemistry* vol. 5, G. Mamantov, C. B. Mamantov eds., Elsevier, Amsterdam, 1983, p. 185.
438.   Carlin R. T., Wilkes J. S., "Chemistry and Speciation in Room-Temperature Chloroaluminate Molten Salts" in *Chemistry of Nonaqueous Solutions. Current Progress*, G. Mamantov, A. I. Popov eds., VCH, New York, 1994, p. 277.
439.   Øye H. A., Jagtoyen M., Oksefjell T., Wilkes J. S., *Mater. Sci. Forum.*, 1991, 73–75, 183.
440.   Gale R. J., Osteryoung R. A., *Inorg. Chem.*, 1979, 18, 1603.
441.   Karpinski Z. J., Osteryoung R. A., *Inorg. Chem.*, 1984, 23, 1491.
442.   Hussey C. L., Scheffler T. B., Wilkes J. S., Fannin Jr., A. A., *J. Electrochem. Soc.*, 1986, 133, 1389.
443.   Torsi G., Mamantov G., *Inorg. Chem.*, 1971, 10, 1900.
444.   Boxall L. G., Jones H. L., Osteryoung R. A., *J. Electrochem. Soc.*, 1973, 120, 223.
445.   Hussey C. L., Øye H. A., *J. Electrochem. Soc.*, 1984, 131, 1621.
446.   Dymek Jr. C. J., King L. A., *J. Electrochem. Soc.*, 1985, 132, 1375.
447.   Hussey C. L., Sanders J. R., Øye H. A., *J. Electrochem. Soc.*, 1985, 132, 2156.
448.   Wilkes J. S., Levisky J. A., Wilson R. A., Hussey C. L., *Inorg. Chem.*, 1982, 21, 1263.
449.   Hussey C. L., "The Electrochemistry of Room-Temperature Haloaluminate Molten Salts" in *Chemistry of Nonaqueous Solutions. Current Progress*, G. Mamantov, A. I. Petrov eds., VCH, New York, 1994, p. 227.
450.   Lipsztajn M., Osteryoung R. A., *J. Electrochem. Soc.*, 1983, 130, 1968.
451.   Scordilis-Kelley C., Fuller J., Carlin R. T., Wilkes J. S., *J. Electrochem. Soc.*, 1992, 139, 694.
452.   Melton T. J., Joyce J., Maloy J. T., Boon J. A., Wilkes J. S., *J. Electrochem. Soc.*, 1990, 137, 3865.
453.   Gifford P. R., Palmisano J. P., *J. Electrochem. Soc.*, 1987, 134, 610.
454.   Reichel T. L., Wilkes J. S., *J. Electrochem. Soc.*, 1992, 139, 977.
455.   Wilkes J. S., Zavorotko M. J., *J. Chem. Soc. Chem. Commun.*, 1992, 965.
456.   Koch V. R., Dominey L. A., Nanjundiah C., Ondrechen M. J., *J. Electrochem. Soc.*, 1996, 143, 798.
457.   Suarez P. A. Z., Selbach V. M., Dullius J. E. L., Einloft S., Piatnicki C. M. S., Azambuja D. S., de Souza R. F., Dupont J., *Electrochim. Acta*, 1997, 42, 2533.
458.   Robinson J., Osteryoung R. A., *J. Electrochem. Soc.*, 1980, 127, 122.
459.   Auborn J. J., Barbeiro Y. L., *J. Electrochem. Soc.*, 1985, 132, 598.
460.   Chryssoulakis Y., Poignet J. C., Manoli G., *J. Appl. Electrochem.*, 1987, 17, 857.
461.   Lai P. K., Skyllas-Kazakos M., *J. Electroanal. Chem.*, 1988, 248, 431.
462.   Carlin R. T., Osteryoung R. A., *J. Electrochem. Soc.*, 1989, 136, 1409.
463.   Lai P. K., Skyllas-Kazakos M., *Electrochim. Acta.*, 1987, 32, 1443.

464. Carlin R. T., Crawford W., Bersch M., *J. Electrochem. Soc.*, 1992, 139, 2720.
465. Quing Liao, Pitner W. R., Stewart G., Hussey C. L., Stafford G. R., *J. Electrochem. Soc.*, 1997, 144, 936.
466. Gray G. E., Winnick J., Kohl P. A., *J. Electrochem. Soc.*, 1996, 143, 2262.
467. Scordilis-Kelley C., Carlin R. T., *J. Electrochem. Soc.*, 1993, 140, 1606.
468. Campbell J. L., Johnson K. E., *J. Electrochem. Soc.*, 1994, 141, L 19.
469. Scordilis-Kelley C., Carlin R. T., *J. Electrochem. Soc.*, 1994, 141, 873.
470. Gray G. E., Kohl P. A., Winnick J., *J. Electrochem. Soc.*, 1995, 142, 3636.
471. Fuller J., Osteryoung R. A., Carlin R. T., *J. Electrochem. Soc.*, 1995, 142, 3632.
472. Gray G. E., Winnick J., Kohl P. A., *J. Electrochem. Soc.*, 1996, 143, 3820.
473. Piersma B. J., Ryan D. M., Schumacher E. R., Riechel T. L., *J. Electrochem. Soc.*, 1996, 143, 908.
474. Carlin R. T., De Long H. C., Fuller J., Trulove P. C., *J. Electrochem. Soc.*, 1994, 141, L 73.
475. Yu C. L., Winnick J., Kohl P. A., *J. Electrochem. Soc.*, 1991, 138, 339.

# Index